Applied Probability
Control
Economics
Information and Communication
Modeling and Identification
Numerical Techniques
Optimization

Applications of Mathematics

5

R. S. Liptser
A. N. Shiryayev

Statistics of Random Processes I

General Theory

Translated by A. B. Aries

Springer-Verlag
New York Heidelberg Berlin

R. S. Liptser
Institute for Problems of
Control Theory
Moscow
Profsojuznaja 81
U.S.S.R.

A. N. Shiryayev
Institute of Control Sciences
Moscow 117806
U.S.S.R.

Title of the Russian Original Edition: *Statistika sluchaĭnyk protsessov.* Nauka, Moscow, 1974.

AMS Subject Classifications: 60Gxx, 60Hxx, 60Jxx, 62Lxx, 62Mxx, 62Nxx, 93Exx, 94A05

Library of Congress Cataloging in Publication Data

Liptser, Robert Shevilevich.
Statistics of random processes.
(Applications of mathematics ; v. 5)
Revised and expanded translation of the authors' Statistika sluchaĭnyk protsessov, originally published in 1974.
Bibliography: p.
1. Stochastic processes. 2. Mathematical statistics. I. Shiriaev, Al'bert Nikolaevich, joint author. II. Title.
QA274.S53 519.2 76-49817

Printed in the United States of America.

9 8 7 6 5 4 3 2 1

ISBN 0-387-90226-0 Springer-Verlag New York

ISBN 3-540-90226-0 Springer-Verlag Berlin Heidelberg

Contents of volume I

Contents of volume II

Introduction

A considerable number of problems in the statistics of random processes are formulated within the following scheme.

On a certain probability space $(\Omega, \mathscr{F}, P)$ a partially observable random process $(\theta, \xi) = (\theta_t, \xi_t)$, $t \geq 0$, is given with only the second component $\xi = (\xi_t)$, $t \geq 0$, observed. At any time t it is required, based on $\xi_0^t = \{\xi_s, 0 \leq s \leq t\}$, to estimate the unobservable state θ_t. This problem of estimating (in other words, the *filtering* problem) θ_t from ξ_0^t will be discussed in this book.

It is well known that if $M(\theta_t^2) < \infty$, then the optimal mean square estimate of θ_t from ξ_0^t is the a posteriori mean $m_t = M(\theta_t | \mathscr{F}_t^\xi)$, where $\mathscr{F}_t^\xi = \sigma\{\omega: \xi_s, s \leq t\}$ is the σ-algebra generated by ξ_0^t. Therefore, the solution of the problem of optimal (in the mean square sense) filtering is reduced to finding the conditional (mathematical) expectation $m_t = M(\theta_t | \mathscr{F}_t^\xi)$.

In principle, the conditional expectation $M(\theta_t | \mathscr{F}_t^\xi)$ can be computed by Bayes' formula. However, even in many rather simple cases, equations obtained by Bayes' formula are too cumbersome, and present difficulties in their practical application as well as in the investigation of the structure and properties of the solution.

From a computational point of view it is desirable that the formulae defining the *filter* m_t, $t \geq 0$, should be of a recurrent nature. Roughly speaking, it means that $m_{t+\Delta}$, $\Delta > 0$, must be built up from m_t and observations $\xi_t^{t+\Delta} = \{\xi_s : t \leq s \leq t + \Delta\}$. In the discrete case $t = 0, 1, 2, \ldots$, the simplest form of such recurrence relations can be, for example, the equation

$$\Delta m_t = a(t, m_t) + b(t, m_t)(\xi_{t+1} - \xi_t), \tag{1}$$

where $\Delta m_t = m_{t+1} - m_t$. In the case of continuous time, $t \geq 0$, stochastic differential equations

$$dm_t = a(t, m_t)dt + b(t, m_t)d\xi_t \tag{2}$$

have such a form.

It is evident that without special assumptions concerning the structure of the process (θ, ξ) it is difficult to expect that optimal values m_t should satisfy recurrence relations of the types given by (1) and (2). Therefore, before describing the structure of the process (θ, ξ) whose filtering problems are investigated in this book, we shall study a few specific examples.

Let θ be a Gaussian random variable with $M\theta = m$, $D\theta = \gamma$, which for short will be written $\theta \sim N(m, \gamma)$. Assume that the sequence

$$\xi_t = \theta + \varepsilon_t, \qquad t = 1, 2, \ldots, \tag{3}$$

is observed, where $\varepsilon_1, \varepsilon_2, \ldots$ is a sequence of mutually independent Gaussian random variables with zero mean and unit dispersion independent also of θ. Using a theorem on normal correlation (Theorem 13.1) it is easily shown that $m_t = M(\theta | \xi_1, \ldots, \xi_t)$. The *tracking* errors $\gamma_t = M(\theta - m_t)^2$ are found by

$$m_t = \frac{m + \sum_{i=1}^t \xi_i}{1 + \gamma t}, \qquad \gamma_t = \frac{\gamma}{1 + \gamma t}. \tag{4}$$

From this we obtain the following recurrence equations for m_t and γ_t:

$$\Delta m_t = \frac{\gamma_t}{1 + \gamma t}[\xi_{t+1} - m_t], \tag{5}$$

$$\Delta \gamma_t = -\frac{\gamma_t^2}{1 + \gamma_t}, \tag{6}$$

where $\Delta m_t = m_{t+1} - m_t$, $\Delta \gamma_t = \gamma_{t+1} - \gamma_t$.

Let us make this example more complicated. Let θ and $\xi_1, \xi_2, \ldots$ be the same as in the previous example, and let the observable process $\xi_t, t = 1, 2, \ldots$, be defined by the relations

$$\xi_{t+1} = A_0(t, \xi) + A_1(t, \xi)\theta + \varepsilon_{t+1}, \tag{7}$$

where functions $A_0(t, \xi)$ and $A_1(t, \xi)$ are assumed to be $\mathscr{F}_t^\xi$-measurable (i.e., $A_0(t, \xi)$ and $A_1(t, \xi)$ at any time depend only on the values $(\xi_0, \ldots, \xi_t)$), $\mathscr{F}_t^\xi = \sigma\{\omega : \xi_0, \ldots, \xi_t\}$.

The necessity to consider the coefficients $A_0(t, \xi)$ and $A_1(t, \xi)$ for all "past history" values $(\xi_0, \ldots, \xi_1)$ arises, for example, in control problems (Section 14.3), where these coefficients play the role of "controlling" actions, and also in problems of information theory (Section 16.4), where the pair of functions $(A_0(t, \xi), A_1(t, \xi))$, is treated as "coding" using noiseless feedback.

It turns out that for the scheme given by (7) the optimal value $m_t = M(\theta_t|\mathcal{F}_t^{\xi})$ and the conditional dispersion $\gamma_t = M[(\theta - m_t)^2|\mathcal{F}_t^{\xi}]$ also satisfy recurrence equations (see Section 13.5):

$$\Delta m_t = \frac{\gamma_t A_1(t, \xi)}{1 + A_1^2(t, \xi)\gamma_t}(\xi_{t+1} - A_0(t, \xi) - A_1(t, \xi)m_t), \qquad m_0 = m; \tag{8}$$

$$\Delta\gamma_t = -\frac{A_1^2(t, \xi)\gamma_t^2}{1 + A_1^2(t, \xi)\gamma_t}, \qquad \gamma_0 = \gamma. \tag{9}$$

In the schemes given by (3) and (7) the question, in essence, is a traditional problem of mathematical statistics—Bayes' estimation of a random parameter from the observations ξ_0^t. The next step to make the scheme given by (7) more complicated is to consider a random process θ_t rather than a random variable θ.

Assume that the random process $(\theta, \xi) = (\theta_t, \xi_t)$, $t = 0, 1, \ldots$, is described by the recurrence equations

$$\begin{aligned} \theta_{t+1} &= a_0(t, \xi) + a_1(t, \xi)\theta_t + b(t, \xi)\varepsilon_1(t + 1), \\ \xi_{t+1} &= A_0(t, \xi) + A_1(t, \xi)\theta_t + B(t, \xi)\varepsilon_2(t + 1) \end{aligned} \tag{10}$$

where $\varepsilon_1(t)$, $\varepsilon_2(t)$, $t = 1, 2, \ldots$, the sequence of independent variables, is normally distributed, $N(0, 1)$, and also independent of (θ_0, ξ_0). The coefficients $a_0(t, \xi), \ldots, B(t, \xi)$ are assumed to be $\mathcal{F}_t^{\xi}$-measurable for any $t = 0, 1\ldots$

In order to obtain recurrence equations for estimating $m_t = M(\theta_t|\mathcal{F}_t^{\xi})$ and conditional dispersion $\gamma_t = M\{[\theta_t - m_t]^2|\mathcal{F}_t^{\xi}\}$, let us assume that the conditional distribution $P(\theta_0 \le x|\xi_0)$ is (for almost all ξ_0) normal, $N(m, \gamma)$. The essence of this assumption is that it permits us to prove (see Chapter 13) that the sequence (θ, ξ) satisfying (10) is conditionally Gaussian. This means, in particular, that the conditional distribution $P(\theta_t \le x|\mathcal{F}_t^{\xi})$ is (almost surely) Gaussian. But such a distribution is characterized only by its two conditional moments m_t and γ_t, leading to the following closed system of equations:

$$\begin{aligned} m_{t+1} &= a_0 + a_1 m_t + \frac{a_1 A_1 \gamma_t}{B^2 + A_1^2\gamma_t}[\xi_{t+1} - A_0 - A_1 m_t], \qquad m_0 = m; \\ \gamma_{t+1} &= [a_1^2\gamma_t + b^2] - \frac{(a_1 A_1 \gamma_t)^2}{B^2 + A_1^2\gamma_t}, \qquad \gamma_0 = \gamma \end{aligned} \tag{11}$$

(in the coefficients $a_0, \ldots, B$, for the sake of simplicity, arguments t and ξ are omitted).

The equations in (11) are deduced (in a somewhat more general framework) in Chapter 13. Their deduction does not need anything except the theorem of normal correlation. In this chapter, equations for optimal estimation in extrapolation problems (estimating θ_τ from ξ_0^t, when $\tau > t$) and interpolation problems (estimating θ_τ from ξ_0^t when $\tau < t$) are derived. Chapter 14 deals with applications of these equations to various statistical problems of random sequences, to control problems, and to problems of constructing pseudosolutions to linear algebraic systems.

These two chapters can be read independently of the rest of the book, and this is where the reader should start if he is interested in nonlinear filtering problems but is not sufficiently acquainted with the general theory of random processes.

The main part of the book concerns problems of optimal filtering and control (and also related problems of interpolation, extrapolation, sequential estimation, testing of hypotheses, etc.) in the case of *continuous* time. These problems are interesting per se; and, in addition, easy formulations and compact formulae can be obtained for them. It should be added that often it is easier, at first, to study the continuous analog of problems formulated for discrete time, and use the results obtained in the solution of the latter.

The simplicity of formulation in the case of continuous time is, however, not easy to achieve—rather complicated techniques of the theory of random processes have to be invoked. Later on, we will discuss the methods and the techniques used in this book in more detail, but here we consider particular cases of the filtering problem for the sake of illustration.

Assume that the partially observable random process $(\theta, \xi) = (\theta_t, \xi_t)$, $t \geq 0$, is Gaussian, governed by stochastic differential equations (compare with the system (10)):

$$d\theta_t = a(t)\theta_t\,dt + b(t)dw_1(t), \qquad d\xi_t = A(t)\theta_t\,dt + B(t)dw_2(t), \qquad \theta_0 \equiv 0, \tag{12}$$

where $w_1(t)$ and $w_2(t)$ are standard Wiener processes, mutually independent and independent of (θ_0, ξ_0), and $B(t) \geq C > 0$. Let us consider the component $\theta = (\theta_t)$, $t \geq 0$, as unobservable. The filtering problem is that of optimal estimation of θ_t from ξ_0^t in the mean square sense for any $t \geq 0$.

The process (θ, ξ), according to our assumption, is Gaussian; hence the optimal estimate $m_t = M(\theta_t | \mathscr{F}_t^\xi)$ depends linearly on $\xi_0^t = \{\xi_s : s \leq t\}$. More precisely, there exists (Lemma 10.1) a function $G(t, s)$, with $\int_0^t G^2(t, s)ds < \infty$, $t > 0$, such that (almost surely):

$$m_t = \int_0^t G(t, s)d\xi_s. \tag{13}$$

If this expression is formally differentiated, we obtain

$$dm_t = G(t, t)d\xi_t + \left(\int_0^t \frac{\partial G(t, s)}{\partial t}\,d\xi_s\right)dt. \tag{14}$$

The right side of this equation can be transformed using the fact that the function $G(t, s)$ satisfies the Wiener–Hopf equation (see (10.25)), which in our case reduces to

$$\frac{\partial G(t,s)}{\partial t} = \left[a(t) - \gamma_t \frac{A^2(t)}{B^2(t)}\right] G(t,s), \qquad t > s, \tag{15}$$

$$G(s,s) = \frac{\gamma_s A(s)}{B^2(s)}, \qquad \gamma_s = M[\theta_s - m_s]^2. \tag{16}$$

Taking into account (15) and (14), we infer that the optimal estimate m_t, $t > 0$, satisfies a linear stochastic differential equation,

$$dm_t = a(t)m_t\,dt + \frac{\gamma_t A(t)}{B^2(t)}[d\xi_t - A(t)m_t\,dt]. \tag{17}$$

This equation includes the tracking error $\gamma = M[\theta_t - m_t]^2$, which in turn is the solution of the Ricatti equation

$$\dot{\gamma}_t = 2a(t)\gamma_t - \frac{A^2(t)\gamma_t^2}{B^2(t)} + b^2(t). \tag{18}$$

(Equation (18) is easy to obtain applying the Ito formula for substitution of variables to the square of the process $[\theta_t - m_t]$ with posterior averaging.)

Let us discuss Equation (17) in more detail taking, for simplicity, $\xi_0 \equiv 0$. Denote

$$\overline{w} = \int_0^t \frac{d\xi_s - A(s)m_s\,ds}{B(s)}. \tag{19}$$

Then Equation (17) can be rewritten:

$$dm_t = a(t)m_t\,dt + \frac{\gamma_t A(t)}{B(t)}\,d\overline{w}_t. \tag{20}$$

The process $(\overline{w}_t)$, $t \geq 0$, is rather remarkable and plays a key role in filtering problems. The point is that, first, this process turns out to be a Wiener process (with respect to the σ-algebras $(\mathscr{F}_t^\xi)$, $t \geq 0$), and secondly, it contains the same *information* as the process ξ does. More precisely, it means that for all $t \geq 0$, the σ-algebras $\mathscr{F}_t^{\overline{w}} = \sigma\{\omega : \overline{w}_s, s \leq t\}$ and $\mathscr{F}_t^\xi = \sigma\{\omega : \xi_s, s \leq t\}$ coincide:

$$\mathscr{F}_t^{\overline{w}} = \mathscr{F}_t^\xi, \qquad t \geq 0 \tag{21}$$

(see Theorem 7.16). By virtue of these properties of the process, $\overline{w}$ it is referred to as the *innovation* process.

The equivalence of σ-algebras $\mathscr{F}_t^\xi$ and $\mathscr{F}_t^{\overline{w}}$ suggests that for m_t not only is Equation (13) justified but also the representation

$$m_t = \int_0^t F(t,s)\,d\overline{w}_s \tag{22}$$

where $\overline{w} = (\overline{w}_t)$, $t \geq 0$ is the innovation process, and functions $F(t,s)$ are such that $\int_0^t F^2(t,s)\,ds < \infty$. In the main part of the text (Theorem (7.16)) it is shown that the representation given by (22) can actually be obtained from results on the structure of functionals of diffusion type processes. Equation (20) can be deduced in a simpler way from the representation given by (22) than the representation given by (13). It should be noted, however, that the proof of (22) is more difficult than that of (13).

In this example, the optimal (Kalman–Bucy) filter was linear because of the assumption that the process (θ, ξ) is Gaussian. Let us take now an example where the optimal filter is nonlinear.

Let (θ_t), $t \geq 0$, be a Markov process starting at zero with two states 0 and 1 and the only transition $0 \to 1$ at a random moment σ, distributed (due to assumed Markov behavior) exponentially: $P(\sigma > t) = e^{-\lambda t}$, $\lambda > 0$. Assume that the observable process $\xi = (\xi_t)$, $t \geq 0$, has a differential

$$d\xi_t = \theta_t\, dt + dw_t \qquad \xi_0 = 0, \tag{23}$$

where $w = (w_t)$, $t \geq 0$, is a Wiener process independent of the process $\theta = (\theta_t)$, $t \geq 0$.

We shall interpret the transition of the process θ from the "zero" state into the unit state as *the occurrence of discontinuity* (at the moment σ). There arises the following problem: to determine at any $t > 0$ from observations ξ_0^t whether or not discontinuity has occurred before this moment.

Denote $\pi_t = P(\theta_t = 1 | \mathscr{F}_t^\xi) = P(\sigma \leq t | \mathscr{F}_t^\xi)$. It is evident that $\pi_t = m_t = M(\theta_t | \mathscr{F}_t^\xi)$. Therefore, the a posteriori probability π_t, $t \geq 0$, is the optimal (in the mean square sense) state estimate of an unobservable process $\theta = (\theta_t)$, $t \geq 0$.

For the a posteriori probability π_t, $t \geq 0$, we can deduce (using, for example, Bayes' formula and results with respect to a derivative of the measure corresponding to the process ξ, with respect to the Wiener measure) the following stochastic differential equation:

$$d\pi_t = \lambda(1 - \pi_t)dt + \pi_t(1 - \pi_t)[d\xi_t - \pi_t\, dt], \qquad \pi_0 = 0. \tag{24}$$

It should be emphasized that whereas in the Kalman–Bucy scheme the optimal filter is linear, Equation (24) is essentially nonlinear. Equation (24) defines the *optimal nonlinear filter*.

As in the previous example, the (innovation) process

$$\bar{w}_t = \int_0^t [d\xi_s - \pi_t\, ds], \qquad t \geq 0,$$

turns out to be a Wiener process and $\mathscr{F}_t^{\bar{w}} = \mathscr{F}_t^\xi$, $t \geq 0$. Therefore, Equation (24) can be written in the following equivalent form:

$$d\pi_t = \lambda(1 - \pi_t)dt + \pi_t(1 - \pi_t)d\bar{w}_t, \qquad \pi_0 = 0. \tag{25}$$

It appears that all these examples are within the following general scheme adopted in this book.

Let $(\Omega, \mathscr{F}, P)$ be a certain probability space with a distinguished non-decreasing set of σ-algebras $(\mathscr{F}_t)$, $t \geq 0 (\mathscr{F}_s \subseteq \mathscr{F}_t \subseteq \mathscr{F}, s \leq t)$. In this probability space we are given a partially observable process (θ_t, ξ_t), $t \geq 0$, and an estimated process (h_t), $t \geq 0$, dependent, generally speaking, on both the unobservable process θ_t, $t \geq 0$, and the observable component (ξ_t), $t \geq 0$.

As to the observable process[1] $\xi = (\xi_t, \mathcal{F}_t)$ it will be assumed that it permits a stochastic differential

$$d\xi_t = A_t(\omega)dt + dw_t, \qquad \xi_0 = 0, \tag{26}$$

where $w = (w_t, \mathcal{F}_t)$, $t \geq 0$, is a standard Wiener process (i.e., a square integrable martingale with continuous trajectories with $M[(w_t - w_s)^2 | \mathcal{F}_s] = t - s$, $t \geq s$, and $w_0 = 0$), and $A = (A_t(\omega), \mathcal{F}_t)$, $t \geq 0$, is a certain integrable random process.[2]

The structure of the unobservable process $\theta = (\theta_t, \mathcal{F}_t)$, $t \geq 0$, is not directly concretized, but it is assumed that the estimated process $h = (h_t, \mathcal{F}_t)$, $t \geq 0$, permits the following representation:

$$h_t = h_0 + \int_0^t a_s(\omega)ds + x_t, \qquad t \geq 0, \tag{27}$$

where $a = (a_t(\omega), \mathcal{F}_t)$, $t \geq 0$, is some integrable process, and $x = (x_t, \mathcal{F}_t)$, $t \geq 0$, is a square integrable martingale.

For any integrable process $g = (g_t, \mathcal{F}_t)$, $t \geq 0$, write $\pi_t(g) = M[g_t | \mathcal{F}_t^\xi]$. Then, if $Mg_t^2 < \infty$, $\pi_t(g)$ is the optimal (in the mean square sense) estimate of g_t from $\xi_0^t = \{\xi_s : s \leq t\}$.

One of the main results of this book (Theorem 8.1) states that for $\pi_t(h)$ the following representation is correct:

$$\pi_t(h) = \pi_0(h) + \int_0^t \pi_s(a)ds + \int_0^t \pi_s(D)d\bar{w}_s + \int_0^t [\pi_s(hA) - \pi_s(h)\pi_s(A)]d\bar{w}_s \tag{28}$$

Here $\bar{w} = (\bar{w}_t, \mathcal{F}_t^\xi)$, $t \geq 0$, is a Wiener process (compare with the innovation processes in the two previous examples), and the process $D = (D_t, \mathcal{F}_t)$, $t \geq 0$, characterizes *correlation* between the Wiener process $w = (w_t, \mathcal{F}_t)$, $t \geq 0$, and the martingale $x = (x_t, \mathcal{F}_t)$, $t \geq 0$. More precisely, the process

$$D_t = \frac{d\langle x, w\rangle_t}{dt}, \qquad t \geq 0, \tag{29}$$

where $\langle x, w\rangle_t$ is a random process involved in Doob–Meyer decomposition of the product of the martingales x and w:

$$M[x_t w_t - x_s w_s | \mathcal{F}_s] = M[\langle x, w\rangle_t - \langle x, w\rangle_s | \mathcal{F}_s] \tag{30}$$

We call the representation in (28) the *main equation* of (optimal nonlinear) *filtering*. Most of known results (within the frame of the assumptions given by (26) and (27)) can be deduced from this equation.

[1] $\xi = (\xi_t, \mathcal{F}_t)$ suggests that values ξ_t are $\mathcal{F}_t$-measurable for any $t \geq 0$.

[2] Actually, this book examines processes ξ of a somewhat more general kind (see Chapter 8).

Let us show, for example, in what way the filtering Equations (17) and (18) in the Kalman–Bucy scheme are deduced from (28), taking, for simplicity, $b(t) \equiv B(t) \equiv 1$.

Comparing (12) with (26) and (27), we see that $A_t(\omega) = A(t)\theta_t$, $w_t = w_2(t)$. Assume $h_t = \theta_t$. Then, due to (12),

$$h_t = h_0 + \int_0^t a(s)\theta_s\, ds + w_1(t). \tag{31}$$

The processes $w_1 = (w_1(t))$ and $w_2 = (w_2(t))$, $t \geq 0$, are independent square integrable martingales, hence for them $D_t \equiv 0$(P-a.s.). Then, due to (28), $\pi_t(\theta)$ has a differential

$$d\pi_t(\theta) = a(t)\pi_t(\theta)dt + A(t)[\pi_t(\theta^2) - \pi_t^2(\theta)]d\overline{w}_t, \tag{32}$$

i.e.,

$$dm_t = a(t)m_t\, dt + A(t)\gamma_t\, d\overline{w}_t \tag{33}$$

where we have taken advantage of the Gaussian behavior of the process (θ, ξ); (P-a.s.)

$$\pi_t(\theta^2) - \pi_t^2(\theta) = M[(\theta_t - m_t)^2 | \mathscr{F}_t^\xi] = M[\theta_t - m_t]^2 = \gamma_t.$$

In order to deduce an equation for γ_t from (28), we take $h_t = \theta_t^2$. Then, from the first equation of the system given by (12) we obtain by the Ito formula for substitution of variables (Theorem 4.4),

$$\theta_t^2 = \theta_0^2 + \int_0^t a_s(\omega)ds + x_t, \tag{34}$$

where

$$a_s(\omega) = 2a(s)\theta_s^2 + b^2(s)$$

and

$$x_t = \int_0^t \theta_s\, dw_1(s).$$

Therefore, according to (28)

$$d\pi_t(\theta^2) = [2a(t)\pi_t(\theta^2) + b^2(s)(t)]dt + A(t)[\pi_t(\theta^3) - \pi_t(\theta)\pi_t(\theta^2)]d\overline{w}_t. \tag{35}$$

From (32) and (35) it is seen that in using the main filtering equation, (28), we face the difficulty that for finding conditional lower moments a knowledge of higher moments is required. Thus, for finding equations for $\pi_t(\theta^2)$ the knowledge of the third a posteriori moment $\pi_t(\theta^3) = M(\theta_t^3 | \mathscr{F}_t^\xi)$ is required. In the case considered this difficulty is easy to overcome, since, due to the Gaussian behavior of the process (θ, ξ), the moments $\pi_t(\theta^n) = M(\theta_t^n | \mathscr{F}_t^\xi)$ for all $n \geq 3$ are expressed through $\pi_t(\theta)$ and $\pi_t(\theta^2)$. In particular, $\pi_t(\theta^3) - \pi_t(\theta)\pi_t(\theta^2) = M[\theta_t^2(\theta_t - m_t) | \mathscr{F}_t^\xi] = 2m_t\gamma_t$ and, therefore,

$$d\pi_t(\theta^2) = [2a(t)\pi_t(\theta^2) + b^2(t)]dt + 2A(t)m_t\gamma_t\, d\overline{w}_t. \tag{36}$$

By the Ito formula for substitution of variables, from (33) we find that

$$dm_t^2 = 2m_t[a(t)m_t\,dt + A(t)\gamma_t m_t\,d\overline{w}_t] + A^2(t)\gamma^2(t)dt.$$

Together with Equation (36) this relation provides the required equation for $\gamma_t = \pi_t(\theta^2) - m_t^2$.

The deduction above of Equations (17) and (18) shows that in order to obtain a closed system of equations defining optimal filtering some supplementary knowledge on the ratios between higher conditional moments is needed.

This book deals mainly with the so-called *conditionally Gaussian* processes (θ, ξ) for which it appears possible to obtain closed system of equations for the optimal nonlinear filter. Therefore, a wide class of random processes (including processes described by the Kalman–Bucy scheme) is described, i.e., random processes for which it has been possible to solve effectively the problem of constructing an optimal nonlinear filter. This class of processes (θ, ξ) is described in the following way.

Assume that the process (θ, ξ) is a diffusion type process with a differential

$$\begin{aligned} d\theta_t &= [a_0(t, \xi) + a_1(t, \xi)\theta_t]dt + b_1(t, \xi)dw_1(t) + b_2(t, \xi)dw_2(t), \\ d\xi_t &= [A_0(t, \xi) + A_1(t, \xi)\theta_t]dt + B_1(t, \xi)dw_1(t) + B_2(t, \xi)dw_2(t), \end{aligned} \tag{37}$$

where each of the functionals $a_0(t, \xi), \ldots, B_2(t, \xi)$ is $\mathscr{F}_t^\xi$-measurable at any $t \geq 0$ (compare with the system given by (10)). We emphasize the fact that the unobservable component θ_t enters into (37) in a linear way, whereas the observable process ξ can enter into the coefficients in any $\mathscr{F}_t^\xi$-measurable way. The Wiener processes $w_1 = (w_1(t))$, $w_2 = (w_2(t))$, $t \geq 0$, and the random vector (θ_0, ξ_0) included in (37) are assumed to be independent.

It will be proved (Theorem 11.1) that if the conditional distribution $P(\theta_0 \leq x \mid \xi_0)$ (for almost all ξ_0) is Gaussian, $N(m_0, \gamma_0)$, where $m_0 = M(\theta_0 \mid \xi_0)$, $\gamma_0 = M[(\theta_0 - m_2)^2 \mid \xi_0]$, then the process (θ, ξ) governed by (37) will be conditionally Gaussian in the sense that for any $t \geq 0$ the conditional distributions $P(\theta_{t_0} \leq x_0, \ldots, \theta_{t_k} \leq x_k \mid \mathscr{F}_t^\xi)$, $0 \leq t_0 < t_1 < \cdots < t_k \leq k$, are Gaussian. Hence, in particular, the distribution $P(\theta_t \leq x \mid \mathscr{F}_t^\xi)$ is also (almost surely) Gaussian, $N(m_t, \gamma_t)$, with parameters $m_t = M(\theta_t \mid \mathscr{F}_t^\xi)$, $\gamma_t = M[(\theta_t - m_t)^2 \mid \mathscr{F}_t^\xi]$.

For the conditionally Gaussian case (as well as in the Kalman–Bucy schemes), the higher moments $M(\theta_t^n \mid \mathscr{F}_t^\xi)$ are expressed in terms of m_t and γ_t. This allows us (from the main filter equation) to obtain a closed system of equations (Theorem 12.1) for m_t and γ_t:

$$\begin{aligned} dm_t &= [a_0(t, \xi) + a_1(t, \xi)m_t]dt \\ &\quad + \frac{\sum_{i=1}^2 b_i(t, \xi)B_i(t, \xi) + \gamma_t A_1(t, \xi)}{\sum_{i=1}^2 B_i^2(t, \xi)}[d\xi_t - (A_0(t, \xi) + A_1(t, \xi)m_t)dt], \end{aligned} \tag{38}$$

$$\dot{\gamma}_t = 2a_1(t, \xi)\gamma_t + \sum_{i=1}^2 b_i^2(t, \xi) - \frac{[\sum_{i=1}^2 b_i(t, \xi)B_i(t, \xi) + \gamma_t A_1(t, \xi)]^2}{\sum_{i=1}^2 B_i^2(t, \xi)}. \tag{39}$$

Note that, unlike (18), Equation (39) for γ_t is an equation with random coefficients dependent on observable data.

Chapters 10, 11 and 12 deal with optimal linear filtering (according to the scheme given by (12)) and optimal nonlinear filtering for conditionally Gaussian processes (according to the scheme given by (37)). Here, besides filtering, the pertinent results for interpolation and extrapolation problems are also given.

The examples and results given in Chapters 8–12 show how extensively we use in this book such concepts of random processes theory as the Wiener process, stochastic differential equations, martingale, square integrable martingales, etc. The desire to provide sufficient proofs of all given results in the nonlinear theory neccesitated a detailed discussion of the theory of martingales and stochastic differential equations (Chapters 2–6). We hope that the material of these chapters may also be useful to those readers who simply wish to become familiar with results in the theory of martingales and stochastic differential equations.

At the same time we would like to emphasize the fact that without this material it does not seem possible to give a satisfactory description of the theory of optimal nonlinear filtering and related problems.

Chapter 7 discusses results, many used later on, on absolute continuity of measures satisfying Ito processes and diffusion type processes.

Chapters 15–17 concern applications of filtering theory to various problems of statistics of random processes. Here, the problems of linear estimation are considered in detail (Chapter 15) and applications to certain control problems and information theory are discussed (Chapter 16).

Applications to nonBayes problems of statistics (maximal likelihood estimation of coefficients of linear regression, sequential estimation and sequential testing of statistical hypotheses) are given in Chapter 17.

Chapters 18 and 19 deal with point (counting) processes. A typical example of such a process is the Poisson process with constant or variable intensity. The presentation is patterned, to a large extent, along the lines of the treatment in Volume I of Ito processes and diffusion type processes. Thus we study the structure of martingales of point processes, of related innovation processes, and the structure of Radon–Nikodym derivatives. Applications to problems of filtering and estimation of unknown parameters from the observations of point processes are included.

Notes at the end of each chapter contain historical and related background materials as well as references to the results discussed in that chapter.

In conclusion the authors wish to thank their colleagues and friends for assistance and recommendations, especially A. V. Balakrishnan, R. Z. Khasminsky and M. P. Yershov. They made some essential suggestions that we took into consideration.

Essentials of probability theory and mathematical statistics 1

1.1 Main concepts of probability theory

1.1.1 Probability spaces

According to Kolmogorov's axiomatics the primary object of probability theory is the *probability* space $(\Omega, \mathscr{F}, P)$. Here $(\Omega, \mathscr{F})$ denotes measurable space, i.e., a set Ω consisting of elementary events ω, with a distinguished system $\mathscr{F}$ of its subsets (events), forming a σ-algebra, and P denotes a probability measure (probability) defined on sets in $\mathscr{F}$.

We recall that the system $\mathscr{F}$ of subsets of the space Ω forms an *algebra* if:

(1) $\Omega \in \mathscr{F}$ implies $\bar{A} \equiv \Omega - A \in \mathscr{F}$;
(2) $A \cup B \in \mathscr{F}$ for any $A \in \mathscr{F}$, $B \in \mathscr{F}$.

The algebra $\mathscr{F}$ forms a *σ-algebra* if given any sequence of subsets $A_1, A_2, \ldots,$ belonging to $\mathscr{F}$, the union $\bigcup_{i=1}^{\infty} A_i \in \mathscr{F}$. A function $P(A)$, defined on sets A from the σ-algebra $\mathscr{F}$, is called a *probability measure* if it has the following properties:

$$P(A) \geq 0 \quad \text{for all } A \in \mathscr{F} \qquad \text{(nonnegativity);}$$

$$P(\Omega) = 1 \qquad \text{(normalization);}$$

$$P\left(\bigcup_{i=1}^{\infty} A_i\right) = \sum_{i=1}^{\infty} P(A_i) \qquad \text{(denumerable or } \sigma\text{-additivity),}$$

where $A_i \in \mathscr{F}$ and $A_i \cap A_j = \varnothing$ for $i \neq j$ ($\varnothing$ = the empty set).

A system of sets $\mathscr{F}^P$ is called an *augmentation* of the σ-algebra $\mathscr{F}$ with respect to measure P, if $\mathscr{F}^P$ contains all sets $A \in \Omega$, for which there exist sets $A_1, A_2 \in \mathscr{F}$ such that $A_1 \subseteq A \subseteq A_2$ and $P(A_2 - A_1) = 0$. The system of sets $\mathscr{F}^P$ is a σ-algebra, and the measure P extends uniquely to $\mathscr{F}^P$. The probability space $(\Omega, \mathcal{F}, P)$ is *complete* if $\mathcal{F}^P$ coincides with $\mathcal{F}$. According to the general custom in probability theory which ignores events of zero probability, all probability spaces $(\Omega, \mathcal{F}, P)$, considered from now on, are assumed (even if not stated specifically) to be complete.

1.1.2 Random elements and variables

Let $(\Omega, \mathcal{F})$ and $(E, \mathcal{B})$ be two measurable spaces. The function $\xi = \xi(\omega)$ defined on $(\Omega, \mathcal{F})$ with values on E, is called *$\mathcal{F}/\mathcal{B}$-measurable* if the set $\{\omega: \xi(\omega) \in B\} \in \mathcal{F}$ for any $B \in \mathcal{B}$. In probability theory such functions are called *random functions with values in E*. In the case where $E = \mathbb{R}$, the real line, and $\mathcal{B}$ is the σ-algebra of Borel subsets of $\mathbb{R}$, $\mathcal{F}/\mathcal{B}$-measurable functions, the $\xi = \xi(\omega)$ are called *(real) random variables*. In this special case $\mathcal{F}/\mathcal{B}$-measurable functions are called simply *$\mathcal{F}$-measurable functions*.

We say that two random variables ξ and η *coincide with probability* 1, or *almost surely* (a.s), if $P(\xi = \eta) = 1$. In this case we shall write: $\xi = \eta(P\text{-a.s.})$. Similarly, the notation $\xi \geq \eta(P\text{-a.s.})$ implies that $P(\xi \geq \eta) = 1$. The notation $\xi = \eta(A; P\text{-a.s.})$ is used for denoting $\xi = \eta$ almost surely on the set A with respect to measure P, i.e.,

$$P(A \cap \{\omega: \xi(\omega) \neq \eta(\omega)\}) = 0.$$

Similar meaning is given to the expression $\xi \geq \eta(A; P\text{-a.s.})$.

Later on the words (P-a.s.) will often be omitted for brevity.

1.1.3 Mathematical expectation

Let $(\Omega, \mathcal{F}, P)$ be a probability space and let $\xi = \xi(\omega)$ be a nonnegative random variable. Its *mathematical expectation* (denoted $M\xi$) is the Lebesgue integral[1] $\int_\Omega \xi(\omega)P(d\omega)$, equal by definition, to

$$\lim_{n\to\infty}\left[\sum_{i=1}^{n\cdot 2^n} i\cdot 2^{-n}P\{i\cdot 2^{-n} < \xi \leq (i+1)2^{-n}\} + nP\{\xi > n\}\right],$$

where $\{i\cdot 2^{-n} < \xi \leq (i+1)2^{-n}\}$ denotes the set of points $\omega \in \Omega$ for which $i\cdot 2^{-n} < \xi(\omega) \leq (i+1)\cdot 2^{-n}$. The set $\{\xi > n\}$ is similarly defined. Due to the assumption $\xi(\omega) \geq 0$ for all $\omega \in \Omega$, the integral $\int_\Omega \xi(\omega)P(d\omega)$ is defined, although it can take on the magnitude $+\infty$.

In the case of the arbitrary random variable $\xi = \xi(\omega)$ the mathematical expectation (also denoted $M\xi$) is defined only when one of the mathematical expectations $M\xi^+$ or $M\xi^-$ is finite (here $\xi^+ = \max(\xi, 0)$, $\xi^- = -\min(\xi, 0)$) and is defined to be equal to $M\xi^+ - M\xi^-$.

The random variable $\xi = \xi(\omega)$ is said to be *integrable* if $M|\xi| = M\xi^+ + M\xi^- < \infty$.

[1] For this integral the notations $\int_\Omega \xi(\omega)dP$, $\int_\Omega \xi\, dP$, $\int \xi(\omega)dP$, $\int \xi\, dP$ will also be used.

Let $\Omega = \mathbb{R}^1$ be the real line and let $\mathscr{F}$ be the system of Borel sets on it. Assume that measure P on $\mathscr{F}$ is generated by a certain distribution function $F(\lambda)$ (i.e., nondecreasing, right continuous, and such that $F(-\infty) = 0$ and $F(\infty) = 1$) according to the law $P\{(a, b]\} = F(b) - F(a)$. Then the integral $\int_a^b \xi(x)P(dx)$ is denoted $\int_a^b \xi(x)dF(x)$ and is called a *Lebesgue–Stieltjes integral.* This integral can be reduced to an integral with respect to Lebesgue measure $P(dt) = dt$ by letting $\xi(x) \geq 0$ and $c(t) = \inf(x : F(x) > t)$. Then

$$\int_a^b \xi(x)dF(x) = \int_{F(a)}^{F(b)} \xi(c(t))dt.$$

1.1.4 *Conditional mathematical expectations and probabilities*

Let $\mathscr{G}$ be a sub-σ-algebra of $\mathscr{F}$ (i.e., $\mathscr{G} \subseteq \mathscr{F}$) and let $\xi = \xi(\omega)$ be a nonnegative random variable. The *conditional mathematical expectation* of ξ with respect to $\mathscr{G}$ (denoted $M(\xi \mid \mathscr{G})$) by definition is any $\mathscr{G}$-measurable function $\eta = \eta(\omega)$, for which $M\eta$ is defined, such that for any $\Lambda \in \mathscr{G}$

$$\int_\Lambda \xi(\omega)P(d\omega) = \int_\Lambda \eta(\omega)P(d\omega).$$

The Lebesgue integral $\int_\Lambda \xi(\omega)P(d\omega)$ with respect to a set $\Lambda \in \mathscr{F}$, is, by definition, $\int_\Omega \xi(\omega)\chi_\Lambda(\omega)P(d\omega)$, where $\chi_\Lambda(\omega)$ is the *characteristic function* of the set Λ:

$$\chi_\Lambda(\omega) = \begin{cases} 1, & \omega \in \Lambda, \\ 0, & \omega \notin \Lambda. \end{cases}$$

The integral $\int_\Lambda \xi(\omega)P(d\omega)$ (if it is defined, i.e., one of the two integrals $\int_\Lambda \xi^+(\omega)P(d\omega)$, $\int_\Lambda \xi^-(\omega)P(d\omega)$ is finite) will be denoted by $M(\xi, \Lambda)$.

Let two probability measures P and Q be given on the measurable space $(\Omega, \mathscr{F})$. We say that measure P is *absolutely continuous with respect to measure* $Q (P \ll Q)$, if $P(A) = 0$ for any $A \in \mathscr{F}$ for which $Q(A) = 0$.

Radon–Nikodym theorem. *If $P \ll Q$, then there exists a nonnegative random variable $\xi = \xi(\omega)$, such that for any $A \in \mathscr{F}$,*

$$P(A) = \int_A \xi(\omega)Q(d\omega).$$

The $\mathscr{F}$-measurable function $\xi = \xi(\omega)$ is unique within stochastic equivalence (i.e., if also $P(A) = \int_A \eta(\omega)Q(d\omega)$ where $A \in \mathscr{F}$, then $\xi = \eta$ (Q-a.s.).

The random variable $\xi(\omega)$ is called the *density of one measure* (P) *with respect to the other* (Q) or *Radon–Nikodym derivative*. Because of this definition the notation

$$\xi(\omega) = \frac{dP}{dQ}(\omega)$$

is used. By the Radon–Nikodym theorem, if $P \ll Q$, the density dP/dQ always exists.

If $\xi(\omega) = \chi_A(\omega)$ is a characteristic function of the set $A \in \mathscr{F}$ (i.e., an *indicator* of the set A), then $M(\chi_A(\omega)|\mathscr{G})$ is denoted by $P(A|\mathscr{G})$ and is called the *conditional probability of the event A with respect to* $\mathscr{G}$. Like $M(\xi|\mathscr{G})$, the conditional probability $P(A|\mathscr{G})$ is defined uniquely within sets of P-measure zero (possibly depending on A).

The function $P(A, \omega)$, $A \in \mathscr{F}$, $\omega \in \Omega$, satisfying the conditions

(1) at any fixed ω it is a probability measure on sets $A \in \mathscr{F}$;
(2) for any $A \in \mathscr{F}$ it is $\mathscr{G}$-measurable;
(3) with probability 1 $P(A, \omega) = P(A|\mathscr{G})$ for any $A \in \mathscr{F}$;

is called the *conditional probability distribution with respect to* $\mathscr{G}$, or *a regular conditional probability*.

The existence of such a function means that conditional probabilities can be defined so that for any ω they would prescribe a probability measure on $A \in \mathscr{F}$.

In the regular case conditional mathematical expectations can be found as integrals with respect to conditional probabilities:

$$M(\xi|\mathscr{G}) = \int_\Omega \xi(\omega)P(d\omega|\mathscr{G}).$$

If $\xi = \xi(\omega)$ is an arbitrary random variable for which $M\xi$ exists (i.e., $M\xi^+ < \infty$ or $M\xi^- < \infty$), then a conditional mathematical expectation is found by the formula

$$M(\xi|\mathscr{G}) = M(\xi^+|\mathscr{G}) - M(\xi^-|\mathscr{G}).$$

If $\mathscr{A}$ is a system of subsets of space Ω, then $\sigma(\mathscr{A})$ denotes the σ-algebra generated by the system $\mathscr{A}$, i.e., the smallest σ-algebra containing $\mathscr{A}$. If $\eta = \eta(\omega)$ is a particular $\mathscr{F}/\mathscr{B}$-measurable function with values on E, then $\sigma(\eta)$ (or $\mathscr{F}^\eta$) denotes the smallest σ-algebra with respect to which the random element $\eta(\omega)$ is measurable. In other words, $\sigma(\eta)$ is a σ-algebra, consisting of sets of the form: $\{\omega:\eta^{-1}(B), B \in \mathscr{B}\}$. For brevity, the conditional mathematical expectation $M(\xi|\mathscr{F}^\eta)$ is denoted by $M(\xi|\eta)$. Similarly, for $P(A|\mathscr{F}^\eta)$ the notation $P(A|\eta)$ is used. In particular, if a random element $\eta(\omega)$ is an n-dimensional vector of random variables $(\eta_1, \ldots, \eta_n)$, then for $M(\xi|\mathscr{F}^\eta)$ the notation $M(\xi|\eta_1, \ldots, \eta_n)$ is used.

Note the basic properties of conditional mathematical expectations:

(1) $M(\xi|\mathscr{G}) \geq 0$, if $\xi \geq 0$ (P-a.s.);
(2) $M(1|\mathscr{G}) = 1$ (P-a.s.);
(3) $M(\xi + \eta|\mathscr{G}) = M(\xi|\mathscr{G}) + M(\eta|\mathscr{G})$ (P-a.s.), assuming the expression $M(\xi|\mathscr{G}) + M(\eta|\mathscr{G})$ is defined;
(4) $M(\xi\eta|\mathscr{G}) = \xi M(\eta|\mathscr{G})$ if $M\xi\eta$ exists and ξ is $\mathscr{G}$-measurable;
(5) If $\mathscr{G}_1 \subseteq \mathscr{G}_2$, then ($P$-a.s.) $M(\xi|\mathscr{G}_1) = M[M(\xi|\mathscr{G}_2)\mathscr{G}_1]$;

(6) If σ-algebras of $\mathcal{G}$ and $\mathcal{F}^{\xi}$ are independent (i.e., $P(A \cap B) = P(A)P(B)$ for any $A \in \mathcal{G}$, $B \in \mathcal{F}^{\xi}$), then (P-a.s.) $M(\xi | \mathcal{G}) = M\xi$. In particular, if $\mathcal{G} = \{\varnothing, \Omega\}$ is a trivial σ-algebra, then $M(\xi | \mathcal{G}) = M\xi$ (P-a.s.).

1.1.5 Convergence of random variables and theorems of the passage to the limit under the sign of mathematical expectation

We say that the sequence of random variables ξ_n, $n = 1, 2, \ldots$, *converges in probability to a random variable* ξ (using in this case $\xi_n \xrightarrow{P} \xi$ or $\xi = P\text{-}\lim_n \xi_n$) if, for any $\varepsilon > 0$, $\lim_{n \to \infty} P\{|\xi_n - \xi| > \varepsilon\} = 0$.

The sequence of random variables ξ_n, $n = 1, 2, \ldots$, is called *convergent to a random variable with probability* 1, *or almost surely* (and is written: $\xi_n \to \xi$ or $\xi_n \to \xi$ (P-a.s.)), if the set $\{\omega : \xi_n(\omega) \nrightarrow \xi(\omega)\}$ has P-measure zero. Note that

$$\{\omega : \xi_n \to \xi\} = \bigcap_{r=1}^{\infty} \bigcup_{n=1}^{\infty} \bigcap_{k=n}^{\infty} \left\{ |\xi_k - \xi| < \frac{1}{r} \right\},$$

from which, in particular, it follows that convergence with probability 1 implies convergence in probability.

We shall write $\xi_n \uparrow \xi$ or $\xi_n \uparrow \xi$ (P-a.s.) if $\xi_n \to \xi$ (P-a.s.) and $\xi_n \leq \xi_{n+1}$ (P-a.s.) for all $n = 1, 2, \ldots$. Convergence $\xi_n \downarrow \xi$ is defined in a similar way. We also say that $\xi_n \to \xi$ on the set $A \in \mathcal{F}$, if $P(A \cap (\xi_n \nrightarrow \xi)) = 0$.

The sequence of random variables ξ_n, $n = 1, 2, \ldots$, is called *convergent in mean square to* ξ (denoted: $\xi = \text{l.i.m.}_{n \to \infty} \xi_n$), if $M\xi_n^2 < \infty$, $M\xi^2 < \infty$ and $M|\xi_n - \xi|^2 \to 0$ as $n \to \infty$.

The sequence of random variables ξ_n, $n = 1, 2, \ldots$, with $M|\xi_n| < \infty$, is called *weakly convergent to a random variable* ξ *with* $M|\xi| < \infty$ if, for any bounded random variable $\eta = \eta(\omega)$,

$$\lim_{n \to \infty} M\xi_n \eta = M\xi\eta.$$

We now state the basic theorems of the passage to the limit under the sign of conditional mathematical expectation. These will be used often later on.

Theorem 1.1 (Monotone convergence). *Let a σ-algebra $\mathcal{G} \subseteq \mathcal{F}$.*

If $\xi_n \uparrow \xi$ (P-a.s.) *and* $M\xi_1^- < \infty$, *then* $M(\xi_n | \mathcal{G}) \uparrow M(\xi | \mathcal{G})$ (P-a.s.).

If $\xi_n \downarrow \xi$ (P-a.s.) *and* $M\xi_1^+ < \infty$, *then* $M(\xi_n | \mathcal{G}) \downarrow M(\xi | \mathcal{G})$ (P-a.s.).

For formulating other criteria the concept of uniform integrability has to be introduced. The set of random variables $\{\xi_\alpha : \alpha \in \mathscr{U}\}$ is called *uniformly integrable* if

$$\lim_{x \to \infty} \sup_{\alpha \in \mathscr{U}} \int_{\{|\xi_\alpha| > x\}} |\xi_\alpha| \, dP = 0. \tag{1.1}$$

Condition (1.1) is equivalent to the two following conditions:

$$\sup_\alpha M|\xi_\alpha| < \infty \quad \text{and} \quad \lim_{P(A) \to 0} \sup_\alpha \int_A |\xi_\alpha| \, dP = 0, \qquad A \in \mathscr{F}.$$

Theorem 1.2 (Fatou's lemma). *If the sequence of random variables* ξ_n^+, $n = 1, 2, \ldots$, *is uniformly integrable and* $M(\limsup_n \xi_n)$ *exists, then*

$$M(\limsup_n \xi_n | \mathscr{G}) \geq \limsup_n M(\xi_n | \mathscr{G}) \qquad (P\text{-a.s.}) \tag{1.2}$$

where[2]

$$\limsup_n \xi_n = \inf_n \sup_{m \geq n} \xi_m.$$

In particular, if for the sequence $\xi_n, n = 1, 2, \ldots,$ *there exists an integrable random variable* ξ *such that* $\xi_n \leq \xi$, *then the inequality given by* (1.2) *holds.*

Theorem 1.3. *Let* $0 \leq \xi_n \to \xi$ $(P\text{-a.s.})$ *and* $M\xi_n < \infty$, $n = 1, 2, \ldots$. *In order that*

$$M(\xi_n | \mathscr{G}) \to M(\xi | \mathscr{G}) < \infty \qquad (P\text{-a.s.}) \tag{1.3}$$

it is necessary and sufficient that the sequence ξ_n, $n = 1, 2, \ldots$, *be uniformly integrable.*

From Theorems 1.2 and 1.3 we have the following useful

Corollary. *If* $\xi_n \to \xi$ $(P\text{-a.s.})$ *and the sequence* ξ_n, $n = 1, 2, \ldots$, *is uniformly integrable, then*

$$M(|\xi_n - \xi| | \mathscr{G}) \to 0 \qquad (P\text{-a.s.}), \qquad n \to \infty. \tag{1.4}$$

Theorem 1.4 (Lebesgue's dominated convergence theorem). *Let* $\xi_n \to \xi$ $(P\text{-a.s.})$, *and let there exist an integrable random variable* η, *such that* $|\xi_n| \leq \eta$. *Then*

$$M(|\xi_n - \xi| | \mathscr{G}) \to 0 \qquad (P\text{-a.s.}), \qquad n \to \infty.$$

Note 1. Theorem 1.3, its corollary, and Theorem 1.4 hold true if the convergence $\xi_n \to \xi$ $(P\text{-a.s.})$ is replaced by a convergence in probability: $\xi = P\text{-}\lim_n \xi_n$.

[2] For the limit superior, $\limsup_n \xi_n$, the notation $\overline{\lim}_n \xi_n$ is also used. Similarly, the limit inferior, $\liminf_n \xi_n$, is denoted $\underline{\lim}_n \xi_n$.

Note 2. Taking in Theorems 1.1–1.4 the trivial algebra $\{\varnothing, \Omega\}$ as $\mathscr{G}$, we obtain the usual theorems of the passage to the limit under the sign of Lebesgue's integral, since in this case $M(\eta|\mathscr{G}) = M\eta$.

Now, let $\ldots, \mathscr{F}_{-2}, \mathscr{F}_{-1}, \mathscr{F}_0, \mathscr{F}_1, \mathscr{F}_2, \ldots$ be a nondecreasing $(\ldots, \mathscr{F}_{-1} \subseteq \mathscr{F}_0 \subseteq \mathscr{F}_1 \subseteq \mathscr{F}_2 \subseteq \ldots)$ sequence of sub-σ-algebras $\mathscr{F}$. Denote a minimal σ-algebra $\mathscr{F}_\infty$ containing the algebra of events $\bigcup_n \mathscr{F}_n$ by $\sigma(\bigcup_n \mathscr{F}_n)$, and assume $\mathscr{F}_{-\infty} = \bigcap_n \mathscr{F}_n$.

Theorem 1.5 (Levy). *Let ξ be a random variable with $M|\xi| < \infty$. Then with probability* 1

$$\begin{aligned} M(\xi|\mathscr{F}_n) &\to M(\xi|\mathscr{F}_\infty), \qquad n \to \infty, \\ M(\xi|\mathscr{F}_n) &\to M(\xi|\mathscr{F}_{-\infty}), \qquad n \to -\infty. \end{aligned} \tag{1.5}$$

The next assumption holds an assertion both of Theorems 1.4 and 1.5.

Theorem 1.6 *Let $\xi_m \to \xi$ (P-a.s.), and let there exist an integrable random variable η, such that $|\xi_m| \le \eta$. Let, moreover, $\ldots, \mathscr{F}_{-2} \subseteq \mathscr{F}_{-1} \subseteq \mathscr{F}_0 \subseteq \mathscr{F}_1 \subseteq \mathscr{F}_2 \ldots$ be a nondecreasing sequence of sub-σ-algebras $\mathscr{F}$, $\mathscr{F}_\infty = \sigma(\bigcup_n \mathscr{F}_n)$, $\mathscr{F}_{-\infty} = \bigcap_n \mathscr{F}_n$. Then with probability* 1

$$\begin{aligned} \lim_{n,m\to\infty} M(\xi_m|\mathscr{F}_n) &= M(\xi|\mathscr{F}_\infty), \\ \lim_{n,m\to\infty} M(\xi_m|\mathscr{F}_{-n}) &= M(\xi|\mathscr{F}_{-\infty}). \end{aligned} \tag{1.6}$$

Theorem 1.7 (Dunford–Pettis compactness criterion). *In order that a family of random variables $\{\xi_\alpha : \alpha \in \mathscr{U}\}$ with $M|\xi_\alpha| < \infty$ be weakly compact,[3] it is necessary and sufficient that it be uniformly integrable.*

To conclude this topic we give one necessary and sufficient condition for uniform integrability.

Theorem 1.8 (Vallée-Poussin). *In order that the sequence $\xi_1, \xi_2, \ldots$ of integrable random variables be uniformly integrable, it is necessary and sufficient that there be a function $G(t)$, $t \ge 0$, which is positive, increasing and convex downward, such that*

$$\lim_{t\to\infty} \frac{G(t)}{t} = \infty, \tag{1.7}$$

$$\sup_n MG(|\xi_n|) < \infty. \tag{1.8}$$

[3] We recall that the weak compactness of the family $\{\xi_\alpha : \alpha \in \mathscr{U}\}$ means that each sequence ξ_{α_i}, $\alpha_i \in \mathscr{U}$, $i = 1, 2, \ldots$, contains a weakly convergent subsequence.

1.1.6 The main inequalities for mathematical expectations

Hölder inequality. If $p > 1$, $(1/p) + (1/q) = 1$, then

$$M|\xi\eta| \leq (M|\xi|^p)^{1/p}(M|\eta|^q)^{1/q}. \tag{1.9}$$

As particular cases of (1.9) we obtain the following inequalities:

(1) Cauchy–Buniakowski inequality:

$$M|\xi\eta| \leq \sqrt{M\xi^2 M\eta^2}; \tag{1.10}$$

(2) Minkowski inequality: if $p \geq 1$, then

$$(M|\xi + \eta|^p)^{1/p} \leq (M|\xi|^p)^{1/p} + (M|\eta|^p)^{1/p}. \tag{1.11}$$

(3) Jensen's inequality: let $f(x)$ be a continuous convex (downward) function of one variable and ξ be an integrable random variable ($M|\xi| < \infty$) such that $M|f(\xi)| < \infty$. Then

$$f(M\xi) \leq Mf(\xi). \tag{1.12}$$

Note. All the above inequalities remain correct if the operation of mathematical expectation $M(\cdot)$ is replaced by the conditional mathematical expectation $M(\cdot|\mathscr{G})$, where $\mathscr{G}$ is a sub-σ-algebra of the main probability space $(\Omega, \mathscr{F}, P)$.

(4) Chebyshev inequality: if $M|\xi| < \infty$, then for all $a > 0$

$$P\{|\xi| > a\} \leq \frac{M|\xi|}{a}.$$

1.1.7 The Borel–Cantelli lemma

The Borel–Cantelli lemma is the main tool in the investigation of properties that hold "with probability 1." Let $A_1, A_2, \ldots,$ be a sequence of sets from $\mathscr{F}$. A set A^* is called an *upper limit of the sequence of sets* $A_1, A_2, \ldots,$ and is denoted by $A^* = \lim_n \sup A_n$, if A^* consists of points ω, each of which belongs to an infinite number of A_n. Starting from this definition, it is easy to show that

$$A^* = \bigcap_{n=1}^{\infty} \bigcup_{m=n}^{\infty} A_m.$$

Often it is also written $A^* = \{A_n \text{ i.o.}\}$

A set A_* is called the *lower limit of the sequence of sets* $A_1, A_2, \ldots,$ and is denoted by $A_* = \lim_n \inf A_n$, if A_* consists of points ω, each of which belongs to all A_n, with the exception of a finite number at the most. According to this definition

$$A_* = \bigcup_{n=1}^{\infty} \bigcap_{m=n}^{\infty} A_m.$$

Borel–Cantelli lemma. *If* $\sum_{n=1}^{\infty} P(A_n) < \infty$, *then* $P(A^*) = 0$. *But if* $\sum_{n=1}^{\infty} PA_n = \infty$ *and sets* $A_1, A_2, \ldots$ *are independent (i.e.,* $P(A_{i_1}, \ldots, A_{i_k}) = P(A_{i_1}) \cdots P(A_{i_k})$ *for any different* $i_1, \ldots, i_k$*), then* $P(A^*) = 1$.

1.1.8 Gaussian systems

A random variable $\xi = \xi(\omega)$, defined on the probability space $(\Omega, \mathscr{F}, P)$, is called *Gaussian* (or *normal*) if its characteristic function

$$\varphi(t) \equiv Me^{it\xi} = e^{itm-(\sigma^2/2)t^2}, \tag{1.13}$$

where $-\infty < m < \infty$, $\sigma^2 < \infty$. In the nondegenerate case ($\sigma^2 > 0$) the distribution function

$$F_\xi(x) = P\{\omega : \xi(\omega) \le x\} \tag{1.14}$$

has the density

$$f_\xi(x) = \frac{1}{\sqrt{2\pi}\sigma} e^{-(x-m)^2/2\sigma^2}, \qquad -\infty < x < \infty. \tag{1.15}$$

In the degenerate case ($\sigma^2 = 0$), evidently, $P\{\xi = m\} = 1$.

Parameters m and σ^2 of the normal distribution given by the characteristic function appearing in (1.13) have simple meaning: $m = M\xi$, $\sigma^2 = D\xi$, where $D\xi = M(\xi - M\xi)^2$ is the dispersion of the random variable ξ. If $m = 0$, then $M\xi^{2n} = (2n-1)!!\,\sigma^{2n}$.

Further on, the notation[4] $\xi \sim N(m, \sigma^2)$ will often be used, noting that ξ is a Gaussian variable with parameters m and σ^2.

A random vector $\xi = (\xi_1, \ldots, \xi_n)$, consisting of random variables $\xi_1, \ldots, \xi_n$, is called *Gaussian* (or *normal*), if its characteristic function

$$\varphi(t) = Me^{i(t,\xi)}, \qquad t = (t_1, \ldots, t_n), \qquad t_j \in \mathbb{R}^1, \qquad (t, \xi) = \sum_{j=1}^{n} t_j \xi_j,$$

is given by a formula

$$\varphi(t) = e^{i(t,m)-(1/2)(Rt,t)}, \tag{1.16}$$

where

$$m = (m_1, \ldots, m_n), \qquad |m_i| < \infty, \qquad (Rt, t) = \sum_{k,j} r_{k,j} t_k t_j,$$

[4] Note that usually it is written $\xi \sim N(m, \sigma)$. We find it, however, convenient to use the notation $\xi \sim N(m, \sigma^2)$.

and $R = \|r_{kj}\|$ is a nonnegative definite symmetric matrix: $\sum_{k,j} r_{kj} t_k t_j \geq 0$, $t_j \in \mathbb{R}^1$, $r_{kj} = r_{jk}$

In the nondegenerate case (when matrix R is positive definite and, therefore, $|R| = \det R > 0$) the distribution function $F_\xi(x_1, \ldots, x_n) = P\{\omega: \xi_1 \leq x_1, \ldots, \xi_n \leq x_n\}$ of the vector $\xi = (\xi_1, \ldots, \xi_n)$ has the density

$$f_\xi(x_1, \ldots, x_n) = \frac{|A|^{1/2}}{(2\pi)^{n/2}} \exp\left\{ -\frac{1}{2} \sum_{i,j} a_{ij}(x_i - m_i)(x_j - m_j) \right\}, \tag{1.17}$$

where

$$A = \|a_{ij}\| \text{ is a matrix reciprocal to } R \qquad (A = R^{-1}, |A| = \det A).$$

Making use of the notations introduced above, the density $f_\xi(x_1, \ldots, x_n)$ can (in the nondegenerate case) be rewritten in the following form:[5]

$$f_\xi(x_1, \ldots, x_n) = \frac{|A|^{1/2}}{(2\pi)^{n/2}} \exp\{ -\tfrac{1}{2}(A(x - m), (x - m))\},$$

where $x = (x_1, \ldots, x_n)$, $m = (m_1, \ldots, m_n)$.

As in the one-dimensional case ($n = 1$), a vector $m = (m_1, \ldots, m_n)$ and a matrix $R = \|r_{ij}\|$ allow a simple and obvious interpretation:

$$m_i = M\xi_i, \qquad r_{ij} = \operatorname{cov}(\xi_i, \xi_j) = M(\xi_i - m_i)(\xi_j - m_j). \tag{1.18}$$

In other words, m is the mean value vector, and R is the covariance matrix of the vector $\xi = (\xi_1, \ldots, \xi_n)$.

The system of random variables $\xi = \{\xi_\alpha, \alpha \in \mathscr{U}\}$, where $\mathscr{U}$ is a finite or infinite set, is called *Gaussian*, if any linear combination

$$c_{\alpha_1}\xi_{\alpha_1} + \cdots + c_{\alpha_n}\xi_{\alpha_n}, \qquad \alpha_i \in \mathscr{U}, \qquad c_{\alpha_i} \in \mathbb{R}^1, \qquad i = 1, 2, \ldots, n,$$

is a Gaussian random variable. Sometimes it is convenient to use another, equivalent, definition of a Gaussian system. According to this definition a system of random variables $\xi = \{\xi_\alpha, \alpha \in \mathscr{U}\}$ is called *Gaussian* if, for all n and for all $\alpha_1, \ldots, \alpha_n \in \mathscr{U}$, the random vector $(\xi_{\alpha_1}, \ldots, \xi_{\alpha_n})$ is Gaussian.

1.2 Random processes: basic notions

1.2.1 Definitions: measurability

Let $(\Omega, \mathscr{F}, P)$ be a probability space and $T = [0, \infty)$. The family $X = (\xi_t)$, $t \in T$, of random variables $\xi_t = \xi_t(\omega)$ is called a (*real*) *random process with continuous time* $t \in T$. In the case where the time parameter t is confined to the set $\mathbb{N} = \{0, 1, \ldots,\}$, the family $X = (\xi_t)$, $t \in N$, is called a random sequence or a random process with discrete time.

With $\omega \in \Omega$ fixed, the time function $\xi_t(\omega)$ ($t \in T$ or $t \in N$) is called a *trajectory* or *realization* (or *sample function*) *corresponding to an elementary event* ω.

[5] As in (1.16), $(\cdot, \cdot)$ denotes a scalar product.

The σ-algebras $\mathscr{F}_t^\xi = \sigma\{\xi_s : s \le t\}$, being the smallest σ-algebras with respect to which the random variables ξ_s, $s \le t$, are measurable, are naturally associated with any random process $X = (\xi_t)$, $t \in Z$ (where $Z = T$ in the case of continuous time and $Z = \mathbb{N}$ in the case of discrete time). For the conditional mathematical expectations $M(\eta | \mathscr{F}_t^\xi)$ we shall also, sometimes, use the following notations: $M(\eta | \xi_s, s \le t)$ and $M(\eta | \xi_0^t)$. For the conditional probabilities $P(A | \mathscr{F}_t^\xi)$ similar notations are used: $P(A | \xi_s, s \le t)$ and $P(A | \xi_0^t)$.

The random process $X = (\xi_t)$, $t \in T$, is called *measurable* if, for all Borel sets $B \in \mathscr{B}$ of the real line $\mathbb{R}^1$,

$$\{(\omega, t) : \xi_t(\omega) \in B\} \in \mathscr{F} \times \mathscr{B}(T),$$

where $\mathscr{B}(T)$ is a σ-algebra of Borel sets on $T = [0, \infty)$.

The next theorem illustrates the significance of the concept of process measurability, given in the complete probability space $(\Omega, \mathscr{F}, P)$.

Theorem 1.9 (Fubini). *Let $X = (\xi_t)$, $t \in T$, be a measurable random process. Then:*

(1) *almost all trajectories of this process are measurable (relative to Borel) functions of $t \in T$;*
(2) *if $M\xi_t$ exists for all $t \in T$, then $m_t = M\xi_t$ is a measurable function of $t \in T$;*
(3) *if S is a measurable set in $T = [0, \infty)$ and $\int_S M|\xi_t| dt < \infty$, then*

$$\int_S |\xi_t| dt < \infty \qquad (P\text{-a.s.})$$

i.e., almost all the trajectories $\xi_t = \xi_t(\omega)$ are integrable on the set S and

$$\int_S M\xi_t \, dt = M \int_S \xi_t \, dt.$$

Let $F = (\mathscr{F}_t)$, $t \in T$, be a nondecreasing family of σ-algebras, $\mathscr{F}_s \subseteq \mathscr{F}_t \subseteq \mathscr{F}$, $s \le t$. We say that a (measurable) random process $X = (\xi_t)$, $t \in T$, is *adapted to a family of σ-algebras $F = (\mathscr{F}_t)$*, $t \in T$, if for any $t \in T$ the random variables ξ_t are $\mathscr{F}_t$-measurable. For brevity, such a random process will be denoted $X = (\xi_t, \mathscr{F}_t)$, $t \in T$, or simply $X = (\xi_t, \mathscr{F}_t)$ and called *F-adapted* or *nonanticipative*.

The random process $X = (\xi_t, \mathscr{F}_t)$, $t \in T$, is called progressively measurable if, for any $t \in T$,

$$\{(\omega, s \le t) : \xi_s(\omega) \in B\} \in \mathscr{F}_t \times \mathscr{B}([0, t]),$$

where B is a Borel set on $\mathbb{R}^1$, and $\mathscr{B}([0, t])$ is a σ-algebra of Borel sets on $[0, t]$.

It is evident that any progressively measurable random process $X = (\xi_t, \mathscr{F}_t)$, $t \in T$, is measurable and adapted to $F = (\mathscr{F}_t)$, $t \in T$.

Any (right, or left) continuous random process $X = (\xi_t, \mathscr{F}_t)$, $t \in T$, is progressively measurable (see [126]).

Two random processes $X = (\xi_t(\omega))$, $t \in T$, and $X' = (\xi_t'(\omega))$, $t \in T$, given, perhaps, on different probability spaces $(\Omega, \mathscr{F}, P)$ and $(\Omega', \mathscr{F}', P')$, will be called *weakly equivalent* if

$$P\{\omega:\xi_{t_1} \in A_1, \ldots, \xi_{t_n} \in A_n\} = P'\{\omega':\xi'_{t_1} \in A_1, \ldots, \xi'_{t_n} \in A_n\}$$

for any $t_1, \ldots, t_n \in T$ and Borel sets $A_1, \ldots, A_n$ of a real line $\mathbb{R}^1$.

Random processes $X = (\xi_t(\omega))$ and $X' = (\xi_t'(\omega))$, $t \in T$, given on the same probability space $(\Omega, \mathscr{F}, P)$ are called *stochastically equivalent* if $P(\xi_t \neq \xi_t') = 1$ for all $t \in T$.

The process $X' = (\xi_t'(\omega))$, $t \in T$, being stochastically equivalent to $X = (\xi_t(\omega))$, $t \in T$, is called the *modification* of the process X.

It is known that if the process $X = (\xi_t(\omega))$, $t \in T$, is measurable and is adapted to (with $F = (\mathscr{F}_t)$, $t \in T$), then it has a progressively measurable modification (see [126]).

Let $\xi = \xi(\omega)$ and $\eta = \eta(\omega)$ be two random variables defined on $(\Omega, \mathscr{F})$, η being $\mathscr{F}^\xi$-measurable, where $\mathscr{F}^\xi = \sigma(\xi)$. Then there exists a Borel function $Y = Y(x)$, $x \in \mathbb{R}^1$, such that $\eta(\omega) = Y(\xi(\omega))$ (P-a.s.). Later on the following generalization of this fact will be often used (see [46], p. 543).

Let $\xi(\omega) = (\xi_t(\omega))$, $0 \le t \le T$, be a random process defined on $(\Omega, \mathscr{F})$, $\mathscr{F}_T^\xi = \sigma\{\omega:\xi_t(\omega), t < T\}$ and $\mathscr{B}_T$ be the smallest σ-algebra on the space $\mathbb{R}^T$ of all real functions $x = (x_t)$, $0 \le t \le T$, containing sets of the form $\{x:x_{t_1} \in A_1, \ldots, x_{t_n} \in A_n\}$, where $0 \le t_i \le T$ and A_i are Borel sets on the real line, $i = 1, \ldots, n$, $n = 1, 2, \ldots$. If the random variable $\eta = \eta(\omega)$ is $\mathscr{F}_T^\xi$-measurable, then a $\mathscr{B}_T$-measurable function $Y = Y(x)$, $x \in \mathbb{R}^T$, can be found such that $\eta(\omega) = Y(\xi(\omega))$, ($P$-a.s.).[6] Moreover, there exists at most a countable number of points $s_1, s_2, \ldots$ belonging to the interval $[0, T]$, and a (measurable) function $Y = Y(z)$, defined for $z = (z_1, z_2, \ldots) \in \mathbb{R}^\infty$, such that

$$\eta(\omega) = Y(\xi_{s_1}(\omega), \xi_{s_2}(\omega), \ldots) \qquad (P\text{-a.s.}).$$

The following assumption will often be used in the book. Let $X = (\xi_t)$, $t \in T$, be the measurable random process on $(\Omega, \mathscr{F}, P)$ with $M|\xi_t| < \infty$, $t \in T$, and let $F = (\mathscr{F}_t)$, $t \in T$, be the family of nondecreasing sub-σ-algebras $\mathscr{F}$. Then the conditional mathematical expectations $\eta_t = M(\xi_t|\mathscr{F}_t)$ can be chosen so that the process $\eta = (\eta_t)$, $t \in T$, is measurable (see [126], [52]).

In accord with this result, from now on (even if it is not specially mentioned) it will be always assumed that the conditional mathematical expectations $M(\xi_t|\mathscr{F}_t)$, $t \in T$, have been defined so that the process $\eta_t = M(\xi_t|\mathscr{F}_t)$, $t \in T$, is measurable.

[6] For random variables η which are $\mathscr{F}_T^\xi$-measurable the notations $\eta = \eta_T(\xi)$, $\eta = \eta(T, \xi)$ will also be used often.

1.2.2 Continuity

The random process $X = (\xi_t)$, $t \in T$, is called *stochastically continuous at the point* $t_0 \in T$ if, for any $\varepsilon > 0$,

$$P\{|\xi_s - \xi_{t_0}| > \varepsilon\} \to 0, \qquad s \to t_0. \tag{1.19}$$

If (1.19) holds for all $t_0 \in S \subseteq T$, then the process X is called *stochastically continuous* (*on the set* S).

The random process $X = (\xi_t)$, $t \in T$, is called *continuous* (*right continuous, left continuous*) on $S \subseteq T$ if almost all its trajectories are continuous (right continuous, left continuous) for $t \in S \subseteq T$. In other words, there must exist a set such that $N \in \mathscr{F}$ with $P(N) = 0$ such that for all $\omega \notin N$ the trajectories $\xi_t(\omega)$, $t \in S$, are continuous (right continuous, left continuous) functions.

The following theorem provides the conditions for the existence of a continuous modification of the process $X = (\xi_t(\omega))$, $t \in [a, b]$.

Theorem 1.10 (Kolmogorov's criterion). *In order that the random process $X = (\xi_t)$, $t \in [a, b]$, permit a continuous modification $X^* = (\xi_t^*)$, $t \in [a, b]$, it is sufficient that there are constants $a > 0$, $\varepsilon > 0$, and C such that*

$$M|\xi_{t+\Delta} - \xi_t|^a \leq C|\Delta|^{1+\varepsilon} \tag{1.20}$$

for all $t, t + \Delta \in [a, b]$.

The random process $X = (\xi_t)$, $t \in T$, is called *continuous in the mean square at the point* $t_0 \in T$, if

$$M|\xi_s - \xi_{t_0}|^2 \to 0, \qquad s \to t_0 \tag{1.21}$$

If (1.21) holds for all the points $t_0 \in S \subseteq T$, then the process X will be called *continuous in the mean square* (*on the set* S).

1.2.3 Some classes of processes

We will now consider the main classes of random processes.

(1) *Stationary processes.* The random process $X = (\xi_t(\omega))$, $t \in T = [0, \infty)$, is said to be *stationary* (or *stationary in a narrow sense*), if for any real Δ the finite dimensional distributions do not change with the shift on Δ:

$$P\{\xi_{t_1} \in A_1, \ldots, \xi_{t_n} \in A_n\} = P\{\xi_{t_1+\Delta} \in A_1, \ldots, \xi_{t_n+\Delta} \in A_n\},$$

for $t_1, \ldots, t_n, t_1 + \Delta, \ldots, t_n + \Delta \in T$.

The random process $X = (\xi_t(\omega))$, $t \in T = [0, \infty)$, is called *stationary in a wide sense* if

$$M\xi_t^2 < \infty \qquad (t \in T) \quad \text{and} \quad M\xi_t = M\xi_{t+\Delta}, \qquad M\xi_s\xi_t = M\xi_{s+\Delta}\xi_{t+\Delta},$$

i.e., if the first and second moments do not change with the shift.

(2) *Markov processes.* The real random process $X = (\xi_t, \mathscr{F}_t)$, $t \in T$, given on $(\Omega, \mathscr{F}, P)$ is called *Markov with respect to a nondecreasing system of σ-algebras* $F = (\mathscr{F}_t)$, $t \in T$, if (P-a.s.).[7]

$$P(A \cap B \mid \xi_t) = P(A \mid \xi_t) P(B \mid \xi_t) \tag{1.22}$$

for any $t \in T$, $A \in \mathscr{F}_t$, $B \in \mathscr{F}^{\xi}_{[t,\infty)} = \sigma(\xi_s, s \geq t)$.

The real random process $X = (\xi_t)$, $t \in T$, is called (*simply*) *Markov*, if it is Markov with respect to a system of σ-algebras $\mathscr{F}_t = \mathscr{F}^{\xi}_t = \sigma(\xi_s, s \leq t)$.

The following statements provide different but equivalent definitions of Markov behavior of the process $X = (\xi_t, \mathscr{F}_t)$, $t \in T$.

Theorem 1.11. *The following conditions are equivalent:*

(1) $X = (\xi_t, \mathscr{F}_t)$, $t \in T$, *is a Markov process with respect to* $F = (\mathscr{F}_t)$, $t \in T$;

(2) *for each* $t \in T$ *and any bounded* $\mathscr{F}^{\xi}_{[t,\infty)}$*-measurable random variable* η,

$$M(\eta \mid \mathscr{F}_t) = M(\eta \mid \xi_t) \qquad (P\text{-a.s.}); \tag{1.23}$$

(3) *for* $t \geq s \geq 0$ *and any (measurable) function* $f(x)$ *with* $\sup_x |f(x)| < \infty$,

$$M[f(\xi_t) \mid \mathscr{F}_s] = M[f(\xi_t) \mid \xi_s]. \tag{1.24}$$

For deciding when the process $X = (\xi_t)$, $t \in T$, is Markov, the following criterion is useful.

Theorem 1.12. *In order that the random process* $X = (\xi_t)$, $t \in T$, *be Markov, it is necessary and sufficient that for each (measurable) function* $f(x)$ *with* $\sup_x |f(x)| < \infty$ *and any collection* t_n *where* $0 \leq t_1 \leq t_2 \leq \cdots \leq t_n \leq t$,

$$M[f(\xi_t) \mid \xi_{t_1}, \ldots, \xi_{t_n}] = M[f(\xi_t) \mid \xi_{t_n}]. \tag{1.25}$$

Processes with independent increments are a significant special case of Markov processes. One can say that the process $X = (\xi_t)$, $t \in T$, is a *process with independent increments* if, for any $t_n > t_{n-1} > \cdots > t_1 > 0$, the increments $\xi_{t_2} - \xi_{t_1}, \ldots, \xi_{t_n} - \xi_{t_{n-1}}$ yield a system of independent random variables.

A process with independent increments is called *homogeneous* (*relative to time*) if the distribution of the probabilities of the increments $\xi_t - \xi_s$ depends only on the difference $t - s$. Often such processes are also called *processes with stationary independent increments.*

(3) *Martingales.* The random process $X = (\xi_t, \mathscr{F}_t)$, $t \in T$, is called a *martingale* (*with respect to the system* $F = (\mathscr{F}_t)$, $t \in T$) if $M|\xi_t| < \infty$, $t \in T$ and

$$M(\xi_t \mid \mathscr{F}_s) = \xi_s \qquad (P\text{-a.s.}), \qquad t \geq s. \tag{1.26}$$

[7] In accord with previous conventions, $P(\cdot \mid \xi_t)$ denotes the conditional probability $P(\cdot \mid \sigma(\xi_t))$.

A considerable part of this book deals with martingales (and a closely related concept—semimartingales).

1.3 Markov times

1.3.1 Definitions

Let $(\Omega, \mathscr{F}, P)$ be a probability space, and let $F = (\mathscr{F}_t)$, $t \in T$ where $T = [0, \infty)$, be a nondecreasing sequence of sub-σ-algebras ($\mathscr{F}_s \subseteq \mathscr{F}_t \subseteq \mathscr{F}, s \leq t$). As noted in Section 1.1, the σ-algebra $\mathscr{F}$ is assumed to be augmented relative to the measure $P(\mathscr{F} = \mathscr{F}^P)$. From now on, it will be assumed that the σ-algebras $\mathscr{F}_t$, $t \in T$, are augmented by the sets from $\mathscr{F}$ which have P-measure zero.

The random variable (i.e., $\mathscr{F}$-measurable function) $\tau = \tau(\omega)$, taking on the values in $\bar{T} = [0, \infty]$, is called a *Markov time (relative to the system $F = (\mathscr{F}_t)$, $t \in T$)* if, for all $t \in T$,

$$\{\omega : \tau(\omega) \leq t\} \in \mathscr{F}_t. \tag{1.27}$$

The Markov times (m.t.), called also *random variables*, are independent of the future. If $P\{\tau(\omega) < \infty\} = 1$, then m.t. is called a *stopping time* (s.t.).

With every m.t. $\tau = \tau(\omega)$ (relative to the system $F = (\mathscr{F}_t)$, $t \in T$) is adapted to the σ-algebra $\mathscr{F}_\tau$-union of those sets $A \subseteq \{\omega : \tau < \infty\}$ for which $A \cap \{\tau \leq t\} \in \mathscr{F}_t$ for all $t \in T$.

If $\mathscr{F}_t$ denotes the totality of the events observed before time t, then $\mathscr{F}_\tau$ consists of the events observed before random time τ.

Techniques based on Markov times will be rather extensively used in this book.

1.3.2 Properties of Markov times

For any $t \in T$ we set[8]

$$\mathscr{F}_{t+} = \bigcap_{s>t} \mathscr{F}_s, \qquad \mathscr{F}_{t-} = \sigma\left(\bigcup_{s<t} \mathscr{F}_s\right), \qquad \mathscr{F}_{0-} = \mathscr{F}_0$$

and

$$\mathscr{F}_\infty = \sigma\left(\bigcup_{s \geq 0} \mathscr{F}_s\right).$$

The sequence of σ-algebras $F = (\mathscr{F}_t)$, $t \in T$, is called *right continuous* if $\mathscr{F}_t = \mathscr{F}_{t+}$ for all $t \in F$. Note that the family $F_+ = (\mathscr{F}_{t+})$ is always *right continuous*.

Lemma 1.1. *Let $\tau = \tau(\omega)$ be a* m.t. *Then $\{\tau < t\} \in \mathscr{F}_t$, and, consequently, $\{\tau = t\} \in \mathscr{F}_t$.*

[8] The smallest σ-algebra $\sigma(\bigcup_{s<t} \mathscr{F}_s)$ is sometimes denoted by $\bigvee_{s<t} \mathscr{F}_s$.

PROOF. The lemma follows from

$$\{\tau < t\} = \bigcup_{k=1}^{\infty} \left\{\tau \le t - \frac{1}{k}\right\} \quad \text{and} \quad \left\{\tau \le t - \frac{1}{k}\right\} \in \mathscr{F}_{t-(1/k)} \subseteq \mathscr{F}_t. \qquad \square$$

The converse of Lemma 1.1 is wrong. But the following is correct.

Lemma 1.2. *If the family $F = (\mathscr{F}_t)$, $t \in T$, is right continuous and $\tau = \tau(\omega)$ is the random variable with values in $[0, \infty]$ such that $\{\tau < t\} \in \mathscr{F}_t$ for all $t \in T$, then τ is a Markov time, i.e., $\{\tau \le t\} \in \mathscr{F}_t$, $t \in T$.*

PROOF. Since $\{\tau < t\} \in \mathscr{F}_t$, then $\{\tau \le t\} \in \mathscr{F}_{t+\varepsilon}$ for any $\varepsilon > 0$. Consequently, $\{\tau \le t\} \in \mathscr{F}_{t+} = \mathscr{F}_t$. $\square$

Lemma 1.3. *If τ_1, τ_2 are Markov times, then $\tau_1 \wedge \tau_2 \equiv \min(\tau_1, \tau_2)$, $\tau_1 \vee \tau_2 \equiv \max(\tau_1, \tau_2)$ and $\tau_1 + \tau_2$ are also Markov times.*

PROOF. The lemma follows directly from the relations

$$\begin{aligned}
\{\tau_1 \wedge \tau_2 \le t\} &= \{\tau_1 \le t\} \cup \{\tau_2 \le t\}, \\
\{\tau_1 \vee \tau_2 \le t\} &= \{\tau_1 \le t\} \cap \{\tau_2 \le t\}, \\
\{\tau_1 + \tau_2 \le t\} &= \{\tau_1 = 0, \tau_2 = t\} \cup \{\tau_1 = t, \tau_2 = 0\} \\
&\quad \cup \left(\bigcup_{\substack{a+b<t \\ a,b \ge 0}} [\{\tau_1 < a\} \cap \{\tau_2 < b\}]\right),
\end{aligned}$$

where a, b are rational numbers. $\square$

Lemma 1.4. *Let $\tau_1, \tau_2, \ldots$ be a sequence of Markov times. Then* $\sup \tau_n$ *is also a Markov time. If, further, the family $F = (\mathscr{F}_t)$, $t \in T$, is right continuous, then* $\inf \tau_n$, $\lim_n \sup \tau_n$ and $\lim_n \inf \tau_n$ *are also Markov times.*

PROOF. This follows from

$$\left\{\sup_n \tau_n \le t\right\} = \bigcap_n \{\tau_n \le t\} \in \mathscr{F}_t, \qquad \left\{\inf_n \tau_n < t\right\} = \bigcup_n \{\tau_n < t\} \in \mathscr{F}_t$$

and, for $\limsup_n \tau_n = \inf_{n \ge 1} \sup_{m \ge n} \tau_m$, $\liminf_n \tau_n = \sup_{n \ge 1} \inf_{m \ge n} \tau_m$,

$$\begin{aligned}
\left\{\limsup_n \tau_n < t\right\} &= \bigcup_{k=1}^{\infty} \bigcup_{n=1}^{\infty} \bigcap_{m=n}^{\infty} \left\{\tau_m < t - \frac{1}{k}\right\}, \\
\left\{\liminf_n \tau_n > t\right\} &= \bigcup_{k=1}^{\infty} \bigcap_{n=1}^{\infty} \bigcup_{m=n}^{\infty} \left\{\tau_m > t + \frac{1}{k}\right\}.
\end{aligned} \qquad \square$$

Lemma 1.5. *Any Markov time $\tau = \tau(\omega)$ (relative to $F = (\mathscr{F}_t)$, $t \in T$) is a $\mathscr{F}_\tau$-measurable random variable. If τ and σ are two Markov times and $\tau(\omega) \le \sigma(\omega)$ (P-a.s.), then $\mathscr{F}_\tau \subseteq \mathscr{F}_\sigma$*

Proof. Let $A = \{\tau \le s\}$. It should be shown that $A \cap \{\tau \le t\} \in \mathscr{F}_t$, $t \in T$. We have

$$\{\tau \le s\} \cap \{\tau \le t\} = \{\tau \le t \wedge s\} \in \mathscr{F}_{t \wedge s} \subseteq \mathscr{F}_t.$$

Therefore, m.t. τ is $\mathscr{F}_\tau$-measurable. □

Now let $A \subseteq \{\omega: \sigma < \infty\}$ and $A \in \mathscr{F}_\tau$. Then, since $P\{\tau \le \sigma\} = 1$ and σ-algebras $\mathscr{F}_t$ are augmented, the set $A \cap \{\sigma \le t\}$ corresponds to the set $A \cap \{\tau \le t\} \cap \{\sigma \le t\}$ which belongs to $\mathscr{F}_t$, within the sets of zero probability. Therefore, the set $A \cap \{\sigma \le t\} \in \mathscr{F}_t$, and hence $A \in \mathscr{F}_\sigma$

Lemma 1.6. *Let $\tau_1, \tau_2, \ldots$ be a sequence of Markov times relative to the non-decreasing continuous to the right system of σ-algebras $F = (\mathscr{F}_t)$, $t \in T$, and let $\tau = \inf_n \tau_n$. Then $\mathscr{F}_\tau = \bigcap_n \mathscr{F}\tau_n$.*

Proof. According to Lemma 1.4, τ is a m.t. Hence, by Lemma 1.5, $\mathscr{F}_\tau \subseteq \bigcap_n \mathscr{F}_{\tau_n}$. On the other hand, let $A \in \bigcap_n \mathscr{F}_{\tau_n}$. Then

$$A \cap \{\tau < t\} = A \cap \left(\bigcup_n (\tau_n < t)\right) = \bigcup_n (A \cap \{\tau_n < t\}) \in \mathscr{F}_t.$$

From this, due to the continuity to the right ($\mathscr{F}_t - \mathscr{F}_{t+}$), it follows that $A \in \mathscr{F}_\tau$. □

Lemma 1.7. *Let τ and σ be Markov times relative to $F = (\mathscr{F}_t)$, $t \in T$. Then each of the events $\{\tau < \sigma\}$, $\{\tau > \sigma\}$, $\{\tau \le \sigma\}$, $\{\tau \ge \sigma\}$ and $\{\tau = \sigma\}$ belongs at the same time to $\mathscr{F}_\tau$ and $\mathscr{F}_\sigma$.*

Proof. For all $t \in T$,

$$\{\tau < \sigma\} \cap \{\sigma \le t\} = \bigcup_{r<t} (\{\tau < r\} \cap \{r < \sigma \le t\}) \in \mathscr{F}_t,$$

where the r are rational numbers. Hence $\{\tau < \sigma\} \in \mathscr{F}_\sigma$. Further,

$$\{\tau < \sigma\} \cap (\tau \le t) = \bigcup_{r<t} [(\{\tau \le r\} \cap \{r < \sigma\}) \cup (\{\tau \le t\} \cap \{t < \sigma\})] \in \mathscr{F}_t,$$

i.e., $\{\sigma < \tau\} \in \mathscr{F}_\tau$.

Analogously, it can be established that $\{\sigma < \tau\} \in \mathscr{F}_\tau$ and $\{\sigma < \tau\} \in \mathscr{F}_\sigma$ Consequently, $\{\tau \le \sigma\}$, $\{\sigma \le \tau\}$ and $\{\sigma = \tau\}$ belong to both $\mathscr{F}_\tau$ and $\mathscr{F}_\sigma$ □

The advantages of the concept of a progressively measurable random process, introduced in Section 1.2, are illustrated by the following.

Lemma 1.8. *Let $X = \{\xi_t, \mathscr{F}_t\}$, $t \in T$, be a real progressively measurable process and let $\tau = \tau(\omega)$ be a Markov time (relative to $F = (\mathscr{F}_t)$, $t \in T$) such that $P(\tau < \infty) = 1$. Then the function $\xi_\tau = \xi_{\tau(\omega)}(\omega)$ is $\mathscr{F}_\tau$-measurable.*

Proof. Let $\mathscr{B}$ be a system of Borel sets of the real line $\mathbb{R}^1$ and $t \in T$. We must show that for all $B \in \mathscr{B}$,

$$\{\xi_{\tau(\omega)}(\omega) \in B\} \cap \{\tau \le t\} \in \mathscr{F}_t.$$

Set $\sigma = \tau \wedge t$. Then

$$\begin{aligned}\{\xi_\tau \in B\} \cap \{\tau \le t\} &= \{\xi_\tau \in B\} \cap [\{\tau < t\} \cup \{\tau = t\}] \\ &= [\{\xi_\sigma \in B\} \cap \{\sigma < t\}] \cup [\{\xi_\tau \in B\} \cap \{\tau = t\}].\end{aligned}$$

It is clear that $\{\xi_\tau \in B\} \cap \{\tau = t\} \in \mathscr{F}_t$. If it is shown that ξ_σ is a $\mathscr{F}_t$-measurable function, then the event $\{\xi_\sigma \in B\} \cap \{\sigma < t\}$ will also belong to $\mathscr{F}_t$. Note now that the mapping $\omega \to (\omega, \sigma(\omega))$ is the measurable mapping $(\Omega, \mathscr{F}_t)$ in $(\Omega \times [0, t], \mathscr{F}_t \times \mathscr{B}([0, t]))$, and that the mapping $(\omega, s) \to \xi_s(\omega))$ of the space $(\Omega \times [0, t], \mathscr{F}_t \times \mathscr{B}([0, t]))$ in $(\mathbb{R}^1, \mathscr{B})$ is also measurable because of the progressive measurability of the process $X = (\xi_t, \mathscr{F}_t)$, $t \in T$. Therefore, the mapping $(\Omega, \mathscr{F}_t)$ in $(\mathbb{R}^1, \mathscr{B})$, given by $\xi_{\sigma(\omega)}(\omega)$, is measurable as a result of application of two measurable mappings. □

Corollary. *If $X = (\xi_t, \mathscr{F}_t)$, $t \in T$, is a right (or left) continuous process, then ξ_τ is $\mathscr{F}_\tau$-measurable.*

Lemma 1.9. *Let $\varkappa = \varkappa(\omega)$ be the integrable random variable ($M|\varkappa| < \infty$) and let τ be a Markov time relative to the system $F = (\mathscr{F}_t)$, $t \in T$. Then on the set $\{\omega : \tau = t\}$ the conditional mathematical expectation $M(\varkappa | \mathscr{F}_\tau)$ coincides with $M(\varkappa | \mathscr{F}_t)$ i.e.,*

$$M(\varkappa | \mathscr{F}_\tau) = M(\varkappa | \mathscr{F}_t), \qquad (\{\tau = t\}; (P\text{-a.s.}))$$

Proof. It must be shown that

$$P[\{\tau = t\} \cap \{M(\varkappa | \mathscr{F}_\tau) \ne M(\varkappa | \mathscr{F}_t)\}] = 0$$

or, what is equivalent,

$$\chi M(\varkappa | \mathscr{F}_\tau) = \chi M(\varkappa | \mathscr{F}_t) \qquad (P\text{-a.s.})$$

where $\chi = \chi_{\{\tau = t\}}$ is the characteristic function of the set $\{\tau = t\}$. Since the random variable χ is $\mathscr{F}_\tau$- and $\mathscr{F}_t$-measurable (Lemma 1.7), then

$$\chi M(\varkappa | \mathscr{F}_\tau) = M(\varkappa\chi | \mathscr{F}_\tau) \quad \text{and} \quad \chi M(\varkappa | \mathscr{F}_t) = M(\varkappa\chi | \mathscr{F}_t).$$

We shall show that $M(\varkappa\chi | \mathscr{F}_\tau) = M(\varkappa\chi | \mathscr{F}_t)$ (P-a.s.). First of all, note that the random variable $M(\varkappa\chi | \mathscr{F}_t)$ is $\mathscr{F}_\tau$-measurable. Actually, let $s \in T$ and $a \in \mathbb{R}^1$. Then, if $t \le s$, evidently $\{M(\varkappa\chi | \mathscr{F}_t) \le a\} \cap \{\tau \le s\} \in \mathscr{F}_s$. But if $t > s$, then the set

$$\begin{aligned}\{M(\varkappa\chi | \mathscr{F}_t) \le a\} \cap \{\tau \le s\} &= \{\chi M(\varkappa | \mathscr{F}_t) \le a\} \cap \{\tau \le s\} \\ &\subseteq \{\tau \le s\} \in \mathscr{F}_s.\end{aligned}$$

Further, according to the definition of the conditional mathematical expectation for all $A \in \mathscr{F}_\tau$,

$$\int_A M(\varkappa\chi|\mathscr{F}_\tau)dP = \int_A \varkappa\chi\, dP = \int_{A \cap \{\tau = t\}} \varkappa\, dP. \tag{1.28}$$

The set $A \cap \{\tau = t\} \in \mathscr{F}_t$. Hence,

$$\int_{A \cap \{\tau = t\}} \varkappa\, dP = \int_{A \cap \{\tau = t\}} M(\varkappa|\mathscr{F}_t)dP = \int_A \chi M(\varkappa|\mathscr{F}_t)dP = \int_A M(\varkappa\chi|\mathscr{F}_t)dP. \tag{1.29}$$

Since $M(\varkappa\chi|\mathscr{F}_t)$ is $\mathscr{F}_\tau$-measurable because of the arbitrariness of the set $A \in \mathscr{F}_\tau$, from (1.28) and (1.29) it follows that $M(\varkappa\chi|\mathscr{F}_\tau) = M(\varkappa\chi|\mathscr{F}_t)$ (P-a.s.). □

1.3.3 Examples

The following lemma provides examples of the most commonly used Markov times.

Lemma 1.10. *Let $X = (\xi_t, t \in T)$ be a real process, right continuous, let $F = (\mathscr{F}_t)$, $t \in T$, be the nondecreasing family of right continuous σ-algebras $\mathscr{F}_t, = \mathscr{F}_{t+}$, and let C be an open set in $\overline{\mathbb{R}}^1 = [-\infty, \infty]$. Then the times*

$$\sigma_C = \inf\{t \geq 0 : \xi_t \in C\}, \qquad \tau_C = \inf\{t > 0 : \xi_t \in C\}$$

of the first and the first after $+0$ entries into the set C are Markov.

Proof. Let $D = \overline{\mathbb{R}}^1 - C$. Then, because of the right continuity of the trajectories of the process X and the closure of the set D,

$$\{\omega : \sigma_C \geq t\} = \{\omega : \xi_s \in D, s < t\} = \bigcap_{r<t} \{\xi_r \in D\},$$

where the r are rational numbers. Therefore,

$$\{\sigma_C < t\} = \bigcup_{r<t} \{\xi_r \in C\} \in \mathscr{F}_t.$$

Because of the assumption $\mathscr{F}_t = \mathscr{F}_{t+}$ and Lemma 1.2, it follows that σ_C is m.t. In similar fashion one can prove the Markov behavior of the time τ_C. □

The following frequently used lemma can be demonstrated by the same type of proof as the one given above.

Lemma 1.11. *Let $X = (\xi_t)$, $t \in T$, be the real continuous random process, let $\mathscr{F}_t^\xi = \sigma\{\omega : \xi_s, s \leq t\}$, and let D be a closed set in $\overline{\mathbb{R}}^1$. Then the time $\sigma_D = \inf(t \geq 0 : \xi_t \in D)$ is Markov with respect to the system $F^\xi = (\mathscr{F}_t^\xi)$, $t \in T$.*

1.4 Brownian motion processes

1.4.1 Definition

In the class of processes with stationary independent increments the process of Brownian motion plays the key role. We define this process and list its well-known properties.

The random process $\beta = (\beta_t)$, $0 \le t \le T$, given on the probability space $(\Omega, \mathcal{F}, P)$, is called a *Brownian motion process*[9] if:

(1) $\beta_0 = 0$ (P-a.s.);
(2) β is a process with stationary independent increments;
(3) increments $\beta_t - \beta_s$ have a Gaussian normal distribution with

$$M[\beta_t - \beta_s] = 0, \qquad D[\beta_t - \beta_s] = \sigma^2|t - s|;$$

(4) for almost all $\omega \in \Omega$ the functions $\beta_t = \beta_t(\omega)$ are continuous on $0 \le t \le T$.

In the case $\sigma^2 = 1$ the process β is often called the *standard Brownian motion process.*

The existence of such a process on (fairly "rich") probability spaces may be established in a constructive way. Thus let $\eta_1, \eta_2, \ldots$ be a sequence of independent Gaussian, $N(0, 1)$, random variables and $\varphi_1(t), \varphi_2(t), \ldots, 0 \le t \le T$ be an arbitrary complete orthonormal sequance in $L_2[0, T]$. Assume $\Phi_j(t) = \int_0^t \varphi_j(s)ds, j = 1, 2, \ldots$.

Theorem 1.13. *For each* t, $0 \le t \le T$, *the series*

$$\beta_t = \sum_{j=1}^{\infty} \eta_j \Phi_j(t)$$

converges (P-a.s.) *and defines the Brownian motion process on* $[0, T]$.

From its definition the following properties of the (standard) Brownian motion are easily found:

$$M\beta_t = 0, \qquad \operatorname{cov}(\beta_s, \beta_t) = M\beta_s\beta_t = \min(s, t);$$

$$P(\beta_t \le x) = \frac{1}{\sqrt{2\pi t}} \int_{-\infty}^{x} e^{-y^2/2t}\, dy, \qquad M|\beta_t| = \sqrt{\frac{2t}{\pi}}.$$

Let $\mathcal{F}_t^\beta = \sigma(\beta_s, s \le t)$. It is easy to check that the Brownian motion process is a martingale (relative to $(\mathcal{F}_t^\beta)$, $0 \le t \le T$):

$$M(\beta_t | \mathcal{F}_s^\beta) = \beta_s \qquad (P\text{-a.s.}), \qquad t \ge s, \tag{1.30}$$

$$M[(\beta_t - \beta_s)^2 | \mathcal{F}_s^\beta] = t - s \qquad (P\text{-a.s.}), \qquad t \ge s. \tag{1.31}$$

[9] The process of Brownian motion is also called a *Wiener process.* W reserve the term "Wiener" for processes defined somewhat differently (for details, see Section 4.2).

Like any process with independent increments, the Brownian motion process is Markovian, i.e.,

$$M[f(\beta_{t+s})|\mathscr{F}_t^\beta] = M[f(\beta_{t+s})|\beta_t] \qquad (P\text{-a.s.}), \qquad s \geq 0, \tag{1.32}$$

for any measurable function $f(x)$ with $\sup_x |f(x)| < \infty$. In particular, for any Borel set $B \in \mathscr{B}$ on $\mathbb{R}^1$,

$$P(\beta_t \in B|\mathscr{F}_s^\beta) = P(\beta_t \in B|\beta_s) \qquad (P\text{-a.s.}), \qquad t \geq s. \tag{1.33}$$

The important property of the process of Brownian motion $\beta = (\beta_t)$, $0 \leq t \leq T$, is that it is strong Markov in the following sense: for any Markov time $\tau = \tau(\omega)$ (relative to $(\mathscr{F}_t^\beta)$, $0 \leq t \leq T$) with $P(\tau(\omega) \leq T) = 1$, the following extension of the relationship given by (1.32) may be carried out:

$$M[f(\beta_{s+\tau})|\mathscr{F}_{\tau+}^\beta] = M[f(\beta_{s+\tau})|\beta_\tau] \qquad (P\text{-a.s.}), \tag{1.34}$$

where s is such that $P(s + \tau \leq T) = 1$.

The strong Markov property of the Brownian motion process can be given the following form: if the initial process $\beta = (\beta_t)$ is defined for all $t \geq 0$, then, for all Markov times $\tau = \tau(\omega)$ (relative to $(\mathscr{F}_t^\beta)$, $t \geq 0$) with $P(\tau < \infty) = 1$, the process

$$\tilde{\beta}_t = \beta_{t+\tau} - \beta_\tau$$

will also be a Brownian motion independent of the events of the σ-algebra $\mathscr{F}_{\tau+}^\beta$.

1.4.2 Properties of the trajectories of Brownian motion $\beta = (\beta_t)$, $t \geq 0$

The *law of the iterated logarithm* states that

$$P\left\{\limsup_{t\to\infty} \frac{|\beta_t|}{\sqrt{2t \ln \ln t}} = 1\right\} = 1, \tag{1.35}$$

the *local law of the iterated logarithm* states that

$$P\left\{\limsup_{t\to 0} \frac{|\beta_t|}{\sqrt{2t \ln \ln(1/t)}} = 1\right\} = 1, \tag{1.36}$$

and the *Hölder condition of Levy* states that

$$P\left\{\limsup_{0 \leq t-s=h\downarrow 0} \frac{|\beta_t - \beta_s|}{\sqrt{2h \ln(1/h)}} = 1\right\} = 1. \tag{1.37}$$

From (1.37) it follows that with probability 1 the trajectories of a Brownian motion process satisfy the Hölder condition with any exponent $\alpha < \frac{1}{2}$ (and do not satisfy the Hölder condition with the exponent $\alpha = \frac{1}{2}$; see (1.36)).

From (1.35)–(1.37) the following properties of a Brownian motion process result: with probability 1 its trajectories have an arbitrary number of "large" zeroes, are nondifferentiable for all $t > 0$, and are of unbounded variation on any (arbitrarily small) interval.

The set $\mathscr{z}(\omega) = \{t \leq 1, \beta_t(\omega) = 0\}$ of the roots of the equation $\beta_t(\omega) = 0$ has the following properties: $P(\mathscr{z}$ is not bounded$) = 1$; with probability 1, $\mathscr{z}(\omega)$ is close and has no isolated points; $P(\text{mes } \mathscr{z}(\omega) = 0) = 1$ where mes $\mathscr{z}(\omega)$ is the Lebesgue measure of the set $\mathscr{z}(\omega)$.

1.4.3 Certain distributions related to the Brownian motion process $\beta = (\beta_t)$, $t \geq 0$

Let

$$p(s, x, t, y) = \frac{\partial P_{s,x}(t, y)}{\partial y}$$

denote the density of the probability of the conditional distribution $P_{s,x}(t, y) = P\{\beta_t \leq y | \beta_s = x\}$. In the case of the standard Brownian motion process $(\sigma^2 = 1)$ the density

$$p(s, x, t, y) = \frac{1}{\sqrt{2\pi(t - s)}} e^{-(y-x)^2/2(t-s)} \tag{1.38}$$

satisfies the equations (which can be verified readily)

$$\frac{\partial p(s, x, t, y)}{\partial s} = -\frac{1}{2} \frac{\partial^2 p(s, x, t, y)}{\partial x^2}, \qquad s < t, \tag{1.39}$$

$$\frac{\partial p(s, x, t, x)}{\partial t} = \frac{1}{2} \frac{\partial^2 p(s, x, t, y)}{\partial x^2}, \qquad t > s. \tag{1.40}$$

Equations (1.39) and (1.40) are called *Kolmogorov's backward and forward equation.* (The forward equation, (1.40), is also called the *Fokker–Planck equation.*)

From the strictly Markov behavior of the process β, the relation

$$P\left(\max_{0 \leq s \leq t} \beta_s \geq x\right) = 2P(\beta_t \geq x) = \frac{2}{\sqrt{2\pi t}} \int_x^\infty e^{-y^2/2t}\, dy \tag{1.41}$$

is deduced (a *reflection principle*).

Denote the time of the first crossing of the level $a \geq 0$ by the process β by $\tau = \inf\{t \geq 0; \beta_t = a\}$. This is a Markov time (Lemma 1.11). Since

$$P(\tau \leq t) = P\left(\max_{0 \leq s \leq t} \beta_s \geq a\right),$$

then, because of (1.41),

$$P(\tau \leq t) = \frac{2}{\sqrt{2\pi t}} \int_a^\infty e^{-y^2/2t}\, dy = \sqrt{\frac{2}{\pi}} \int_{a/\sqrt{t}}^\infty e^{-y^2/2}\, dy; \tag{1.42}$$

from this we find that the density $p_\tau(t) = \partial P(\tau \le t)/\partial t$ exists and is given by the formula

$$p_\tau(t) = \frac{a}{\sqrt{2\pi}\, t^{3/2}} e^{-a^2/2t}. \tag{1.43}$$

From (1.43) it follows that $p_\tau(t) \sim (a/\sqrt{2\pi})t^{-3/2}$, as $t \to \infty$, and consequently, if $a > 0$, then $M\tau = \infty$.

Let now

$$\tau = \inf\{t \ge 0 : \beta_t = a - bt\}, \qquad a > 0, \qquad 0 \le b < \infty,$$

be the first crossing time of the line $a - bt$ by a Brownian motion process. It is known that in this case the density $p_\tau(t) = \partial P(\tau \le t)/\partial t$ is defined by the formula

$$p_\tau(t) = \frac{a}{\sqrt{2\pi}\, t^{3/2}} e^{-(bt - a)^2/2t}. \tag{1.44}$$

1.4.4 Transformations of the Brownian motion process $\beta = (\beta_t)$, $t \ge 0$

It may be readily verified that

$$y_t(\omega) = \begin{cases} 0, & t = 0, \\ t\beta_{1/t}(\omega), & t > 0, \end{cases}$$

and that

$$z_t(\omega) = c\beta_{t/c^2}(\omega), \qquad c > 0,$$

are also Brownian motion processes.

1.5 Some notions from mathematical statistics

1.5.1

In mathematical statistics the concept of the sample space $(X, \mathscr{A})$ consisting of the set of all sample points X and a σ-algebra $\mathscr{A}$ of its subsets is primary. Usually X is the space of sequences $x = (x_1, x_2, \ldots)$, where $x_i \in \mathbb{R}^k$, or the space of functions $x = (x_t)$, $t \ge 0$. In the problems of diffusion-type processes, discussed later on, the space of continuous functions is a sample space.

Let $(U, \mathscr{B})$ be another measure space. Any measurable (more precisely, $\mathscr{A}/\mathscr{B}$-measurable) mapping $y = y(x)$ of the space X into U is called a *statistic*. If the sample $x = (x_1, x_2, \ldots)$ is the result of observations (for example, the result of independent observations of some random variable $\xi = \xi(\omega)$), then $y = y(x)$ is a function of the observations.

Examples of statistics are:

$$m_n(x) = \frac{1}{n}\sum_{i=1}^{n} x_i \qquad \text{(random sample mean);}$$

$$S_n(x) = \frac{1}{n}\sum_{i=1}^{n} (x_i - m_n)^2 \qquad \text{(random sample dispersion).}$$

1.5.2

The theory of estimation is one of the most important parts of mathematical statistics. We present now some concepts which are used in this book.

Assume that on the sample space $(X, \mathscr{A})$ is given the family $\mathscr{P} = \{P_\theta : \theta \in \Theta\}$ of probability measures depending on the parameter θ, which belongs to a certain parametric set Θ.

The statistic $y = y(x)$ is called an *unbiased parameter estimate of* $\theta \in \Theta$ if $M_\theta y(x) = \theta$ for all $\theta \in \Theta$ (M_θ-averaging on the measure P_θ). The statistic $y = y(x)$ is called *sufficient for* θ (or *for the family* $\mathscr{P}$) if for each $A \in \mathscr{A}$ a version of the conditional probability $P_\theta(A \mid y(x))$ not depending on θ can be chosen.

The following factorization theorem provides necessary and sufficient conditions for a certain statistic $y = y(x)$ to be sufficient.

Theorem 1.14. *Let the family $\mathscr{P} = \{P_\theta, \theta \in \Theta\}$, be dominated by a certain σ-finite measure λ (i.e., $P_\theta \ll \lambda, \theta \in \Theta$). The statistic $y = y(x)$ will be sufficient if and only if there exists a $\mathscr{B}$-measurable (with each $\theta \in \Theta$) function $g(y, \theta)$ such that*

$$dP_\theta(x) = g(y(x), \theta)d\lambda(x).$$

The sequence of the statistics $y_n(x)$, $n = 1, 2, \ldots$, is called a *consistent parameter estimate of* $\theta \in \Theta$, if $y_n(x) \to \theta$, $n \to \infty$, in P-probability for all $\theta \in \Theta$, i.e.,

$$P_\theta\{|y_n(x) - \theta| > \varepsilon\} \to 0, \qquad n \to \infty, \varepsilon > 0.$$

The sequence of statistics $y_n(x)$, $n = 1, 2, \ldots$, is called a *strongly consistent estimate of the parameter* $\theta \in \Theta$ if $y_n(x) \to \theta$ with P_θ-probability one for all $\theta \in \Theta$.

Let the family $\mathscr{P}$ be dominated by a certain σ-finite measure λ. The function

$$L_x(\theta) = \frac{dP_\theta(x)}{d\lambda(x)},$$

considered (with a fixed x) as a function of θ, is called a *likelihood function.* The statistic $\hat{y} = \hat{y}(x)$ that maximizes the likelihood function $L_x(\theta)$ is called a maximal likelihood estimate.

To compare various estimates $y = y(x)$ of the unknown parameter $\theta \in \Theta$ we introduce (nonnegative) loss functions $W(\theta, y)$ and average loss

$$R(\theta, y) = M_\theta W(\theta, y(x)). \tag{1.45}$$

In those cases where $\theta \in \mathbb{R}^1$, $y \in \mathbb{R}^1$, the most commonly used function is

$$W(\theta, y) = |\theta - y|^2. \tag{1.46}$$

While investigating the quality of parameter estimates $\theta = (\theta_1, \ldots, \theta_k) \in \mathbb{R}^k$, the *Fisher information matrix* $I(\theta) = \|I_{ij}(\theta)\|$, where

$$I_{ij}(\theta) = M_\theta \left\{ \frac{\partial}{\partial \theta_i} \ln \frac{dP_\theta}{d\lambda}(x) \right\} \left\{ \frac{\partial}{\partial \theta_j} \ln \frac{dP_\theta}{d\lambda}(x) \right\} \tag{1.47}$$

plays an essential role.

In the one-dimensional case ($\theta \in \mathbb{R}^1$) the value

$$I(\theta) = M_\theta \left\{ \frac{\partial}{\partial \theta} \ln \frac{dP_\theta}{d\lambda}(x) \right\}^2 \tag{1.48}$$

is called the *Fisher information quantity*.

For unbiased estimates $y = y(x)$ of the parameter $\theta \in \Theta \subseteq \mathbb{R}^1$ the Rao–Cramer inequality is true (under certain conditions of regularity; see [128], [138]):

$$M_\theta[\theta - y(x)]^2 \geq \frac{I}{I(\theta)}, \qquad \theta \in \Theta. \tag{1.49}$$

In a multivariate case ($\theta \in \Theta \subseteq \mathbb{R}^k$, $y \in \mathbb{R}^k$) the inequality given by (1.49) becomes the Rao–Cramer matrix inequality[10]

$$M_\theta[\theta - y(x)][\theta - y(x)]^* \geq I^{-1}(\theta), \qquad \theta \in \Theta. \tag{1.50}$$

(For details see [128], [138], and also Section 7.8.)

The unbiased estimate $y(x) \in \mathbb{R}^k$ of the parameter $\theta \in \mathbb{R}^k$ is called *efficient* if, for all $\theta \in \Theta$,

$$M_\theta[\theta - y(x)][\theta - y(x)]^* = I^{-1}(\theta),$$

i.e., if in the Rao–Cramer inequality the equality is actually attained.

1.5.3

Assume that the parameter $\theta \in \Theta$ is itself a random variable with the distribution $\pi = \pi(d\theta)$. Then, along with the mean loss $R(\theta, y)$, the total mean loss

$$R(\pi, y) = \int_\Theta R(\theta, y)\pi(d\theta)$$

can be considered.

[10] For symmetric nonnegative definite matrices A and B, the inequality $A \geq B$ implies that the matrix $A - B$ is nonnegative definite.

The statistic $y^* = y^*(x)$ is called the *Bayes statistic* with respect to the a priori distribution π, if $R(\pi, y^*) \leq R(\pi, y)$ for any other statistic $y = y(x)$.

The statistic $\tilde{y} = \tilde{y}(x)$ is called *minimax* if

$$\max_{\theta} R(\theta, \tilde{y}) \leq \inf_{y} \max_{\theta} R(\theta, y).$$

Notes and references

1.1. The axiomatic foundation of probability theory is presented in Kolmogorov [86]. The proofs of Theorems 1.1–1.5 can be found in many works. See, for instance, Doob [46], Loeve [120], Kolmogorov and Fomin [89], Meyer [126]. Theorem 1.6 has been proved in [11]. The Fatou lemma formulation (Theorem 1.2) is contained in [160]. The proof of Vallée-Poussin criterion of uniform integrability (Theorem 1.8) is given in [126].

1.2. For more details on measurable, progressively measurable, and stochastically equivalent processes, see [126]. Stationary processes have been discussed in Rozanov [139], Cramer and Leadbetter [91], and in a well-known paper of Yaglom [172]. The modern theory of Markov processes has been dealt with by Dynkin [47] and in Blumental and Getoor [12]. The reader can find the fundamentals of the stationary and Markov process theory in Prokhorov and Rozanov [135].

1.3. Our discussion of Markov time properties follows Meyer [126], Blumental and Getoor [12], and Shiryayev [169].

1.4. A large amount of information about a Brownian motion process is available in Levy [100], Ito and McKean [61], Doob [46], and Gykhman and Škorokhod [34], [36].

1.5. For more details on the concepts of mathematical statistics used here, see Lynnik [106], Cramer [90], and Ferguson [153].

Martingales and semimartingales: discrete time 2

2.1 Semimartingales on the finite time interval

2.1.1

Let $(\Omega, \mathscr{F}, P)$ be a probability space, and let $\mathscr{F}_1 \subseteq \mathscr{F}_2 \subseteq \cdots \subseteq \mathscr{F}_N \subseteq \mathscr{F}$ be a nondecreasing family of sub-σ-algebras $\mathscr{F}$.

Definition 1. The sequence $X = (x_n, \mathscr{F}_n)$, $n = 1, \ldots, N$, is called respectively a *supermartingale* or *submartingale*, if $M|x_n| < \infty$, $n = 1, \ldots, N$, and

$$M(x_n|\mathscr{F}_m) \le x_m \qquad (P\text{-a.s.}), \qquad n \ge m, \tag{2.1}$$

or

$$M(x_n|\mathscr{F}_m) \ge x_m \qquad (P\text{-a.s.}), \qquad n \ge m. \tag{2.2}$$

The supermartingales and submartingales are together referred to as *semimartingales*.

If $X = (x_n, \mathscr{F}_n)$ is a supermartingale, then $Y = (-x_n, \mathscr{F}_n)$ is a submartingale. Therefore, the investigation of the supermartingales (or submartingales for convenience) is sufficient for the investigation of properties of semimartingales.

It is clear that a sequence $X = (x_n, \mathscr{F}_n)$ which is both a supermartingale and a submartingale is a *martingale*:

$$M(x_n|\mathscr{F}_m) = x_m \qquad (P\text{-a.s.}), \qquad n \ge m. \tag{2.3}$$

For a supermartingale the mathematical expectation Mx_n does not increase: $Mx_n \le Mx_m$, $n \ge m$. For the martingale the mathematical expectation is a constant: $Mx_n = Mx_1$, $n \le N$.

2.1.2

EXAMPLE 1. Let $z = z(\omega)$ be a random variable with $M|z| < \infty$ and $x_n = M(z|F_n)$. The sequence $(x_n, \mathscr{F}_n)$ is a martingale.

EXAMPLE 2. Let $\eta_1, \eta_2, \ldots$ be a sequence of integrable independent random variables with $M\eta_i = 0, i = 1, 2, \ldots, S_n = \eta_1 + \cdots \eta_n, \mathscr{F}_n = \sigma\{\omega: \eta_1, \ldots, \eta_n\}$. Then $S = (S_n, \mathscr{F}_n)$ is a martingale.

EXAMPLE 3. If $X = (x_n, \mathscr{F}_n)$ and $Y = (y_n, \mathscr{F}_n)$ are two supermartingales, then the sequence $z = (x_n \wedge y_n, \mathscr{F}_n)$ is also a supermartingale.

EXAMPLE 4. If $X = (x_n, \mathscr{F}_n)$ is a martingale and $f(x)$ is a function convex downward such that $M|f(x)| < \infty$, then the sequence $F = (f(x_n), \mathscr{F}_n)$ is a submartingale. This follows immediately from Jensen's inequality. In particular, the sequences

$$(|x_n|^\alpha, \mathscr{F}_n), \alpha \geq 1, \qquad (|x_n| \log^+ |x_n|, \mathscr{F}_n),$$

where $\log^+ a = \max(0, \log a)$, are submartingales.

2.1.3

We shall now formulate and prove the main properties of semimartingales.

Theorem 2.1. *Let $X = (x_n, \mathscr{F}_n), n = 1, \ldots, N$, be a supermartingale. Then for any two Markov times τ and σ (with respect to $F = (\mathscr{F}_n), n = 1, \ldots, N$) such that $P(\tau \leq N) = P(\sigma \leq N) = 1$,*

$$x_\sigma \geq M(x_\tau | \mathscr{F}_\sigma) \qquad (\{\tau \geq \sigma\}, \qquad P\text{-a.s.}) \tag{2.4}$$

or, what is equivalent,

$$x_{\tau \wedge \sigma} \geq M(x_\tau | \mathscr{F}_\sigma) \qquad (P\text{-a.s.}). \tag{2.5}$$

PROOF. First of all note that $M|x_\tau| < \infty$. Actually,

$$M|x_\tau| = \sum_{n=1}^N \int_{\{\tau = n\}} |x_\tau| dP = \sum_{n=1}^N \int_{\{\tau = n\}} |x_n| \, dP \leq \sum_{n=1}^N M|x_n| < \infty.$$

Consider the set $\{\sigma = n\}$ and show that on the set $\{\sigma = n\} \cap \{\tau \geq \sigma\} = \{\sigma = n\} \cap \{\tau \geq n\}$ the inequality given by (2.4) is valid. On this set $x_\sigma = x_n$, according the Lemma 1.9,

$$M(x_\tau | \mathscr{F}_\sigma) = M(x_\tau | \mathscr{F}_n) \qquad (\{\sigma = n\}, \quad P\text{-a.s.})$$

So it is sufficient to establish that on $\{\sigma = n\} \cap \{\tau \geq n\}$, ($P$-a.s.)

$$x_n \geq M(x_\tau | \mathscr{F}_n).$$

Let $A \in \mathscr{F}_n$. Then,

$$\int_{A \cap \{\sigma = n\} \cap \{\tau \geq n\}} (x_n - x_\tau) dP = \int_{A \cap \{\sigma = n\} \cap \{\tau = n\}} (x_n - x_\tau) dP + \int_{A \cap \{\sigma = n\} \cap \{\tau > n\}} (x_n - x_\tau) dP$$

$$= \int_{A \cap \{\sigma = n\} \cap \{\tau > n\}} (x_n - x_\tau) dP$$

$$\geq \int_{A \cap \{\sigma = n\} \cap \{\tau \geq n+1\}} (x_{n+1} - x_\tau) dP, \qquad (2.6)$$

where the last inequality holds due to the fact that $x_n \geq M(x_{n+1} | \mathscr{F}_n)$ (P-a.s.) and that the set $A \cap \{\sigma = n\} \cap \{\tau > n\} \in \mathscr{F}_n$.

Continuing the inequality given by (2.6) we find

$$\int_{A \cap \{\sigma = n\} \cap \{\tau \geq n\}} (x_n - x_\tau) dP \geq \int_{A \cap \{\sigma = n\} \cap \{\tau \geq n+1\}} (x_{n+1} - x_\tau) dP \geq \dots$$

$$\geq \int_{A \cap \{\sigma = n\} \cap \{\tau = N\}} (x_N - x_\tau) dP = 0. \qquad (2.7)$$

Since $\Omega - \bigcup_{n=1}^N \{\sigma = n\}$ is a set of measure zero, (2.4) follows from (2.7). □

Corollary 1. *Let* $X = (x_n, \mathscr{F}_n)$, $n = 1, \dots, N$, *be a supermartingale. If* $P(\tau \geq \sigma) = 1$, *then* $Mx_1 \geq Mx_\sigma \geq Mx_\tau \geq Mx_N$.

Corollary 2. *Let* $X = (x_n, \mathscr{F}_n)$, $n = 1, \dots, N$, *be a submartingale. If* $P(\tau \geq \sigma) = 1$, *then* $Mx_1 \leq Mx_\sigma \leq Mx_\tau \leq Mx_N$.

Corollary 3. *Let* $X = (x_n, \mathscr{F}_n)$, $n = 1, \dots, N$, *be a supermartingale. Then, if* τ *is a Markov time and* $P(\tau \leq N) = 1$, *we have*

$$M|x_\tau| \leq Mx_1 + 2Mx_N^- \leq 3 \sup_{n \leq N} M|x_n|.$$

Actually, $|x_\tau| = x_\tau + 2x_\tau^-$ and, by Corollary 1, $M|x_\tau| = Mx_\tau + 2Mx_\tau^- \leq Mx_1 + 2Mx_\tau^-$. Since $(x_n \wedge 0, \mathscr{F}_n)$, $n = 1, \dots, N$, is a supermartingale (Example 3), the sequence $(x_n^-, \mathscr{F}_n)$, where $x_n^- = -x_n \wedge 0$, forms a submartingale and, by Corollary 2, $Mx_\tau^- \leq Mx_N^-$. Hence,

$$M|x_\tau| \leq Mx_1 + 2Mx_\tau^- \leq Mx_1 + 2Mx_N^- \leq Mx_1 + 2M|x_N| \leq 3 \sup_{n \leq N} M|x_n|.$$

Surveying the proof of Theorem 2.1, we note that if $X = (x_n, \mathscr{F}_n)$, $n = 1, \ldots, N$, is a martingale, then in (2.6), (2.7) the inequalities become equalities. Therefore, we have:

Theorem 2.2 *Let $X = (x_n, \mathscr{F}_n)$, $n = 1, \ldots, N$, be a martingale. Then, for any two Markov times τ and σ, such that*

$$x_\sigma = M(x_\tau | \mathscr{F}_\sigma) \qquad \{\tau \geq \sigma\}, \qquad (P\text{-a.s.}) \tag{2.8}$$

or, equivalently, $x_{\sigma \wedge \tau} = M(x_\tau | \mathscr{F}_\sigma)$ (P-a.s.).

Corollary 1. *If $P(\tau \geq \sigma) = 1$, then, $Mx_1 = Mx_\sigma = Mx_\tau = Mx_N$.*

2.1.4

Theorem 2.3. *Let $X = (x_n, \mathscr{F}_n)$, $n = 1, \ldots, N$, be a submartingale. Then for any $\lambda > 0$,*

$$\lambda \cdot P\left\{\max_{n \leq N} x_n \geq \lambda\right\} \leq \int_{\{\max_{n \leq N} x_n \geq \lambda\}} x_N \, dP \leq Mx_N^+, \tag{2.9}$$

$$\lambda \cdot P\left\{\min_{n \leq N} x_n \leq -\lambda\right\} \leq -Mx_1 + \int_{\{\max_{n \leq N} x_n \geq -\lambda\}} x_N \, dP \tag{2.10}$$

Proof. Introduce the Markov time $\tau = \min\{n \leq N : x_n \geq \lambda\}$, assuming $\tau = N$, if $\max_{n \leq N} x_n < \lambda$. Then by Corollary 2 of Theorem 2.1,

$$Mx_N \geq Mx_\tau = \int_{\{\max_{n \leq N} x_n \geq \lambda\}} x_\tau \, dP + \int_{\{\max_{n \leq N} x_n < \lambda\}} x_\tau \, dP$$

$$\geq \lambda \int_{\{\max_{n \leq N} x_n \geq \lambda\}} dP + \int_{\{\max_{n \leq N} x_n < \lambda\}} x_N \, dP.$$

From this we obtain

$$\lambda P\left\{\max_{n \leq N} x_n \geq \lambda\right\} \leq Mx_N - \int_{\{\max_{n \leq N} x_n < \lambda\}} x_N \, dP$$

$$= \int_{\{\max_{n \leq N} x_n \geq \lambda\}} x_N \, dP \leq \int_{\{\max_{n \leq N} x_n \geq \lambda\}} x_N^+ \, dP \leq Mx_N^+,$$

which proves (2.9).

Analogously, (2.10) follows. It need only be assumed that $\tau = \min\{n \leq N : x_n \leq -\lambda\}$, with $\tau = N$, if $\min_{n \leq N} x_n > -\lambda$. □

Corollary (Kolmogorov's inequality). *Let $X = (x_n, \mathscr{F}_n)$, $n = 1, \ldots, N$ be a square integrable martingale (i.e. a martingale with $Mx_n^2 < \infty$, $n = 1, \ldots, N$). Then the sequence $(x_n^2, \mathscr{F}_n)$ will be a submartingale (Example 4) and from (2.9) we obtain the inequality*

$$P\left\{\max_{n \leq N} |x_n| \geq \lambda\right\} \leq \frac{Mx_N^2}{\lambda^2}. \tag{2.11}$$

Theorem 2.4. *Let $X = (x_n, \mathscr{F}_n)$, $n = 1, \ldots, N$, be a nonnegative submartingale. Let $Mx_N^p < \infty$ $(1 < p < \infty)$. Then $M[\max_{n \leq N} x_n]^p < \infty$ and*

$$M\left[\max_{n \leq N} x_n\right]^p \leq \left(\frac{p}{p-1}\right)^p Mx_N^p. \tag{2.12}$$

Proof. Denote $y = \max_{n \leq N} x_n$ and $F(\lambda) = P\{y > \lambda\}$. Then, due to (2.9),

$$\lambda F(\lambda) \leq \int_{(y \geq \lambda)} x_N \, dP. \tag{2.13}$$

To deduce (2.12) we estimate, first $M(y \wedge L)^p$, where $L \geq 0$. Making use of (2.13) we find that

$$\begin{aligned} M(y \wedge L)^p &= L^p F(\lambda) - \int_0^L \lambda^p F(d\lambda) = \int_0^L F(\lambda) d(\lambda^p) \\ &\leq \int_0^L \frac{1}{\lambda}\left(\int_{(y \geq \lambda)} x_N \, dP\right) d(\lambda^p) \\ &= \int_\Omega x_N \left[\int_0^{y \wedge L} \frac{d(\lambda^p)}{\lambda}\right] dP = \frac{p}{p-1} M[x_N (y \wedge L)^{p-1}]. \end{aligned}$$

By the Hölder inequality ($q = p(p-1)^{-1}$),

$$\begin{aligned} M[x_N(y \wedge L)^{p-1}] &\leq [Mx_N^p]^{1/p} M[(y \wedge L)^{(p-1)q}]^{1/q} \\ &= [Mx_N^p]^{1/p} [M(y \wedge L)^p]^{1/q}. \end{aligned}$$

Thus,

$$M(y \wedge L)^p \leq q[M(y \wedge L)^p]^{1/q}[Mx_N^p]^{1/p}$$

and, since $M(y \wedge L)^p \leq L^p < \infty$,

$$M(y \wedge L)^p \leq q^p M x_N^p. \tag{2.14}$$

By Theorem 1.1, $My^p = \lim_{L \uparrow \infty} M(y \wedge L)^p$. Hence from (2.14) we have the desired estimate:

$$My^p \leq q^p M x_N^p < \infty. \qquad \square$$

Corollary. *Let $X(x_n, \mathscr{F}_n)$, $n = 1, \ldots, N$, be a square integrable martingale. Then*

$$M\left[\max_{n \le N} x_n^2\right] \le 4Mx_N^2.$$

2.1.5

For investigating the asymptotic properties of the semimartingales $X = (x_n, \mathscr{F}_n)$, $n = 1, 2, \ldots$, Doob's inequalities on the number of crossings of the interval (a, b) (see Theorem 2.5) play a significant role. For formulating these inequalities we introduce some necessary definitions.

Let $X = (x_n, \mathscr{F}_n)$, $n = 1, \ldots, N$, be a submartingale and let (a, b) be a nonempty interval. We need to define the "number of up-crossings of the interval (a, b) by the submartingale X." For this purpose denote:

$$\begin{aligned}
\tau_0 &= 0,\\
\tau_1 &= \min\{0 < n \le N : x_n \le a\},\\
\tau_2 &= \min\{\tau_1 < n \le N : x_n \ge b\},\\
&\cdots\cdots\cdots\cdots\cdots\cdots\\
\tau_{2m-1} &= \min\{\tau_{2m-2} < n \le N : x_n \le a\},\\
\tau_{2m} &= \min\{\tau_{2m-1} < n \le N : x_n \ge b\},\\
&\cdots\cdots\cdots\cdots\cdots\cdots
\end{aligned}$$

In this case, if $\inf_{n \le N} x_n \ge a$, then τ_1 is assumed to be equal to N, and the times $\tau_2, \tau_3, \ldots$ are not defined. This also applies to the subsequent times.

Definition 2. The maximal m for which τ_{2m} is defined is called the *number of up-crossings of the interval* (a, b), and is denoted $\beta(a, b)$.

Theorem 2.5. *If $X = (x_n, \mathscr{F}_n)$, $n = 1, \ldots, N$, is a submartingale, then*

$$M\beta(a, b) \le \frac{M[x_N - a]^+}{b - a} \le \frac{Mx_N^+ + |a|}{b - a}. \tag{2.15}$$

PROOF. Since the number of crossings of the interval (a, b) by the submartingale $X = (x_n, \mathscr{F}_n)$, $n \le N$, corresponds to the number of crossings of the interval $(0, b - a)$ by the nonnegative submartingale $X^+ = ((x_n - a)^+, \mathscr{F}_n)$, $n \le N$, we see that the initial submartingale is nonnegative and $a = 0$. Thus we need to show that for $b > 0$,

$$M\beta(0, b) \le \frac{Mx_N}{b}. \tag{2.16}$$

Assume $x_0 \equiv 0$, and for $i = 1, \ldots,$ let

$$\chi_i = \begin{cases} 1, & \text{if } \tau_m < i \le \tau_{m+1} \quad \text{for some odd } m,\\ 0, & \text{if } \tau_m < i \le \tau_{m+1} \quad \text{for some even } m. \end{cases}$$

Then (P-a.s.)

$$b\beta(0, b) \leq \sum_{i=1}^{N} \chi_i[x_i - x_{i-1}]$$

and

$$\{\chi_i = 1\} = \bigcup_{m \text{ is odd}} [\{\tau_m < i\} - \{\tau_{m+1} < i\}].$$

Hence,

$$\begin{aligned} bM\beta(0, b) &\leq M \sum_{i=1}^{N} \chi_i[x_i - x_{i-1}] \\ &= \sum_{i=1}^{N} \int_{\{\chi_i = 1\}} (x_i - x_{i-1}) dP \\ &= \sum_{i=1}^{N} \int_{\{\chi_i = 1\}} M(x_i - x_{i-1} | \mathscr{F}_{i-1}) dP \\ &= \sum_{i=1}^{N} \int_{\{\chi_i = 1\}} [M(x_i | \mathscr{F}_{i-1}) - x_{i-1}] dP \\ &\leq \sum_{i=1}^{N} \int_{\Omega} [M(x_i | \mathscr{F}_{i-1}) - x_{i-1}] dP = Mx_N. \end{aligned}$$ □

Note. By analogy with $\beta(a, b)$, the number of down-crossings $\alpha(a, b)$ of the interval (a, b) can be found. For $M\alpha(a, b)$ (in the same way as one deduces (2.15)) the following estimate can be obtained:

$$M\alpha(a, b) \leq \frac{M(x_N - a)^+}{b - a} \leq \frac{M[x_N^+ + |a|]}{b - a}. \tag{2.17}$$

2.2 Semimartingales on an infinite time interval, and the theorem of convergence

In this section it will be assumed that the semimartingales $X = (x_n, \mathscr{F}_n)$ are defined for $n = 1, 2, \ldots$

Theorem 2.6. *Let $X = (x_n, \mathscr{F}_n)$, $n < \infty$ be a submartingale such that*

$$\sup_n Mx_n^+ < \infty. \tag{2.18}$$

Then $\lim_n x_n (= x_\infty)$ exists with probability 1, and $Mx_\infty^+ < \infty$.

Proof. Let $x^* = \lim_n \sup x_n$, $x_* = \lim_n \inf x_n$. Assume that

$$P\{x^* > x_*\} > 0. \tag{2.19}$$

Then, since $\{x^* > x_*\} = \bigcup_{a<b} \{x^* > b > a > x_*\}$ (a, b are rational numbers), there exist a and b such that

$$P\{x^* > b > a > x_*\} > 0. \tag{2.20}$$

Let $\beta_N(a, b)$ be the number of crossings of the interval (a, b) by the submartingale $(x_n, \mathscr{F}_n)$, $n \le N$, and $\beta_\infty(a, b) = \lim_N \beta_N(a, b)$. Then according to (2.15),

$$M\beta_N(a, b) \le \frac{Mx_N^+ + |a|}{b - a}$$

and, due to (2.18),

$$M\beta_\infty(a, b) = \lim_N M\beta_N(a, b) \le \frac{\sup_N Mx_N^+ + |a|}{b - a} < \infty.$$

This, however, contradicts the assumption in (2.20), from which it follows that with positive probability $\beta_\infty(a, b) = \infty$. Thus, $P(x^* = x_*) = 1$, and, therefore, $\lim_n x_n$ exists with probability 1. □

This variable will henceforth be denoted by x_∞. Note that, due to Fatou's lemma, $Mx_\infty^+ \le \sup_n Mx_n^+$.

Corollary 1. *If $X = (x_n, \mathscr{F}_n)$, $n \ge 1$, is a negative submartingale (or a positive supermartingale) then with probability 1 $\lim_n x_n$ exists.*

Corollary 2. *Let $X = (x_n, \mathscr{F}_n)$, $n \ge 1$, be a negative submartingale (or a positive supermartingale). Then the sequence $\bar{X} = (x_n, \mathscr{F}_n)$, $n = 1, 2, \ldots, \infty$, with $x_\infty = \lim_n x_n$ and $\mathscr{F}_\infty = \sigma(\bigcup_{n=1}^\infty \mathscr{F}_n)$, forms a negative submartingale (a positive supermartingale).*

Actually, if $X = (x_n, \mathscr{F}_n)$, $n = 1, 2, \ldots$, is a negative submartingale, then, by Fatou's lemma,

$$Mx_\infty = M \lim_n x_n \ge \overline{\lim_n}\, Mx_n \ge Mx_1 > -\infty$$

and

$$M(x_\infty | \mathscr{F}_m) = M\left(\lim_n x_n | \mathscr{F}_m\right) \ge \overline{\lim_n}\, M(x_n | \mathscr{F}_m) \ge x_m \qquad (P\text{-a.s.}).$$

Corollary 3. *If $X = (x_n, \mathscr{F}_n)$, $n \ge 1$, is a martingale, then (2.18) is equivalent to the condition*

$$\sup_n M|x_n| < \infty. \tag{2.21}$$

Actually,

$$M|x_n| = Mx_n^+ + Mx_n^- = 2Mx_n^+ - Mx_n = 2Mx_n^+ - Mx_1.$$

Hence, $\sup_n M|x_n| = 2 \sup_n Mx_n^+ - Mx_1$.

2.3 Regular martingales: Levy's theorem

2.3.1

The generalization of Theorems 2.1 and 2.2 to the case of an infinite sequence requires some additional assumptions on the structure of martingales and semimartingales. A crucial concept is:

Definition 3. The martingale $X = (x_n, \mathscr{F}_n)$, $n \ge 1$, is called *regular* if there exists an integrable random variable $\eta = \eta(\omega)$, such that

$$x_n = M(\eta \mid \mathscr{F}_n) \qquad (P\text{-a.s.}), \qquad n \ge 1.$$

Note that in the case of finite time, $1 \le n \le N$, any martingale is regular, since $x_n = M(x_N \mid \mathscr{F}_n)$, $1 \le n \le N$.

Theorem 2.7. *The following conditions on the martingale* $X = (x_n, \mathscr{F}_n)$, $n \ge 1$, *are equivalent*:

(A) *regularity, i.e., feasibility of representation in the form* $x_n = M(\eta \mid \mathscr{F}_n)$ (P-a.s.) *with* $M|\eta| < \infty$;
(B) *uniform integrability of the variables* $x_1, x_2, \ldots$;
(C) *convergence of the sequence* $x_1, x_2, \ldots$ *in* L^1:

$$\lim_n M|x_\infty - x_n| = 0;$$

(D) $\sup_n M|x_n| < \infty$ *and the variable* $x_\infty = \lim_n x_n$ *is such that* $x_n = M(x_\infty \mid \mathscr{F}_n)$ (P-a.s.), *i.e., the sequence* $X = (x_\infty, \mathscr{F}_n)$, $1 \le n \le \infty$, *is a martingale.*

PROOF

(A) $\Rightarrow$ (B). It must be shown that the variables $x_n = M(\eta \mid \mathscr{F}_n)$, $n \ge 1$, are uniformly integrable. We have

$$|x_n| \le M(|n| \mid \mathscr{F}_n), \qquad M|x_n| \le M|\eta|, \qquad \sup_n M|x_n| \le M|\eta| < \infty.$$

From this, for $c > 0$, $b > 0$ we obtain

$$\begin{aligned}
\int_{\{|x_n| \ge c\}} |x_n| \, dP &\le \int_{\{|x_n| \ge c\}} |\eta| \, dP \\
&= \int_{\{|x_n| \ge c\} \cap \{|\eta| \ge b\}} |\eta| \, dP + \int_{\{|x_n| \ge c\} \cap \{|\eta| < b\}} |\eta| \, dP \\
&\le bP\{|x_n| \ge c\} + \int_{\{|\eta| \ge b\}} |\eta| \, dP \\
&\le \frac{b}{c} M|x_n| + \int_{\{|\eta| \ge b\}} |\eta| \, dP.
\end{aligned}$$

Consequently,

$$\sup_n \int_{\{|x_n| \geq c\}} |x_n| dP \leq \frac{b}{c} M|\eta| + \int_{\{|\eta| \geq b\}} |\eta| dP,$$

$$\lim_{c \uparrow \infty} \sup_n \int_{\{|x_n| \geq c\}} |x_n| dP \leq \int_{\{|\eta| \geq b\}} |\eta| dP.$$

But $b > 0$ arbitrarily; therefore,

$$\lim_{c \uparrow \infty} \sup_n \int_{\{|x_n| \geq c\}} |x_n| dP = 0,$$

which proves statement (B).

(B) $\Rightarrow$ (C). Since $x_n = M(\eta | \mathscr{F}_n)$ are uniformly integrable, then: first, $\sup_n M|x_n| < \infty$ and therefore $\lim_n x_n (= x_\infty)$ exists (Corollary 3 of Theorem 2.6); second, by the corollary of Theorem 1.3, $M|x_n - x_\infty| \to 0$, $n \to \infty$, i.e., the sequence $x_1, x_2, \ldots$ converges (to x_∞) in L^1.

(C) $\Rightarrow$ (D). If the sequence of the random variables $x_1, x_2, \ldots$ converges in L^1 (let us say, to a random variable y), then $\sup_n M|x_n| < \infty$. Then, on the basis of Corollary 3 of Theorem 2.6, $\lim_n x_n$ $(= x_\infty)$ exists and therefore $M|x_n - y| \to 0, x_n \to x_\infty$ (P-a.s.), $n \to \infty$. Hence, $y = x_\infty$ (P-a.s.). Consequently, $x_n \to x_\infty$, i.e., $M|x_n - x_\infty| \to 0$, $n \to \infty$, and $M|x_n|\mathscr{F}_m) \xrightarrow{L^1} M(x_\infty | \mathscr{F}_m)$, if $m \leq n \to \infty$. But $M(x_n | \mathscr{F}_m) = x_m$ (P-a.s.), and therefore $x_m = (x_\infty | \mathscr{F}_m)$ (P-a.s.).

(D) $\Rightarrow$ (A). Denoting $\eta = x_\infty$, we immediately obtain statement (A). ☐

From this theorem it follows that any of properties (B), (C), (D) can be taken for the definition of a regular martingale.

2.3.2

As a corollary of Theorems 2.6 and 2.7 we may deduce the following useful result (P. Levy), mentioned in Section 1.1.

Theorem 2.8. *Let $\eta = \eta(\omega)$ be an integrable ($M|\eta| < \infty$) random variable and let $\mathscr{F}_1 \subseteq \mathscr{F}_2 \subseteq \cdots$ be the nondecreasing family of sub-σ-algebras $\mathscr{F}$. Then, with $n \to \infty$.* (P-a.s.)

$$M(\eta | \mathscr{F}_n) \to M(\eta | \mathscr{F}_\infty) \tag{2.22}$$

where

$$\mathscr{F}_\infty = \sigma\left(\bigcup_{n=1}^\infty \mathscr{F}_n\right).$$

PROOF. Denote $x_n = M(\eta | \mathscr{F}_n)$. The sequence $X = (x_n, \mathscr{F}_n)$, $n \geq 1$, is a regular martingale. According to Theorem 2.6 $\lim x_n (= x_\infty)$ exists, and by Fatou's lemma $M|x_\infty| \leq M|\eta|$. Further, if $A \in \mathscr{F}_n$ and $m \geq n$, then

$$\int_A x_m \, dP = \int_A x_n \, dP = \int_A M(\eta | \mathscr{F}_n) dP = \int_A \eta \, dP.$$

By Theorem 2.7 the sequence $\{x_m, m \geq 1\}$ is uniformly integrable. Hence $M\chi_A|x_m - x_\infty| \to 0$, $m \to \infty$, and, therefore,

$$\int_A x_\infty \, dP = \int_A \eta \, dP. \tag{2.23}$$

Equation (2.23) is satisfied for any $A \in \mathscr{F}_n$ and, consequently, for any set A from the algebra $\bigcup_{n=1}^\infty \mathscr{F}_n$. The left and the right sides in (2.23) represent σ-additive signed measures (which may take on negative values, but are finite), agreeing on the algebra $\bigcup_{n=1}^\infty \mathscr{F}_n$. Hence, because of uniqueness of extension of σ-additive finite measures from the algebra $\bigcup_{n=1}^\infty \mathscr{F}_n$ to the smallest σ-algebra $\mathscr{F}_\infty = \sigma(\bigcup_{n=1}^\infty \mathscr{F}_n)$ which contains it, Equation (2.23) remains correct also for $A \in \mathscr{F}_\infty = \sigma(\bigcup_{n=1}^\infty \mathscr{F}_n)$. Thus,

$$\int_A x_\infty \, dP = \int_A \eta \, dP = \int_A M(\eta|\mathscr{F}_\infty) dP, \qquad A \in \sigma\left(\bigcup_{n=1}^\infty \mathscr{F}_n\right).$$

But x_∞ and $M(\eta|\mathscr{F}_\infty)$ are $\mathscr{F}_\infty$-measurable. Consequently, $x_\infty = M(\eta|\mathscr{F}_\infty)$ (P-a.s.). □

Note (an example of a martingale that is not regular). Let $x_n = \exp[S_n - \frac{1}{2}n]$, where $S_n = y_1 + \cdots + y_n$, $y_i \sim N(0, 1)$ and independent, and $\mathscr{F}_n = \sigma\{\omega:(y_1, \ldots, y_n)\}$. Then, $X = (x_n, \mathscr{F}_n)$, $n \geq 1$, is a martingale and, because of the strong law of large numbers,

$$x_\infty = \lim_n x_n = \lim_n \exp\left\{n\left[\frac{S_n}{n} - \frac{1}{2}\right]\right\} = 0 \qquad (P\text{-a.s.}).$$

Consequently, $x_n \neq M(x_\infty|\mathscr{F}_n) = 0$ (P-a.s.).

2.3.3

The result of Theorem 2.2 extends to regular martingales.

Theorem 2.9. *Let $X = (x_n, \mathscr{F}_n)$, $n \geq 1$, be a regular martingale and let τ, σ be Markov times with $P(\tau \geq \sigma) = 1$. Then*

$$x_\sigma = M(x_\tau|\mathscr{F}_\sigma). \tag{2.24}$$

PROOF. Note first that since the martingale X is regular, $\lim_n x_n$ exists and in (2.24) x_∞ is understood to have the value $\lim_n x_n$. Further, to have $M(x_\tau|\mathscr{F}_\sigma)$ defined, it must be shown that $M|x_\tau| < \infty$. But $x_n = M(\eta|\mathscr{F}_n)$ and $x_\tau = M(\eta|\mathscr{F}_\tau)$ (since $x_\tau = x_n$ on the sets $\{\tau = n\}$ by definition, and $M(\eta|\mathscr{F}_\tau) = M(\eta|\mathscr{F}_n)$ because of Lemma 1.9). Hence $M|x_\tau| \leq M|\eta|$. For the proof of (2.24) it need only be noted that, since $\mathscr{F}_\tau \supseteq \mathscr{F}_\sigma$,

$$M(x_\tau|\mathscr{F}_\sigma) = M(M(\eta|\mathscr{F}_\tau)|\mathscr{F}_\sigma) = M(\eta|\mathscr{F}_\sigma) = x_\sigma \qquad (P\text{-a.s.}). \qquad \square$$

Corollary. *If $X = (x_n, \mathscr{F}_n)$, $n \geq 1$, is a regular martingale, then, for any Markov time σ,*

$$x_\sigma = M(x_\infty | \mathscr{F}_\sigma).$$

Note. For the uniformly integrable martingale $X = (x_n, \mathscr{F}_n)$, $n \geq 1$, the property given in (2.24) holds without the assumption that $P(\tau \geq \sigma) = 1$:

$$x_\sigma = M(x_\tau | \mathscr{F}_\sigma) \qquad (\{\tau \geq \sigma\}, \qquad P\text{-a.s.})$$

that is.

$$x_{\sigma \wedge \tau} = M(x_\tau | \mathscr{F}_\sigma) \qquad (P\text{-a.s.}). \tag{2.25}$$

2.4 Invariance of the supermartingale property for Markov times: Riesz and Doob decompositions

2.4.1

Consider the analog of Theorem 2.1 for semimartingales.

Theorem 2.10. *Let $X = (x_n, \mathscr{F}_n)$, $n \geq 1$, be a supermartingale majorizing a certain regular martingale, i.e., for some random variable η with $M|\eta| < \infty$ let*

$$x_n \geq M(\eta | \mathscr{F}_n) \qquad (P\text{-a.s.}) \qquad n \geq 1. \tag{2.26}$$

Then, if $P(\sigma \leq \tau < \infty) = 1$,

$$x_\sigma \geq M(x_\tau | \mathscr{F}_\sigma) \qquad (P\text{-a.s.}). \tag{2.27}$$

Note. The statement of the theorem is also valid without the assumption that $P(\tau < \infty) = 1$. The corresponding generalization will be given in Theorem 2.12.

PROOF. Since $x_n = M(\eta | \mathscr{F}_n) + [x_n - M(\eta | \mathscr{F}_n)]$ and $(z_n, \mathscr{F}_n)$, $z_n = x_n - M(\eta | \mathscr{F}_n)$, $n \geq 1$, is a nonnegative supermartingale, taking Theorem 2.9 into consideration we see that it suffices to prove (2.27) for the case where $x_n \geq 0$ (P-a.s.).

We now show that $Mx_\tau < \infty$. For this, assume $\tau_k = \tau \wedge k$. Then $Mx_{\tau_k} \leq Mx_1$ (Corollary 1 of Theorem 2.1), and since $P(\tau < \infty) = 1$,

$$x_\tau = x_\tau \cdot \chi_{\{\tau < \infty\}} = \lim_k [x_{\tau_k} \cdot \chi_{\{\tau < \infty\}}].$$

Hence, by Fatou's lemma,

$$Mx_\tau \leq \underline{\lim}_k \, Mx_{\tau_k} = Mx_1 < \infty.$$

Now consider the times $\tau_k = \tau \wedge k$, $\sigma_k = \sigma \wedge k$. For them, according to Theorem 2.1 $x_{\sigma_k} \geq M(x_{\tau_k} | \mathscr{F}_{\sigma_k})$ and, consequently, if $A \in \mathscr{F}_\sigma$,

$$\int_{A \cap \{\sigma \leq k\}} x_{\sigma_k}\, dP \geq \int_{A \cap \{\sigma \leq k\}} x_{\tau_k}\, dP,$$

since $A \cap \{\sigma \leq \kappa\} \in \mathscr{F}_{\sigma_\kappa}$.

The event $\{\sigma \leq \kappa\} \supseteq \{\tau \leq \kappa\}$ and $x_n \geq 0$ (P-a.s.). Therefore,

$$\int_{A \cap \{\sigma \leq k\}} x_{\sigma_k}\, dP \geq \int_{A \cap \{\tau \leq k\}} x_{\tau_k}\, dP. \tag{2.28}$$

But $x_{\sigma_\kappa} = x_\sigma$ on the set $\{\sigma \leq k\}$ and $x_{\tau_\kappa} = x_\tau$ on $\{\tau \leq k\}$. From this and from (2.28) we find

$$\int_{A \cap \{\sigma \leq k\}} x_\sigma\, dP \geq \int_{A \cap \{\tau \leq k\}} x_\tau\, dP. \tag{2.29}$$

Assuming in (2.29) that $k \to \infty$, we obtain

$$\int_{A \cap \{\sigma < \infty\}} x_\sigma\, dP \geq \int_{A \cap \{\tau < \infty\}} x_\tau\, dP,$$

since $P(\sigma < \infty) = P(\tau < \infty) = 1$. □

2.4.2

Definition 4. The nonnegative supermartingale $\Pi = (\pi_n, \mathscr{F}_n)$, $n \geq 1$, is called a *potential* if

$$M\pi_n \to 0, \qquad n \to \infty.$$

Note that since $\sup_n M\pi_n \leq M\pi_1 < \infty$, $\lim_n \pi_n (= \pi_\infty)$ exists and $M\pi_\infty \leq \lim_n M\pi_n = 0$. It follows that $\pi_\infty = 0$ (P-a.s.).

Theorem 2.11 (Riesz decomposition). *If the supermartingale $X = (x_n, \mathscr{F}_n)$, $n \geq 1$, majorizes a certain submartingale $Y = (y_n, \mathscr{F}_n)$, $n \geq 1$, then there exists a martingale $M = (m_n, \mathscr{F}_n)$, $n \geq 1$, and a potential $\Pi = (\pi_n, \mathscr{F}_n)$, such that, for each n,*

$$x_n = m_n + \pi_n. \tag{2.30}$$

The decomposition in (2.30) *is unique (to within a stochastic equivalence).*

Proof. Assume, for each $n \geq 1$,

$$x_{n,p} = M(x_{n+p} | \mathscr{F}_n), \qquad p = 0, 1, \ldots$$

Then

$$x_{n,p+1} = M(x_{n+p+1} | \mathscr{F}_n) \leq M(x_{n+p} | \mathscr{F}_n) = x_{n,p},$$

i.e., for each $n \geq 1$ the sequence $\{x_{n,p}, p = 0, 1, \ldots\}$ is nondecreasing. Since, moreover, $x_{n,p} = M(x_{n+p}|\mathscr{F}_n) \geq M(y_{n+p}|\mathscr{F}_n) \geq y_n$, $\lim_{p\to\infty} x_{n,p}(=m_n)$ exists and $x_n \geq m_n \geq y_n$ (P-a.s.). Therefore, $M|m_n| < \infty$ and

$$\begin{aligned} M(m_{n+1}|\mathscr{F}_n &= M\left(\lim_{p\to\infty} x_{n+1,p}|\mathscr{F}_n\right) = \lim_{p\to\infty} M(x_{n+1,p}|\mathscr{F}_n) \\ &= \lim_{p\to\infty} M(x_{n+1+p}|\mathscr{F}_n) = \lim_{p\to\infty} M(x_{n,p+1}|\mathscr{F}_n) \\ &= M\left(\lim_{p\to\infty} x_{n,p+1}|\mathscr{F}_n\right) = M(m_n|\mathscr{F}_n) = m_n. \end{aligned}$$

Thus $(m_n, \mathscr{F}_n)$, $n \geq 1$, is a martingale.

Assume now that $\pi_n = x_n - m_n$. Since $x_n \geq m_n$, then $\pi_n \geq 0$. It is also clear that $\Pi = (\pi_n, \mathscr{F}_n)$, $n \geq 1$, is a supermartingale. It need only be shown that $\lim_n M\pi_n = 0$.

By definition of m_n, $n \geq 1$, (P-a.s.)

$$\begin{aligned} M(\pi_{n+p}|\mathscr{F}_n) &= M[x_{n+p} - m_{n+p}|\mathscr{F}_n] \\ &= M[x_{n+p}|\mathscr{F}_n] - m_n = x_{n,p} - m_n \downarrow 0, \qquad p \to \infty. \end{aligned}$$

Hence, by Theorem 1.3,

$$\lim_{p\to\infty} M\pi_{n+p} = \lim_{p\to\infty} \int_\Omega \pi_{n+p}\, dP = \lim_{p\to\infty} \int_\Omega M(\pi_{n+p}|\mathscr{F}_n) dP = 0.$$

Let us now prove the uniqueness of the decomposition given by (2.30). Let $x_n = \tilde{m}_n + \tilde{\pi}_n$ be another decomposition of the same type. Then

$$M[x_{n+p}|\mathscr{F}_n] = M[\tilde{m}_{n+p}|\mathscr{F}_n] + M[\tilde{\pi}_{n+p}|\mathscr{F}_n] = \tilde{m}_n + M[\tilde{\pi}_{n+p}|\mathscr{F}_n].$$

But with $p \to \infty$, (P-a.s.)

$$M[x_{n+p}|\mathscr{F}_n] \to m_n, \qquad M[\tilde{\pi}_{n+p}|\mathscr{F}_n] \to 0.$$

Hence, $m_n = \tilde{m}_n$, and $\pi_n = \tilde{\pi}_n$ (P-a.s.) for all $n \geq 1$. □

2.4.3

Now we shall prove the generalization of Theorem 2.10.

Theorem 2.12. *Let $X = (x_n, \mathscr{F}_n)$, $n \geq 1$, be the supermartingale majorizing a certain regular martingale ($x_n \geq M(\eta|\mathscr{F}_n)$ for some random variable η with $M|\eta| < \infty$, $n \geq 1$, (P-a.s.)). If $P(\tau \geq \sigma) = 1$, then (P-a.s.)*

$$x_\sigma \geq M(x_\tau|\mathscr{F}_\sigma). \tag{2.31}$$

Proof. Represent x_n in the form $x_n = M(\eta|\mathscr{F}_n) + z_n$, where $z_n = x_n - M(\eta|\mathscr{F}_n)$. The supermartingale $Z = (z_n, \mathscr{F}_n)$, $n \geq 1$, has the decomposition $z_n = m_n + \pi_n$, where $m_n = M(z_\infty|\mathscr{F}_n)$, $\pi_n = z_n - M(z_\infty|\mathscr{F}_n)$ and $z_\infty = \lim_n z_n$. Hence $x_n = M(\eta + z_\infty|\mathscr{F}_n) + \pi_n$.

The martingale $(M(\eta + \varkappa_\infty)|\mathscr{F}_n)$, $n \geq 1$ is regular, and Theorem 2.9 can be applied to it. Hence it is enough to establish that $\pi_\sigma \geq M(\pi_\tau|\mathscr{F}_\sigma)$. As shown in Theorem 2.10, for any $A \in \mathscr{F}_\sigma$

$$\int_{A \cap \{\sigma < \infty\}} \pi_\sigma \, dP \geq \int_{A \cap \{\tau < \infty\}} \pi_\tau \, dP.$$

Considering now that $\pi_\infty = \lim_n [\varkappa_n - M(\varkappa_\infty|\mathscr{F}_n)] = \varkappa_\infty - M(\varkappa_\infty|\mathscr{F}_\infty) = 0$ (P-a.s.) we obtain

$$\int_A \pi_\sigma \, dP \geq \int_A \pi_\tau \, dP.$$

Together with Theorem 2.9 this inequality proves (2.31). □

2.4.4

Definition 5. The random process A_n, $n = 0, 1, \ldots$, given on the probability space $(\Omega, \mathscr{F}, P)$ with a distinguished nondecreasing family of σ-algebras $\mathscr{F}_0 \subseteq \mathscr{F}_1 \subseteq \cdots \mathscr{F}$, is called *increasing* if

(1) $0 = A_0 \leq A_1 \leq \cdots$ (P-a.s.)

and *natural*, if

(2) A_{n+1} is $\mathscr{F}_n$-measurable, $n = 0, 1, \ldots$

Theorem 2.13 (Doob decomposition). *Any supermartingale*[1] $X = (x_n, \mathscr{F}_n)$, $n \geq 0$, *permits the unique (to within a stochastic equivalence) decomposition*

$$x_n = m_n - A_n, \qquad n \geq 0, \tag{2.32}$$

where $M = (m_n, \mathscr{F}_n)$, $n \geq 0$ *is a martingale and* A_n, $n \geq 0$, *is a natural increasing process.*

PROOF. A decomposition of the type given by (2.32) is obtained if we put

$$\begin{aligned} m_0 &= x_0, \qquad m_{n+1} - m_n = x_{n+1} - M(x_{n+1}|\mathscr{F}_n), \\ A_0 &= 0, \qquad A_{n+1} - A_n = x_n - M(x_{n+1}|\mathscr{F}_n). \end{aligned} \tag{2.33}$$

Let there be another decomposition: $x_n = m'_n - A'_n$, $n \geq 0$. Then

$$A'_{n+1} - A'_n = (m'_{n+1} - m'_n) + (x_{n+1} - x_n). \tag{2.34}$$

From this, taking into account that A'_n and A'_{n+1} are $\mathscr{F}_n$-measurable we find (taking in (2.34) the conditional mathematical expectation $M(\cdot|\mathscr{F}_n)$)

$$A'_{n+1} - A'_n = x_n - M(x_{n+1}|\mathscr{F}_n) = A_{n+1} - A_n.$$

But $A'_0 = A_0 = 0$, hence $A'_n = A_n$, $m'_n = m_n$, $n \geq 0$ (P-a.s.). □

[1] Here it is more convenient (keeping in mind forthcoming applications to the case of continuous time) to examine the supermartingales defined for $n \geq 0$ (and not for $n \geq 1$ as before).

Corollary 1. *If* $\Pi = (\pi_n, \mathscr{F}_n)$, $n \geq 0$, *is a potential, then there exists a natural increasing process* A_n, $n = 0, 1, \ldots$ *such that*

$$\pi_n = M(A_\infty | \mathscr{F}_n) - A_n,$$

where $A_\infty = \lim_n A_n$.

Actually, according to the theorem, $\pi_n = m_n - A_n$, where $(m_n, \mathscr{F}_n)$ is a certain martingale. We can show that $m_n = M(A_\infty | \mathscr{F}_n)$. We have $0 \leq A_n = m_n - \pi_n \leq m_n$ and $0 \leq A_n \leq A_\infty$, where $MA_\infty = \lim_n MA_n = \lim_n [Mm_0 - M\pi_n] = Mm_0$. Hence, the sequence $A_0, A_1, \ldots$ is uniformly integrable. The variables $\pi_0, \pi_1, \ldots$ are also uniformly integrable since $\pi_n \geq 0$ and $M\pi_n \to 0$, $n \to \infty$. From this it follows that the sequence $m_0, m_1, \ldots$ is the same. From Theorem 2.7 we find that $\lim_n m_n = m_\infty$ exists whereas $m_n = M(m_\infty | \mathscr{F}_n)$. Denote $\pi_\infty = \lim_n \pi_n$. Then $\pi_\infty = \lim_n [m_n - A_n] = m_\infty - A_\infty$. But $\pi_\infty = 0$ (P-a.s.), hence $m_\infty = A_\infty$ (P-a.s.). Therefore,

$$\pi_n = m_n - A_n = M(m_\infty | \mathscr{F}_n) - A_n = M(A_\infty | \mathscr{F}_n) - A_n.$$

Corollary 2. *If the supermartingale* $X = (x_n, \mathscr{F}_n)$, $n \geq 0$, *majorizes a certain submartingale* $Y = (y_n, \mathscr{F}_n)$, $n \geq 0$, *then there exists a natural increasing process* A_n, $n \geq 0$, *and a martingale* $(m_n, \mathscr{F}_n)$, $n \geq 0$, *such that*

$$x_n = m_n + M(A_\infty | \mathscr{F}_n) - A_n \qquad (P\text{-a.s.}) \qquad n \geq 0. \tag{2.35}$$

The proof follows immediately from the Riesz decomposition, given by (2.30), and the preceding corollary.

2.4.5

The natural process A_n, $n = 0, 1, \ldots$, by definition is $\mathscr{F}_{n-1}$-measurable (and not only $\mathscr{F}_n$-measurable) for every $n \geq 1$. This assumption can be given a somewhat different but equivalent formulation, which is more convenient in the case of continuous time (see Section 3.3). Namely, let $0 = A_0 \leq A_1 \leq, \ldots,$ where the random variables A_n are $\mathscr{F}_n$-measurable and $MA_\infty < \infty$.

Theorem 2.14. *In order that* A_n *be* $\mathscr{F}_{n-1}$*-measurable,* $n \geq 1$, *it is necessary and sufficient that for each bounded martingale* $Y = (y_n, \mathscr{F}_n)$, $n = 0, 1, \ldots,$

$$M \sum_{n=1}^{\infty} y_{n-1}(A_n - A_{n-1}) = My_\infty A_\infty, \tag{2.36}$$

where $y_\infty = \lim_n y_n$.

PROOF. Necessity: let A_n be $\mathscr{F}_{n-1}$-measurable, $MA_\infty < \infty$. Since

$$My_n A_n = My_{n-1} A_n, \qquad n \geq 1, \tag{2.37}$$

therefore

$$M \sum_{n=1}^{\infty} y_{n-1}(A_n - A_{n-1}) = \lim_{N\to\infty} M \sum_{n=1}^{N} y_{n-1}(A_n - A_{n-1})$$
$$= \lim_{N\to\infty} \sum_{n=1}^{N} [My_nA_n - My_{n-1}A_{n-1}]$$
$$= \lim_{N\to\infty} My_NA_N = My_\infty A_\infty .$$

Sufficiency: let (2.36) be satisfied. Then

$$M \sum_{n=1}^{\infty} A_n[y_{n-1} - y_n] = 0 \tag{2.38}$$

for any bounded martingale $Y = (y_n, \mathscr{F}_n)$, $n \geq 0$. Now make use of the fact that if $Y = (y_n, \mathscr{F}_n)$, $n \geq 0$, is a martingale, then the "stopped" sequence $(y_{n \wedge \tau}, \mathscr{F}_n)$, $n \geq 0$ will be also a martingale for any Markov time τ (see, further, Theorem 2.15). Taking $\tau \equiv 1$ and applying (2.38) to the martingale $(y_{n \wedge 1}, \mathscr{F}_n)$, we infer that

$$MA_1(y_0 - y_1) = 0. \tag{2.39}$$

Similar considerations on $\tau \equiv 2$, $\tau \equiv 3$, etc., lead to the fact that if (2.38) is correct, then, there exist the equalities given by Equation (2.37) for any bounded martingale $Y = (y_n, \mathscr{F}_n)$, $n \geq 0$.

From (2.37) it follows that

$$M\{[y_n - y_{n-1}][A_n - M(A_n|\mathscr{F}_{n-1})]\} = 0. \tag{2.40}$$

Let $y_{n+m} = y_n, m \geq 0, y_n = \text{sign}[A_n - M(A_n|\mathscr{F}_{n-1})], y_k = M(y_n|\mathscr{F}_k), k < n$. Then, from (2.40), we find

$$0 = M\{\text{sign}[A_n - M(A_n|\mathscr{F}_{n-1})] - y_{n-1}\}\{A_n - M(A_n|\mathscr{F}_{n-1})\}$$
$$= M\{\text{sign}[A_n - M(A_n|\mathscr{F}_{n-1})]\}\{A_n - M(A_n|\mathscr{F}_{n-1})\}$$
$$= M|A_n - M(A_n|\mathscr{F}_{n-1})|,$$

from which $A_n = M(A_n|\mathscr{F}_{n-1})$ (P-a.s.), i.e., the A_n are $\mathscr{F}_{n-1}$-measurable. □

2.4.6

Theorem 2.15. *Let $X = (x_n, \mathscr{F}_n)$, $n \geq 1$, be a martingale (semimartingale) and let $\tau = \tau(\omega)$ be a* m.t. *with respect to the system $(\mathscr{F}_n)$, $n \geq 1$. Then the "stopped" sequence $(x_{n \wedge \tau}, \mathscr{F}_n)$, $n \geq 1$, is also a martingale (semimartingale).*

Proof. It is sufficient to prove the theorem for the case where X is a supermartingale. From the equality

$$x_{\tau \wedge n} = \sum_{m<n} x_m \chi_{\{\tau=m\}} + x_n \chi_{\{\tau \geq n\}}$$

it follows that the variables $x_{\tau \wedge n}$ are $\mathscr{F}_n$-measurable integrable with any n, $n = 1, 2, \ldots$, and $x_{\tau \wedge (n+1)} - x_{\tau \wedge n} = \chi_{(\tau > n)}(x_{n+1} - x_n)$. Hence,

$$M\{_{\tau \wedge (n+1)} - x_{\tau \wedge n} | \mathscr{F}_n\} = \chi_{\{\tau > n\}} M\{x_{n+1} - x_n | \mathscr{F}_n\} \leq 0,$$

from which the theorem follows readily. □

Also note that this could be deduced immediately from (2.5) (for supermartingales). Actually, taking $\sigma = m$ in (2.5) and instead of τ taking $\tau \wedge n$, we find ($n \geq m$) that (P-a.s.)

$$x_{\tau \wedge m} = x_{(\tau \wedge n) \wedge m} \geq M(x_{\tau \wedge n} | \mathscr{F}_m).$$

Notes and references

2.1–2.4. The theory of martingales and semi-martingales for the case of discrete time is presented in Doob [46], Meyer [126], Neveu [130], and Gykhman and Skorokhod [37].

Martingales and semimartingales: continuous time 3

3.1 Right continuous semimartingales

3.1.1

Let $(\Omega, \mathscr{F}, P)$ be a probability space and let $F = (\mathscr{F}_t)$, $t \geq 0$, be a non-decreasing family of sub-σ-algebras of $\mathscr{F}$.

Definition 1. The supermartingale

$$X = (x_t, \mathscr{F}_t),\ t \geq 0\ (M|x_t| < \infty,\ M(x_t|\mathscr{F}_s) \leq x_s,\ t \leq s),$$

is said to be right continuous if:

(1) The trajectories x_t are right continuous (P-a.s.);
(2) the family $(\mathscr{F}_t)$, $t \geq 0$, is right continuous, i.e.,

$$\mathscr{F}_t = \mathscr{F}_{t+} = \bigcap_{s>t} \mathscr{F}_s, \qquad t \geq 0$$

Many of the results of the previous chapter are extended to right continuous supermartingales and submartingales (i.e., to semimartingales).

First of all, we prove a useful result on the conditions for the existence of a right continuous modification of the supermartingale $X = (x_t, \mathscr{F}_t)$, $t \geq 0$.

Theorem 3.1. *Let the family $F = (\mathscr{F}_t)$, $t \geq 0$, be right continuous and each of the σ-algebras $\mathscr{F}_t$ be completed by the P-nullsets from F. In order that the supermartingale $X = (x_t, \mathscr{F}_t)$, $t \geq 0$, permit a right continuous modification, it is necessary and sufficient that the function $m_t = MX_t$, $t \geq 0$, be right continuous.*

For the proof we need the following.

Lemma 3.1. *Let $X = (x_t, \mathscr{F}_t)$, $t \geq 0$, be a supermartingale for which there exists an integrable random variable y such that $x_s \leq M(y|\mathscr{F}_s)$ (P-a.s.), $s \geq 0$. Let $\tau_1 \geq \tau_2 \geq \cdots$ be a nonincreasing sequence of Markov times. Then the family of the random variables $\{x_{\tau_n}, n = 1, 2, \ldots\}$ is uniformly integrable.*

PROOF. Assume $y_n = x_{\tau_n}$, $\mathscr{G}_n = \mathscr{F}_{\tau_n}$. Then by Theorem 2.10, $x_{\tau_n} \geq M(x_{\tau_{n-1}}|\mathscr{F}_{\tau_n})$ or, in new notation,

$$y_n \geq M(y_{n-1}|\mathscr{G}_n). \tag{3.1}$$

Note further that $Mx_0 \geq My_n \geq My_{n-1} \geq My$.

Take now $\varepsilon > 0$ and find $k = k(\varepsilon)$ such that $\lim_n My_n - My_k \leq \varepsilon$. Then for all $n \geq k$, $My_n - My_k < \varepsilon$. Next, by (3.1) for $n \geq k$,

$$\begin{aligned}
\int_{\{|y_n| > \lambda\}} |y_n| \, dP &= \int_{\{y_n > \lambda\}} y_n \, dP - \int_{\{y_n < -\lambda\}} y_n \, dP \\
&= My_n - \int_{\{y_n \leq \lambda\}} y_n \, dP - \int_{\{y_n < -\lambda\}} y_n \, dP \\
&\leq My_n - \int_{\{y_n \leq \lambda\}} y_k \, dP - \int_{\{y_n < -\lambda\}} y_k \, dP \\
&\leq \varepsilon + My_k - \int_{\{y_n \leq \lambda\}} y_k \, dP - \int_{\{y_n < -\lambda\}} y_k \, dP \\
&\leq \varepsilon + \int_{\{|y_n| \geq \lambda\}} |y_k| \, dP.
\end{aligned} \tag{3.2}$$

But

$$P\{|y_n| \geq \lambda\} = \frac{M|y_n|}{\lambda} = \frac{My_n + 2My_n^-}{\lambda} \leq \frac{Mx_0 + 2M|y|}{\lambda} \to 0$$

with $\lambda \to \infty$. Hence

$$\sup_{n \geq k} \int_{\{|y_n| \geq \lambda\}} |y_k| \, dP \to 0, \qquad \lambda \to \infty,$$

and therefore, according to (3.2),

$$\lim_{\lambda \to \infty} \sup_{n \geq k} \int_{\{|y_n| \geq \lambda\}} |y_n| \, dP \leq \varepsilon. \tag{3.3}$$

Since the variables $y_1, \ldots, y_k$ are integrable, given $\varepsilon > 0$ we can find $L > 0$ such that

$$\max_{i \leq k} \int_{\{|y_i| \geq L\}} |y_i| \, dP \leq \varepsilon.$$

Together with (3.3) this result leads to uniform integrability of the sequence $y_1, y_2, \ldots$. □

Note. If $P(\tau_1 \leq N) = 1$, $N < \infty$, then the lemma holds true without the assumption $x_s \geq M(y|\mathscr{F}_s)$, $s \geq 0$, since then it will be sufficient to consider only $s \in [0, N]$, and for such s $x_s \geq M(y|\mathscr{F}_s)$ with $y = x_N$, $M|x_N| < \infty$.

3.1.2

PROOF OF THEOREM 3.1. Let S be a countable dense set on $[0, \infty)$. For any rational a and b $(-\infty < a < b < \infty)$ denote by $\beta(a, b; n; S)$ the number of up-crossings of the interval (a, b) by the supermartingale $x = (x_s, \mathscr{F}_s)$, $s \in [0, n] \cap S$. Using Theorem 2.5 we conclude that $M(\beta(a, b; n; S) < \infty$, and therefore the set

$$A(a, b; n; S) = \{\omega | \beta(a, b; n; S) = \infty\}$$

has P-measure zero. Hence the set $A = \cup A(a, b; n; S)$, where the union is taken over all pairs (a, b) of rational numbers and $n = 1, 2, \ldots$, also has P-measure zero.

For every elementary event $\omega \in \bar{A}$ we have $\beta(a, b; n; S) < \infty$ and therefore, as is well known from analysis, the function $x_s = x_s(\omega)$, $s \in S$, has left and right limits,

$$x_{t-}(\omega) - \lim_{\substack{s \uparrow t \\ s \in S}} x_s(\omega) \quad \text{and} \quad x_{t+}(\omega) = \lim_{\substack{s \downarrow t \\ s \in S}} x_s(\omega)$$

respectively, for each $t \in [0, \infty)$.

For every $\omega \in A$ let us now set

$$x_{t+}(\omega) = 0, \qquad t \in [0, \infty).$$

Clearly, for all $\omega \in \Omega$ the trajectories x_{t+}, $t \geq 0$, are continuous on the right and for each $t \geq 0$ the variables x_{t+} are $\mathscr{F}_{t+}$-measurable by the construction and by noting that according to the assumption every set $A \in F$ with $P(A) = 0$ belongs to $\mathscr{F}_t$ for all $t \geq 0$.

Finally, if $s_n \downarrow t$, $s_n \in S$, then by Lemma 3.1 the variables $(x_{s_n}, n = 1, 2, \ldots)$ are uniformly integrable and therefore the inequality

$$x_t \geq M(x_{s_n}|\mathscr{F}_t) \qquad (P\text{-a.e.}) \tag{3.4}$$

implies (see Theorem 1.3)

$$x_t \geq M(x_{t+}|\mathscr{F}_t) \qquad (P\text{-a.e.}) \tag{3.5}$$

According to the assumption we have $\mathscr{F}_t = \mathscr{F}_{t+}$ and x_{t+} are $\mathscr{F}_{t+}$-measurable. Therefore (3.5) implies $P(x_t \geq x_{t+}) = 1$. Assume that $m_t = m_{t+}$, i.e., $Mx_t = Mx_{t+}$. Then from the equality $P(x_t \geq x_{t+}) = 1$ it immediately follows that $P(x_t = x_{t+}) = 1$. In this case the supermartingale $X = (x_t, \mathscr{F}_t)$, $t \geq 0$, has the modification $X^* = (x_{t+}, \mathscr{F}_t)$, $t \geq 0$, whose trajectories are obviously right continuous with probability 1.

Suppose now the supermartingale $X = (x_t, \mathscr{F}_t)$, $t \geq 0$, has the right continuous modification $Y = (y_t, \mathscr{F}_t)$, $t \geq 0$. Then since $P(x_t = y_t) = 1$, $t \geq 0$, $Mx_t = My_t$, and by Lemma 3.1

$$\lim_{s \downarrow t} My_s = M \lim_{s \downarrow t} y_s = My_{t+} = My_t.$$

In other words, the mathematical expectation $m_t = Mx_t (= My_t)$ is right continuous. □

Corollary. *Any martingale $X = (x_t, \mathscr{F}_t)$, $\mathscr{F}_t = \mathscr{F}_{t+}$, $t \geq 0$, permits a right continuous modification.*

Note. In Theorem 3.1 the assumption of right continuity of the family $F = \{\mathscr{F}_t\}$, $t \geq 0$, is essential. Another sufficient condition for the existence of a right continuous modification of the supermartingale $X = (x_t, \mathscr{F}_t)$, $t \geq 0$, is, for example, that the process x_t, $t \geq 0$, be right continuous in probability at each point t, i.e., $P\text{-}\lim_{s \downarrow t} x_s = x_t$.

3.2 Basic inequalities, the theorem of convergence, and invariance of the supermartingale property for Markov times

3.2.1

Theorem 3.2. *Let $X = (x_t, \mathscr{F}_t)$, $t \leq T$, be a submartingale with right continuous trajectories. The following inequalities are valid:*

$$\lambda P\left\{\sup_{t \leq T} x_t \geq \lambda\right\} \leq \int_{\{\sup_{t \leq T} x_t \geq \lambda\}} x_T \, dP \leq Mx_T^+, \tag{3.6}$$

$$\lambda P\left\{\inf_{t \leq T} x_t \leq -\lambda\right\} \leq -Mx_0 + \int_{\{\inf_{t \leq T} x_t \geq -\lambda\}} x_T \, dP. \tag{3.7}$$

If X is a nonnegative submartingale with $Mx_T^p < \infty$ for $1 < p < \infty$, then

$$M\left[\sup_{t \leq T} x_t\right]^p \leq \left(\frac{p}{p-1}\right)^p Mx_T^p. \tag{3.8}$$

If $\beta_T(a, b)$ is the number of up-crossings of the interval (a, b) by the submartingale $X = (x_t, \mathscr{F}_t)$, $t \leq T$, then

$$M\beta_T(a, b) \leq \frac{M[x_T - a]^+}{b - a} \leq \frac{Mx_T^+ + |a|}{b - a}. \tag{3.9}$$

PROOF. Since the trajectories x_t, $t \geq 0$, are right continuous, then the events

$$\left\{\inf_{t \leq T} x_t \leq -\lambda\right\} = \left\{\inf_{r \leq T} x_r \leq -\lambda\right\} \quad \text{and} \quad \left\{\sup_{t \leq T} x_t \geq \lambda\right\} = \left\{\sup_{r \leq T} x_r \geq \lambda\right\}$$

belong to $\mathscr{F}$ (the r are rational numbers). Hence (3.6)–(3.9) are easily obtained from the corresponding inequalities for the case of discrete time, discussed in the preceding chapter. □

Corollary 1. *If $X = (x_t, \mathscr{F}_t)$, $t \geq 0$, is a submartingale (or a supermartingale) with right continuous trajectories, then for each $t > 0$ (P-a.s.) $x_{t-} = \lim_{s \uparrow t} x_s$ exists.*

Actually, if with positive probability this limit did not exist, then (compare with the assumptions used for proving Theorem 2.6) for some $a < b$, $M\beta_t(a, b) = \infty$. But this contradicts the estimate in (3.9).

Corollary 2. *Let $X = (x_t, \mathscr{F}_t)$, $t \geq 0$, be a martingale with $x_t = M(\xi | \mathscr{F}_t)$, $M|\xi| < \infty$, and let the family $(\mathscr{F}_t)$, $t \geq 0$, be right continuous. Then the process x_t, $t \geq 0$, has the modification $\tilde{x}_t$, $t \geq 0$, with trajectories right continuous (P-a.s.) and having the limit to the left (at each point $t > 0$).*

Actually, from Theorem 1.5 it follows that for each $t \geq 0$ there exists

$$x_{t+} = \lim_{s \downarrow t} M(\xi | \mathscr{F}_s) = M(\xi | \mathscr{F}_{t+}) = M(\xi | \mathscr{F}_t) = x_t.$$

Hence if we put $\tilde{x}_t \equiv x_{t+}$, then we obtain the right continuous modification (P-a.s.). Because of the previous corollary, the process $\tilde{x}_t$, $t \geq 0$, has for each $t > 0$ the limits to the left $\tilde{x}_{t-} = \lim_{s \uparrow t} \tilde{x}_s$ (P-a.s.).

3.2.2

Theorem 3.3. *Let $X = (x_t, \mathscr{F}_t)$, $t \geq 0$, be a submartingale with right continuous trajectories x_t, $t \geq 0$, such that*

$$\sup_t M x_t^+ < \infty. \tag{3.10}$$

Then with probability 1 $\lim_{t \to \infty} x_t (= x_\infty)$ exists and $M x_\infty^+ < \infty$.

PROOF. The proof follows from (3.9) by means of the assertions used for proving Theorem 2.6. □

3.2.3

Analogous to the case of discrete time we introduce the concept of the potential $\Pi = (\pi_t, \mathscr{F}_t)$, $t \geq 0$—a nonnegative supermartingale with $\lim_{t \to \infty} M\pi_t = 0$—and prove the following result.

Theorem 3.4 (Riesz decomposition). *If the supermartingale $X = (x_t, \mathscr{F}_t)$, $t \geq 0$, with right continuous trajectories $x_t, t \geq 0$ majorizes some submartingale $Y = (y_t, \mathscr{F}_t)$, $t \geq 0$, then there exists a martingale $M = (m_t, \mathscr{F}_t), t \geq 0$, and a potential $\Pi = (\pi_t, \mathscr{F}_t), t \geq 0$, such that, for each $t \geq 0$,*

$$x_t = m_t + \pi_t \qquad (P\text{-a.s.}). \tag{3.11}$$

The decomposition in (3.11) is unique (to within a stochastic equivalence).

3.2.4

Theorem 3.5. *Let $X = (x_t, \mathscr{F}), t \geq 0$, be a supermartingale with right continuous trajectories, such that, for a certain random variable η with $M|\eta| < \infty$,*

$$x_t \geq M(\eta | \mathscr{F}_t) \qquad (P\text{-a.s.}), \qquad t \geq 0.$$

If τ and σ are Markov times and $P(\sigma \leq \tau) = 1$ then

$$x_\sigma \geq M(x_\tau | \mathscr{F}_\sigma). \tag{3.12}$$

Proof. For each n, $n = 1, 2, \ldots$, let $\tau_n = \tau_n(\omega)$ where

$$\tau_n(\omega) = \frac{k}{2^n} \quad \text{on } \left\{\omega: \frac{k-1}{2^n} \leq \tau(\omega) < \frac{k}{2^n}\right\}, \qquad k = 1, 2, \ldots,$$

and $\tau_n(\omega) = +\infty$ on $\{\omega: \tau(\omega) = \infty\}$. Analogously define the times σ_n, $n = 1, 2, \ldots$. Assume that $P(\sigma_n \leq \tau_n) = 1$ for each n, $n = 1, 2, \ldots$ (otherwise, $\sigma_n \wedge \tau_n$ should be considered instead of σ_n).

By Theorem 2.12,

$$x_{\sigma_n} \geq M(x_{\tau_n} | \mathscr{F}_{\sigma_n}) \qquad (P\text{-a.s.}), \qquad n = 1, 2, \ldots$$

Take the set $A \in \mathscr{F}_\sigma$. Then since $\mathscr{F}_\sigma \subseteq \mathscr{F}_{\sigma_n}$, $A \in \mathscr{F}_{\sigma_n}$, and from the preceding inequality we obtain

$$\int_A x_{\sigma_n}\, dP \geq \int_A x_{\tau_n}\, dP. \tag{3.13}$$

Note now that the random variables $(x_{\sigma_n}, n = 1, 2, \ldots)$ and $(x_{\tau_n}, n = 1, 2, \ldots)$ are uniformly integrable (Lemma 3.1) and $\tau_n(\omega) \downarrow \tau(\omega)$, $\sigma_n(\omega) \downarrow \sigma(\omega)$ for all ω. Hence passing to the limit in (3.13) with $n \to \infty$ it is found (Theorem 1.3) that

$$\int_A x_\sigma\, dP \geq \int_A x_\tau\, dP. \tag{3.14}$$

Hence, $x_\sigma \geq M[x_\tau | \mathscr{F}_\sigma]$ (P-a.s.). □

Note 1. From Theorem 3.5 it is seen that (3.12) holds true for the supermartingales with continuous trajectories $X = (x_t, \mathscr{F}_t)$, $0 \leq t \leq T < \infty$, and the m.t. τ and σ such that $P(\sigma \leq \tau \leq T) = 1$.

Note 2. If $X = (x_t, \mathscr{F}_t)$, $t \geq 0$, is a nonnegative supermartingale and $x_\tau = 0$, then $x_t = 0(\{t \geq \tau\}, (P\text{-a.s.}))$.

3.2.5

The above proof shows that if the supermartingale $X = (x_t, \mathscr{F}_t)$, $t \geq 0$, is a uniformly integrable martingale, then the inequality given by (3.12) turns into an equality. To make this statement analogous in its form to the corresponding statement (Theorem 2.9) for discrete time, we introduce such a definition.

Definition 2. The martingale $X = (x_t, \mathscr{F}_t)$, $t \geq 0$, is called *regular* if there exists an integrable random variable $\eta(M|\eta| < \infty)$ such that

$$x_t = M(\eta | \mathscr{F}_t) \qquad (P\text{-a.s.}), \qquad t \geq 0.$$

As in Theorem 2.7, it can be shown that the regularity of the martingale $X = (x_t, \mathscr{F}_t)$, $t \geq 0$, is equivalent to uniform integrability of the family of random variables $(x_t, t \geq 0)$.

Theorem 3.6. *Let $X = (x_t, \mathscr{F}_t)$, $t \geq 0$ be a regular martingale with right continuous trajectories. Then if τ and σ are Markov times and $P(\sigma \leq \tau) = 1$, then*

$$x_\sigma = M(x_\tau | \mathscr{F}_\sigma) \qquad (P\text{-a.s.}). \tag{3.15}$$

PROOF. This follows from the proof of Theorem 3.5, noting that for a regular martingale families of the random variables $\{x_{\sigma_n}: n = 1, 2, \ldots\}$ and $\{x_{\tau_n}: n = 1, 2, \ldots\}$ are uniformly integrable. □

Note 1. Since for the martingale $X = (x_t, \mathscr{F}_t)$, $t \geq 0$, $m_t = Mx_t \equiv \text{const.}$ for right continuity of its trajectories (in accordance with Theorem 3.1) it is sufficient to require only right continuity of the family $(\mathscr{F}_t)$, $t \geq 0$. More precisely, in this case there exists a martingale $Y = (y_t, \mathscr{F}_t)$, $t \geq 0$, such that its trajectories y_t, $t \geq 0$, are right continuous and $P(x_t = y_t) = 1$, $t \geq 0$.

Note 2. Statement (3.15) of Theorem 3.6 remains correct for the martingale $X = (x_t, \mathscr{F}_t)$ with right continuous trajectories over the finite time interval $0 \leq t \leq T$ and Markov times τ and σ such that $P(\sigma \leq \tau \leq T) = 1$.

Note 3. If in Theorem 3.6 we omit the condition that $P(\sigma \leq \tau) = 1$, then (3.15) must be modified as follows:

$$x_{\sigma \wedge \tau} = M(x_\tau | \mathscr{F}_\sigma) \qquad (P\text{-a.s.}) \tag{3.16}$$

(compare with (2.25)). From this it follows in particular that the "stopped" process $X^* = (x_{t \wedge \tau}, \mathscr{F}_t)$, $t \geq 0$, will also be a martingale. For proving (3.16) note that, according to (2.25),

$$x_{\sigma_n \wedge \tau_k} = M(x_{\tau_k} | \mathscr{F}_{\sigma_n}) \qquad (P\text{-a.s.})$$

for all $k \geq n$. From this, because of the uniform integrability of the variables $\{x_{\tau_k}, k = 1, 2, \ldots\}$ with $k \to \infty$, we find that

$$x_{\sigma_n \wedge \tau} = M(x_\tau | \mathscr{F}_{\sigma_n}).$$

Allowing $n \to \infty$, we arrive at the necessary equality in (3.16).

3.3 Doob–Meyer decomposition for supermartingales

3.3.1

In this section the analog of Theorem 2.13 (Doob decomposition) for the case of continuous time is considered. We introduce some preliminary necessary concepts.

Definition 3. The supermartingale $X = (x_t, \mathscr{F}_t)$, $t \geq 0$, with right continuous trajectories $x_t = x_t(\omega)$, $t \geq 0$ *belongs to class* D if the family of random variables $(x_\tau, \tau \in \mathscr{T})$, where $\mathscr{T}$ is the set of the Markov times τ with $P(\tau < \infty) = 1$, is uniformly integrable.

Definition 4. The supermartingale $X = (x_t, \mathscr{F}_t)$, $t \geq 0$, with right continuous trajectories $x_t = x_t(\omega)$, $t \geq 0$, *belongs to class* DL, if for any a, $0 \leq a < \infty$, the family of random variables $(x_\tau, \tau \in \mathscr{T}_a)$, where $\mathscr{T}_a$ is the set of the Markov times τ with $P(\tau \leq a) = 1$, is uniformly integrable.

It is clear that class $DL \supseteq D$. The next theorem gives criteria for membership in classes D and DL.

Theorem 3.7

(1) *Any martingale* $X = (x_t, \mathscr{F}_t)$, $t \geq 0$, *with right continuous trajectories, belongs to class* DL.
(2) *Any uniformly integrable martingale* $X = (X_t, \mathscr{F}_t)$, $t \geq 0$ *with right continuous trajectories, belongs to class* D.
(3) *Any negative supermartingale* $X = (x_t, \mathscr{F}_t)$, $t \geq 0$, *with right continuous trajectories, belongs to class* DL.

PROOF. Let $P(\tau \leq a) = 1$, $a < \infty$. Then according to Note 2 to Theorem 3.6, $x_\tau = M(x_a | \mathscr{F}_\tau)$ (P-a.s.). But the family $(x_\tau, \tau \in \mathscr{T}_a)$ of such random variables is uniformly integrable as can be proved in the same way as the implication (A) $\Rightarrow$ (B) in Theorem 2.7. The second statement is proved in a similar manner. Let us next prove the last statement.

Let $P(\tau \leq a) = 1$. Then according to Note 1 to Theorem 3.5, for $\lambda > 0$

$$\int_{\{|x_\tau| > \lambda\}} |x_\tau| \, dP = -\int_{\{|x_\tau| > \lambda\}} x_\tau \, dP \leq -\int_{\{|x_\tau| > \lambda\}} x_a \, dP$$

and also $M|x_\tau| \leq M|x_a|$. Hence, by Chebyshev's inequality,

$$\lambda P\{|x_\tau| > \lambda\} \leq M|x_\tau| \leq M|x_a|.$$

Therefore, $P\{|x_\tau| > \lambda\} \to 0$, $\lambda \to \infty$ and consequently,

$$\sup_{\tau \in \mathcal{T}_a} \int_{\{|x_\tau| > \lambda\}} |x_\tau| dP \leq \sup_{\tau \in \mathcal{T}_a} \left[- \int_{\{|x_\tau| > \lambda\}} x_a \, dP \right] \to 0, \qquad \lambda \to \infty. \qquad \square$$

3.3.2

Definition 5. Let $(\Omega, \mathcal{F}, P)$ be a probability space and let $F = (\mathcal{F}_t)$, $t \geq 0$, be a nondecreasing family of right continuous sub-σ-algebras $\mathcal{F}$. The right continuous random process A_t, $t \geq 0$, is called *increasing*, if the values A_t are $\mathcal{F}_t$-measurable, $A_0 = 0$ and $A_s \leq A_t$ (P-a.s.), $s \leq t$. The increasing process $A = (A_t, \mathcal{F}_t)$, $t \geq 0$, is called a *natural increasing process*, if for any bounded positive right continuous martingale $Y = (y_t, \mathcal{F}_t)$, $t \geq 0$, having the limits to the left,

$$M \int_0^\infty y_{s-} \, dA_s = M y_\infty A_\infty. \tag{3.17}$$

The increasing process A_t, $t \geq 0$, is called *integrable* if $MA_\infty < \infty$.

Lemma 3.2. *The integrable increasing process $A = (A_t, \mathcal{F}_t)$, $t \geq 0$, is natural if and only if for any bounded martingale, right continuous and having limits to the left, $Y = (y_t, \mathcal{F}_t)$, $t \geq 0$,*

$$M \int_0^T y_s \, dA_s = M \int_0^T y_{s-} \, dA_s \tag{3.18}$$

for any $T > 0$.

Proof. Let us first show that for any increasing process $A = (A_t, \mathcal{F}_t)$, $t \geq 0$, with $A_0 = 0$, $MA_\infty < \infty$ and the martingale $Y = (y_t, \mathcal{F}_t)$, $t \geq 0$, having right continuous trajectories

$$M \int_0^T y_s \, dA_s = M y_T A_T. \tag{3.19}$$

Set $c_t(\omega) = \inf\{s : A_s(\omega) > t\}$ and use the fact that for almost all ω the Lebesgue–Stieltjes integral can be reduced to a Lebesgue integral (Section 1.1)

$$\int_0^T y_s \, dA_s = \int_0^{A_T(\omega)} y_{c_t(\omega)} \, dt = \int_0^\infty y_{c_t(\omega)} \chi_{\{t : t < A_T(\omega)\}} \, dt,$$

where, according to the corollary of Lemma 1.8, $y_{c_t(\omega)}$ is a $\mathcal{F}_{c_t}$-measurable variable. But (P-a.s.)

$$\{t : t < A_T(\omega)\} = \{t : c_t(\omega) < T\}.$$

Hence,

$$\int_0^T y_s\,dA_s = \int_0^\infty y_{c_t(\omega)}\chi_{\{t:c_t(\omega)<T\}}\,dt,$$

and by Fubini's theorem

$$M\int_0^T y_s\,dA_s = \int_0^\infty M[y_{c_t(\omega)}\chi_{\{t:c_t(\omega)<T\}}]dt.$$

Fix $t \geq 0$ and note that the random time $\tau(\omega) = c_t(\omega)$ is Markov. Then, since the event $\{\omega:\tau(\omega) < T\} \in \mathcal{F}_\tau$ (Lemma 1.7) and $Y = (y_t, \mathcal{F}_t)$, $t \geq 0$, is a martingale, by Note 2 to Theorem 3.6

$$\begin{aligned} M[y_{\tau(\omega)}\chi_{\{t:c_t(\omega)<T\}}] &= M[y_{\tau(\omega)}\chi_{(\omega:\tau(\omega)<T\}}] \\ &= M[\chi_{\{\omega:\tau(\omega)<T\}}M(y_T|\mathcal{F}_\tau)] = M[\chi_{\{\omega:\tau(\omega)<T\}}y_T]. \end{aligned}$$

Therefore,

$$\begin{aligned} M\int_0^T y_s\,dA_s &= \int_0^\infty M\{\chi_{\{t:c_t(\omega)<T\}}y_T\}dt \\ &= M\left[y_T\int_0^\infty \chi_{\{t:c_t(\omega)<T\}}\,dt\right] = My_T A_T. \end{aligned}$$

Hence if (3.18) is satisfied for any $T > 0$, then $M\int_0^T y_{s-}\,dA_s = My_T A_T$, and, taking limits as $T \to \infty$, we obtain (3.17).

Suppose next that (3.17) is satisfied. Since

$$M\int_0^\infty y_s\,dA_s = MA_\infty y_\infty$$

then

$$M\int_0^\infty y_s\,dA_s = M\int_0^\infty y_{s-}\,dA_s.$$

Let now $y_s^* = y_s\chi_{\{s<T\}} + y_T\chi_{\{s\geq T\}}$. The process $Y^* = (y_s^*, \mathcal{F}_s)$, $s \geq 0$, is a martingale (right continuous, bounded, as is easily verified[1]) and the equality

$$M\int_0^\infty y_s^*\,dA_s = M\int_0^\infty y_{s-}^*\,dA_s$$

turns into Equation (3.18), as required. □

Let us consider now the analog of Theorem 2.13 (Doob decomposition), restricting ourselves first to nonnegative supermartingales which are potentials.

[1] A more general result of this kind is given in Lemma 3.3.

Theorem 3.8 (Doob–Meyer decomposition). *Let the right continuous potential $\Pi = (\pi_t, \mathscr{F}_t)$, $0 \le t \le \infty$, belong to class D. Then there exists an integrable increasing process $A = (A_t, \mathscr{F}_t)$, $t \ge 0$, such that*

$$\pi_t = M(A_\infty | \mathscr{F}_t) - A_t \qquad (P\text{-a.s.}), \qquad t \ge 0. \tag{3.20}$$

In the expansion given in (3.20) *the process A_t, $t \ge 0$, can be taken as natural.*

The expansion given in (3.20) *with a natural increasing process is unique.*

PROOF. For any n, $n = 0, 1, \ldots$, the sequence $(\pi_{i \cdot 2^{-n}}, \mathscr{F}_{i \cdot 2^{-n}})$, $i = 0, 1, \ldots$, is a potential (with discrete times $0, 2^{-n}, 2 \cdot 2^{-n}, \ldots$). According to Corollary 1 of Theorem 2.13, for any n

$$\pi_{i \cdot 2^{-n}} = M[A_\infty(n) | \mathscr{F}_{i \cdot 2^{-n}}] - A_{i \cdot 2^{-n}}(n), \qquad i = 0, 1, \ldots, \tag{3.20$'$}$$

where the variables $A_{i \cdot 2^{-n}}(n)$ are $\mathscr{F}_{(i-1) \cdot 2^{-n}}$-measurable, constitute an increasing process and

$$A_\infty(n) = \lim_{i \to \infty} A_{i \cdot 2^{-n}}(n). \tag{3.21}$$

Assume now that the values $A_\infty(n)$, $n = 0, 1, \ldots$, are uniformly integrable (it will be shown further that for this it is necessary and sufficient that the potential Π should belong to class D). Then, according to Theorem 1.7, a sequence of integers $n_1, n_2, \ldots \to \infty$ and an integrable function A_∞ can be found such that, for any limited random variable ξ,

$$\lim_{i \to \infty} MA_\infty(n_i)\xi = MA_\infty \xi. \tag{3.22}$$

Denote by m_t the right continuous modification $M(A_\infty | \mathscr{F}_t)$, existing because of Corollary 2 of Theorem 3.2. Let $r \le s$ be the numbers of the form $i \cdot 2^{-n}$, $i = 0, 1, \ldots$. Then $A_r(n) \le A_s(n)$, and together with (3.20$'$)) this yields

$$M[A_\infty(n) | \mathscr{F}_r] - \pi_r \le M[A_\infty(n) | \mathscr{F}_s] - \pi_s. \tag{3.23}$$

From this, with $n = n_i \to \infty$, we obtain

$$m_r - \pi_r \le m_s - \pi_s. \tag{3.24}$$

Set $A_t = m_t - \pi_t$. This function is (P-a.s.) right continuous and since, according to (3.24), it does not decrease on a binary rational sequence, A_t is an increasing process. Further, $\pi_t \to 0$(P-a.s.), $t \to \infty$, and $m_t = M(A_\infty | \mathscr{F}_t) \to M(A_\infty | \mathscr{F}_\infty) = A_\infty$, $t \to \infty$. Hence, (P-a.s.) $\lim_{t \to \infty} A_t$ yields the variable A_∞, introduced before.

Let us show now that the process A_t, $t \ge 0$, is natural. Let $Y = (y_t, \mathscr{F}_t)$, $t \ge 0$, be a bounded nonnegative martingale, having (P-a.s.) the limits to the left $y_{t-} = \lim_{s \uparrow t} y_s$ at each point $t > 0$. Since the process A_t, $t \ge 0$, is right continuous, and the process y_{t-}, $t > 0$, is left continuous, then, by the Lebesgue bounded convergence theorem (Theorem 1.4),

$$M \int_0^\infty y_{s-}\, dA_s = \lim_{n \to \infty} \sum_{i=0}^{\infty} M[y_{i \cdot 2^{-n}}(A_{(i+1)2^{-n}} - A_{i \cdot 2^{-n}})]. \tag{3.25}$$

But the $y_{i\cdot 2^{-n}}$ are $\mathscr{F}_{i\cdot 2^{-n}}$-measurable. Hence

$$\sum_{i=0}^{\infty} M[y_{i\cdot 2^{-n}}(A_{(i+1)\cdot 2^{-n}} - A_{i\cdot 2^{-n}})]$$

$$= \sum_{i=0}^{\infty} M[y_{i\cdot 2^{-n}} M(A_{(i+1)\cdot 2^{-n}} - A_{i\cdot 2^{-n}} | \mathscr{F}_{i\cdot 2^{-n}})]$$

$$= \sum_{i=0}^{\infty} M[y_{i\cdot 2^{-n}} M((m_{(i+1)\cdot 2^{-n}} - \pi_{(i+1)\cdot 2^{-n}}) - (m_{i\cdot 2^{-n}} - \pi_{i\cdot 2^{-n}}) | \mathscr{F}_{i\cdot 2^{-n}})]$$

$$= \sum_{i=0}^{\infty} M[y_{i\cdot 2^{-n}} M(\pi_{i\cdot 2^{-n}} - \pi_{(i+1)\cdot 2^{-n}} | \mathscr{F}_{i\cdot 2^{-n}})]$$

$$= \sum_{i=0}^{\infty} M[y_{i\cdot 2^{-n}}(A_{(i+1)\cdot 2^{-n}}(n) - A_{i\cdot 2^{-n}}(n))]. \tag{3.26}$$

Note now that the $A_{(i+1)2^{-n}}(n)$ are $\mathscr{F}_{i\cdot 2^{-n}}$-measurable, and therefore

$$M[y_{i\cdot 2^{-n}} A_{(i+1)\cdot 2^{-n}}(n)] = M[y_{(i+1)\cdot 2^{-n}} A_{(i+1)\cdot 2^{-n}}]. \tag{3.27}$$

From (3.25)–(3.27) we find that

$$M \int_0^{\infty} y_{s-}\, dA_s = \lim_n M[A_{\infty}(n) y_{\infty}]. \tag{3.28}$$

According to (3.22),

$$\lim_{n_i \to \infty} M[A_{\infty}(n_i) y_{\infty}] = M[A_{\infty} y_{\infty}]. \tag{3.29}$$

From the comparison of (3.28) with (3.29) we conclude that

$$M \int_0^{\infty} y_{s-}\, dA_s = M A_{\infty} y_{\infty}, \tag{3.30}$$

i.e., the process A_t, $t \geq 0$, is a natural one.

Assume now that along with $\pi_t = M(A_{\infty} | \mathscr{F}_t) - A_t$ there also exists an expansion $\pi_t = M(B_{\infty} | \mathscr{F}_t) - B_t$ with a natural increasing process $(B_t, t \geq 0)$. We will show that $A_t = B_t$ (P-a.s.) for any $t \geq 0$. To see this it is enough to show that for any fixed t and any bounded $\mathscr{F}_t$-measurable random variable η,

$$M[\eta A_t] = M[\eta B_t]. \tag{3.31}$$

Let η_s, $s \leq t$, be a right continuous modification of the conditional expectation $M(\eta | \mathscr{F}_s)$, $s \leq t$. (3.19) and (3.18) imply that

$$\begin{aligned} M[\eta A_t] &= M\left[\int_0^t \eta_s\, dA_s\right] = M\left[\int_0^t n_{s-}\, dA_s\right], \\ M[\eta B_t] &= M\left[\int_0^t \eta_s\, dB_s\right] = M\left[\int_0^t n_{s-}\, dB_s\right]. \end{aligned} \tag{3.32}$$

Since $(A_s - B_s, \mathscr{F}_s)$, $s \le t$, is a martingale, we have

$$M[\eta_{i\cdot 2^{-n}}(B_{(i+1)\cdot 2^{-n}} - B_{i\cdot 2^{-n}})] = M[\eta_{i\cdot 2^{-n}}(A_{(i+1)\cdot 2^{-n}} - A_{i\cdot 2^{-n}})]$$

and hence (see (3.25))

$$\begin{aligned} M\left[\int_0^t \eta_{s-}\, dB_s\right] &= \lim_{n\to\infty} \sum_{\{i:i\cdot 2^{-n}\le t\}} M[\eta_{i\cdot 2^{-n}}(B_{(i+1)\cdot 2^{-n}} - B_{i\cdot 2^{-n}})] \\ &= \lim_{n\to\infty} \sum_{\{i:1\cdot 2^{-n}\le t\}} M[\eta_{i\cdot 2^{-n}}(A_{(i+1)\cdot 2^{-n}} - A_{i\cdot 2^{-n}})] \\ &= M\left[\int_0^t \eta_{s-}\, dA_s\right]. \end{aligned}$$

This and (3.22) prove (3.31), as required.

To complete the proof it also has to be established that for uniform integrability of the sequence $\{A_\infty(n), n = 0, 1, \ldots\}$ it is necessary and sufficient that the potential $\pi = (\pi_t, \mathscr{F}_t)$, $t \ge 0$, belongs to class D.

If the family $\{A_\infty(n), n = 0, 1, \ldots\}$ is uniformly integrable, then, as already established, $\pi_t = M[A_\infty | \mathscr{F}_t] - A_t$. Therefore $\pi_\tau \le M[A_\infty | \mathscr{F}_\tau]$. But the family $\{M[A_\infty | \mathscr{F}_\tau], \tau \in \mathscr{F}\}$ is uniformly integrable (Theorem 3.7); hence the family $\{\pi_\tau, \tau \in \mathscr{F}\}$ has the same property, i.e., the potential Π belongs to class D.

Suppose $\Pi \in D$. Then according to the Doob decomposition, for each $n = 0, 1, \ldots$, (P-a.s.)

$$\pi_{i\cdot 2^{-n}} = M[A_\infty(n) | \mathscr{F}_{i\cdot 2^{-n}}] - A_{i\cdot 2^{-n}}(n). \tag{3.33}$$

Since the $A_{(i+1)\cdot 2^{-n}}(n)$ are $\mathscr{F}_{i\cdot 2^{-n}}$-measurable for each $\lambda > 0$, the time

$$\tau_{n,\lambda} = \inf\{i\cdot 2^{-n} : A_{(i+1)\cdot 2^{-n}}(n) > \lambda\} \tag{3.34}$$

($\tau_{n,\lambda} = \infty$, if the set $\{\cdot\}$ in (3.34) is empty) will be a Markov time with respect to the family $\{\mathscr{F}_{i\cdot 2^{-n}}, i = 0, 1, \ldots\}$.

It is clear that $\{\omega : A_\infty(n) > \lambda\} = \{\omega : \tau_{n,\lambda} < \infty\}$, and by (3.33)

$$\pi_{\tau_{n,\lambda}} = M[A_\infty(n) | \mathscr{F}_{\tau_{n,\lambda}}] - A_{\tau_{n,\lambda}}(n) \qquad (P\text{-a.s.}). \tag{3.35}$$

From this we find

$$\begin{aligned} M[A_\infty(n); \{A_\infty(n) > \lambda\}] &= M[A_{\tau_{n,\lambda}}(n); \{\tau_{n,\lambda} < \infty\}] + M[\pi_{\tau_{n,\lambda}}; \{\tau_{n,\lambda} < \infty\}] \\ &\le \lambda P\{A_\infty(n) > \lambda\} + M[\pi_{\tau_{n,\lambda}}; \{\tau_{n,\lambda} < \infty\}], \end{aligned} \tag{3.36}$$

since from (3.34) $A\tau_{n,\lambda}(n) \le \lambda$.

From (3.36) we obtain

$$\begin{aligned} M[A_\infty(n) - \lambda; \{A_\infty(n) > 2\lambda\}] &\le M[A_\infty(n) - \lambda; \{A_\infty(n) > \lambda\}] \\ &\le M[\pi_{\tau_{n,\lambda}}; \{\tau_{n,\lambda} < \infty\}]. \end{aligned} \tag{3.37}$$

Therefore

$$\lambda P\{A_\infty(n) > 2\lambda\} \le M[\pi_{\tau_{n,\lambda}}; \{\tau_{n,\lambda} < \infty\}]. \tag{3.38}$$

From (3.36) (with substitution of λ for 2λ) and (3.38) we find

$$\begin{aligned} &M[A_\infty(n); \{A_\infty(n) > 2\lambda\}] \\ &\le 2\lambda P\{A_\infty(n) > 2\lambda\} + M[\pi_{\tau_{n,2\lambda}}; \{\tau_{n,2\lambda} < \infty\}] \\ &\le 2M[\pi_{\tau_{n,\lambda}}; \{\tau_{n,\lambda} < \infty\}] + M[\pi_{\tau_{n,2\lambda}}; \{\tau_{n,2\lambda} < \infty\}]. \end{aligned} \tag{3.39}$$

Note now that

$$P\{\tau_{n,\lambda} < \infty\} = P\{A_\infty(n) > \lambda\} \le \frac{MA_\infty(n)}{\lambda} = \frac{M\pi_0}{\lambda} \to 0, \qquad \lambda \to \infty.$$

From this and the assumption that $\Pi \in D$ it follows that as $\lambda \to \infty$ the right side in (3.39) converges to zero uniformly in n, $n = 0, 1, \ldots$.

Hence, uniformly in all n, $n = 0, 1, \ldots$,

$$\int_{\{A_\infty(n) > 2\lambda\}} A_\infty(n)dP \to 0, \qquad \lambda \to \infty,$$

which proves uniform integrability of the variables $\{A_\infty(n), n = 0, 1, \ldots\}$. □

Corollary. *Let $X = (x_t, \mathscr{F}_t)$, $t \ge 0$, be a right continuous supermartingale, belonging to class D. Then there exists a right continuous uniformly integrable martingale $M = (m_t, \mathscr{F}_t)$, $t \ge 0$, and an integrable natural increasing process $A = (A_t, \mathscr{F}_t)$, such that*

$$x_t = m_t - A_t \qquad (P\text{-a.s.}), \qquad t \ge 0. \tag{3.40}$$

This decomposition (with the natural process A_t, $t \ge 0$) is unique to within a stochastic equivalence.

Proof. Since $X \in D$, in particular $\sup_t M|x_t| < \infty$ and $\sup_t Mx_t^- < \infty$. Consequently, by Theorem 3.3 there exists $x_\infty = \lim_{t\to\infty} x_t$ with $M|x_\infty| < \infty$.

Let $\tilde{m}_t$ be a right continuous modification of the martingale $M(x_\infty | \mathscr{F}_t)$, $t \ge 0$. Then if $\pi_t = x_t - \tilde{m}_t$, the process $\Pi = (\pi_t, \mathscr{F}_t)$, $t \ge 0$, will form a right continuous potential belonging to class D, since $X \in D$ and the martingale $\tilde{m}_t = (M(x_\infty | \mathscr{F}_t), \mathscr{F}_t)$, $t \ge 0$, also belongs to class D (Theorem 3.7). Applying now the Doob–Meyer decomposition to the potential $\Pi = (\pi_t, \mathscr{F}_t)$, $t \ge 0$, we find that

$$x_t = M(x_\infty | \mathscr{F}_t) + M(A_\infty | \mathscr{F}_t) - A_t, \tag{3.41}$$

where A_t, $t \ge 0$, is a certain integrable natural increasing process. □

Note. Theorem 3.8 and its corollary remain correct also for the right continuous supermartingales $X = (x_t, \mathscr{F}_t)$, $t \ge 0$, belonging to class DL, with the only difference being that the natural increasing process A_t, $t \ge 0$ is such that, generally speaking, $MA_\infty \le \infty$ (see [126]).

3.3.3

In Theorem 3.8 and in its note it was assumed that the supermartingale $\Pi = (\pi_t, \mathscr{F}_t)$, $0 \le t \le T \le \infty$, belongs to class D or class DL. Let us now look at the analog of the Doob–Meyer decomposition without the assumption that $\Pi \in D$ or $\Pi \in DL$.

Definition 6. The random process $M = (m_t, \mathscr{F}_t)$, $t \ge 0$, is called a *local martingale*, if there exists an increasing sequence of the Markov times τ_n, $n = 1, 2, \ldots$ (with respect to $F = (\mathscr{F}_t)$, $t \ge 0$), such that:

(1) $P(\tau_n \le n) = 1$, $P(\lim \tau_n = \infty) = 1$;
(2) for any n, $n = 1, 2, \ldots$, the sequences $(m_{t \wedge \tau_n}, \mathscr{F}_t)$, $t \ge 0$, are uniformly integrable martingales.

In connection with this definition we note that any martingale is a local martingale.

Lemma 3.3. *Let $X = (x_t, \mathscr{F}_t)$, $t \ge 0$, be a martingale with right continuous trajectories and let $\tau = \tau(\omega)$ be a Markov time with respect to the system $F = (\mathscr{F}_t)$, $t \ge 0$. Then the process $(x_{t \wedge \tau}, \mathscr{F}_t)$, $t \ge 0$, is also a martingale.*

PROOF. Put

$$\tau_n = \frac{k}{2^n} \qquad \text{on} \qquad \left\{\omega: \frac{k-1}{2^n} \le \tau < \frac{k}{2^n}\right\},$$

taking $\tau_n = \infty$ on $\{\omega: \tau = \infty\}$. Fix two numbers s and t, $s \le t$, and let $t_n = k/2^n$ if $(k-1)/2^n \le t \le k/2^n$, and $s_n = k/2^n$ if $(k-1)/2^n < s \le k/2^n$. With sufficiently large n, obviously, $s_n \le t_n$.

According to Theorem 2.15, for any $A \in \mathscr{F}_s$

$$\int_A x_{\tau_n \wedge t_n}\, dP = \int_A x_{\tau_n \wedge s_n}\, dP.$$

Since the variables $x_{\tau_n \wedge t_n}$ and $x_{\tau_n \wedge s_n}$ $(n = 1, 2, \ldots)$ are uniformly integrable (Lemma 3.1), passing to the limit $(n \to \infty)$ in the preceding equality we obtain $M(x_{\tau \wedge t} \mid \mathscr{F}_s) = x_{\tau \wedge s}$ (P-a.s.). □

Note. The statement of the lemma is valid also for the supermartingales having right continuous trajectories and majorizing some regular martingale (compare with Theorem 3.5).

Theorem 3.9. *Let $X = (x_t, \mathscr{F}_t)$, $t \ge 0$, be a right continuous nonnegative supermartingale. Then, there exists a right continuous process $M = (m_t, \mathscr{F}_t)$, $t \ge 0$, which is a local martingale, and a natural integrable increasing process $A = (A_t, \mathscr{F}_t)$, $t \ge 0$, such that*

$$x_t = m_t - A_t \qquad (P\text{-a.s.}), \qquad t \ge 0. \tag{3.42}$$

This decomposition is unique.

PROOF. From the analog of the inequality given by (3.6) for the nonnegative supermartingale $X = (x_t, \mathscr{F}_t)$, $t \geq 0$, we find that

$$P\left\{\sup_t x_t \geq \lambda\right\} \leq \frac{Mx_0}{\lambda}.$$

From this it follows that

$$P\left\{\sup_t x_t < \infty\right\} = 1. \tag{3.43}$$

Set $\tau_n = \inf\{t : x_t \geq n\} \wedge n$. Then $P[\tau_n \leq n\} = 1$, $P(\tau_n \leq \tau_{n+1}\} = 1$ and, because of (3.43), $P\{\lim_n \tau_n = \infty\} = 1$. Now set $x_n(t) = x_{t \wedge \tau_n}$. It is clear that $x_{\tau \wedge \tau_n} \leq \max\{n, x_{\tau_n}\}$, from which it follows that for any n, $n = 1, 2, \ldots$, the supermartingale $X_n = (x_n(t), \mathscr{F}_t)$, $t \geq 0$, belongs to class D. Hence, according to the corollary of Theorem 3.8,

$$x_n(t) = m_n(t) - A_n(t), \tag{3.44}$$

where $M_n = (m_n(t), \mathscr{F}_t)$, $t \geq 0$, is a uniformly integrable martingale, and $A_n(t)$, $t \geq 0$ is a natural increasing process.

Note that $x_{n+1}(\tau_n \wedge t) = x_n(t)$. Further, since $\{m_{n+1}(t), t \geq 0\}$ is uniformly integrable, the family $\{m_{n+1}(t \wedge \tau_n), t \geq 0\}$ is also integrable. The process $A_{n+1}(\tau_n \wedge t)$, $t \geq 0$, which is obtained from the natural increasing process $A_{n+1}(t)$, $t \geq 0$ by "stopping" at the time τ_n, will be also natural and increasing, as can easily be proved.

Because of the uniqueness of the Doob–Meyer decomposition.

$$m_{n+1}(\tau_n \wedge t) = m_n(t), \qquad t \geq 0,$$
$$A_{n+1}(\tau_n \wedge t) = A_n(t), \qquad t \geq 0.$$

Hence the processes $(m_t, t \geq 0)$ and $(A_t, t \geq 0)$ are defined, where

$$m_t = m_n(t) \qquad \text{for} \quad t \leq \tau_n,$$
$$A_t = A_n(t) \qquad \text{for} \quad t \leq \tau_n.$$

It is clear that the process $M = (m_t, \mathscr{F}_t)$, $t \geq 0$, is a local martingale, and that A_t, $t \geq 0$, is an increasing process.

Since for $A_t^N = A_t \wedge N$

$$\begin{aligned} MA_t^N &= \lim_{n\to\infty} M(A_t^N; \tau_n \geq t) = \lim_{n\to\infty} M(A_n^N(t); \tau_n \geq t) \\ &\leq \lim_{n\to\infty} MA_n^N(t) \leq \lim_{n\to\infty} [Mx_n(0) - Mx_n(t)] \leq \lim_{n\to\infty} Mx_n(0) = Mx_0 < \infty, \end{aligned}$$

it follows that the variables A_t^N, $t \geq 0$, are integrable, and, by Fatou's lemma, $MA_t < \infty$ and $MA_\infty < \infty$.

Let now $Y = (y_t, \mathscr{F}_t)$, $t \geq 0$, be a positive bounded martingale, having the limits to the left $y_{t-} = \lim_{s \uparrow t} y_s$ (P-a.s.). Then applying Lemma 3.2 to the processes $A_n(t)$, $t \geq 0$, $n = 1, 2, \ldots$, we obtain

$$\begin{aligned} M \int_0^t y_s \, dA_s &= \lim_{n \to \infty} M\left[\int_0^t y_s \, dA_s; \tau_n \geq t\right] \\ &= \lim_{n \to \infty} M\left[\int_0^t y_s \, dA_n(s); \tau_n \geq t\right] \\ &= \lim_{n \to \infty} M\left[\int_0^t y_{s-} \, dA_n(s), \tau_n \geq t\right] \\ &= \lim_{n \to \infty} M\left[\int_0^t y_{s-} \, dA_s; \tau_n \geq t\right] = M \int_0^t y_{s-} \, dA_s. \end{aligned}$$

From the equality

$$M \int_0^t y_s \, dA_s = M \int_0^t y_{s-} \, dA_s$$

and Lemma 3.2 it follows that the process A_t, $t \geq 0$, is natural. Uniqueness of the expansion given in (3.42) follows from uniqueness of the Doob–Meyer expansion. □

3.4 Some properties of natural increasing processes

3.4.1

In the case of discrete time $n = 0, 1, \ldots$, the increasing process $A = (A_n, \mathscr{F}_n)$, $n = 0, 1, \ldots$, was called natural, if the values A_{n+1} were $\mathscr{F}_n$-measurable. It would be natural to expect that in the case of continuous time the definition of the natural increasing process $A = (A_t, \mathscr{F}_t)$, $t \geq 0$, given in the previous section (see (3.17)), leads to the fact that at each $t \geq 0$ the random variables A_t are actually $\mathscr{F}_{t-}$-measurable. We shall show now that this is really so.

Theorem 3.10. *Let $A = (A_t, \mathscr{F}_t)$, $t \geq 0$, be a right continuous integrable increasing process, $\mathscr{F}_t = \mathscr{F}_{t+}$, $t \geq 0$. Then for each $t > 0$ the variables A_t are $\mathscr{F}_{t-}$-measurable.*

PROOF. Form the potential

$$\pi_t = M[A_\infty | \mathscr{F}_t] - A_t, \tag{3.45}$$

taking as $M[A_\infty | \mathscr{F}_t]$ a right continuous modification. Using the same notation as in proving Theorem 3.8, we have

$$\pi_{(i+1)\cdot 2^{-n}} = M[A_\infty(n) | \mathscr{F}_{(i+1)\cdot 2^{-n}}] - A_{(i+1)\cdot 2^{-n}}(n). \tag{3.46}$$

Fix $t > 0$ and set $t_n = (i + 1) \cdot 2^{-n}$ if $i \cdot 2^{-n} < t \leq (i + 1) \cdot 2^{-n}$. Then from (3.46), because of $\mathcal{F}_{i \cdot 2^{-n}}$-measurability of the variable $A_{(i+1) \cdot 2^{-n}}(n)$, we obtain

$$M[\pi_{(i+1) \cdot 2^{-n}} | \mathcal{F}_t] = M[A_\infty(n) | \mathcal{F}_t] - A_{(i+1) \cdot 2^{-n}}(n). \tag{3.47}$$

Using the variables $\pi_{(i+1) \cdot 2^{-n}}$, from (3.45) we find

$$M[A_\infty(n) | \mathcal{F}_t] = M[A_\infty - A_{t_n} | \mathcal{F}_t] + A_{t_n}(n), \qquad t_n = (i + 1) \cdot 2^{-n}. \tag{3.48}$$

Since the decomposition given by (3.45) with the natural process $A = (A_t, \mathcal{F}_t)$, $t \geq 0$, is unique, by Theorem 3.8., we can find a subsequence $\{n_j, j = 1, 2, \ldots\}$ such that the $A_\infty(n_j)$ converge weakly to A_∞. Then obviously $M[A_\infty(n_j) | \mathcal{F}_t]$ also converges weakly to $M[A_\infty | \mathcal{F}_t]$. Note also that because of continuity to the right of the process A_t, $t \geq 0$,

$$M | M(A_{t_{n_j}} | \mathcal{F}_t) - A_t | \to 0, \qquad n_j \to \infty.$$

Taking all this into account, from (3.48) we infer that

$$A_{t_{n_j}}(n_j) \text{ converges weakly to } A_t, \text{ as } n_j \to \infty.$$

The variables

$$A_{t_{n_j}}(n_j)$$

are $\mathcal{F}_{i \cdot 2^{-n_j}}$-measurable and, since $i \cdot 2^{-n_j} < t \leq t_{n_j}$, they are also $\mathcal{F}_t$-measurable. □

We shall show now that the weak limit A_t will be also $\mathcal{F}_{t-}$-measurable. This follows from the following more general result.

Lemma 3.4. *On the probability space $(\Omega, \mathcal{F}, P)$ let there be given the sequence of random variables ξ_i, $i = 1, 2, \ldots$, with $M|\xi_i| < \infty$, weakly converging to the random variables ξ, i.e., for any bounded $\mathcal{F}$-measurable variable η, let*

$$M\xi_i \eta \to M\xi\eta, \qquad i \to \infty. \tag{3.49}$$

Assume that the random variables ξ_i are $\mathcal{G}$-measurable, where $\mathcal{G}$ is the sub-σ-algebra $\mathcal{F}$. Then the random variable ξ is also $\mathcal{G}$-measurable.

PROOF. According to Theorem 1.7, the sequence of the random variables $\xi_1, \xi_2, \ldots$ is uniformly integrable. This sequence will continue to be uniformly integrable, with respect to the new probability space $(\Omega, \mathcal{G}, P)$. Therefore, using Theorem 1.7 once more, we infer that there will be a subsequence $\xi_{n_1}, \xi_{n_2}, \ldots$ and a $\mathcal{G}$-measurable random variable $\tilde{\xi}$, such that for any bounded $\mathcal{G}$-measurable variable $\tilde{\eta}$,

$$M\xi_{n_i}\tilde{\eta} \to M\tilde{\xi}\tilde{\eta}, \qquad i \to \infty. \tag{3.50}$$

According to (3.49), $M\xi_{n_i}\eta \to M\xi\eta$, and, on the other hand, because of (3.50),

$$M\xi_{n_i}\eta = M\{\xi_{n_i} M(\eta|\mathscr{G})\} \to M\{\tilde{\xi} M(\eta|\mathscr{G})\} = M\tilde{\xi}\eta.$$

Consequently, $M\xi\eta = M\tilde{\xi}\eta$, and $\xi = \tilde{\xi}$ (P-a.s.); therefore ξ is $\mathscr{G}$-measurable. □

Note. If τ is a Markov time, then the random variable $A_\tau = A_{\tau(\omega)}(\omega)$ is $\mathscr{F}_{\tau-}$-measurable. Recall that $\mathscr{F}_{\tau-}$ is the σ-algebra generated by sets of the form $\{\tau > t\} \cap \Lambda_t$, where $\Lambda_t \in \mathscr{F}_t$, $t \geq 0$.

3.4.2

In the following theorem are given the conditions under which the natural process A_t, corresponding to the potential π_t, is continuous. We introduce first:

Definition 7. The potential π_t, $t \geq 0$, is *regular*, if for any sequence $\{\tau_n, n = 1, 2, \ldots\}$ of Markov times such that $\tau_n \uparrow \tau$, $P(\tau < \infty) = 1$,

$$M\pi_{\tau_n} \to M\pi_\tau.$$

Theorem 3.11. *Let $\Pi = (\pi_t, \mathscr{F}_t)$, $t \geq 0$, be a right continuous potential belonging to class D. In order that the natural increasing process A_t, $t \geq 0$, corresponding to this potential be (P-a.s.) continuous (more precisely: should have a continuous modification), it is necessary and sufficient that the potential be regular.*

Proof of necessity. Let A_t, $t \geq 0$, be a (P-a.s.) continuous process. Then, if $\tau_n \uparrow \tau$, by Lebesgue Theorem 1.4 $\lim_{n\to\infty} MA_{\tau_n} = MA\tau$. Hence

$$\lim_{n\to\infty} M\pi_{\tau_n} = \lim_{n\to\infty} M[A_\infty - A_{\tau_n}] = M[A_\infty - A_\tau] = M\pi_\tau. \quad \square \tag{3.51}$$

Proof of sufficiency is more complicated and will be divided into several stages.

3.4.3

Lemma 3.5. *Let $\Pi = (\pi_t, \mathscr{F}_t)$, $t \geq 0$, be a right continuous potential and let*

$$\pi_t = M[A_\infty | \mathscr{F}_t] - A_t, \tag{3.52}$$

where A_t, $t \geq 0$ is a natural integrable increasing process. Then

$$MA_\infty^2 = M\int_0^\infty [\pi_t + \pi_{t-}]dA_t, \tag{3.53}$$

where the limit $\pi_{t-} = \lim_{s\uparrow t} \pi_s$ exists according to Corollary 1 of Theorem 3.2.

PROOF

(a) Assume first that $MA_\infty^2 < \infty$; under this assumption we will establish (3.53).

Let $m_t, t \geq 0$ be right continuous and have limits to the left modification $M(A_\infty | \mathscr{F}_t)$ (see Corollary 2 of Theorem 3.2). Then, because of uniform integrability of the family of the values $\{m_t, t \geq 0\}$,

$$M\left[\int_0^\infty m_t \, dA_t\right] = M\left[\int_0^\infty m_{t+} \, dA_t\right] = \lim_{k\to\infty} M\left[\sum_{i=0}^\infty m_{(i+1)/k}(A_{(i+1)/k} - A_{i/k})\right]$$

$$= \lim_{k\to\infty} \sum_{i=0}^\infty M[m_{i+1/k}(A_{(i+1)/k} - A_{i/k})]$$

$$= \lim_{k\to\infty} \sum_{i=0}^\infty [Mm_{i+1/k} A_{(i+1)/k} - Mm_{i/k} A_{i/k}]$$

$$= Mm_\infty A_\infty = MA_\infty^2. \tag{3.54}$$

Now make use of the fact that the process $A_t, t \geq 0$, is natural. If $m_t^N = M(A_\infty \wedge N | \mathscr{F}_t), t \geq 0$, then

$$M \int_0^\infty m_{t-}^N \, dA_t = Mm_\infty^N A_\infty.$$

Letting $N \to \infty$ we have

$$M \int_0^\infty m_{t-} \, dA_t = Mm_\infty A_\infty = MA_\infty^2. \tag{3.55}$$

Note also that

$$M \int_0^\infty (A_t + A_{t-}) dA_t = \lim_{k\to\infty} M\left[\sum_{i=0}^\infty (A_{(i+1)/k} + A_{i/k})(A_{(i+1)/k} - A_{i/k})\right]$$

$$= \lim_{k\to\infty} M \sum_{i=0}^\infty [A_{(i+1)/k}^2 - A_{i/k}^2] = MA_\infty^2. \tag{3.56}$$

From (3.54)–(3.56) obtain

$$M \int_0^\infty (\pi_t + \pi_{t-}) dA_t = M \int_0^\infty (m_t + m_{t-}) dA_t - M \int_0^\infty (A_t + A_{t-}) dA_t$$

$$= 2MA_\infty^2 - MA_\infty^2 = MA_\infty^2.$$

(b) Assume now that

$$M \int_0^\infty [\pi_t + \pi_{t-}] dA_t < \infty. \tag{3.57}$$

Then, if we can prove that in this case also $MA_\infty^2 < \infty$, Equation (3.53) will follow from the preceding considerations.

For proving the inequality $MA_\infty^2 < \infty$ it is sufficient to establish that for all n, larger than some $N_0 < \infty$,

$$MA_\infty^2(n) \le C < \infty. \tag{3.58}$$

But this follows from the fact that A_∞ is the weak limit of some sequence $\{A_\infty(n_i), i = 1, 2, \ldots\}$ and from the following:

Lemma 3.6. *Let ξ_i, $i = 1, 2, \ldots$, be a sequence of the random variables $M|\xi_i| < \infty$, $i = 1, 2, \ldots$, weakly converging to some variable ξ, i.e., for any bounded random variable η let*

$$M\xi_i\eta \to M\xi\eta, \qquad i \to \infty. \tag{3.59}$$

Assume that $\sup_i M\xi_i^2 \le C < \infty$. *Then* $M\xi^2 \le C$.

PROOF. Denote

$$\xi_{(n)} = \begin{cases} \xi, & \text{if } |\xi| \le n, \\ 0, & \text{if } |\xi| > n. \end{cases}$$

Then assuming in (3.59) that $\eta = \xi_{(n)}$, and taking into account that $\xi\xi_{(n)} = \xi_{(n)}^2$ (P-a.s.), we obtain

$$M\xi_{(n)}^2 = M\xi\xi_{(n)} = \lim_{i\to\infty} M\xi_i\xi_{(n)} \le \left[\sup_i M\xi_i^2 \cdot M\xi_{(n)}^2\right]^{1/2} = C^{1/2}(M\xi_{(n)}^2)^{1/2}. \tag{3.60}$$

But $M\xi_{(n)}^2 \le n \le \infty$, and hence (3.60) leads to the inequality $M\xi_{(n)}^2 \le C$. Finally, by Fatou's lemma, $M\xi^2 = M \lim \xi_{(n)}^2 \le C < \infty$, which proves Lemma 3.6. □

Thus, returning to the proof of Lemma 3.5, we need to establish (3.58).

From (3.57) it follows that we can find $N_0 < \infty$, such that for all $n \ge N_0$,

$$M\sum_{i=0}^{\infty} \pi_{i\cdot 2^{-n}}[A_{(i+1)\cdot 2^{-n}} - A_{i\cdot 2^{-n}}] \le C < \infty \tag{3.61}$$

or, what is equivalent (see (3.26)),

$$M\sum_{i=0}^{\infty} \pi_{i\cdot 2^{-n}}[A_{(i+1)\cdot 2^{-n}}(n) - A_{i\cdot 2^{-n}}(n)] \le C < \infty.$$

Let $a^N = \min(a, N)$ and $\pi_{i\cdot 2^{-n}}^N = M(A_\infty^N(n)|\mathscr{F}_{i\cdot 2^{-n}}) - A_{i\cdot 2^{-n}}^N(n)$. Since $A_{i\cdot 2^{-n}}^N(n) \le N < \infty$, the results of (a) can be applied, according to which

$$M[A_\infty^N(n)]^2 = M\sum_{i=0}^{\infty} (\pi_{(i+1)\cdot 2^{-n}}^N + \pi_{i\cdot 2^{-n}}^N)(A_{(i+1)\cdot 2^{-n}}^N(n) - A_{i\cdot 2^{-n}}^N(n)).$$

Note that

$$A^N_{(i+1)\cdot 2^{-n}}(n) - A^N_{i\cdot 2^{-n}}(n) \le (A_{(i+1)\cdot 2^{-n}}(n) - A_{i\cdot 2^{-n}}(n))^N$$

and

$$\begin{aligned}\pi^N_{i\cdot 2^{-n}} &= M[A^N_\infty(n) - A^N_{i\cdot 2^{-n}}(n) \mid \mathscr{F}_{i\cdot 2^{-n}}]\\ &\le M[(A_\infty(n) - A_{i\cdot 2^{-n}}(n))^N \mid \mathscr{F}_{i\cdot 2^{-n}}] \le \pi_{i\cdot 2^{-n}}.\end{aligned}$$

Hence, according to (3.61),

$$M[A^N_\infty(n)]^2 \le M\sum_{i=0}^{\infty}(\pi_{(i+1)\cdot 2^{-n}} + \pi_{i\cdot 2^{-n}})(A_{(i+1)\cdot 2^{-n}}(n) - A_{i\cdot 2^{-n}}(n)) \le 2C,$$

where we used the fact that

$$\begin{aligned}& M\sum_{i=0}^{\infty}\pi_{(i+1)\cdot 2^{-n}}(A_{(i+1)\cdot 2^{-n}}(n) - A_{i\cdot 2^{-n}}(n))\\ &= M\sum_{i=0}^{\infty}M\{\pi_{(i+1)\cdot 2^{-n}}[A_{(i+1)\cdot 2^{-n}}(n) - A_{i\cdot 2^{-n}}(n)] \mid \mathscr{F}_{i\cdot 2^{-n}}\}\\ &= M\sum_{i=0}^{\infty}M(\pi_{(i+1)\cdot 2^{-n}} \mid \mathscr{F}_{i\cdot 2^{-n}})(A_{(i+1)\cdot 2^{-n}}(n) - A_{i\cdot 2^{-n}}(n))\\ &\le M\sum_{i=0}^{\infty}\pi_{i\cdot 2^{-n}}(A_{(i+1)\cdot 2^{-n}}(n) - A_{i\cdot 2^{-n}}(n)) \le C.\end{aligned}$$

Thus $M[A^N_\infty(n)]^2 \le 2C < \infty$ and, by Fatou's lemma, $M[A_\infty(n)]^2 \le 2C$ for all $n \ge N_0$. □

3.4.4

For formulating two additional results needed for proving Theorem 3.11 we introduce additional notation.

Using the process A_t, $t \ge 0$, construct the submartingale $(A_n(t), \mathscr{F}_t)$, $t \ge 0$ by:

$$A_n(t) = M[A_{\varphi_n}(t) \mid \mathscr{F}_t], \tag{3.62}$$

where $\varphi_n(t) = (k+1)2^{-n}$, if $k2^{-n} \le t < (k+1)2^{-n}$. According to Theorem 3.1 we may assume that the trajectories $A_n(t)$, $t \ge 0$, are right continuous (P-a.s.) and have limits to the left at each point $t \ge 0$.

Let τ be a m.t. (relative to $(\mathscr{F}_t)$, $t \ge 0$). Then from Lemma 1.9 and the definition of conditional mathematical expectation it is easy to deduce that

$$A_n(\tau) = M[A_{\varphi_n}(\tau) \mid \mathscr{F}_\tau]. \tag{3.63}$$

For each $\varepsilon > 0$ define

$$\tau_{n,\varepsilon} = \inf\{t : A_n(t) - A_t \ge \varepsilon\}, \tag{3.64}$$

taking $\tau_{n,\varepsilon} = +\infty$ if the set $\{\cdot\}$ in (3.64) is empty. It is clear that $\tau_{n,\varepsilon} \le \tau_{n+1,\varepsilon}$ (P-a.s.). Put $\tau_\varepsilon = \lim_{n\to\infty} \tau_{n,\varepsilon}$.

Lemma 3.7. *For all n, $n = 1, 2, \ldots$,*

$$M[A_{\tau_\varepsilon} - A_{\tau_{n,\varepsilon}}] \ge \varepsilon P(\tau_\varepsilon < \infty) + M[A_{\tau_\varepsilon} - A_{\varphi_n(\tau_\varepsilon)}]. \tag{3.65}$$

PROOF. We have

$$A_{\tau_\varepsilon} - A_{\tau_{n,\varepsilon}} = [A_{\tau_\varepsilon} - A_{\varphi_n(\tau_\varepsilon)}] + [A_{\varphi_n(\tau_\varepsilon)} - A_{\tau_{n,\varepsilon}}]$$

and

$$\begin{aligned} MA_{\varphi_n(\tau_\varepsilon)} &= MM[A_{\varphi_n(\tau_\varepsilon)} | \mathscr{F}_{\tau_{n,\varepsilon}}] \\ &\ge MM[A_{\varphi_n(\tau_{n,\varepsilon})} | \mathscr{F}_{\tau_{n,\varepsilon}}] = MA_n(\tau_{n,\varepsilon}). \end{aligned}$$

Hence, taking into account that $A_n(t) \ge A_t$ (P-a.s.), $t \ge 0$, we obtain

$$\begin{aligned} M[A_{\tau_\varepsilon} - A_{\tau_{n,\varepsilon}}] &\ge M[A_{\tau_\varepsilon} - A_{\varphi_n(\tau_\varepsilon)}] + M[A_n(\tau_{n,\varepsilon}) - A_{\tau_{n,\varepsilon}}] \\ &\ge M[A_{\tau_\varepsilon} - A_{\varphi_n(\tau_\varepsilon)}] + \int_{\{\tau_\varepsilon < \infty\}} [A_n(\tau_{n,\varepsilon}) - A_{\tau_{n,\varepsilon}}] dP \\ &\ge M[A_{\tau_\varepsilon} - A_{\varphi_n(\tau_\varepsilon)}] + \varepsilon P(\tau_\varepsilon < \infty), \end{aligned}$$

where we made use of the fact that, because of right continuity of the processes $A_n(t)$ and A_t, $t \ge 0$, on the set $\{\tau_\varepsilon < \infty\}$,

$$A_n(\tau_{n,\varepsilon}) - A_{\tau_{n,\varepsilon}} \ge \varepsilon. \qquad \square$$

Lemma 3.8. *Let A_t, $t \ge 0$, be a natural process corresponding to the regular potential $\Pi = (\pi_t, \mathscr{F}_t)$, and let $MA_\infty^2 < \infty$. Then for all n, $n = 1, 2, \ldots$, and any $\varepsilon > 0$,*

$$M\int_0^\infty [A_t - A_{t-}] dA_t \le \lim_{n\to\infty} \{\varepsilon MA_{\tau_{n,\varepsilon}} + M[A_\infty(A_\infty - A\tau_{n,\varepsilon})]\}. \tag{3.66}$$

PROOF. Put $\Delta_{n,k} = \{t : k \cdot 2^{-n} \le t < (k+1)2^{-n}\}$. Since for $t \in \Delta_{n,k}$ the process $(A_n(t), \mathscr{F}_t)$ forms a martingale, and the process $(A_t, \mathscr{F}_t)$, $t \ge 0$, is natural, it is not difficult to deduce from Lemma 3.2 that

$$M\int_{\Delta_{n,k}} A_n(t) dA_t = M\int_{\Delta_{n,k}} A_n(t-) dA_t.$$

Consequently

$$M\int_0^\infty A_n(t-) dA_t = M\int_0^\infty A_n(t) dA_t. \tag{3.67}$$

On the other hand (compare with (3.54))

$$M\int_{\Delta_{n,k}} A_n(t) dA_t = \lim_{\varepsilon\downarrow 0} M\{A_n((k+1)\cdot 2^{-n} - \varepsilon)[A_{(k+1)\cdot 2^{-n} - \varepsilon} - A_{k\cdot 2^{-n}}]\}. \tag{3.68}$$

But with $\varepsilon \downarrow 0$,

$$A_n((k+1)\cdot 2^{-n} - \varepsilon) = M[A_{(k+1)\cdot 2^{-n}} | \mathscr{F}_{(k+1)\cdot 2^{-n}-\varepsilon}] \\ \to M[A_{(k+1)\cdot 2^{-n}} | \mathscr{F}_{(k+1)\cdot 2^{-n}-}] = A_{(k+1)\cdot 2^{-n}},$$

because the variable $A_{(k+1)\cdot 2^{-n}}$ is $\mathscr{F}_{((k+1)\cdot 2^{-n})}$-measurable according to Theorem 3.10.

Since the potential $\Pi = (\pi_t, \mathscr{F}_t)$, $t \geq 0$, is regular, $MA_t = MA_\infty - M\pi_t$ is a continuous function, and therefore, for each $t > 0$, $P(A_t = A_{t-}) = 1$. (Note that $A_{t-} = \lim_{s \uparrow t} A_s$ exists for each $t > 0$, since $A_s = M[A_\infty | \mathscr{F}_s] - \pi_s$, and $\pi_{t-} = \lim_{s \uparrow t} \pi_s$ and $M[A_\infty | \mathscr{F}_{t-}] = \lim_{s \uparrow t} M[A_\infty | \mathscr{F}_s]$ exist by Corollary 1 of Theorem 3.2 and by Theorem 1.5 respectively.)

Further,

$$A_{(k+1)\cdot 2^{-n}-\varepsilon} \to A_{((k+1)\cdot 2^{-n})-}, \qquad \varepsilon \to 0,$$

where, according to what was said,

$$P(A_{((k+1)\cdot 2^{-n})-} = A_{((k+1)\cdot 2^{-n})}) = 1.$$

Hence if $MA_\infty^2 < \infty$, then from (3.68) it follows that

$$M \int_{\Delta_{n,k}} A_n(t) dA_t = M\{A_{(k+1)\cdot 2^{-n}}[A_{(k+1)\cdot 2^{-n}} - A_{k\cdot 2^{-n}}]\},$$

and therefore

$$M \int_0^\infty A_n(t) dA_t = \sum_{k=0}^\infty M\{A_{(k+1)\cdot 2^{-n}}[A_{(k+1)\cdot 2^{-n}} - A_{k\cdot 2^{-n}}]\}. \tag{3.69}$$

From this, taking into account (3.67), we obtain

$$M \int_0^\infty A_t dA_t = \lim_{n\to\infty} \sum_{k=0}^\infty M\{A_{(k+1)\cdot 2^{-n}}[A_{(k+1)\cdot 2^{-n}} - A_{k\cdot 2^{-n}}]\}$$

$$= \lim_{n\to\infty} M \int_0^\infty A_n(t) dA_t = \lim_{n\to\infty} M \int_0^\infty A_n(t-) dA_t, \tag{3.70}$$

and consequently

$$M \int_0^\infty [A_t - A_{t-}] dA_t = \lim_{n\to\infty} M \int_0^\infty [A_n(t-) - A_{t-}] dA_t. \tag{3.71}$$

To obtain the inequality given by (3.66), transform the right side in (3.71). We have

$$M \int_0^\infty [A_n(t-) - A_{t-}] dA_t = M \int_0^{\tau_{n,\varepsilon}} [A_n(t-) - A_{t-}] dA_t$$

$$+ M \int_{\tau_{n,\varepsilon}}^\infty [A_n(t-) - A_{t-}] dA_t$$

$$\leq \varepsilon M A_{\tau_{n,\varepsilon}} + M \int_{\tau_{n,\varepsilon}}^\infty A_n(t-) dA_t. \tag{3.72}$$

Put $B_t = M(A_\infty | \mathscr{F}_t)$. Then, obviously, $B_{t-} \geq A_n(t-)$, and therefore (see (3.19))

$$M \int_{\tau_{n,\varepsilon}}^{\infty} A_n(t-)dA_t \leq M \int_{\tau_{n,\varepsilon}}^{\infty} B_{t-}\,dA_t = M[A_\infty(A_\infty - A_{\tau_{n,\varepsilon}})]. \quad (3.73)$$

From (3.72) and (3.73) it follows that

$$M \int_0^{\infty} [A_n(t-) - A_{t-}]dA_t \leq \varepsilon M A_{\tau_{n,\varepsilon}} + M[A_\infty(A_\infty - A_{\tau_{n,\varepsilon}})]. \quad (3.74)$$

This together with (3.71), in an obvious manner leads to the inequality given by (3.66). □

3.4.5

Proof of Theorem 3.11: sufficiency. Assume first that $MA_\infty^2 < \infty$. Since the potential $\Pi = (\pi_t, \mathscr{F}_t)$, $t \geq 0$, is regular, then

$$M[A_{\tau_\varepsilon} - A_{\tau_{\varepsilon,n}}] = M[\pi_{\tau_{n,\varepsilon}} - \pi_{\tau_\varepsilon}] \to 0, \qquad n \to \infty. \quad (3.75)$$

Because of the right continuity of the process A_t, $t \geq 0$,

$$M[A_{\tau_\varepsilon} - A_{\varphi_n(\tau_\varepsilon)}] \to 0, \qquad n \to \infty, \quad (3.76)$$

since $\varphi_n(\tau_\varepsilon) \downarrow \tau_\varepsilon$, $n \to \infty$.

From (3.75), (3.76) and the inequality given by (3.65) of Lemma 3.7 we infer that $P(\tau_\varepsilon < \infty) = 0$ for any $\varepsilon > 0$. But then (see (3.66))

$$\varliminf_{n\to\infty} \{\varepsilon M A_{\tau_{n,\varepsilon}} + M[A_\infty(A_\infty - A_{\tau_{n,\varepsilon}})]\} = \varepsilon M A_\infty,$$

and consequently

$$M \int_0^{\infty} [A_t - A_{t-}]dA_t \leq \varepsilon M A_\infty.$$

Because of arbitrariness of $\varepsilon > 0$,

$$M \int_0^{\infty} [A_t - A_{t-}]dA_t = 0,$$

and therefore (P-a.s.) trajectories of the process are left continuous. Since the trajectories A_t, $t \geq 0$, are also right continuous, the process A_t, $t \geq 0$, is continuous with probability 1.

Let us get rid now of the assumption $MA_\infty^2 < \infty$. Let $\Pi = (\pi_t, \mathscr{F}_t)$, $t \geq 0$, be a right continuous regular potential of class D and let

$$\pi_t = M(A_\infty | \mathscr{F}_t) - A_t, \quad (3.77)$$

where A_t, $t \geq 0$, is a natural increasing process. For $n = 1, 2, \ldots$, set

$$A_t^{(n)} = A_t \wedge n, \qquad B_t^{(n)} = A_t^{(n+1)} - A_t^{(n)}$$

and

$$\pi_t^{(n)} = M[B_\infty^{(n)} | \mathscr{F}_t] - B_t^{(n)}. \tag{3.78}$$

It is clear that for each $t \geq 0$,

$$\pi_t = \sum_{n=1}^{\infty} \pi_t^{(n)}, \tag{3.79}$$

where the potentials $\pi_t^{(n)}$, $t \geq 0$, are bounded and right continuous. Let us show that each of them is regular, if the potential $\Pi = (\pi_t, \mathscr{F}_t)$ is regular.

From (3.77) and (3.78) it follows that from each n, $n = 1, 2, \ldots$,

$$\pi_t = \pi_t^{(n)} + z_t,$$

where the potential

$$z_t = M[A_\infty - B_\infty^{(n)} | \mathscr{F}_t] - (A_t - B_t^{(n)}).$$

Let the sequence of Markov times be such that $\tau_m \uparrow \tau$. Then by Theorem 3.5,

$$M\pi_{\tau_m}^{(n)} \geq M\pi_\tau^{(n)}, \qquad Mz_{\tau_m} \geq Mz_\tau,$$

and consequently

$$\lim_{m\to\infty} M\pi_{\tau_m}^{(n)} \geq M\pi_\tau^{(n)}, \qquad \lim_{m\to\infty} Mz_{\tau_m} = Mz_\tau. \tag{3.80}$$

Actually, both of these inequalities are equalities since the potential π_t, $t \geq 0$, is regular:

$$\lim_{m\to\infty} M\pi_{\tau_m} = M\pi_\tau.$$

Thus each of the potentials $\pi_t^{(n)}$, $n = 1, 2, \ldots$, is regular, limited, and, according to the proof given above, the natural increasing process $B_t^{(n)}$, $t \geq 0$, corresponding to them are continuous with probability 1 ($M(B_\infty^{(n)})^2 < \infty$ by Lemma 3.5).

For the potential $\sum_{n=1}^{\infty} \pi_t^{(n)}$ the corresponding natural process is the process $B_t = \sum_{n=1}^{\infty} B_t^{(n)}$, where each of the processes $B_t^{(n)}$, $t \geq 0$, is continuous. This process is also continuous. Actually,

$$0 \leq B_t - \sum_{n=1}^{N} B_t^{(n)} \leq B_\infty - \sum_{n=1}^{N} B_\infty^{(n)} \qquad (P\text{-a.s.}) \tag{3.81}$$

where, with probability 1, $B_\infty - \sum_{n=1}^{N} B_\infty^{(n)} \to 0$, $N \to \infty$, since $MB_\infty = M\sum_{n=1}^{\infty} B_\infty^{(n)} = MA_\infty < \infty$. From (3.81) it follows that the process B_t, $t \geq 0$, is continuous with probability 1.

To complete the proof it remains only to note that from the uniqueness of the decomposition in (3.77) with the natural process A_t, $t \geq 0$, it follows that $P(A_t = B_t) = 1$, $t \geq 0$. From this it follows that in (3.77) the natural proccss A_t, $t \geq 0$, can bc chosen to be continuous with probability 1. □

Notes and references

3.1,3.2. See also Meyer [126] and Doob [46].

3.3,3.4. The proof of Doob–Meyer decomposition has been copied from Rao's paper [137], (see also Meyer [126]).

4 The Wiener process, the stochastic integral over the Wiener process, and stochastic differential equations

4.1 The Wiener process as a square integrable martingale

4.1.1

Let $(\Omega, \mathcal{F}, P)$ be a probability space and $\beta = (\beta_t)$, $t \geq 0$, be a Brownian motion process (in the sense of the definition given in Section 1.4). Denote $\mathcal{F}_t^\beta = \sigma\{\omega: \beta_s, s \leq t\}$. Then, according to (1.30) and (1.31), (P-a.s.)

$$M(\beta_t | \mathcal{F}_s^\beta) = \beta_s, \qquad t \geq s, \tag{4.1}$$

$$M[(\beta_t - \beta_s)^2 | \mathcal{F}_s^\beta] = t - s, \qquad t \geq s. \tag{4.2}$$

From this it follows that the Brownian motion process β is a square integrable ($M\beta_t^2 < \infty$, $t \geq 0$) martingale (with respect to the system of σ-algebras $F^\beta = (\mathcal{F}_t^\beta)$, $t \geq 0$) with continuous (P-a.s.) trajectories.

In a certain sense the converse is also correct; to formulate this we introduce the following:

Definition 1. Let $(\Omega, \mathcal{F}, P)$ be a probability space and $F = (\mathcal{F}_t)$, $t \geq 0$, be a nondecreasing family of sub-σ-algebras of $\mathcal{F}$. The random process $W = (W_t, \mathcal{F}_t)$, $t \geq 0$, is called a *Wiener process* (*relative to the family* $F = (\mathcal{F}_t)$. $t \geq 0$) if:

(1) the trajectories W_t, $t \geq 0$, are continuous (P-a.s.) over t;
(2) $W = (W_t, \mathcal{F}_t)$, $t \geq 0$, is a square integrable martingale with $W_0 = 0$ and

$$M[(W_t - W_s)^2 | \mathcal{F}_s] = t - s, \qquad t \geq s.$$

Theorem 4.1 (Levy). *Any Wiener process* $W = (W_t, \mathcal{F}_t)$, $t \geq 0$, *is a Brownian motion process.*

Note 1. This theorem can be reformulated in the following equivalent way: any continuous square integrable martingale $W = (W_t, \mathscr{F}_t)$, $t \geq 0$, with $W_0 = 0$ and $M[(W_t - W_s)^2|\mathscr{F}_s] = t - s$ is a process with stationary independent Gaussian increments with $M[W_t - W_s] = 0$, $M[W_t - W_s]^2 = t - s$, $t \geq s$.

Note 2. Because of the Levy theorem, from now on we shall not distinguish between Wiener processes and Brownian motion processes $\beta = (\beta_t)$, $t \geq 0$, since the latter are Wiener processes relative to the system of the σ-algebras $F^\beta = (\mathscr{F}_t^\beta)$, $t \geq 0$.

Note 3. A useful generalization of the Levy theorem, which is due to Doob, will be given in Chapter 5 (Theorem 5.12).

Two lemmas will now be proved preparatory to proving Theorem 4.1.

Lemma 4.1. *Let σ be a Markov time (with respect to $F = (\mathscr{F}_t)$, $t \geq 0$), $P(\sigma \leq T) = 1$, $T < \infty$, and $\tilde{W}_t = W_{t \wedge \sigma}$, $\tilde{\mathscr{F}}_t = \mathscr{F}_{t \wedge \sigma}$. Then $\tilde{W} = (\tilde{W}_t, \tilde{\mathscr{F}}_t)$, $t \geq 0$, is a martingale,*

$$M(\tilde{W}_t - \tilde{W}_s|\tilde{\mathscr{F}}_s) = 0, \tag{4.3}$$

and

$$M[(\tilde{W}_t - \tilde{W}_s)^2|\tilde{\mathscr{F}}_s] = M[(t \wedge \sigma) - (s \wedge \sigma)|\tilde{\mathscr{F}}_s], \qquad t \geq s. \tag{4.4}$$

PROOF. It is sufficient to apply Theorem 3.6 to the martingales $W = (W_t, \mathscr{F}_t)$ and $(W_t^2 - t, \mathscr{F}_t)$, $t \geq 0$. □

Lemma 4.2. *Let $X = (x_t, \mathscr{F}_t)$, $0 \leq t \leq T < \infty$, be a continuous bounded ($P\{\sup_{t \leq T}|x_t| \leq K < \infty\} = 1$) martingale and let $f(x)$ be continuous and bounded together with its first and second derivatives $f'(x)$, $f''(x)$.*

If for any s, t, $0 \leq s \leq t \leq T$,

$$M[(x_t - x_s)^2|\mathscr{F}_s] = \int_s^t M[g_u|\mathscr{F}_s]du \tag{4.5}$$

for some measurable function $g_u = g_u(\omega)$, with each u, $0 \leq u \leq T$, being $\mathscr{F}_u$-measurable and such that $M\int_0^T g_u^2\,du < \infty$, then ($P$-a.s.)

$$M[f(x_t)|\mathscr{F}_s] = f(x_s) + \frac{1}{2}\int_s^t M[f''(x_u)g_u|\mathscr{F}_s]du, \qquad s \leq t \leq T. \tag{4.6}$$

PROOF. For given s, t ($0 \leq s \leq t \leq T$), consider the partition of the interval $[s, t]$ into n parts, $s \equiv t_0^{(n)} < t_1^{(n)} < \cdots < t_n^{(n)} \equiv t$, such that $\max_j[t_{j+1}^{(n)} - t_j^{(n)}] \to 0$, $n \to \infty$. Then, obviously,

$$f(x_t) - f(x_s) = \sum_{i=0}^{n-1}[f(x_{t_{j+1}^{(n)}}) - f(x_{t_j^{(n)}})],$$

and, by the theorem on the mean,

$$\begin{aligned} f(x_{t_{j+1}^{(n)}}) - f(x_{t_j^{(n)}}) = {} & f'(x_{t_j^{(n)}})[x_{t_{j+1}^{(n)}} - x_{t_j^{(n)}}] \\ & + \tfrac{1}{2}f''(x_{t_j^{(n)}})[x_{t_{j+1}^{(n)}} - x_{t_j^{(n)}}]^2 + \tfrac{1}{2}\Delta f_j''[x_{t_{j+1}^{(n)}} - x_{t_j^{(n)}}]^2, \end{aligned}$$

where

$$\Delta f''_j = f''(x_{t_j^{(n)}} + \theta[x_{t_{j+1}^{(n)}} - x_{t_j^{(n)}}]) - f''(x_{t_j^{(n)}})$$

and θ is a random variable, $0 \leq \theta \leq 1$. It is clear that

$$M[f(x_{t_{j+1}^{(n)}}) - f(x_{t_j^{(n)}}) | \mathcal{F}_{t_j^{(n)}}] = \tfrac{1}{2} f''(x_{t_j^{(n)}}) \int_{t_j^{(n)}}^{t_{j+1}^{(n)}} M[g_u | \mathcal{F}_{t_j^{(n)}}] du$$
$$+ \tfrac{1}{2} M\{\Delta f''_j [x_{t_{j+1}^{(n)}} - x_{t_j^{(n)}}]^2 | \mathcal{F}_{t_j^{(n)}}\}.$$

Hence

$$M[f(x_t) - f(x_s) | \mathcal{F}_s] = \frac{1}{2} \sum_{j=0}^{n-1} \int_{t_j^{(n)}}^{t_{j+1}^{(n)}} M[f''(x_{t_j^{(n)}}) g_u | \mathcal{F}_s] du$$
$$+ \frac{1}{2} \sum_{j=0}^{n-1} M\{\Delta f''_j [x_{t_{j+1}^{(n)}} - x_{t_j^{(n)}}]^2 | \mathcal{F}_s\}. \quad (4.7)$$

Let us now show that with $n \to \infty$, (P-a.s.).

$$\sum_{j=0}^{n-1} \int_{t_j^{(n)}}^{t_{j+1}^{(n)}} M[f''(x_{t_j^{(n)}}) g_u | \mathcal{F}_s] du \to \int_s^t M[f''(x_u) g_u | \mathcal{F}_s] du \quad (4.8)$$

and

$$\sum_{j=0}^{n-1} M\{\Delta f''_j [x_{t_{j+1}^{(n)}} - x_{t_j^{(n)}}]^2 | \mathcal{F}_s\} \xrightarrow{P} 0. \quad (4.9)$$

For this purpose we define

$$f''_n(u) = f''(x_{t_j^{(n)}}), \qquad t_j^{(n)} \leq u < t_{j+1}^{(n)}.$$

Then, as $n \to \infty$,

$$\sum_{j=0}^{n-1} \int_{t_j^{(n)}}^{t_{j+1}^{(n)}} M[f''(x_{t_j^{(n)}}) g_u | \mathcal{F}_s] = \int_s^t M[f''_n(u) g_u | \mathcal{F}_s] du \to \int_s^t M[f''(x_u) g_u | \mathcal{F}_s] du$$

because of Theorem 1.4 and the fact that $f''_n(u) \to f''(x_u)$ (P-a.s.). Next

$$M \left| \sum_{j=0}^{n-1} M\{\Delta f''_j [x_{t_{j+1}^{(n)}} - x_{t_j^{(n)}}]^2 | \mathcal{F}_s\} \right| \leq \sum_{j=0}^{n-1} M |\Delta f''_j [x_{t_{j+1}^{(n)}} - x_{t_j^{(n)}}]^2 |$$
$$\leq M\left[\max_{j,\theta} |\Delta f''_j| \sum_{j=0}^{n-1} [x_{t_{j+1}^{(n)}} - x_{t_j^{(n)}}]^2\right]$$
$$\leq \left(M\left[\max_{j,\theta} |\Delta f''_j|\right]^2 \cdot M\left[\sum_{j=0}^{n-1} [x_{t_{j+1}^{(n)}} - x_{t_j^{(n)}}]^2\right]^2\right)^{1/2}.$$

But $M[\max_{j,\theta}|\Delta f_j''|]^2 \to 0$ as $n \to \infty$ because of continuity with probability 1 of the process x_t, $0 \le t \le T$, and boundedness of the function $f''(x)$, and

$$\begin{aligned} M\left(\sum_{j=0}^{n-1}[x_{t_{j+1}^{(n)}} - x_{t_j^{(n)}}]^2\right)^2 &= M\left(\sum_{j=0}^{n-1}[x_{t_{j+1}^{(n)}}^2 + x_{t_j^{(n)}}^2 - 2x_{t_j^{(n)}}x_{t_{j+1}^{(n)}}]\right)^2 \\ &= M\left(\sum_{j=0}^{n-1}[x_{t_{j+1}^{(n)}}^2 - x_{t_j^{(n)}}^2] - 2\sum_{j=0}^{n-1} x_{t_j^{(n)}}[x_{t_{j+1}^{(n)}} - x_{t_j^{(n)}}]\right)^2 \\ &\le 2M(x_t^2 - x_s^2)^2 + 8M\left(\sum_{j=0}^{n-1} x_{t_j^{(n)}}[x_{t_{j+1}^{(n)}} - x_{t_j^{(n)}}]\right)^2 \\ &= 2M(x_t^2 - x_s^2)^2 + 8\sum_{j=0}^{n-1} M[x_{t_j^{(n)}}(x_{t_{j+1}^{(n)}} - x_{t_j^{(n)}})]^2 \\ &= 2M(x_t^2 - x_s^2)^2 + 8\sum_{j=0}^{n-1}\int_{t_j^{(n)}}^{t_{j+1}^{(n)}} Mx_{t_j^{(n)}}^2 g_u\,du \\ &\le 8K^2 + 8K^2\int_s^t Mg_u^2\,du < \infty. \end{aligned}$$

This proves (4.9), and therefore Lemma 4.2. □

4.1.2

PROOF OF THEOREM 4.1. Let $\sigma_N = \inf\{t \le T: \sup_{s\le T}|W_s| = N\}$, $\sigma_N = T$ on the set $\{\omega:\sup_{s\le T}|W_s| < N\}$. Denote also $\tilde{W}_N(t) = W_{t\wedge\sigma_N}$ and $\tilde{\mathscr{F}}_t = \mathscr{F}_{t\wedge\sigma_N}$. According to Lemma 4.1, $(\tilde{W}_N(t), \tilde{\mathscr{F}}_t)$, $0 \le t \le T$, is a martingale with

$$\begin{aligned} M([\tilde{W}_N(t) - \tilde{W}_N(s)]^2|\tilde{\mathscr{F}}_s) &= M[(t\wedge\sigma_N - s\wedge\sigma_N)|\tilde{\mathscr{F}}_s] \\ &= \int_s^t M[\chi_N(u)|\tilde{\mathscr{F}}_s]du, \end{aligned}$$

where

$$\chi_N(u) = \begin{cases} 1, & \sigma_N > u, \\ 0, & \sigma_N \le u. \end{cases}$$

Then, by Lemma 4.2, for any function $f(x)$ bounded and continuous (together with its derivatives $f'(x)$ and $f''(x)$),

$$M[f(\tilde{W}_N(t))|\mathscr{F}_{s\wedge\sigma_N}] = f(\tilde{W}_N(s)) + \frac{1}{2}\int_s^t M[f''(\tilde{W}_N(u))\chi_N(u)|\mathscr{F}_{s\wedge\sigma_N}]du. \tag{4.10}$$

Note now that with probability 1 with $N \to \infty$,

$$\tilde{W}_N(u) \to W_u, \qquad \chi_N(u) \to 1, \quad \sigma_N \to T,$$

and $\mathscr{F}_{s \wedge \sigma_N} \uparrow \mathscr{F}_s$. Hence from (4.10), using Theorem 1.6 by means of passage to the limit over $N \to \infty$, we infer that

$$M[f(W_t)|\mathscr{F}_s] = f(W_s) + \frac{1}{2}\int_s^t M[f''(W_u)|\mathscr{F}_s]du. \tag{4.11}$$

Set $f(x) = e^{i\lambda x}$, where $-\infty < \lambda < \infty$. Then from the relation given in (4.11) (applied to the real and imaginary parts of this function), we obtain

$$M[e^{i\lambda W_t}|\mathscr{F}_s] = e^{i\lambda W_s} - \frac{\lambda^2}{2}\int_s^t M[e^{i\lambda W_u}|\mathscr{F}_s]du. \tag{4.12}$$

Let $y_t = M[e^{i\lambda W_t}|\mathscr{F}_s]$, $t \geq s$, with $y_s = e^{i\lambda W_s}$. Then because of (4.12) for $t \geq s$,

$$\frac{dy_t}{dt} = -\frac{\lambda^2}{2} y_t.$$

The unique continuous solution y_t of this equation with the initial condition $y_s = e^{i\lambda W_s}$ is given by the formula

$$y_t = y_s e^{-\lambda^2(t-s)/2},$$

from which we obtain

$$M[e^{i\lambda(W_t - W_s)}|\mathscr{F}_s] = e^{-\lambda^2(t-s)/2}. \tag{4.13}$$

From this formula it is seen that the increments $W_t - W_s$ do not depend on the random variables which are measurable with respect to the σ-algebra $\mathscr{F}_s$, $t \geq s$, and are Gaussian with the mean $M[W_t - W_s] = 0$ and the dispersion $D[W_t - W_s] = t - s$, proving Levy's theorem. □

4.1.3

Let us now consider the multivariate analog of this theorem.

Theorem 4.2. *Let* $W = (W_t, \mathscr{F}_t)$, $t \geq 0$, $W_t = (W_1(t), \ldots, W_n(t))$, *be an n-dimensional continuous martingale with* $P(W_i(0) = 0) = 1$, $i \leq n$, $M[W_i(t)|\mathscr{F}_s] = W_i(s)$, $t \geq s$, (P-a.s.), *and*

$$M[(W_t - W_s)(W_t - W_s)^*|\mathscr{F}_s] = E(t - s), \tag{4.14}$$

where $E = E(n \times n)$ *is a unit square matrix of order* $n \times n$. *Then* $W = (W_t, \mathscr{F}_t)$, $t \geq 0$, *is an n-dimensional Brownian motion process with independent components.*

PROOF. The proof differs little from that for the univariate case. Set

$$\sigma_N = \inf\left\{t \leq T: \sup_{s \leq t} \sum_{j=1}^n |W_j(s)| = N\right\}$$

and $\sigma_N = T$ on the set $\{\omega: \sup_{s \leq T} \sum_{j=1}^n |W_j(s)| < N\}$. First, in the same way as in the univariate case we can establish that for any function $f =$

$f(x_1, \ldots, x_n)$ bounded and continuous together with its first and second partial derivatives f'_{x_i} and $f''_{x_i x_j}$,

$$
\begin{aligned}
M[f(\tilde{W}_1^N(t), \ldots, &\tilde{W}_n^N(t)) \mid \mathscr{F}_{s \wedge \sigma_N}] \\
= f(\tilde{W}_1^N(s), \ldots, &\tilde{W}_n^{(N)}(s)) \\
&+ \frac{1}{2} \int_s^t \sum_{i=1}^n M[f''_{x_i x_i}(\tilde{W}_1^N(u), \ldots, \tilde{W}_n^N(u))\chi_N(u) \mid \mathscr{F}_{s \wedge \sigma_N}]du,
\end{aligned}
\tag{4.15}
$$

where

$$
\tilde{W}_i^N(t) = W_i(t \wedge \sigma_N), \qquad \chi_N(u) = \begin{cases} 1, & \sigma_N > u, \\ 0, & \sigma_N \leq u. \end{cases}
$$

From this, after the passage to the limit with $N \to \infty$, we infer that

$$
\begin{aligned}
&M[f(W_1(t), \ldots, W_n(t)) \mid \mathscr{F}_s] \\
&= f(W_1(s), \ldots, W_n(s)) + \frac{1}{2} \int_s^t \sum_{i=1}^n M[f''_{x_i x_i}(W_1(u), \ldots, W_n(u)) \mid \mathscr{F}_s]du.
\end{aligned}
\tag{4.16}
$$

Taking $f(x_1, \ldots, x_n) = \exp[i \sum_{j=1}^n \lambda_j x_j]$, we find that

$$
M\left\{\exp\left[i \sum_{j=1}^n \lambda_j(W_j(t) - W_j(s))\right] \middle| \mathscr{F}_s\right\} = \exp\left\{-\frac{1}{2} \sum_{j=1}^n \lambda_j^2(t - s)\right\}, \tag{4.17}
$$

which proves the desired result. □

4.1.4

In conclusion of this section which deals with the Wiener process $W = (W_t, \mathscr{F}_t)$, $t \geq 0$, we give one result on the continuity of the family of σ-algebras $\mathscr{F}_t^W$.

Theorem 4.3. *Let $(\Omega, \mathscr{F}, P)$ be a complete probability space, and let $W = (W_t, \mathscr{F}_t)$, $t \geq 0$, be a Wiener process on it. Let $\mathscr{F}_t^W = \sigma\{\omega: W_s, s \leq t\}$, assuming that the $\mathscr{F}_t^W$ are augmented by sets from $\mathscr{F}$ having P-measure zero. Then the family of σ-algebras $(\mathscr{F}_t^W)$, $t \geq 0$, is continuous: for all $t \geq 0$, $\mathscr{F}_{t-}^W = \mathscr{F}_t^W = \mathscr{F}_{t+}^W$, where $\mathscr{F}_{0-}^W = \mathscr{F}_0^W$.*

Proof. Left continuity, $\mathscr{F}_{t-}^W = \mathscr{F}_t^W$, easily follows from continuity of the trajectory of a Wiener process. Actually, $\mathscr{F}_{t-}^W = \sigma(\bigcup_{s<t} \mathscr{F}_s^W)$ and $\mathscr{F}_t^W = \sigma(\bigcup_{s<t} \mathscr{F}_s^W \cup \mathscr{F}^W(t))$, where $\mathscr{F}^W(t) = \sigma\{W_t\}$. But $W_t = \lim_{r \uparrow t} W_r$, where the r are rational numbers. Hence $\mathscr{F}^W(t) \subseteq \sigma(\bigcup_{s<t} \mathscr{F}_s^W)$, and therefore, $\mathscr{F}_{t-}^W = \mathscr{F}_t^W$.

In a somewhat more complicated way right continuity, $\mathscr{F}_{t+}^W = \mathscr{F}_t^W$, is proved.

Let $t > s$. Because of (4.13)

$$
M(e^{izW_t} \mid \mathscr{F}_s^W) = M[M(e^{izW_t} \mid \mathscr{F}_s) \mid \mathscr{F}_s^W] = e^{izW_s - (z^2/2)(t-s)}. \tag{4.18}
$$

Let ε be given such that $0 < \varepsilon < t - s$. Then

$$M(e^{izW_t} | \mathscr{F}_{s+}^W) = M[M(e^{izW_t} | \mathscr{F}_{s+\varepsilon}^W) | \mathscr{F}_{s+}^W]$$

$$= M\left[\exp\left\{izW_{s+\varepsilon} - \frac{z^2}{2}(t - s - \varepsilon)\right\}\Big| \mathscr{F}_{s+}^W\right]. \tag{4.19}$$

Passing to the limit with $\varepsilon \downarrow 0$, we find that

$$M[e^{izW_t} | \mathscr{F}_{s+}^W] = M\left[\exp\left\{izW_s - \frac{z^2}{2}(t - s)\right\}\Big| \mathscr{F}_{s+}^W\right]$$

$$= \exp\left\{izW_s - \frac{z^2}{2}(t - s)\right\}, \tag{4.20}$$

since W_s is measurable relative to $\mathscr{F}_{s+}^W$. Consequently,

$$M[e^{izW_t} | \mathscr{F}_s^W] = M[e^{izW_t} | \mathscr{F}_{s+}^W]. \tag{4.21}$$

From this it follows that, for any bounded measurable function,

$$M[f(W_t) | \mathscr{F}_s^W] = M[f(W_t) | \mathscr{F}_{s+}^W]. \tag{4.22}$$

Let now $s < t_1 < t_2$ and $f_1(x), f_2(x)$ be two bounded measurable functions. Then according to the preceding equality,

$$\begin{aligned} M[f_2(W_{t_2}) f_1(W_{t_1}) | \mathscr{F}_s^W] &= M[M(f_2(W_{t_2}) | W_{t_1}) f_1(W_{t_1}) | \mathscr{F}_s^W] \\ &= M[M(f_2(W_{t_2}) | W_{t_1}) f_1(W_{t_1}) | \mathscr{F}_{s+}^W] \\ &= M[f(W_{t_2}) f(W_{t_1}) | \mathscr{F}_{s+}^W], \end{aligned} \tag{4.23}$$

and analogously

$$M\left[\prod_{j=1}^n f_j(W_{t_j}) | \mathscr{F}_s^W\right] = M\left[\prod_{j=1}^n f_j(W_{t_j}) | \mathscr{F}_{s+}^W\right], \tag{4.24}$$

where $s < t_1 < \cdots < t_n$, and the $f_j(x)$ are bounded measurable functions, $j = 1, \ldots, n$. From this it follows that for $t > s$ and any $\mathscr{F}_t^W$-measurable bounded random variable $\eta = \eta(\omega)$.

$$M[\eta | \mathscr{F}_s^W] = M[\eta | \mathscr{F}_{s+}^W]. \tag{4.25}$$

Taking, in particular, the $\mathscr{F}_{s+}^W$-measurable random variable $\eta = \eta(\omega)$ we find that $M(\eta | \mathscr{F}_s^W) = \eta$ (P-a.s.). Because of completeness of the σ-algebras $\mathscr{F}_s^W$, $\mathscr{F}_{s+}^W$, it follows that η is $\mathscr{F}_s^W$-measurable. Consequently, $\mathscr{F}_s^W \supseteq \mathscr{F}_{s+}^W$. The reverse inclusion, $\mathscr{F}_s^W \subseteq \mathscr{F}_{s+}^W$, is obvious. Hence $\mathscr{F}_s^W = \mathscr{F}_{s+}^W$. □

4.2 Stochastic integrals: Ito processes

4.2.1

We shall consider as given the probability space $(\Omega, \mathscr{F}, P)$ with a distinguished nondecreasing family of sub-σ-algebras $F = (\mathscr{F}_t)$, $t \geq 0$. From now on it will be assumed that each σ-algebra $\mathscr{F}_t$, $t \geq 0$, is augmented[1] by sets from $\mathscr{F}$, having zero P-measure.

Let $W = (W_t, \mathscr{F}_t)$, $t \geq 0$, be a Wiener process. In this section the construction and properties of the stochastic integrals $I_t(f)$ of the form $\int_0^t f(s, \omega)dW_s$ for some class of functions $f = f(s, \omega)$, are given. First of all note that the integrals of this type cannot be defined as Lebesgue–Stieltjes or Riemann–Stieltjes integrals, since realizations of a Wiener process have unbounded variation in any arbitrarily small interval of time (Section 1.4). However, Wiener trajectories have some properties which in some sense are analogous to bounded variation.

Lemma 4.3. *Let* $0 \equiv t_0^{(n)} < t_1^{(n)} < \cdots < t_n^{(n)} \equiv t$ *be the subdivision of the interval* $[0, t]$, *with* $\max_i[t_{i+1}^{(n)} - t_i^{(n)}] \to 0$, $n \to \infty$. *Then*

$$\underset{n}{\text{l.i.m.}} \sum_{i=0}^{n-1} [W_{t_{i+1}^{(n)}} - W_{t_i^{(n)}}]^2 = t, \tag{4.26}$$

and with probability 1

$$\lim_n \sum_{i=0}^{n-1} [W_{t_{i+1}^{(n)}} - W_{t_i^{(n)}}]^2 = t. \tag{4.27}$$

PROOF. Since for any n,

$$M \sum_{i=0}^{n-1} [W_{t_{i+1}^{(n)}} - W_{t_i^{(n)}}]^2 = t,$$

in order to prove (4.26) it is sufficient to check that

$$D \sum_{i=0}^{n-1} [W_{t_{i+1}^{(n)}} - W_{t_i^{(n)}}]^2 \to 0, \qquad n \to \infty.$$

But because of independence and the Gaussian behavior of the Wiener process increments,

$$D \sum_{i=0}^{n-1} [W_{t_{i+1}^{(n)}} - W_{t_i^{(n)}}]^2 = 2 \sum_{i=0}^{n-1} [t_{j+1}^{(n)} - t_j^{(n)}]^2$$
$$\leq 2t \max_i [t_{i+1}^{(n)} - t_i^{(n)}] \to 0, \qquad n \to \infty.$$

Equation (4.27) will be proved assuming that $t_i^{(n)} = (i/n)t$ (the general case is somewhat more complicated). For this we make use of the following known fact.

Let $\{\xi_n, n = 1, 2, \ldots\}$ be a sequence of random variables such that for each $\varepsilon > 0$

$$\sum_{n=1}^{\infty} P\{|\xi_n| > \varepsilon\} < \infty. \tag{4.28}$$

[1] Such an assumption provides the opportunity to choose in the random processes on $(\Omega, \mathscr{F}, P)$ under consideration the modifications with necessary properties of measurability (see, for example the note to Lemma 4.4).

Then $\xi_n \to 0$ with probability 1 as $n \to \infty$.

Actually, let $A_n^\varepsilon = \{\omega : |\xi_n| > \varepsilon\}$ and $B^\varepsilon = \lim_n \sup A_n^\varepsilon = \bigcap_{n=1}^\infty \bigcup_{m=n}^\infty A_m^\varepsilon$. Then $\{\omega : \xi_n(\omega) \nrightarrow 0\} = \bigcup_k B^{1/k}$. But because of (4.28), by the lemma of Borel–Cantelli (Section 1.1) $P(B^\varepsilon) = 0$. Hence $P\{\omega : \xi_n \nrightarrow 0\} = 0$.

Returning to the proof of (4.27), where $t_i^{(n)} = (i/n)t$, put

$$\xi_n = \sum_{i=0}^{n-1} \left\{[W_{((i+1)/n)t} - W_{(i/n)t}]^2 - \frac{t}{n}\right\}.$$

By Chebyshev's inequality,

$$P\{|\xi_n| > \varepsilon\} \le \frac{M|\xi_n|^4}{\varepsilon^4}.$$

Using independence of the Wiener process increments over nonoverlapping intervals and the formula

$$M[W_t - W_s]^{2m} = (2m-1)!!(t-s)^m, \qquad m = 1, 2, \ldots,$$

it is not difficult to calculate that

$$M\xi_n^4 \le C\left(\frac{t}{n}\right)^3 \cdot t,$$

where C is a constant.

Hence the series $\sum_{n=1}^\infty P\{|\xi_n| > \varepsilon\} < \infty$, and according to the above remark $\xi_n \to 0$ (P-a.s.), $n \to \infty$, which proves (4.27) assuming $t_j^{(n)} = (i/n)t$. ☐

Note. Symbolically, Equations (4.26) and (4.27) are often written in the following form:

$$\int_0^t (dW_s)^2 = \int_0^t ds.$$

4.2.2

Let us now define the class of random functions $f = f(t, \omega)$ for which the stochastic integral $\int_0^t f(s, \omega)dW_s$ will be constructed.

Definition 2. The measurable (with respect to a pair of variables (t, ω)) function $f = f(t, \omega)$, $t \ge 0$, $\omega \in \Omega$, is called *nonanticipative with respect to the family* $F = (\mathscr{F}_t)$, $t \ge 0$, if, for each t, it is $\mathscr{F}_t$-measurable.

Definition 3. The nonanticipative function $f = f(t, \omega)$ is said to be *of class* $\mathscr{P}_T$ if

$$P\left\{\int_0^T f^2(t, \omega)dt < \infty\right\} = 1. \tag{4.29}$$

Definition 4. $f = f(t, \omega)$ is said to be *of class* $\mathcal{M}_T$ if

$$M \int_0^T f^2(t, \omega)dt < \infty. \tag{4.30}$$

Note. The nonanticipative functions are often also called *functions independent of the future.*

In accordance with the definitions of Section 1.2 the nonanticipative functions $f = f(t, \omega)$ are measurable random processes, adapted to the family $F = (\mathcal{F}_t)$, $t \leq T$. Obviously for any $T > 0$, $\mathcal{P}_T \supseteq \mathcal{M}_T$.

By analogy with the conventional integration theory it is natural to determine first the stochastic integral $I_t(f)$ for a certain set of "elementary" functions. This set has to be sufficiently "rich": so that, on the one hand, any functions from classes $\mathcal{M}_T$ and $\mathcal{P}_T$ can be "approximated" by functions from this set; and, on the other hand, so that it would be possible to describe properties of stochastic integrals from the representatives of this set.

Such a class of "elementary" functions consists of simple functions introduced in Definition 5.

Definition 5. The function $e = e(t, \omega)$, $0 \leq t \leq T$, is called *simple* if there exists a finite subdivision $0 = t_0 < t_1 < \cdots < t_n = T$ of the interval $[0, T]$, random variables $\alpha, \alpha_0, \ldots, \alpha_{n-1}$, where α is $\mathcal{F}_0$-measurable, and α_i are $\mathcal{F}_{t_i}$-measurable, $i = 0, 1, \ldots, n-1$, such that

$$e(t, \omega) = \alpha\chi_{\{0\}}(t) + \sum_{i=0}^{n-1} \alpha_i \chi_{(t_i, t_{i+1}]}(t)$$

($\chi_{\{0\}}(t)$ is a characteristic function of the "point" $\{0\}$ and $\chi_{(t_i, t_{i+1}]}$ is a characteristic function of the half-closed interval $(t_i, t_{i+1}]$, and $e \in \mathcal{M}_T$).

Note. The simple functions $e = e(t, \omega)$ are defined as left continuous functions. This choice is motivated by the analogy with the usual Stieltjes integral, defined so that if $a = a(t)$ is a nondecreasing right continuous function, then

$$\int_0^\infty \chi_{(u, v]}(t)da(t) = a(v) - a(u).$$

The fact that, when constructing a stochastic integral over a Wiener process, we start from the "elementary" left continuous functions is not essential. Right continuous step functions could have been taken as "elementary." However, this fact becomes essential when constructing stochastic integrals over square integrable martingales (see Section 5.4).

4.2.3

For the simple functions $e = e(t, \omega)$, $0 \leq t \leq T$, the stochastic integral $I_t(e)$ by definition is assumed to satisfy

$$I_t(e) = aW_0 + \sum_{\{0 \leq i \leq m,\, t_{m+1} < t\}} a_i[W_{t_{i+1}} - W_{t_i}] + a_{m+1}[W_t - W_{t_{m+1}}]$$

or, since $P(W_0 = 0) = 1$,

$$I_t(e) = \sum_{\{0 \le i \le m,\, t_{m+1} < t\}} a_i[W_{t_{i+1}} - W_{t_i}] + a_{m+1}[W_t - W_{t_{m+1}}]. \tag{4.31}$$

For brevity, instead of the sums in (4.31) we shall use the following (integral) notation:

$$I_t(e) = \int_0^t e(s, \omega)dW_s. \tag{4.32}$$

The integral $\int_s^t e(u, \omega)dW_u$ will be understood to be the integral $I_t(\tilde{e})$, where $\tilde{e}(u, \omega) = e(u, \omega)\chi(u > s)$.

Note the main properties of stochastic integrals from simple functions following immediately from (4.31):

$$I_t(ae_1 + be_2) = aI_t(e_1) + bI_1(e_2), \qquad a, b = \text{const.}; \tag{4.33}$$

$$\int_0^t e(s, \omega)dW_s = \int_0^u e(s, \omega)dW_s + \int_0^t e(s, \omega)dW_s \qquad (P\text{-a.s.}); \tag{4.34}$$

$$I_t(e) \text{ is a continuous function over } t, \qquad 0 \le t \le T; \tag{4.35}$$

$$M\left(\int_0^t e(u, \omega)dW_u | \mathscr{F}_s\right) = \int_0^s e(u, \omega)dW_u \qquad (P\text{-a.s.}); \tag{4.36}$$

$$M\left(\int_0^t e_1(u, \omega)dW_u\right)\left(\int_0^t e_2(u, \omega)dW_u\right) = M\int_0^t e_1(u, \omega)e_2(u, \omega)du. \tag{4.37}$$

If $e(s, \omega) = 0$ for all s, $0 \le s \le T$, and $\omega \in A \subseteq \mathscr{F}_T$, then $\int_0^t e(s, \omega)dW_s = 0$, $t \le T$, $\omega \in A$.

The process $I_t(e)$, $0 \le t \le T$, is progressively measurable and, in particular, the $I_t(e)$ are $\mathscr{F}_t$-measurable at each t, $0 \le t \le T$.

From (4.36), in particular, it follows that

$$M\int_0^t e(u, \omega)dW_u = 0. \tag{4.38}$$

4.2.4

Starting from integrals $I_t(e)$ of simple functions, define now the stochastic integrals $I_t(f)$, $t \le T$, for the functions $f \in \mathscr{M}_T$. The possibility of such a definition is based on the following lemma:

Lemma 4.4. *For each function $f \in \mathscr{M}_T$, we can find a sequence of simple functions f_n, $n = 1, 2, \ldots$, such that*

$$M\int_0^T [f(t, \omega) - f_n(t, \omega)]^2\, dt \to 0, \qquad n \to \infty. \tag{4.39}$$

PROOF

(a) First of all, note that without restricting generality the function $f(t, \omega)$ can be considered as bounded:

$$|f(t, \omega)| \leq C < \infty, \qquad 0 \leq t \leq T, \omega \in \Omega.$$

(Otherwise, one can go over from $f(t, \omega)$ to the function $f^{(N)}(t, \omega) = f(t, \omega)\chi^N(t, \omega)$, where

$$\chi^N(t, \omega) = \begin{cases} 1, & |f(t, \omega)| \leq N, \\ 0, & |f(t, \omega)| > N, \end{cases}$$

and make use of the fact that $M \int_0^T |f(t, \omega) - f^{(N)}(t, \omega)|^2\, dt \to 0$ with $N \to \infty$). Next, if $T = \infty$, then it can be immediately assumed that the function $f(t, \omega)$ vanishes outside a certain finite interval.

Thus, let $|f(t, \omega)| \leq C < \infty$, $T < \infty$.

(b) If the function $f(t, \omega)$ is continuous over t(P-a.s.), then a sequence of simple functions is easy to construct. For example, we can take:

$$f_n(t, \omega) = f\left(\frac{kT}{n}, \omega\right), \qquad \frac{kT}{n} < t \leq \frac{(k+1)T}{n}.$$

Then (4.39) is satisfied by the Lebesgue bounded convergence theorem.

(c) If the function $f(t, \omega)$, $0 \leq t \leq T$, $\omega \in \Omega$, is progressively measurable, then the sequence of approximating functions can be constructed in the following way. Let $F(t, \omega) = \int_0^t f(s, \omega)ds$, where the integral is understood as a Lebesgue integral. Because of the progressive measurability of the functions $f(s, \omega)$, the process $F(t, \omega)$, $0 \leq t \leq T$, is measurable, and for each t the random variables $F(t, \omega)$ are $\mathscr{F}_t$-measurable.

Assume

$$\tilde{f}_m(t, \omega) = m \int_{(t-(1/m)) \vee 0}^{t} f(s, \omega)ds\left(= \left[F(t, \omega) - F\left(\left(t - \frac{1}{m}\right) \vee 0, \omega\right)\right] \Big/ \frac{1}{m}\right).$$

The random process $\tilde{f}_m(t, \omega)$, $0 \leq t \leq T$, $\omega \in \Omega$, is measurable, nonanticipative, and has (P-a.s.) continuous trajectories. Hence, according to (b), for each m there exists a sequence of nonanticipative step functions $\tilde{f}_{m,n}(t, \omega)$, $n = 1, 2, \ldots$, such that

$$M \int_0^T [\tilde{f}_m(t, \omega) - \tilde{f}_{m,n}(t, \omega)]^2\, dt \to 0, \qquad n \to \infty.$$

Note now that (P-a.s.) for almost all $t \leq T$ there exists a derivative $F'(t, \omega)$, and $F'(t, \omega) = f(t, \omega)$. But at those points where the derivative $F'(t, \omega)$ exists,

$$F'(t, \omega) = \lim_{m \to \infty} m\left[F(t, \omega) - F\left(\left(t - \frac{1}{m}\right) \vee 0, \omega\right)\right] = \lim_{m \to \infty} \tilde{f}_m(t, \omega).$$

Hence for almost all (t, ω) (on measure $dt\, dP$), $\lim_{m\to\infty} \tilde{f}_m(t, \omega) = f(t, \omega)$ and by the Lebesgue bounded convergence theorem

$$M \int_0^T [\tilde{f}_m(t, \omega) - f(t, \omega)]^2\, dt \to 0, \qquad m \to \infty.$$

By this, the statement of the lemma is proved in the case where the functions $f(t, \omega)$, $0 \le t \le T$, $\omega \in \Omega$, are progressively measurable.

(d) In the general case we proceed in the following way. Complete a definition of the function $f(t, \omega)$ for negative t, taking $f(t, \omega) = f(0, \omega)$. We shall consider the function $f(t, \omega)$ to be bounded and finite. Set

$$\psi_n(t) = \frac{j}{2^n}, \qquad \frac{j}{2^n} < t \le \frac{j+1}{2^n}, \qquad j = 0, \pm 1, \ldots,$$

and note that the function $f_n(t, \omega) = f[\psi_n(t - \tilde{s}) + \tilde{s}, \omega]$ is simple for every fixed $\tilde{s}$. The lemma will have been proved if it is shown that the point $\tilde{s}$ can be chosen so that (4.39) is satisfied.

For this we use the following: if $f = f(t, \omega)$, $t \ge 0$, and $\omega \in \Omega$ is a measurable bounded function with finite support, then

$$\lim_{h\to 0} M \int_0^\infty [f(s + h, \omega) - f(s, \omega)]^2\, ds = 0. \tag{4.40}$$

Actually, according to (c), for any $\varepsilon > 0$ there will be a (P-a.s.) continuous function $f_\varepsilon(t, \omega)$ such that

$$M \int_0^\infty [f_\varepsilon(s, \omega) - f(s, \omega)]^2\, ds \le \varepsilon^2$$

But because of the Minkowski inequality,

$$\varlimsup_{h\to 0}\left[M \int_0^\infty [f(s + h, \omega) - f(s, \omega)]^2\, ds\right]^{1/2}$$
$$\le \varlimsup_{h\to 0}\left[M \int_0^\infty [f_\varepsilon(s + h, \omega) - f_\varepsilon(s, \omega)]^2\, ds\right]^{1/2} + 2\varepsilon,$$

from which, because of the arbitrariness of $\varepsilon > 0$, (4.40) follows.

From (4.40) it also follows that for any $t \ge 0$,

$$\lim_{h\to 0} M \int_0^\infty [f(s + t + h, \omega) - f(s + t, \omega)]^2\, ds = 0$$

and, in particular,

$$\lim_{n\to 0} M \int_0^\infty [f(s + \psi_n(t), \omega) - f(s + t, \omega)]^2\, ds = 0$$

and

$$\lim_{n\to\infty} M \int_0^\infty \int_0^\infty [f(s + \psi_n(t), \omega) - f(s + t, \omega)]^2\, ds\, dt = 0.$$

From the last equality it follows that there exists a sequence of numbers n_i, $i = 1, 2, \ldots, n_i$, such that for almost all (s, t, ω) (over measure $ds\,dt\,dP$)

$$[f(s + \psi_{n_i}(t), \omega) - f(s + t, \omega)]^2 \to 0, \qquad n_i \to \infty.$$

From this, going over to new variables $u = s$, $v = s + t$, we infer that for almost all (u, v, ω) (over measure $du\,dv\,dP$)

$$[f(u + \psi_{n_i}(v - u), \omega) - f(v, \omega)]^2 \to 0, \qquad n_i \to \infty,$$

and, therefore, there will be a point $\tilde{u} = \tilde{s}$ such that

$$\lim_{n_i \to \infty} M \int_0^\infty [f(\tilde{u} + \psi_{n_i}(v - \tilde{u}), \omega) - f(v, \omega)]^2\, dv$$

$$= \lim_{n_i \to \infty} M \int_0^\infty [f(\tilde{s} + \psi_{n_i}(t - \tilde{s}), \omega) - f(t, \omega)]^2\, dt = 0. \quad \square$$

Note. If the random process $f(t, \omega)$, $0 \le t \le T < \infty$, is progressively measurable and $P(\int_0^T |f(t, \omega)| dt < \infty) = 1$, then the process $F(t, \omega) = \int_0^t f(s, \omega) ds$, $0 \le t \le T$, where the integral is understood as a Lebesgue integral, is measurable and $\mathscr{F}$-adapted and, more than this, is progressively measurable (with (P-a.s.) continuous trajectories).

If the measurable random process $f(t, \omega)$, $0 \le t \le T < \infty$,

$$P\left(\int_0^T |f(t, \omega)| dt < \infty\right) = 1$$

is $\mathscr{F}_t$-measurable for every t, $0 \le t \le T$, then it has (Subsection 1.2.1) a progressively measurable modification $\tilde{f}(t, \omega)$ and the process $\tilde{F}(t, \omega) = \int_0^t \tilde{f}(s, \omega) ds$ is also progressively measurable. We can show that $\tilde{F}(t, \omega)$ is a (progressively measurable) modification of the process $F(t, \omega)$.

Actually, let $\chi_s(\omega) = \chi_{\{\omega : f(s, \omega) \ne \tilde{f}(s, \omega)\}}$. Then by the Fubini theorem, $M \int_0^T \chi_s(\omega) ds = \int_0^T M \chi_s(\omega) ds = 0$, and, consequently, ($P$-a.s.) $\int_0^T \chi_s(\omega) ds = 0$, and, therefore, $P(F(t, \omega) = \tilde{F}(t, \omega)) = 1$, $t \le T$.

As it was noted at the beginning of this section, it is assumed that the σ-algebras $\mathscr{F}_t$ are augmented by sets from $\mathscr{F}$, having P-measure zero. Hence, from the fact that the $\tilde{F}(t, \omega)$ are $\mathscr{F}_t$-measurable for each $t \le T$, it follows that the $F(t, \omega)$ are also $\mathscr{F}_t$-measurable for each $t \le T$.

Taking into consideration the fact that the process $F(t, \omega)$, $t \le T$, is continuous, we infer that the integral $F(t, \omega) = \int_0^t f(s, \omega) ds$, $t \le T$, from the nonanticipative process $f(s, \omega)$, $s \le t$, is a progressively measurable random process.

The considerations described above can be used (in the case where the σ-algebras $\mathscr{F}_t$ are augmented by sets from $\mathscr{F}$, having P-measure zero) for proving (d) of Lemma 4.4 by reducing to the case examined in (c). However, the proof given in (d) is valuable since it shows the way of constructing simple functions $f_n(t, \omega)$ directly from the functions $f(t, \omega)$.

4.2.5

Thus let $f \in \mathcal{M}_T$. According to the lemma which has been proved, there exists a sequence of the simple functions $f_n(t, \omega)$, for which (4.39) is satisfied. But then obviously

$$\lim_{\substack{n \to \infty \\ m \to \infty}} M \int_0^T [f_n(t, \omega) - f_m(t, \omega)]^2 \, dt = 0,$$

and consequently (by the property (4.37))

$$\lim_{\substack{n \to \infty \\ m \to \infty}} M \left[\int_0^T f_n(t, \omega) dW_t - \int_0^T f_m(t, \omega) dW_t \right]^2$$

$$= \lim_{\substack{n \to \infty \\ m \to \infty}} M \int_0^T [f_n(t, \omega) - f_m(t, \omega)]^2 \, dt = 0. \quad (4.41)$$

Thus the sequence of random variables $I_T(f_n)$ is fundamental in the sense of convergence in the mean square, and, therefore, converges to some limit which will be denoted $I_T(f)$ or $\int_0^T f(t, \omega) dW_t$:

$$I_T(f) = \underset{n}{\text{l.i.m.}}\, I_T(f_n). \quad (4.42)$$

The value (to within stochastic equivalence) of this limit, it is not difficult to show, does not depend on the choice of the approximating sequence $\{f_n, n = 1, 2, \ldots\}$. Therefore, the stochastic integral $I_T(f)$ is well defined.

Note 1. Since the value of the stochastic integral $I_T(f)$ is determined to within stochastic equivalence, we agree to consider $I_T(f) = 0$ for all those ω for which $f(t, \omega) = 0$ for all $0 \le t \le T$ (compare with the properties of stochastic integrals of simple functions, Subsection 4.2.3).

Again let $f \in \mathcal{M}_T$. Define the family of stochastic integrals $I_t(f)$, with $0 \le t \le T$, assuming $I_t(f) = I_T(f\chi_t)$, i.e.,

$$I_t(f) = \int_0^T f(s, \omega)\chi_t(s) dW_s. \quad (4.43)$$

For $I_t(f)$ it is natural to use also the notation

$$I_t(f) = \int_0^t f(s, \omega) dW_s. \quad (4.44)$$

Consider the basic properties of the stochastic integrals $I_t(f)$, $0 \le t \le T$, of the functions f, $f_i \in \mathcal{M}_T$, $i = 1, 2$:

$$I_t(af_1 + bf_2) = aI_t(f_1) + bI_t(f_2), \qquad a, b = \text{const.}; \quad (4.45)$$

$$\int_0^t f(s, \omega) dW_s = \int_0^u f(s, \omega) dW_s + \int_u^t f(s, \omega) dW_s, \quad (4.46)$$

where

$$\int_u^t f(s, \omega) dW_s = \int_0^T f(s, \omega) \chi_{[u,t]}(s) dW_s,$$

and $\chi_{[u,t]}(s)$ is a characteristic function of the set $u \leq s \leq t$;

$$I_t(f) \text{ is a continuous function over } t, 0 \leq t \leq T; \tag{4.47}$$

$$M\left[\int_0^t f(u, \omega) dW_u | \mathscr{F}_s\right] = \int_0^s f(u, \omega) dW_u, \qquad s \leq t;^2 \tag{4.48}$$

$$M\left[\int_0^t f_1(u, \omega) dW_u\right]\left[\int_0^t f_2(u, \omega) dW_u \cdot\right] = M \int_0^t f_1(u, \omega) f_2(u, \omega) du. \tag{4.49}$$

If $f(s, \omega) = 0$ for all s, $0 \leq s \leq T$, and $\omega \in A \in \mathscr{F}_T$, then

$$\int_0^t f(s, \omega) dW_s = 0, \qquad t \leq T, \qquad \omega \in A. \tag{4.50}$$

The process $I_t(f)$, $0 \leq t \leq T$, $f \in \mathscr{M}_T$, is progressively measurable, and, in particular, $I_t(f)$ are $\mathscr{F}_t$-measurable for each t, $0 \leq t \leq T$.

For proving (4.45) it is sufficient to choose sequences of the simple functions $f_1^{(n)}$ and $f_2^{(n)}$ such that

$$M \int_0^T (f_i - f_i^{(n)})^2 \, ds \to 0, \qquad n \to \infty, \qquad i = 1, 2,$$

and then to pass to the limit in the equality

$$I_t(af_1^{(n)} + bf_2^{(n)}) = aI_t(f_1^{(n)}) + bI_t(f_2^{(n)}).$$

(4.46) is proved analogously. (4.48) and (4.49) follow from (4.36) and (4.37) since from the convergence of random variables in the mean square there follows the convergence of their moments of the first two orders.

(4.49) can be somewhat generalized:

$$M\left\{\int_s^t f_1(u, \omega) dW_u \int_s^t f_2(u, \omega) dW_u | \mathscr{F}_s\right\} = M\left\{\int_s^t f_1(u, \omega) f_2(u, \omega) du | \mathscr{F}_s\right\}.$$

This property is verified in the usual way: first its correctness for simple functions is established, and second, the corresponding passage to the limit is made.

(4.50) follows from Note 1 given above. Let us show now that the process $I_t(f)$, $0 \leq t \leq T$, is progressively measurable, and, what is more, has (P-a.s.) continuous trajectories (more precisely, it has a modification with these two properties).

[2] (4.48) is true also for functions $f \in \mathscr{P}_t$ with $M\sqrt{\int_0^t f^2(u, \omega) du} < \infty$ (cf. [132]).

For proving this we note that for the simple functions f_n the process $(I_t(f_n), \mathscr{F}_t)$, $0 \le t \le T$, forms a continuous (P-a.s.) martingale (with the properties given in (4.35) and (4.36)). Hence by Theorem 3.2,

$$P\left\{\sup_{t\le T}\left|\int_0^t f_n(s,\omega)dW_s - \int_0^t f_m(s,\omega)dW_s\right| > \lambda\right\}$$
$$\le \frac{1}{\lambda^2} M\left\{\int_0^T [f_n(s,\omega) - f_m(s,\omega)]dW_s\right\}^2$$
$$= \frac{1}{\lambda^2} M \int_0^T [f_n(s,\omega) - f_m(s,\omega)]^2\, ds. \tag{4.51}$$

Choose the sequence of simple functions f_n converging to $f \in \mathscr{M}_T$ so that $f_0 \equiv 0$,

$$M \int_0^T [f(s,\omega) - f_n(s,\omega)]^2\, ds \to 0, \qquad n \to \infty,$$

and

$$M \int_0^T [f_{n+1}(s,\omega) - f_n(s,\omega)]^2\, ds \le \frac{1}{2^n}. \tag{4.52}$$

Note now that the series

$$\int_0^t f_1(s,\omega)dW_s + \left[\int_0^t (f_2(s,\omega) - f_1(s,\omega))dW_s\right] + \cdots$$
$$+ \left[\int_0^t (f_{n+1}(s,\omega) - f_n(s,\omega))dW_s\right] + \cdots$$

converges in the mean square to $\int_0^t f(s,\omega)dW_s$ and that the terms of this series are (P-a.s.) continuous over t, $0 \le t \le T$. Further, according to (4.51) and (4.52),

$$\sum_{n=0}^{\infty} P\left\{\sup_{t\le T}\left|\int_0^t (f_{n+1}(s,\omega) - f_n(s,\omega))dW_s\right| > \frac{1}{n^2}\right\} \le \sum_{n=0}^{\infty} \frac{n^4}{2^n} < \infty.$$

Hence, because of the Borel–Cantelli lemma, with probability 1 there will be a (random) number $N = N(\omega)$ starting from which

$$\sup_{t\le T}\left|\int_0^t (f_{n+1}(s,\omega) - f_n(s,\omega))dW_s\right| \le \frac{1}{n^2}, \qquad n \ge N.$$

Consequently, the series of continuous functions

$$\int_0^t f_1(s,\omega)dW_s + \sum_{n=1}^{\infty}\left[\int_0^t (f_{n+1}(s,\omega) - f_n(s,\omega))dW_s\right]$$

converges uniformly with probability 1 and defines the continuous function (P-a.s.) which, for each t, is $\mathscr{F}_t$-measurable.[3] From these two properties it follows that the random process defined by this series is progressively measurable (Subsection 1.2.1).

Thus, we see that the sequence of simple functions f_n satisfying (4.52) can be chosen so that the integrals $I_t(f)$, $0 \le t \le T$, constructed with their help will be continuous over t, $0 \le t \le T$, with probability 1. Since, to within stochastic equivalence, the values of $I_t(f)$ do not depend on the choice of an approximating sequence, if follows that the integrals $I_t(f)$ have a continuous modification. From now on, while considering the integrals $I_t(f)$, $f \in \mathscr{M}_T$, it will be assumed that the $I_t(f)$ have continuous (P-a.s.) trajectories.

Note 2. Observe that from the construction of the approximating sequence $\{f_n, n = 1, 2, \dots\}$ satisfying (4.52) it follows, in particular, that with probability 1

$$\sup_{0 \le t \le T} \left| \int_0^t f(s, \omega) dW_s - \int_0^t f_n(s, \omega) dW_s \right| \to 0, \qquad n \to \infty.$$

In other words, uniformly over t, $0 \le t \le T$, with probability 1

$$\int_0^t f_n(s, \omega) dW_s \to \int_0^t f(s, \omega) dW_s, \qquad n \to \infty.$$

Observe two more useful properties of the stochastic integrals $I_t(f)$, $f \in \mathscr{M}_T$, which follow immediately from Theorem 3.2 and from the note that $(I_t(f), \mathscr{F}_t)$, $0 \le t \le T$, is a square integrable martingale with continuous trajectories:

$$P\left\{ \sup_{0 \le t \le T} \left| \int_0^t f(s, \omega) dW_s \right| > \lambda \right\} \le \frac{1}{\lambda^2} \int_0^T M f^2(s, \omega) ds; \qquad (4.53)$$

$$M \sup_{0 \le t \le T} \left| \int_0^t f(s, \omega) dW_s \right|^2 \le 4 \int_0^T M f^2(s, \omega) ds. \qquad (4.54)$$

From the last property, in particular, it follows that if $f \in \mathscr{M}_T$ and the sequence of the functions $\{f_n, n = 1, 2, \dots\}$ is such that $f_n \in \mathscr{M}_T$ and

$$M \int_0^T [f(t, \omega) - f_n(t, \omega)]^2 \, dt \to 0,$$

then

$$\underset{n}{\text{l.i.m.}} \int_0^t f_n(s, \omega) dW_s = \int_0^t f(s, \omega) dW_s.$$

Note 3. The construction of the stochastic integrals $I_t(f)$, $0 \le t \le T$, carried out above, and their basic properties carry over to the case $T = \infty$.

[3] Note that the σ-algebras $\mathscr{F}_t$ are assumed to be augmented by sets from $\mathscr{F}$ of zero probability.

It is enough if $f \in \mathcal{M}_\infty$, where $\mathcal{M}_\infty$ is the class of nonanticipative functions $f = f(s, \omega)$ with the property

$$\int_0^\infty Mf^2(s, \omega)ds < \infty.$$

4.2.6

Let us now construct the stochastic integrals $I_t(f)$, $t \le T$, for functions f from class $\mathcal{P}_T$, satisfying the condition

$$P\left\{\int_0^T f^2(s, \omega)ds < \infty\right\} = 1. \tag{4.55}$$

For this purpose we establish first the following lemma.

Lemma 4.5. *Let $f \in \mathcal{P}_T$, $T \le \infty$. Then we can find a sequence of functions $f_n \in \mathcal{M}_T$, such that in probability*

$$\int_0^T [f(t, \omega) - f_n(t, \omega)]^2\, dt \to 0, \qquad n \to \infty. \tag{4.56}$$

There exists a sequence of simple functions $f_n(t, \omega)$, for which (4.56) *is satisfied both in the sense of convergence in probability and with probability* 1.

PROOF. Let $f \in \mathcal{P}_T$. Put

$$\tau_N(\omega) = \begin{cases} \inf\left\{t \le T: \int_0^t f^2(s, \omega)ds \ge N\right\}, \\ T, \quad \text{if} \quad \int_0^T f^2(s, \omega)ds < N, \end{cases}$$

and

$$f_N(s, \omega) = f(s, \omega)\chi_{\{s \le \tau_N(\omega)\}}. \tag{4.57}$$

Since it is assumed that the σ-algebras $\mathcal{F}_t$, $0 \le t \le T$, are augmented by sets from $\mathcal{F}$ of zero probability, then, according to the note to Lemma 4.4., the process $\int_0^t f^2(s, \omega)ds$, $t \le T$, is progressively measurable. From this it follows that the moments $\tau_N(\omega)$ are Markov (relative to the family $(\mathcal{F}_t)$, $0 \le t \le T$). Hence the functions $f_N(s, \omega)$, $N = 1, 2, \ldots$, are nonanticipative and belong to class $\mathcal{M}_T$ since

$$\int_0^T Mf_N^2(s, \omega)ds \le N < \infty.$$

To prove the final part of the lemma we make use of Lemma 4.4, according to which for each N, $N = 1, 2, \ldots$, there exists a sequence of simple functions $f_N^{(n)}$, $n = 1, 2, \ldots$, such that

$$M\int_0^T [f_N^{(n)}(t, \omega) - f_N(t, \omega)]^2\, dt \to 0, \qquad n \to \infty,$$

and (because of (4.57))

$$P\left\{\int_0^T [f(t,\omega) - f_N(t,\omega)]^2\,dt > 0\right\} \le P\left\{\int_0^T f^2(t,\omega)dt > N\right\}. \quad (4.58)$$

Then

$$\begin{aligned} &P\left\{\int_0^T [f(t,\omega) - f_N^{(n)}(t,\omega)]^2\,dt > \varepsilon\right\} \\ &\le P\left\{\int_0^T [f(t,\omega) - f_N(t,\omega)]^2\,dt > 0\right\} \\ &+ P\left\{\int_0^T [f_N(t,\omega) - f_N^{(n)}(t,\omega)]^2\,dt > \frac{\varepsilon}{2}\right\} \\ &\le P\left\{\int_0^T f^2(t,\omega)dt > N\right\} + \frac{2}{\varepsilon} M \int_0^T [f_N(t,\omega) - f_N^{(n)}(t,\omega)]^2\,dt, \end{aligned}$$

which proves the existence of the sequence of simple functions $f_n(t,\omega)$ approximating the function f in the sense of (4.56) with convergence in probability.

Without loss of generality we may assume that the functions f_n have been chosen so that

$$P\left\{\int_0^T [f(t,\omega) - f_n(t,\omega)]^2\,dt > 2^{-n}\right\} \le 2^{-n}.$$

(Otherwise we may work with an appropriate subsequence of the sequence $\{f_n\}, n = 1, 2, \ldots$) Hence, by the Borel–Cantelli lemma, for almost all ω there will be numbers $N(\omega)$ such that for all $n \ge N(\omega)$,

$$\int_0^T [f(t,\omega) - f_n(t,\omega)]^2\,dt < 2^{-n}.$$

In particular, with probability 1,

$$\lim_{n\to\infty} \int_0^T [f(t,\omega) - f_n(t,\omega)]^2\,dt = 0. \qquad \square$$

Note 4. If the nonanticipative function $f = f(t,\omega)$ is such that, with probability 1,

$$\int_0^T |f(t,\omega)|dt < \infty,$$

then there will be a sequence of simple functions $\{f_n(t,\omega), n = 1, 2, \ldots\}$ such that, with probability 1,

$$\lim_n \int_0^T |f(t,\omega) - f_n(t,\omega)|dt = 0.$$

Proving this is analogous to the case where

$$P\left(\int_0^T f^2(t,\omega)dt < \infty\right) = 1.$$

From now on we shall also need the following.

Lemma 4.6. *Let $f \in \mathcal{M}_T$ and the event $A \in \mathcal{F}_T$. Then for any $N > 0$, $C > 0$,*

$$P\left\{A \cap \left(\sup_{0 \le t \le T}\left|\int_0^t f(s, \omega)dW_s\right| > C\right)\right\}$$
$$\le \frac{N}{C^2} + P\left\{A \cap \left(\int_0^T f^2(s, \omega)ds > N\right)\right\} \tag{4.59}$$

and, in particular,

$$P\left\{\sup_{0 \le t \le T}\left|\int_0^t f(s, \omega)dW_s\right| > C\right\} \le \frac{N}{C^2} + P\left\{\int_0^T f^2(s, \omega)ds > N\right\}. \tag{4.60}$$

PROOF. Let the functions $f_N(s, \omega)$ be defined by (4.57). Then, by (4.49),

$$M\left(\int_0^T f_N(s, \omega)dW_s\right)^2 = \int_0^T Mf_N^2(s, \omega)ds \le N < \infty.$$

In accordance with properties of the stochastic integrals,

$$\left\{\omega: \sup_{0 \le t \le T}\left|\int_0^t [f(s, \omega) - f_N(s, \omega)]dW_s\right| = 0\right\} \supseteq \left\{\omega: \int_0^T f^2(s, \omega)ds \le N\right\}.$$

Hence

$$A \cap \left\{\omega: \sup_{0 \le t \le T}\left|\int_0^t [f(s, \omega) - f_N(s, \omega)]dW_s\right| > 0\right\}$$
$$\subseteq A \cap \left\{\omega: \int_0^T f^2(s, \omega)ds > N\right\},$$

and therefore

$$P\left\{A \cap \left(\sup_{0 \le t \le T}\left|\int_0^t f(s, \omega)dW_s\right| > C\right)\right\}$$
$$= P\left\{A \cap \left(\sup_{0 \le t \le T}\left|\int_0^t f_N(s, \omega)dW_s + \int_0^t [f(s, \omega) - f_N(s, \omega)]dW_s\right| > C\right)\right\}$$
$$\le P\left\{A \cap \left(\sup_{0 \le t \le T}\left|\int_0^t f_N(s, \omega)dW_s\right| > C\right)\right\}$$
$$+ P\left\{A \cap \left(\sup_{0 \le t \le T}\left|\int_0^t [f(s, \omega) - f_N(s, \omega)]dW_s\right| > 0\right)\right\}$$
$$\le P\left\{\sup_{0 \le t \le T}\left|\int_0^t f_N(s, \omega)dW_s\right| > C\right\} + P\left\{A \cap \left(\int_0^t f^2(s, \omega)ds > N\right)\right\}$$
$$\le \frac{1}{C^2} M\left(\int_0^T f_N(s, \omega)dW_s\right)^2 + P\left\{A \cap \left(\int_0^T f^2(s, \omega)ds > N\right)\right\}$$
$$\le \frac{N}{C^2} + P\left\{A \cap \left(\int_0^T f^2(s, \omega)ds > N\right)\right\}.$$

□

Corollary. *If $f \in \mathscr{M}_T$, then*

$$P\left\{\left(\int_0^T f^2(s, \omega)ds \le N\right) \cap \left(\sup_{0 \le t \le T}\left|\int_0^t f(s, \omega)dW_s\right| > C\right)\right\} \le \frac{N}{C^2}.$$

Note 5. The statement of the lemma remains correct if in its formulation the time T is replaced with the Markov time σ; in this case it is required that $f \in \mathscr{M}_\sigma$, $A \in \mathscr{F}_\sigma$.

Let us now go directly to the construction of the integral $I_T(f)$ for $f \in \mathscr{P}_T$, $T \le \infty$. Let $f_n = f_n(t, \omega)$, $n = 1, 2, \ldots$, be a sequence of functions from class $\mathscr{M}_T$, approximating the function $f(t, \omega)$ in the sense of convergence of (4.56). Then, obviously, for any $\varepsilon > 0$,

$$\lim_{n, m \to \infty} P\left\{\int_0^T [f_n(t, \omega) - f_m(t, \omega)]^2\, dt > \varepsilon\right\} = 0$$

and, according to Lemma 4.6, for any $\varepsilon > 0$, $\delta > 0$,

$$\overline{\lim_{n, m \to \infty}}\, P\left\{\left|\int_0^T f_n(t, \omega)dW_t - \int_0^T f_m(t, \omega)dW_t\right| > \delta\right\}$$

$$\le \frac{\varepsilon}{\delta^2} + \lim_{n, m \to \infty} P\left\{\int_0^T [f_n(t, \omega) - f_m(t, \omega)]^2\, dt > \varepsilon\right\} = \frac{\varepsilon}{\delta^2}.$$

From this, because of the arbitrariness of $\varepsilon > 0$, we obtain

$$\lim_{n, m \to \infty} P\left\{\left|\int_0^T f_n(t, \omega)dW_t - \int_0^T f_m(t, \omega)dW_t\right| > \delta\right\} = 0.$$

Thus the sequence of random variables $I_T(f_n) - \int_0^T f_n(t, \omega)dW_t$ converges in probability to some random variable which we denote as $I_T(f)$ or $\int_0^T f(t, \omega)dW_t$ and is called a **stochastic integral** (of the function $f \in \mathscr{P}_T$ with respect to the Wiener process $W = (W_t, \mathscr{F}_t)$, $t \le T$).

The value $I_T(f)$ (to within equivalence) does not depend on the choice of approximating sequences (say, $\{f_n\}$ and $\{g_n\}$, $n = 1, 2, \ldots$). Actually, joining the sequences into one, $\{h_n\}$, we can establish the existence of the limit in probability of the sequence of the variables $I_T(h_n)$, $n \to \infty$. Consequently, the limits over the subsequences $\lim I_T(f_n)$, $\lim I_T(g_n)$ will coincide.

The construction of the stochastic integrals $I_T(f)$ for $t \le T$ in the case of the functions $f \in \mathscr{P}_T$ is accomplished in the same way as for $f \in \mathscr{M}_T$. Namely, we define the integrals $I_t(f) = \int_0^t f(s, \omega)dW_s$ with the help of the equalities

$$I_t(f) = \int_0^T f(s, \omega)\chi_t(s)dW_s, \qquad 0 \le t \le T, \tag{4.61}$$

where $\chi_t(s)$ is the characteristic function of the set $0 \le s \le t$.

Since (to within stochastic equivalence) the value of the stochastic integrals $I_t(f)$ does not depend on the choice of the approximating sequence, then while investigating the properties of the process $I_t(f)$, $0 \le t \le T$, particular types of such sequences can be used.

In particular, take as such a sequence the functions $f_N(s, \omega)$ from (4.57). Since $P\{\int_0^T f^2(s, \omega)ds < \infty\} = 1$, then the set $\Omega' = \bigcup_{N=1}^{\infty} \Omega_N$, where $\Omega_N = \{\omega : N - 1 \leq \int_0^T f^2(s, \omega)ds < N\}$, differs from Ω by a subset of P-measure zero.

Note now that on the set Ω_N,

$$f_N(s, \omega) = f_{N+1}(s, \omega) = \cdots = f(s, \omega)$$

for all s, $0 \leq s \leq T$. Consequently, on the set Ω_N,

$$I_t(f) = \int_0^t f(s, \omega)dW_s = \int_0^t f_N(s, \omega)dW_s = I_t(f_N).$$

But $f_N \in \mathscr{M}_T$. Hence the process $I_t(f_N)$ is continuous over t, $0 \leq t \leq T$, with probability 1 (more precisely, it has a continuous modification). From this it follows that on the set Ω_N the stochastic integrals $I_t(f)$, $f \in \mathscr{P}_T$, $0 \leq t \leq T$, form a continuous process.

But, as it has been noted, $\Omega' = \bigcup_{N=1}^{\infty} \Omega_N$ differs from Ω only on the set of P-measure zero, therefore, (P-a.s.) the random process $I_t(f)$, $0 \leq t \leq T$, has continuous trajectories. Due to progressive measurability of the processes $I_t(f_N)$, $0 \leq t \leq T$, the same consideration shows that the process $I_t(f)$, $0 \leq t \leq T$, is also progressively measurable.

Note 6. According to Note 2 above, if $f \in \mathscr{M}_T$, then there exists a sequence $\{f_n, n = 1, 2, \ldots\}$ of simple functions such that, uniformly over t, $0 \leq t \leq T$, with probability 1 $\int_0^t f_n \, dW_s \to \int_0^t f \, dW_s$.

A similar result holds true also for the functions $f \in \mathscr{P}_T$ (see [123]).

Note 7. It is also useful to note that the inequalities given by (4.59) and (4.60) hold true for any function $f \in \mathscr{P}_T$. Actually, let $\{f_n, n = 1, 2, \ldots\}$ be a sequence of simple functions such that

$$|f_n(s, \omega)| \leq |f(s, \omega)|, \qquad 0 \leq s \leq T, \qquad \omega \in \Omega,$$

and

$$\int_0^T [f_n(s, \omega) - f(s, \omega)]^2 \, ds \to 0$$

(in probability) with $n \to \infty$. Then for any $N > 0$, $C > 0$, and $A \in \mathscr{F}_T$,

$$P\left\{A \cap \left(\sup_{t \leq T} \left| \int_0^t f(s, \omega)dW_s \right| > C\right)\right\}$$

$$\leq P\left\{A \cap \left(\sup_{t \leq T} \left| \int_0^t f_n(s, \omega)dW_s \right| > C\right)\right\}$$

$$+ P\left\{A \cap \left(\sup_{t \leq T} \left| \int_0^t [f(s, \omega) - f_n(s, \omega)]dW_s \right| > 0\right)\right\}$$

$$\leq \frac{N}{C^2} + P\left\{A \cap \left(\int_0^T f_n^2(s, \omega)ds > N\right)\right\}$$

$$+ P\left\{\int_0^T [f(s, \omega) - f_n(s, \omega)]^2 \, ds > 0\right\}.$$

From this, passing to the limit with $n \to \infty$, we obtain the desired inequality:

$$P\left\{A \cap \left(\sup_{t \le T}\left|\int_0^t f(s, \omega)dW_s\right| > C\right)\right\} \le \frac{N}{C^2} + P\left\{A \cap \left(\left(\int_0^T f^2(s, \omega)ds > N\right)\right)\right\}.$$

Completing the construction of the stochastic integrals $I_t(f)$ for the functions $f \in \mathcal{P}_T$, we note their properties. The properties given by (4.45)–(4.47) remain valid. However, the properties given by (4.48) and (4.49) can be violated (see below, Note 9 in Subsection 4.2.8). While in the case of $f \in \mathcal{M}_T$ the stochastic integrals $(I_t(f), \mathcal{F}_t)$, $0 \le t \le T$, yield a martingale (and square integrable), for the functions $f \in \mathcal{P}_T$ this is, generally speaking, not true. By the way, in the case of $f \in \mathcal{P}_T(I_t(f)\mathcal{F}_t)$, $t \le T$, is a local martingale (see, further, Subsection 4.2.10).

4.2.7

Let $f \in \mathcal{P}_\infty$, and let $\tau = \tau(\omega)$ be a finite ($P(\tau < \infty) = 1$) Markov time relative to the system $(\mathcal{F}_t)$, $t \ge 0$. Along with the stochastic integrals $I_t(f) = \int_0^t f(s, \omega)dW_s$ we may introduce the stochastic integral with the random upper limit τ.

Set

$$I_\tau(f) = I_t(f) \quad \text{on} \quad \{\omega : \tau(\omega) = t\}. \tag{4.62}$$

Since the stochastic integral $I_t(f)$, $t \ge 0$, is a progressively measurable process, then by Lemma 1.8 $I_\tau(f)$ is a $\mathcal{F}_\tau$-measurable random variable.

By analogy with the notation $I_t(f) = \int_0^t f(s, \omega)dW_s$, we shall use also the notation $I_\tau(f) = \int_0^\tau f(s, \omega)dW_s$.

While operating with the stochastic integrals $I_\tau(f)$ with the random upper limit τ the following equality is useful:

$$I_\tau(f) = I_\infty(\chi \cdot f) \qquad (P\text{-a.s.}), \tag{4.63}$$

where $\chi = \chi_{\{t \le \tau\}}$ is the characteristic function of the set $\{t \le \tau\}$. In other notation (4.63) can be rewritten in the following way:

$$\int_0^\tau f(s, \omega)dW_s = \int_0^\infty \chi_{\{s \le \tau\}} f(s, \omega)dW_s \qquad (P\text{-a.s.}). \tag{4.64}$$

PROOF OF (4.63) (OR (4.64)). For the simple functions $f = \mathcal{M}_\infty$ the equality

$$\int_0^\tau f(s, \omega)dW_s = \int_0^\infty \chi_{\{s \le \tau\}} f(s, \omega)dW_s \tag{4.65}$$

is obvious.

Let $f \in \mathcal{P}_\infty$, and let $f_n \in \mathcal{M}_\infty$, $n = 1, 2, \ldots$, be a sequence of simple functions involved in the construction of the integrals $I_t(f)$, $t \ge 0$. Since (in probability)

$$\int_0^\infty [f_n(s, \omega)\chi_{\{s \le \tau\}} - f(s, \omega)\chi_{\{s \le \tau\}}]^2\, ds$$

$$\le \int_0^\infty [f_n(s, \omega) - f(s, \omega)]^2\, ds \to 0, \qquad n \to \infty,$$

then

$$P\text{-}\lim_{n\to\infty} \int_0^\infty f_n(s, \omega)\chi_{\{s\le\tau\}}\, dW_s = \int_0^\infty f(s, \omega)\chi_{\{s\le\tau\}}\, dW_s. \tag{4.66}$$

Note now that on the set $\{\omega : \tau(\omega) = t\}$,

$$\int_0^\tau f_n(s, \omega) dW_s = \int_0^t f_n(s, \omega) dW_s, \qquad \int_0^\tau f(s, \omega) dW_s = \int_0^t f(s, \omega) dW_s$$

and

$$P\text{-}\lim_{n\to\infty} \int_0^t f_n(s, \omega) dW_s = \int_0^t f(s, \omega) dW_s.$$

Hence on the set $\{\omega : \tau(\omega) = t\}$,

$$P\text{-}\lim_{n\to\infty} \int_0^\tau f_n(s, \omega) dW_s = \int_0^\tau f(s, \omega) dW_s. \tag{4.67}$$

From (4.65)–(4.67) the desired equality, (4.64), follows. □

The following result, which will be used frequently, is a generalization of Lemma 4.6.

Lemma 4.7. *Let $f = f(t, \omega)$, $t \ge 0$, be a nonanticipative (relative to the system $F = (\mathscr{F}_t)$, $t \ge 0$) process. Let $\{\sigma_n, n = 1, 2, \ldots\}$ be a nondecreasing sequence of Markov times, $\sigma = \lim_n \sigma_n$, such that for each $n = 1, 2, \ldots$,*

$$P\left(\int_0^{\sigma_n} f^2(s, \omega) ds < \infty\right) = 1.$$

Then for any event $A \in \mathscr{F}_\sigma$, and $N > 0$, $C > 0$,

$$P\left\{A \cap \left(\sup_n \left| \int_0^{\sigma_n} f(s, \omega) dW_s \right| > C\right)\right\} \le \frac{N}{C^2} + P\left\{A \cap \int_0^\sigma f^2(s, \omega) ds > N\right\}.$$

Proof. Let

$$\tau_N = \begin{cases} \inf\left[t \le \sigma : \int_0^t f^2(s, \omega) ds \ge N\right], \\ \sigma, \quad \text{if } \int_0^\sigma f^2(s, \omega) ds < N, \end{cases}$$

and let $f_N(s, \omega) = f(s, \omega)\chi_{\{s \le \tau_N\}}$. Then, as in the proof of Lemma 4.6, we infer that

$$\begin{aligned} &P\left\{A \cap \left(\sup_n \left| \int_0^{\sigma_n} f(s, \omega) dW_s \right| > C\right)\right\} \\ &\le P\left\{\sup_n \left| \int_0^{\sigma_n} f_N(s, \omega) dW_s \right| > C\right\} \\ &\quad + P\left\{A \cap \left(\sup_n \left| \int_0^{\sigma_n} (f(s, \omega) - f_N(s, \omega)) dW_s \right| > 0\right)\right\} \\ &\le P\left\{\sup_n \left| \int_0^{\sigma_n} f_N(s, \omega) dW_s \right| > C\right\} + P\left\{A \cap \left(\int_0^\sigma f^2(s, \omega) ds > N\right)\right\}. \end{aligned}$$

From Theorem 2.3 and the properties of the stochastic integrals it follows that

$$P\left\{\sup_n \left| \int_0^{\sigma_n} f_N(s, \omega) dW_s \right| > C \right\} \le \frac{1}{C^2} M \int_0^{\sigma} f_N^2(s, \omega) ds \le \frac{N}{C^2};$$

that, together with the previous inequality, proves the lemma. □

Corollary. *Let*

$$A = \left\{\omega: \int_0^{\sigma} f^2(s, \omega) ds < \infty \right\}.$$

Then

$$P\left\{A \cap \left(\sup_n \left| \int_0^{\sigma_n} f(s, \omega) dW_s \right| = \infty \right)\right\} = 0.$$

In other words, on the set A

$$\sup_n \left| \int_0^{\sigma_n} f(s, \omega) dW_s \right| < \infty, \qquad (P\text{-a.s.}).$$

4.2.8

As a corollary of Equation (4.64) we shall now deduce the following formula known as the Wald identity.

Lemma 4.8. *Let $W = (W_t, \mathcal{F}_t)$, $t \ge 0$, be a Wiener process, and let $\tau = \tau(\omega)$ be a Markov time (relative to $(\mathcal{F}_t)$, $t \ge 0$) with $M\tau < \infty$. Then*

$$MW_\tau = 0, \tag{4.68}$$

$$MW_\tau^2 = M\tau. \tag{4.69}$$

PROOF. Consider the nonanticipative function $f(s, \omega) = \chi_{\{s \le \tau(\omega)\}}$. It is clear that

$$P\left\{\int_0^\infty f^2(s, \omega) ds < \infty \right\} = P\left\{\int_0^\infty \chi_{\{s \le \tau(\omega)\}} ds \right\} = P\{\tau < \infty\} = 1,$$

i.e., this function belongs to class $\mathcal{P}_\infty$. We shall show that, for $t \ge 0$,

$$\int_0^t \chi_{\{s \le \tau\}} dW_s = W_{t \wedge \tau} \qquad (P\text{-a.s.}). \tag{4.70}$$

With this purpose introduce for each n, $n = 1, 2, \ldots$, the Markov times

$$\tau_n = \frac{k}{2^n} \quad \text{on} \quad \left\{\omega: \frac{k-1}{2^n} \le \tau(\omega) < \frac{k}{2^n}\right\},$$

$$\tau_n(\omega) = \infty \quad \text{on} \quad \{\omega: \tau(\omega) = \infty\},$$

and consider the integrals

$$\int_0^t \chi_{\{s \le \tau_n\}} \, dW_s = \int_0^\infty \chi_{\{s \le \tau_n \wedge t\}} \, dW_s.$$

If t takes one of the values of the form $k/2^n$, then it is obvious that

$$\int_0^t \chi_{\{s \le \tau_n\}} \, dW_s = \int_0^\infty \chi_{\{s \le \tau_n \wedge t\}} \, dW_s = W_{\tau_n \wedge t}. \tag{4.71}$$

Because of the continuity of the stochastic integrals and the trajectories of the Wiener process in t, Equation (4.71) remains correct also for all $t \ge 0$.

Note now that

$$\int_0^\infty M[\chi_{\{s \le \tau_n\}} - \chi_{\{s \le \tau\}}]^2 \, ds = \int_0^\infty [P(s \le \tau_n) - P(s \le \tau)] ds$$
$$= M\tau_n - M\tau \le \frac{1}{2^n} \to 0, \qquad n \to \infty.$$

Hence

$$\underset{n\to\infty}{\text{l.i.m.}} \int_0^t \chi_{\{s \le \tau_n\}} \, dW_s = \int_0^t \chi_{\{s \le \tau\}} \, dW_s. \tag{4.72}$$

Comparing (4.72) with (4.71) and taking into account that for all $\omega \in \Omega$, $\tau_n(\omega) \downarrow \tau$, we arrive at the desired equation, (4.70).

From (4.70) and (4.64) we find that (P-a.s.)

$$W_\tau = \int_0^\tau \chi_{\{s \le \tau\}} \, dW_s = \int_0^\infty \chi_{\{s \le \tau\}} \, dW_s,$$

since

$$\chi^2_{\{s \le \tau\}} = \chi_{\{s \le \tau\}}.$$

Now make use of the properties given in (4.48) and (4.49), the application of which is valid since, under the conditions of the lemma, $\int_0^\infty M\chi^2_{\{s \le \tau\}} \, ds = M\tau < \infty$. Then

$$MW_\tau = M \int_0^\infty \chi_{\{s \le \tau\}} \, dW_s = 0$$

and

$$MW_\tau^2 = M\left(\int_0^\infty \chi_{\{s \le \tau\}} \, dW_s\right)^2 = \int_0^\infty M\chi^2_{\{s \le \tau\}} \, ds = M\tau. \qquad \square$$

Note 8. The equality $MW_\tau = 0$ remains correct also under the condition $M\sqrt{\tau} < \infty$ (see [130], [132]).

Note 9. The condition $M\tau < \infty$, yielding the equalities $MW_\tau^2 = M\tau$, cannot be weakened, generally speaking, as may be illustrated by an example. Let $\tau = \inf(t \ge 0 : W_t = 1)$. Then $P(\tau < \infty) = 1$, $M\tau = \infty$ (see Subsection 1.4.3) and $1 = MW_\tau^2 \ne M\tau = \infty$.

4.2.9

Let $f = f(t, \omega)$ be an arbitrary nonanticipative function, i.e., such that, generally speaking, $P(\int_0^T f^2(s, \omega)ds = \infty) > 0$.

Set

$$\sigma_n = \inf\left\{t \le T: \int_0^t f^2(s, \omega)ds \ge n\right\},$$

considering $\sigma_n = \infty$ if $\int_0^T f^2(s, \omega)ds < n$, and let $\sigma = \lim_n \sigma_n$. It is clear that on the set $\{\sigma \le T\}$, $\int_0^\sigma f^2(s, \omega)ds = \infty$.

Since

$$P\left(\int_0^{\sigma_n \wedge T} f^2(s, \omega)ds < \infty\right) = 1,$$

we can define the stochastic integrals

$$I_{\sigma_n \wedge T}(f) = \int_0^{\sigma_n \wedge T} f(s, \omega)dW_s = \int_0^T f_n(s, \omega)dW_s,$$

where $f_n(s, \omega) = f(s, \omega)\chi_{\{s \le \sigma_n\}}$. The stochastic integral $I_{\sigma \wedge T}(f)$ is not defined, generally speaking, since $\int_0^\sigma f^2(s, \omega)ds = \infty$ on the set $\{\omega: \sigma \le T\}$ (P-a.s.), and the constructions of the stochastic integrals $I_\sigma(f)$, given above, assumed that $P\{\int_0^\sigma f^2(s, \omega)ds < \infty\} = 1$.

In the case where the condition $P\{\int_0^\sigma f^2(s, \omega)ds < \infty\} = 1$ is violated, one could try to define the integral $I_\sigma(f)$ as the limit (in this or the other sense) of the integrals $I\sigma_n(f)$ with $n \to \infty$. But it is not difficult to give examples where, on the set $\{\sigma \le T\}$, (P-a.s.)

$$\overline{\lim_n}\, I_{\sigma_n}(f) = \infty, \qquad \underline{\lim_n}\, I_{\sigma_n}(f) = -\infty.$$

It is sufficient to assume that $T = \infty$, $f \equiv 1$. Hence $\lim_n I_{\sigma_n}(f)$ does not exist, generally speaking. We can show, however, that there exists[4]

$$P\text{-}\lim_n \chi_{\{\int_0^{\sigma \wedge T} f^2(s, \omega)ds < \infty\}} I_{\sigma \wedge T}(f_n), \tag{4.73}$$

which we shall denote $\Gamma_{\sigma \wedge T}(f)$.

For proving this, note that

$$P\text{-}\lim_n \chi_{\{\int_0^{\sigma \wedge T} f^2(s, \omega)ds < \infty\}} \int_0^{\sigma \wedge T} [f(s, \omega) - f_n(s, \omega)]^2\, ds = 0. \tag{4.74}$$

Denoting

$$\chi_{\sigma \wedge T} = \chi_{\{\int_0^{\sigma \wedge T} f^2(s, \omega)ds > \infty\}},$$

[4] This fact will be used frequently in Chapters 6 and 7.

by Lemma 4.6 (and the note to it) we find that for any $\varepsilon > 0$, $\delta > 0$,

$$P\left\{\chi_{\sigma\wedge T}\left|\int_0^{\sigma\wedge T}(f_n(s,\omega) - f_m(s,\omega))dW_s\right| > \delta\right\}$$
$$\leq \frac{\varepsilon}{\delta^2} + P\left\{\chi_{\sigma\wedge T}\int_0^{\sigma\wedge T}[f_n(s,\omega) - f_m(s,\omega)]^2\,ds > \varepsilon\right\}.$$

From this, because of (4.74), we infer that

$$\lim_{m,n\to\infty} P\left\{\left|\chi_{\sigma\wedge T}\int_0^{\sigma\wedge T} f_n(s,\omega)dW_s - \chi_{\sigma\wedge T}\int_0^{\sigma\wedge T} f_m(s,\omega)dW_s\right| > \delta\right\} = 0. \tag{4.75}$$

Consequently, the sequence of the random variables $\chi_{\sigma\wedge T}\int_0^{\sigma\wedge T} f_n(s,\omega)dW_s$ converges in probability to a certain random variable which is denoted $\Gamma_{\sigma\wedge T}(f)$.

Note that according to our construction, $|\Gamma_{\sigma\wedge T}(f)| < \infty$ (P-a.s.), is valid on the set $\{\omega: \int_0^{\sigma\wedge T} f^2(s,\omega)ds = \infty\}$; and if

$$P\left\{\int_0^{\sigma\wedge T} f^2(s,\omega)ds < \infty\right\} = 1,$$

then

$$\Gamma_{\sigma\wedge T}(f) = I_{\sigma\wedge T}(f) = \int_0^{\sigma\wedge T} f(s,\omega)dW_s.$$

Let now τ be an arbitrary Markov time (not necessarily equal to $\lim_n \sigma_n$, where the σ_n are defined above), and let $\{f_n(s,\omega), n = 1, 2, \ldots\}$ be a sequence of nonanticipative functions such that, for each n, $n = 1, 2, \ldots$,

$$P\left\{\int_0^{\tau\wedge T} f_n^2(s,\omega)ds < \infty\right\} = 1,$$

and approximating the given function f in the sense that

$$P\text{-}\lim_{n\to\infty} \chi_{\{\int_0^{\tau\wedge T} f^2(s,\omega)ds<\infty\}}\int_0^{\tau\wedge T}[f(s,\omega) - f_n(s,\omega)]^2\,ds = 0.$$

The arguments given above in defining the values $\Gamma_{\sigma\wedge T}(f)$ show that in the case considered there also exists

$$P\text{-}\lim_{n\to\infty} \chi_{\{\int_0^{\tau\wedge T} f^2(s,\omega)ds<\infty\}}\int_0^{\tau\wedge T} f_n(s,\omega)dW_s,$$

which we shall denote $\Gamma_{\tau\wedge T}(f)$. It is important to note that, for the given τ and f, this value (to within stochastic equivalence) does not depend on the special form of the approximating sequences $\{f_n(s,\omega), n = 1, 2, \ldots\}$.

Note also that, on the set $\{\omega: \int_0^{T\wedge\tau} f^2(s,\omega)ds < \infty\}$, for the process $\Gamma_t(f)$, considered for $t \leq T \wedge \tau$, there exists (P-a.s.) a continuous modification. Only such modifications will be discussed from now on.

4.2.10

As noted above, the process $(I_t(f), \mathscr{F}_t)$, $t \geq 0$, in the case $f \in \mathscr{P}_\infty$ is, generally speaking, not a martingale. However, this process will be a local martingale.

Actually, let $\tau_n = \inf(t: \int_0^t f^2(s, \omega)ds \geq n) \wedge n$. Then $P(\tau_n \leq n) = 1$, $P(\tau_n \leq \tau_{n+1}) = 1$, and $P\{\lim_n \tau_n = \infty\} = 1$.

Consider for a given n, $n = 1, 2, \dots$, the process

$$I_{t \wedge \tau_n}(f) = \int_0^{t \wedge \tau_n} f(s, \omega) dW_s = \int_0^t f(s, \omega)\chi_{\{s \leq \tau_n\}}\, dW_s.$$

Since

$$\int_0^\infty M[f(s, \omega)\chi_{\{s \leq \tau_n\}}]^2\, ds = \int_0^n M[f(s, \omega)\chi_{\{s \leq \tau_n\}}]^2\, ds \leq n,$$

the process $(I_{t \wedge \tau_n}(f), \mathscr{F}_t)$, $t \geq 0$, is a square integrable martingale. In this case

$$I_{t \wedge \tau_n}(f) = M[I_{\tau_n}(f) | \mathscr{F}_t],$$

where $M|I_{\tau_n}(f)| < \infty$. From this representation it follows that the sequence of the random variables $\{I_{t \wedge \tau_n}(f), t \geq 0\}$ is uniformly integrable (see the proof of Theorem 2.7).

According to Definition 6 in Section 3.3, this proves that the process $(I_t(f), \mathscr{F}_t)$, $t \geq 0$, is a local martingale.

4.2.11

Further on, while considering nonlinear filtering problems, we shall deal with stochastic integrals where integration is carried out not over a Wiener process, but over so-called Ito processes. Let us give some necessary definitions.

Let $(\Omega, \mathscr{F}, P)$ be a probability space, $(\mathscr{F}_t)$, $0 \leq t \leq T$, be a nondecreasing family of sub-σ-algebras, and $W = (W_t, \mathscr{F}_t)$ be a Wiener process.

Definition 6. The continuous random process $\xi = (\xi_t, \mathscr{F}_t)$, $0 \leq t \leq T$, is called an *Ito process* (*relative to the Wiener process* $W = (W_t, \mathscr{F}_t)$, $t \leq T$), if there exist two nonanticipative processes $a = (a_t, \mathscr{F}_t)$ and $b = (b_t, \mathscr{F}_t)$, $0 \leq t \leq T$, such that

$$P\left\{\int_0^T |a_t| dt < \infty\right\} = 1, \tag{4.76}$$

$$P\left\{\int_0^T b_t^2\, dt < \infty\right\} = 1 \tag{4.77}$$

and, with probability 1 for $0 \leq t \leq T$,

$$\xi_t = \xi_0 + \int_0^t a(s, \omega) ds + \int_0^t b(s, \omega) dW_s. \tag{4.78}$$

(For brevity it is said that the process ξ_t has the stochastic differential

$$d\xi_t = a(t, \omega)dt + b(t, \omega)dW_t, \tag{4.79}$$

with (4.79) understood as shorthand for the representation given by (4.78).)

Let now $f = (f(t, \omega), \mathscr{F}_t)$ be a certain nonanticipative function. The stochastic integral $I_t(f) = \int_0^t f(s, \omega)d\xi_s$ of the function $f = f(s, \omega)$ over the process with the differential (4.79) will be understood to be

$$\int_0^t f(s, \omega)a(s, \omega)ds + \int_0^t f(s, \omega)b(s, \omega)dW_s \tag{4.80}$$

under the condition that both of these integrals exist; for which it is sufficient that

$$P\left(\int_0^T |f(s, \omega)a(s, \omega)|ds < \infty\right) = 1,$$

$$P\left(\int_0^T f^2(s, \omega)b^2(s, \omega)ds < \infty\right) = 1.$$

The definition of the integral $\int_0^t f(s, \omega)d\xi_s$ given by (4.80) is not quite convenient, since it does not provide an effective method for calculating $I_t(f)$ immediately over the process $\xi = (\xi_s, \mathscr{F}_s)$, $0 \le s \le t$. It is, however, possible to obtain the integral, defined in this way, as the limit of integral sums of the form

$$I_T(f_n) = \sum_{\{0 \le t \le m, t_{m+1}^{(n)} < T\}} f_n(t_i^{(n)}, \omega)[\xi_{t_{i+1}^{(n)}} - \xi_{t_i^{(n)}}] + f_n(t_{m+1}^{(n)}, \omega)[\xi_T - \xi_{t_{m+1}^{(n)}}] \tag{4.81}$$

(compare with (4.31)), where the $f_n(t, \omega)$ are simple functions approximating $f(t, \omega)$ in the sense that

$$\int_0^T (|a(t, \omega)||f(t, \omega) - f_n(t, \omega)| + b^2(t, \omega))|f(t, \omega) - f_n(t, \omega)|^2)dt \xrightarrow{P} 0, \qquad n \to \infty. \tag{4.82}$$

For the correctness of (4.82) it is sufficient, for example, to require that

$$P\left\{\int_0^T f^2(t, \omega)(|a(t, \omega)| + b^2(t, \omega))dt < \infty\right\} = 1. \tag{4.83}$$

If the condition (4.83) is not satisfied, then take the simple functions $f_n^{(N)}(t, \omega)$, such that for each N, $N = 1, 2, \dots,$

$$\int_0^T [f^{(N)}(t, \omega) - f_n^{(N)}(t, \omega)]^2(|a(t, \omega)| + b^2(t, \omega))dt \xrightarrow{P} 0, \qquad n \to \infty,$$

where

$$f^{(N)}(t, \omega) = \begin{cases} f(t, \omega), & |f(t, \omega)| \le N, \\ 0, & |f(t, \omega)| > N. \end{cases}$$

Then from the sequence $f_n^{(N)}(t, \omega)$ $(n, N = 1, \ldots)$ a subsequence $\tilde{f}_n(t, \omega)$ approximating $f(t, \omega)$ can be chosen in such a way that

$$\int_0^T |f(t, \omega) - \tilde{f}_n(t, \omega)||a(t, \omega)|dt$$
$$+ \int_0^T [f(t, \omega) - \tilde{f}_n(t, \omega)]^2 b^2(t, \omega)dt \xrightarrow{P} 0, \qquad n \to \infty.$$

Proving the existence of the approximizing sequence (under the condition given by (4.83)) and the existence of the limit $P\text{-}\lim_n I_T(f_n)$ is accomplished in the same way as in the case of constructing integrals over a Wiener process. The integrals $I_t(f)$, $0 \le t \le T$, defined by $\int_0^T f(s, \omega)\chi_{\{s \le t\}}\, d\xi_s$, form, as in the case of integration over a Wiener process, a continuous random process (P-a.s.).

4.2.12

The important particular case of Ito processes is processes of the diffusion type.

Definition 7. The Ito process $\xi = (\xi_t, \mathscr{F}_t)$, $0 \le t \le T$, is called *a process of the diffusion type* (*relative to the Wiener process* $W = (W_t, \mathscr{F}_t)$, $0 \le t \le T$), if the functionals $a(s, \omega)$ and $b(s, \omega)$ being a part of (4.78) are $\mathscr{F}_s^\xi$-measurable for almost all s, $0 \le s \le T$.

Denote by $(C_T, \mathscr{B}_T)$ the measure space of functions $x = (x_t)$, $0 \le t \le T$, continuous on $[0, T]$ with σ-algebra $\mathscr{B}_T = \sigma\{x : x_t, t \le T\}$. Let $\mathscr{B}_t = \sigma\{x : x_s, s \le t\}$, and let $\mathscr{B}_{[0,t]}$ be the smallest σ-algebra of the sets on $[0, T]$ containing all the Borel subsets of the interval $[0, t]$.

Lemma 4.9, given below, shows that if ξ is a process of the diffusion type with the coefficients $a(s, \omega)$ and $b(s, \omega)$ then there will be the jointly measurable (s, x) functionals $A(s, x)$ and $B(s, x)$, which are $\mathscr{B}_{s+}$-measurable with each s, such that, for almost all $0 \le s \le T$,

$$A(s, \xi(\omega)) = a(s, \omega), \qquad B(s, \xi(\omega)) = b(s, \omega). \qquad (P\text{-a.s.}).$$

From this it follows that, for the processes of the diffusion type, along with the equalities

$$\xi_t = \xi_0 + \int_0^t a(s, \omega)ds + \int_0^t b(s, \omega)dW_s \qquad (P\text{-a.s.}), \qquad 0 \le t < T,$$

the equalities

$$\xi_t = \xi_0 + \int_0^t A(s, \xi)ds + \int_0^t B(s, \xi)dW_s \qquad (P\text{-a.s.}), \qquad 0 \le t \le T,$$

also hold, where the (measurable) functionals $A(s, x)$ and $B(s, x)$ are $\mathscr{B}_{s+}$-measurable for each s, $0 \le s \le T$, $\mathscr{B}_{T+} = \mathscr{B}_T$.

Lemma 4.9. *Let $\xi = (\xi_t)$, $0 \le t \le T$, be a continuous random process defined on the complete probability space $(\Omega, \mathscr{F}, P)$. Let next the measurable process $\zeta = (\zeta_t)$, $0 \le t \le T$, be adapted to the family of the σ-algebras $F^\xi = (\mathscr{F}_t^\xi)$. Then there exists a measurable functional $\varphi = \varphi(t, x)$ defined on $([0, T] \times C_T, \mathscr{B}_{[0,T]} \times \mathscr{B}_T)$ which is $\mathscr{B}_{t+}$-measurable for each $0 \le t \le T$, and such that*

$$\lambda \times P\{(t, \omega): \zeta(\omega) \neq \varphi(t, \xi(\omega))\} = 0,$$

where λ is the Lebesgue measure on $[0, T]$ and $\lambda \times P$ is the direct product of the measures λ and P.

Proof. Since the process $\zeta = (\zeta_t)$, $0 \le t \le T$, is measurable and adapted to F^ξ, then (see Section 1.2) it has a progressively measurable modification. We shall assume that the process $\zeta = (\zeta_t)$, $0 \le t \le T$, itself has this property. Then for each u, $0 \le u \le T$, the function $\zeta_{t \wedge u}(\omega)$, considered as a function of (t, ω), where $0 \le t \le T$, $\omega \in \Omega$, is measurable relative to $\lambda \times P$ (augmentation of the algebra $\mathscr{B}_{[0,u]} \times \mathscr{F}_u^\xi$). Hence for each u, $0 \le u \le T$, on $([0, T] \times C_T, \mathscr{B}_{[0,u]} \times \mathscr{B}_u)$ there exists a measurable functional $\varphi_u(t, x)$ such that

$$\lambda \times P\{(t, \omega): \zeta_{t \wedge u}(\omega) \neq \varphi_u(t, \xi(\omega))\} = 0.$$

Let $u_{k,n} = T/2^n \cdot k$, $k = 1, 2, \dots, 2^n$, $n = 1, 2, \dots$. Assume

$$\varphi^{(n)}(t, x) = \varphi_0(0, x)\chi_{\{0\}}(t) + \sum_{k=1}^{2^n} \varphi_{u_{k,n}}(t, x)\chi_{(u_{k-1,n}, u_{kn}]}(t)$$

and

$$\varphi(t, x) = \overline{\lim_n}\, \varphi^{(n)}(t, x).$$

The functionals $\varphi^{(n)}(t, x)$ are measurable over (t, x) for each n, and therefore the functional $\varphi(t, x)$ is also measurable. From the constructions of the functionals $\varphi^{(n)}(t, x)$, $n = 1, 2, \dots$, it is also seen that $\varphi(t, x)$ at each t are $\mathscr{B}_{t+}$-measurable. Further by the equation $\lambda \times P\{(t, \omega): \zeta(\omega)_{t \wedge u} \neq \varphi_u(t, \xi(\omega))\} = 0$ and the definition of functionals $\varphi^{(n)}(t, x)$, it follows that for any number $n = 1, 2, \dots$ $\varphi^{(n)}(t, \xi(\omega))$ coincides with $\zeta_t(\omega)$ on the (t, ω)-set of full $\lambda \times P$-measure. Hence $\lambda \times P\{(t, \omega): \zeta_t(\omega) \neq \varphi(t, \xi(\omega))\} = 0$. □

4.2.13

Let $A = (A(t, x), \mathscr{B}_{t+})$, $\tilde{A} = (\tilde{A}(t, x), \mathscr{B}_{t+})$, $B = (B(t, x), \mathscr{B}_{t+})$, $\tilde{B} = (\tilde{B}(t, x), \mathscr{B}_{t+})$ be nonanticipative functionals and $\xi = (\xi_t, \mathscr{F}_t)$, $\tilde{\xi} = (\tilde{\xi}_t, \mathscr{F}_t)$, $0 \le t \le T$, be processes of the diffusion type with

$$d\xi_t = A(t, \xi)dt + B(t, \xi)dW_t,$$

$$d\tilde{\xi}_t = \tilde{A}(t, \tilde{\xi})dt + \tilde{B}(t, \tilde{\xi})dW_t.$$

The functionals A, $\tilde{A}$, B, $\tilde{B}$ are assumed to be such that (P-a.s.)

$$\int_0^T [|A(t, \xi)| + |\tilde{A}(t, \tilde{\xi})| + B^2(t, \xi) + \tilde{B}^2(t, \tilde{\xi})]dt < \infty.$$

(Note that for each s the values $B(s, \xi)$ and $\tilde{B}(s, \tilde{\xi})$ are $\mathscr{F}_{s+}$-measurable and the existence of the stochastic integrals $\int_0^t B(s, \xi)dW_s$, $\int_0^t B(s, \tilde{\xi})dW_s$ follows from the previous inequality and the fact that the process $W_t = (W_t, \mathscr{F}_{t+})$, as well as $W = (W_t, \mathscr{F}_t)$, is also a Wiener process).

Let now $g = (g(t, x), \mathscr{B}_{t+})$, $0 \le t \le T$, be a nonanticipative functional with

$$P\left(\int_0^T |g(t, \xi)|\, dt < \infty\right) = P\left(\int_0^T |g(t, \tilde{\xi})|\, dt < \infty\right) = 1.$$

Consider the (Lebesgue) integrals

$$\int_0^T g(t, \xi)dt, \qquad \int_0^T g(t, \tilde{\xi})dt.$$

Since they are $\mathscr{F}_T^\xi$- and $\mathscr{F}_T^{\tilde{\xi}}$-measurable respectively, then there are $\mathscr{B}_T$-measurable functionals $\psi(x)$ and $\tilde{\psi}(x)$ such that (P-a.s.)

$$\psi(\xi) = \int_0^T g(t, \xi)dt, \qquad \tilde{\psi}(\tilde{\xi}) = \int_0^T g(t, \tilde{\xi})dt.$$

These equalities can determine the functionals $\psi(x)$ and $\tilde{\psi}(x)$ in different ways. Hence, generally speaking,

$$P\{\psi(\tilde{\xi}) \neq \tilde{\psi}(\tilde{\xi})\} \geqslant 0, \qquad P\{\psi(\xi) \neq \tilde{\psi}(\xi)\} \geqslant 0.$$

Consider now the stochastic integrals

$$\int_0^T f(t, \xi)d\xi_t, \qquad \int_0^T f(t, \tilde{\xi})d\tilde{\xi}_t,$$

for the existence of which we suppose that for μ_ξ and $\mu_{\tilde{\xi}}$ almost surely[5]

$$\int_0^T [|f(t, x)|(|A(t, x)| + |\tilde{A}(t, x)|) + f^2(t, x)(B^2(t, x) + \tilde{B}^2(t, x))]dt < \infty.$$

The stochastic integrals

$$\int_0^T f(t, \xi)d\xi_t, \qquad \int_0^T f(t, \tilde{\xi})d\tilde{\xi}_t$$

are $\mathscr{F}_T^\xi$- and $\mathscr{F}_T^{\tilde{\xi}}$-measurable respectively. Hence we can find $\mathscr{B}_T$-measurable functionals $\Phi(x)$ and $\tilde{\Phi}(x)$ such that (P-a.s.)

$$\Phi(\xi) = \int_0^T f(t, \xi)d\xi_t, \qquad \tilde{\Phi}(\tilde{\xi}) = \int_0^T f(t, \tilde{\xi})d\tilde{\xi}_t.$$

[5] μ_ξ and $\mu_{\tilde{\xi}}$ are measures in the space $(C_T, \mathscr{B}_T)$, corresponding to the processes ξ and $\tilde{\xi}$ respectively.

For the functionals $\Phi(x)$ and $\tilde{\Phi}(x)$ the following equalities: $\Phi(\tilde{\xi}) = \tilde{\Phi}(\tilde{\xi})$, $\Phi(\xi) = \tilde{\Phi}(\xi)$ are not necessarily correct.

Actually, let $f(t, x) = x_t$, $\xi_t = W_t$, $\tilde{\xi}_t = 2W_t$. Then

$$\int_0^T W_t\, dW_t = \frac{W_T^2}{2} - \frac{T}{2}, \qquad \int_0^T (2W_t) d(2W_t) = \frac{(2W_T)^2}{2} - 2T.$$

Therefore,

$$\Phi(x) = \frac{x_T^2}{2} - \frac{T}{2}, \qquad \tilde{\Phi}(x) = \frac{x_T^2}{2} - 2T$$

and

$$P(\Phi(\tilde{\xi}) > \tilde{\Phi}(\tilde{\xi})) = 1.$$

Note that in this example the measures μ_ξ and $\mu_{\tilde{\xi}}$ are singular. Hence, it is natural to expect that the equality (P-a.s.) of the functionals $\Phi(\tilde{\xi})$ and $\tilde{\Phi}(\tilde{\xi})$, $\Phi(\xi)$ and $\tilde{\Phi}(\xi)$, and also of the functionals $\psi(\tilde{\xi})$ and $\tilde{\psi}(\tilde{\xi})$, $\psi(\xi)$ and $\tilde{\psi}(\xi)$, depends on the absolute continuity properties of the measures μ_ξ and $\mu_{\tilde{\xi}}$.

Lemma 4.10

(1) *If the measure μ_ξ is absolutely continuous with respect to the measure $\mu_{\tilde{\xi}}(\mu_\xi \ll \mu_{\tilde{\xi}})$, then $\psi(\xi) = \tilde{\psi}(\xi)$, $\Phi(\xi) = \tilde{\Phi}(\xi)$ (P-a.s.).*

(2) *If $\mu_{\tilde{\xi}} \ll \mu_\xi$, then $\psi(\tilde{\xi}) = \tilde{\psi}(\tilde{\xi})$, $\Phi(\tilde{\xi}) = \tilde{\Phi}(\tilde{\xi})$ (P-a.s.)*

PROOF. Let us establish the correctness of the equality $\psi(\xi) = \tilde{\psi}(\xi)$. Let $g_n = (g_n(t, x), \mathscr{B}_{t+})$, $0 \le t \le T$, $n = 1, 2, \ldots$, be a sequence of (simple) functionals such that

$$P\text{-}\lim_{n\to\infty} \int_0^T g_n(t, \tilde{\xi})dt = \int_0^T g(t, \tilde{\xi})dt.$$

Then the functional

$$\tilde{\psi}(x) = \mu_{\tilde{\xi}}\text{-}\lim_{n\to\infty} \int_0^T g_n(t, x)dt.$$

Because of the absolute continuity of $\mu_\xi \ll \mu_{\tilde{\xi}}$,

$$\tilde{\psi}(x) = \mu_\xi\text{-}\lim_{n\to\infty} \int_0^T g_n(t, x)dt.$$

Hence we infer that

$$\tilde{\psi}(\xi) = P\text{-}\lim_{n\to\infty} \int_0^T g_n(t, \xi)dt = \psi(\xi) \qquad (P\text{-a.s.}).$$

For proof of the equality $\Phi(\xi) = \tilde{\Phi}(\xi)$ we consider the density (a Radon–Nikodym derivative)

$$\varkappa(x) = \frac{d\mu_\xi}{d\mu_{\tilde{\xi}}}(x)$$

of the measure μ_{ξ} with respect to the measure $\mu_{\tilde{\xi}}$. On the original probability space $(\Omega, \mathscr{F})$ introduce the new probability measure $\tilde{P}$, and set $\tilde{P}(d\omega) = \varkappa(\tilde{\xi}(\omega))P(d\omega)$. Then, if $\Gamma \in \mathscr{B}_T$,

$$\tilde{P}\{\tilde{\xi} \in \Gamma\} = \int_{\{\omega:\tilde{\xi}(\omega)\in\Gamma\}} \varkappa(\tilde{\xi}(\omega))P(d\omega) = \int_{\Gamma} \varkappa(x)d\mu_{\tilde{\xi}}(x) = \mu_{\xi}(\Gamma) = P\{\xi \in \Gamma\}.$$

Let now $f_n = (f_n(t, x), \mathscr{B}_{t+})$, $0 \le t \le T$, $n = 1, 2, \ldots$, be a sequence of (simple) functionals such that (P-a.s.)

$$\lim_n \int_0^T \{[B^2(t, \tilde{\xi}) + \tilde{B}^2(t, \tilde{\xi})][f(t, \tilde{\xi}) - f_n(t, \tilde{\xi})]^2$$
$$+ (|A(t, \tilde{\xi})| + |\tilde{A}(t, \tilde{\xi})|)(|f(t, \tilde{\xi}) - f_n(t, \tilde{\xi})|)\}dt = 0.$$

Then, since $\tilde{P} \ll P$, this limit is also equal to zero ($\tilde{P}$-a.s.), from which (because of the equality $\tilde{P}\{\tilde{\xi} \in \Gamma\} = P\{\xi \in \Gamma\}$) it follows that

$$P\text{-}\lim_n \int_0^T \{[B^2(t, \xi) + \tilde{B}^2(t, \xi)][f(t, \xi) - f_n(t, \xi)]^2$$
$$+ (|A(t, \xi)| + |\tilde{A}(t, \xi)|)(|f(t, \xi) - f_n(t, \xi)|\})dt = 0.$$

Therefore (see Subsection 4.2.11),

$$P\text{-}\lim_n \int_0^T f_n(t, \xi)d\xi_t = \int_0^T f(t, \xi)d\xi_t = \Phi(\xi),$$

$$\tilde{P}\text{-}\lim_n \int_0^T f_n(t, \tilde{\xi})d\tilde{\xi}_t - \int_0^T f(t, \tilde{\xi})d\tilde{\xi}_t - \tilde{\Phi}(\tilde{\xi}).$$

Because of the definition of stochastic integrals of simple functions and the equality $\tilde{P}\{\tilde{\xi} \in \Gamma\} = P\{\xi \in \Gamma\}$, $\Gamma \in \mathscr{B}_T$,

$$\lim_n P\left\{\left|\tilde{\Phi}(\xi) - \int_0^T f_n(t, \xi)d\xi_t\right| > \varepsilon\right\}$$
$$= \lim_n \tilde{P}\left\{\left|\tilde{\Phi}(\tilde{\xi}) - \int_0^T f_n(t, \tilde{\xi})d\tilde{\xi}_t\right| > \varepsilon\right\} = 0.$$

Then

$$P\{|\tilde{\Phi}(\xi) - \Phi(\xi)| > \varepsilon\} \le P\left\{\left|\tilde{\Phi}(\xi) - \int_0^T f_n(t, \xi)d\xi_t\right| > \frac{\varepsilon}{2}\right\}$$
$$+ P\left\{\left|\Phi(\xi) - \int_0^T f_n(t, \xi)d\xi_t\right| > \frac{\varepsilon}{2}\right\} \to 0, \qquad n \to \infty,$$

Consequently, (P-a.s.) $\tilde{\Phi}(\xi) = \Phi(\xi)$, proving the first statement of the lemma. Similarly the correctness of the second statement is established. □

4.3 Ito's formula

4.3.1

Let $\xi = (\xi_t, \mathscr{F}_t)$, $0 \le t \le T$, be a random process having the stochastic differential

$$d\xi_t = a(t, \omega)dt + b(t, \omega)dW_t, \tag{4.84}$$

where $W = (W_t, \mathscr{F}_t)$ is a Wiener process, and the nonanticipative functions $a(t, \omega)$, $b(t, \omega)$ are such that

$$P\left\{\int_0^T |a(t, \omega)|dt < \infty\right\} = 1, \tag{4.85}$$

$$P\left\{\int_0^T b^2(t, \omega)dt < \infty\right\} = 1. \tag{4.86}$$

Let now $f = f(t, x)$ be a measurable function defined on $[0, T] \times \mathbb{R}^1$. The theorem given below states the conditions under which the random process $f(t, \xi_t)$ also permits a stochastic differential.

Theorem 4.4. *Let the function $f(t, x)$ be continuous and have the continuous partial derivatives $f'_t(t, x)$, $f'_x(t, x)$ and $f''_{xx}(t, x)$. Assume that the random process $\xi = (\xi_t, \mathscr{F}_t)$, $0 \le t \le T$, has the stochastic differential given by* (4.84). *Then the process $f(t, \xi_t)$ also has a stochastic differential and*

$$df(t, \xi_t) = [f'_t(t, \xi_t) + f'_t(t, \xi_t)a(t, \omega) + \tfrac{1}{2}f''_{xx}(t, \xi_t)b^2(t, \omega)]dt$$
$$+ f'_x(t, \xi_t)b(t, \omega)dW_t. \tag{4.87}$$

The formula given by (4.87), *obtained by K. Ito, will be called from now on the Ito formula.*

Proof. First of all let us show that for proving the Ito formula it is sufficient to restrict oneself to considering only simple functions $a(s, \omega)$ and $b(s, \omega)$. Actually, let $a_n(s, \omega)$, $b_n(s, \omega)$, $n = 1, 2, \ldots$, be sequences of simple functions such that with probability 1

$$\int_0^T |a(s, \omega) - a_n(s, \omega)|ds \to 0, \qquad n \to \infty,$$

$$\int_0^T [b(s, \omega) - b_n(s, \omega)]^2 ds \to 0, \qquad n \to \infty$$

(see Lemma 4.5 and Note 4 to it). According to Note 6 (see Section 4.2) the sequence $\{b_n(s, \omega), n = 1, 2, \dots\}$ may be chosen so that uniformly over $t \leq T$ with probability 1

$$\int_0^t b_n(s, \omega) dW_s \to \int_0^t b(s, \omega) dW_s.$$

Then the sequence of processes

$$\xi_t^n = \xi_0 + \int_0^t a_n(s, \omega) ds + \int_0^t b_n(s, \omega) dW_s$$

with probability 1 uniformly over t, $0 \leq t \leq T$, converges to the process ξ_t.

Assume now that the formula given by (4.87) is established for the processes $\xi_t^{(n)}$. In other words, for $0 \leq s \leq T$, let

$$f(s, \xi_s^{(n)}) = f(0, \xi_0) + \int_0^s [f_t'(t, \xi_t^{(n)}) + f_x'(t, \xi_t^{(n)}) a_n(t, \omega) + \tfrac{1}{2} f_{xx}''(t, \xi_t^{(n)}) b_n^2(t, \omega)] dt$$
$$+ \int_0^s f_x'(t, \xi_t^{(n)}) b_n(t, \omega) dW_t \qquad (P\text{-a.s.}). \tag{4.88}$$

Then, since $\sup_{0 \leq t \leq T} |\xi_t^{(n)} - \xi_t| \to 0$, $n \to \infty$, with probability 1, and the functions f, f_t', f_x', f_{xx}'' are continuous, taking the passage to the limit in (4.88) we infer that

$$f(s, \xi_s) = f(0, \xi_0) + \int_0^s [f_t'(t, \xi_t) + f_x'(t, \xi_t) a(t, \omega) + \tfrac{1}{2} f_{xx}''(t, \xi_t) b^2(t, \omega)] dt$$
$$+ \int_0^s f_x'(t, \xi_t) b(t, \omega) dW_t. \tag{4.89}$$

(The stochastic integrals $\int_0^s f_x'(t, \xi_t^{(n)}) b_n(t, \omega) dW_t \to \int_0^s f_x'(t, \xi_t) b(t, \omega) dW_t$ as $n \to \infty$ because of Note 6 from the previous section).

Thus it is sufficient to prove the formula given by (4.89) assuming that the functions $a(t, \omega)$ and $b(t, \omega)$ are simple. In this case, because of the additivity of stochastic integrals, it is sufficient to consider only $t \geq 0$ such that

$$\xi_t = \xi_0 + at + bW_t, \tag{4.90}$$

where $a = a(\omega)$, $b = b(\omega)$ are certain random variables (independent of t).

Let the representation (4.90) be satisfied for $t \leq t_0$, and for simplicity let $\xi_0 = 0$. Then obviously there will be a function $u(t, x)$ of the same degree of smoothness as that of $f(t, x)$ such that

$$u(t, W_t) = f(t, at + bW_t), \qquad t \leq t_0.$$

Hence it is sufficient to establish the Ito formula only for the function $u = u(t, W_t)$, $t \leq t_0$.

Assume $l = [2^{-n}t]$, $\Delta W = W_{k\cdot 2^{-n}} - W_{(k-1)\cdot 2^{-n}}$, $\Delta = 1/2^n$, $n = 1, 2, \ldots$. Then, by the Taylor formula after a number of transformations, we find that

$$
\begin{aligned}
&u(t, W_t) - u(0, 0)\\
&\quad = \sum_{k\leq 1} [u(k\cdot 2^{-n}, W_{k\cdot 2^{-n}}) - u((k-1)\cdot 2^{-n}, W_{k\cdot 2^{-n}})]\\
&\qquad + \sum_{k\leq l}[u((k-1)\cdot 2^{-n}, W_{k\cdot 2^{-n}}) - u((k-1)\cdot 2^{-n}, W_{(k-1)\cdot 2^{-n}})]\\
&\qquad + [u(t, W_t) - u(l\cdot 2^{-n}, W_{l\cdot 2^{-n}})]\\
&\quad = \sum_{k\leq l}[u_t'((k-1)\cdot 2^{-n}, W_{k\cdot 2^{-n}})\Delta + \{u_t'(((k-1)+\theta_k)\cdot 2^{-n}, W_{k\cdot 2^{-n}})\\
&\qquad - u_t'((k-1)\cdot 2^{-n}, W_{k\cdot 2^{-n}})\}\Delta] + \sum_{k\leq l}[u_x'((k-1)\cdot 2^{-n}, W_{(k-1)\cdot 2^{-n}})\Delta W\\
&\qquad + \tfrac{1}{2}u_{xx}''((k-1)\cdot 2^{-n}, W_{(k-1)\cdot 2^{-n}})(\Delta W)^2\\
&\qquad + \tfrac{1}{2}(\Delta W)^2\{u_{xx}''((k-1)\cdot 2^{-n}, W_{(k-1)\cdot 2^{-n}} + \theta_k'\Delta W)\\
&\qquad - u_{xx}''((k-1)\cdot 2^{-n}, W_{(k-1)\cdot 2^{-n}})\}] + \delta_n(\omega),
\end{aligned}
\tag{4.91}
$$

where θ_k, θ_k' are random variables such that $0 \leq \theta_k \leq 1$, $0 \leq \theta_k' \leq 1$, and $\lim_n \delta_n(\omega) = 0$ (P-a.s.).

Note now that the random variables

$$a_n = \sup_{k\leq l} |u_t'(((k-1)+\theta_k)\cdot 2^{-n}, W_{k\cdot 2^{-n}}) - u_t'((k-1)\cdot 2^{-n}, W_{k\cdot 2^{-n}})|$$

and

$$
\begin{aligned}
\beta_n = \sup_{k\leq l} |&u_{xx}''((k-1)\cdot 2^{-n}, W_{(k-1)\cdot 2^{-n}} + \theta_k'\Delta W)\\
&- u_{xx}''((k-1)\cdot 2^{-n}, W_{(k-1)\cdot 2^{-n}})|
\end{aligned}
$$

converge with $n \to \infty$ to zero with probability 1 because of the continuity of the Wiener process and the continuity of the derivatives u_t', u_{xx}''. Hence

$$
\begin{aligned}
u(t, W_t) - u(0, 0) = &\sum_{k\leq l} u_t'((k-1)\cdot 2^{-n}, W_{k\cdot 2^{-n}})\Delta\\
&+ \sum_{k\leq l}(u_x'((k-1)\cdot 2^{-n}, W_{(k-1)\cdot 2^{-n}})\Delta W\\
&+ \tfrac{1}{2}u_{xx}''((k-1)\cdot 2^{-n}, W_{(k-1)\cdot 2^{-n}})\Delta)\\
&+ A_n + B_n + C_n + \delta_n(\omega),
\end{aligned}
\tag{4.92}
$$

where

$$A_n \leq a_n \cdot t, \qquad B_n \leq \frac{1}{2}\beta_n \sum_{k\leq l}(\Delta W)^2,$$

$$C_n = \frac{1}{2}\sum_{k\leq l} u_{xx}''((k-1)\cdot 2^{-n}, W_{(k-1)\cdot 2^{-n}})((\Delta W)^2 - \Delta).$$

It is clear that with probability 1 $A_n \to 0$, $B_n \to 0$, since with probability 1 $\sum_{k \le l} (\Delta W)^2 \to t$ (Lemma 4.3). Let us show that $C_n \to 0$ (in probability) as $n \to \infty$.

Let

$$\chi_k^N = \chi_{\left\{\max\limits_{i \le k} |W_{i \cdot 2^{-n}}| \le N\right\}}.$$

Then

$$M\left[\sum_{k \le l} u''_{xx}((k-1)\cdot 2^{-n}, W_{(k-1)\cdot 2^{-n}})\chi_k^N((\Delta W)^2 - \Delta)\right]^2$$
$$\le \sup_{t \le t_0, |x| \le N} |u''_{xx}(t, x)|^2 \sum_{k \le l} M((\Delta W)^2 - \Delta)^2$$
$$= 2 \sup_{t \le t_0, |x| \le N} |u''_{xx}(t, x)|^2 \sum_{k \le l} (\Delta)^2 \to 0, \qquad n \to \infty. \tag{4.93}$$

Further

$$P\left\{\sum_{k \le l} u''_{xx}((k-1)\cdot 2^{-n}, W_{(k-1)\cdot 2^{-n}})(1 - \chi_k^N)((\Delta W)^2 - \Delta) \ne 0\right\}$$
$$\le P\left\{\sup_{t \le t_0} |W_t| > N\right\} \to 0, \qquad N \to \infty. \tag{4.94}$$

From (4.93) and (4.94) it follows that $P\text{-}\lim_n C_n = 0$. Passing now in (4.92) to the limit as $n \to \infty$, we obtain that (P-a.s.) for all t, $0 \le t \le t_0$,

$$u(t, W_t) - u(0, 0) = \int_0^t u'_t(s, W_s)ds + \int_0^t u'_x(s, W_s)dW_s + \frac{1}{2}\int_0^t u''_{xx}(s, W_s)ds. \tag{4.95}$$

For passing from the function $u(t, W_t)$ to the function $f(t, \xi_t)$, remember that $u(t, W_t) = f(t, at + bW_t)$. Hence

$$u'_t(s, W_s) = f'_s(s, \xi_s) + af'_x(s, \xi_s),$$
$$u'_x(s, W_s) = bf'_x(s, \xi_s),$$
$$u''_{xx}(s, W_s) = b^2 f''_{xx}(s, \xi_s).$$

Substituting these values in (4.95), we obtain the desired result:

$$f(t, \xi_t) = f(0, 0) + \int_0^t [f'_t(s, \xi_s) + af'_x(s, \xi_s) + \tfrac{1}{2}b^2 f''_{xx}(s, \xi_s)]ds$$
$$+ \int_0^t bf'_x(s, \xi_s)dW_s. \qquad \square \tag{4.96}$$

Note. The Ito formula, (4.87), holds true with substitution of t for the Markov time $\tau = \tau(\omega)$ (with respect to $(\mathscr{F}_t)$, $t \ge 0$) if only $P(\tau < \infty) = 1$ and

$$P\left(\int_0^\tau |a(s, \omega)|ds < \infty\right) = 1, \qquad P\left(\int_0^\tau b^2(s, \omega)ds < \infty\right) = 1.$$

4.3.2

We give now a multi-dimensional variant of Ito's formula.

Let $\xi = (\xi_t, \mathscr{F}_t), t \leq T$, be a vectorial random process $\xi_t = (\xi_1(t), \ldots, \xi_m(t))$, having the stochastic differential

$$d\xi_t = a(t, \omega)dt + b(t, \omega)dW_t, \tag{4.97}$$

where $W = (W_t, \mathscr{F}_t)$, $t \geq 0$, is a (vector) Wiener process[6], $W_t = (W_1(t), \ldots, W_m(t))$. The vector $a(t, \omega) = (a_1(t, \omega), \ldots, a_m(t, \omega))$ and matrix $b(t, \omega) = \|b_{ij}(t, \omega)\|$, $i, j = 1, \ldots, m$, consist of nonanticipative functions satisfying the conditions

$$P\left(\int_0^T |a_i(t, \omega)|dt < \infty\right) = 1, \qquad i = 1, \ldots, m,$$

$$P\left(\int_0^T b_{ij}^2(t, \omega)dt < \infty\right) = 1, \qquad i, j = 1, \ldots, m.$$

In the more complete form (4.97) is written as follows:

$$d\xi_i(t) = a_i(t, \omega)dt + \sum_{j=1}^m b_{ij}(t, \omega)dW_j(t), \qquad i = 1, \ldots, m.$$

Theorem 4.5. *Let the function $f(t, x_1, \ldots, x_m)$ be continuous and have the continuous derivatives f_t', f_{x_i}', $f_{x_ix_j}''$. Then the process $f(t, \xi_1(t), \ldots, \xi_m(t))$ has the stochastic differential*

$$\begin{aligned} &df(t, \xi_1(t), \ldots, \xi_m(t)) \\ &= \Bigg[f_t'(t, \xi_1(t), \ldots, \xi_m(t)) + \sum_{i=1}^m f_{x_i}'(t, \xi_1(t), \ldots, \xi_m(t))a_i(t, \omega) \\ &\quad + \frac{1}{2}\sum_{i,j=1}^m f_{x_ix_j}''(t, \xi_1(t), \ldots, \xi_m(t))\sum_{k=1}^m b_{ik}(t, \omega)b_{jk}(t, \omega)\Bigg]dt \\ &\quad + \sum_{i,j=1}^m f_{x_i}'(t, \xi_1(t), \ldots, \xi_m(t))b_{ij}(t, \omega))b_{ij}(t, \omega)dW_j(t). \end{aligned} \tag{4.98}$$

This theorem is proved in the same way as in the case $m = 1$.

4.3.3

Let us consider now a number of examples illustrating the use of Ito's formula, (4.98).

Example 1. Let $X_i = (x_i(t), \mathscr{F}_t)$, $i = 1, 2$, be two random processes with the differentials

$$dx_i(t) = a_i(t, \omega)dt + b_i(t, \omega)dW_t.$$

[6] That is, a vector process the components of which are independent Wiener processes.

It is assumed that $x_1(t) = (x_{11}(t), \ldots, x_{1n}(t))$, $x_2(t) = (x_{21}(t), \ldots, x_{2m}(t))$, $a_1(t) = a_{11}(t), \ldots, a_{1n}(t))$, $a_2(t) = (a_{21}(t), \ldots, a_{2m}(t))$, are all vector functions, the matrices $b_1(t) = \|b_{ij}^1(t)\|$, $b_2(t) = \|b_{ij}^2(t)\|$ have respectively the order $n \times k$, $m \times k$, and the Wiener process $W = (W_t, \mathscr{F}_t)$ has k independent components.

Consider the matrix $Y(t) = x_1(t)x_2^*(t)$. Applying Ito's formula to the elements of the matrix $Y(t)$, we find that

$$dY(t) = [x_1(t)a_2^*(t) + a_1(t)x_2^*(t) + b_1(t)b_2^*(t)]dt + b_1(t)dW_t x_2^*(t) + x_1(t)dW_t^* b_2^*(t). \tag{4.99}$$

In particular, if $n = m = k = 1$,

$$d(x_1(t)x_2(t)) = [x_1(t)a_2(t) + a_1(t)x_2(t) + b_1(t)b_2(t)]dt + [b_1(t)x_2(t) + x_1(t)b_2(t)]dW_t. \tag{4.100}$$

EXAMPLE 2. Let the function $f(t, x_1, \ldots, x_m) = (x, B(t)x)$ where $x = (x_1, \ldots, x_m)$, and $B(t)$ is a matrix (nonrandom) of the order $m \times m$ with differentiable elements. Let $X = (x_t, \mathscr{F}_t)$ be a process with the differential

$$dx_t = a(t)dt + b(t)dW_t,$$

where $x_t = (x_1(t), \ldots, x_m(t))$, $W_t = (W_1(t), \ldots, W_m(t))$ is a Wiener process.

Let us find the differential of the process $(x_t, B(t)x_t)$. Applying the formula given by (4.98) to $y_t = B(t)x_t$, we find

$$dy_t = [\dot{B}(t)x_t + B(t)a(t)]dt + B(t)b(t)dW_t.$$

For computing the differential $d(x_t, B(t)x_t)$ we make use of the formula given by (4.99), according to which

$$d(x_t y_t^*) = [a(t)y_t^* + x_t x_t^* \dot{B}^*(t) + x_t a^*(t)B^*(t) + b(t)b^*(t)B(t)]dt + x_t dW_t^* b^*(t)B^*(t) + b(t)dW_t x_t^* B^*(t).$$

Then

$$\begin{aligned} d(x_t, B(t)x_t) &= \operatorname{Sp} d(x_t y_t^*) \\ &= [\operatorname{Sp} a(t)\, x_t^* B^*(t) + \operatorname{Sp} x_t x_t^* \dot{B}^*(t) + \operatorname{Sp} x_t a^*(t)B^*(t) \\ &\quad + \operatorname{Sp} b(t)b^*(t)B(t)]dt + \operatorname{Sp} x_t\, dW_t^* b^*(t)B^*(t) + \operatorname{Sp} b(t)dW_t x_t^* B^*(t) \\ &= [(x_t, B^*(t)a(t)) + (x_t, B(t)a(t)) + (x_t, \dot{B}(t)x_t) + \operatorname{Sp} b(t)b^*(t)B(t)]dt \\ &\quad + (b^*(t)B^*(t)x_t, dW_t) + (b^*(t)B(t)x_t, dW_t). \end{aligned}$$

Thus

$$d(x_t, B(t)x_t) = \{(x_t, \dot{B}(t)x_t) + (x_t, [B(t) + B^*(t)a(t)]) + \operatorname{Sp} b(t)b^*(t)B(t)\}dt + (b^*(t)[B(t) + B^*(t)]x_t, dW_t). \tag{4.101}$$

In particular, if $x_t \equiv W_t$, and $B(t)$ is a symmetric matrix, then

$$d(W_t, B(t)W_t) = [(W_t, \dot{B}(t)W_t) + \operatorname{Sp} B(t)]dt + 2(B(t)W_t, dW_t). \tag{4.102}$$

EXAMPLE 3. Let $a(t) = a(t, \omega) \in \mathscr{P}_T$ and

$$z_t = \exp\left\{\int_0^t a(s)dW_s - \frac{1}{2}\int_0^t a^2(s)ds\right\}.$$

Denoting $x_t = \int_0^t a(s)dW_s - \frac{1}{2}\int_0^t a^2(s)ds$, we find (from (4.87)), that $z_t = \exp x_t$ has the differential

$$dz_t = z_t a(t)dW_t. \tag{4.103}$$

Also

$$d\left(\frac{1}{z_t}\right) = \frac{a^2(t)}{z_t}dt - \frac{a(t)}{z_t}dW_t. \tag{4.104}$$

(Note that $P\{\inf_{t \le T} z_t > 0\} = 1$, since $P\{\int_0^T a^2(t)dt < \infty\} = 1$.)

EXAMPLE 4. Let $a(t)$, $b(t)$, $0 \le t \le T$, be nonrandom functions with $\int_0^T |a(t)|dt < \infty$, $\int_0^T b^2(t)dt < \infty$.

Using the Ito formula, we find that the random process

$$x_t = \exp\left\{\int_0^t a(s)ds\right\}\left\{\xi + \int_0^t \exp\left[-\int_0^s a(u)du\right]b(s)dW_s\right\}$$

has the stochastic differential

$$dx_t = a(t)x_t\,dt + b(t)dW_t, \qquad x_0 = \xi.$$

4.3.4

Let us apply Ito's formula for deducing useful estimates for the mathematical expectations $M(\int_0^t f(s, \omega)dW_s)^{2m}$ of even degrees of stochastic integrals.

Lemma 4.11. *Let $W = (W_t, \mathscr{F}_t)$, $0 \le t \le T$, be a Wiener process, and let $f(t, \omega)$ be a bounded nonanticipative function, $|f(t, \omega)| \le K$, $0 \le t \le T$. Then*

$$M\left(\int_0^t f(s, \omega)dW_s\right)^{2m} \le K^{2m}t^m(2m - 1)!!.$$

PROOF. Let $x_t = \int_0^t f(s, \omega)dW_s$. Set

$$\tau_N = \inf\left(t: \sup_{s \le t}|x_s| \ge N\right),$$

assuming $\tau_N = T$, if $\sup_{s \le T}|x_s| < N$.

By the Ito formula,

$$x_{t \wedge \tau_N}^{2m} = 2m\int_0^{t \wedge \tau_N} x_s^{2m-1} f(s, \omega)dW_s + m(2m - 1)\int_0^{t \wedge \tau_N} x_s^{2m-2} f^2(s, \omega)ds.$$

From the definition of τ_N, the assumption $|f(s, \omega)| \le K, 0 \le s \le T$, and (4.48), it follows that

$$M \int_0^{t \wedge \tau_N} x_s^{2m-1} f(s, \omega) dW_s = 0.$$

Hence

$$Mx_{t \wedge \tau_N}^{2m} = m(2m-1)M \int_0^{t \wedge \tau_N} x_s^{2m-2} f^2(s, \omega) ds$$

$$\le K^2 m(2m-1)M \int_0^{t \wedge \tau_n} x_s^{2m-2}\, ds \le K^2 m(2m-1)M \int_0^t x_s^{2m-2}\, ds.$$

From this, by Fatou's lemma, we obtain

$$Mx_t^{2m} \le K^2 m(2m-1)M \int_0^t x_s^{2m-2}\, ds.$$

In the above inequality set $m = 1$. Then it follows that $Mx_t^2 \le K^2 t$. Similarly, with $m = 2$, we obtain the estimate $Mx_t^4 \le 3K^4 t^2$. Proof of the desired estimate is now completed by induction: assuming that $Mx_t^{2m} \le K^{2m} t^m (2m-1)!!$, from the inequality given above we easily infer that

$$Mx_t^{2(m+1)} \le K^{2(m+1)} t^{m+1} (2m+1)!!. \qquad \square$$

Let us relax now the assumption on the boundeness of the function $f(t, \omega)$, replacing it with the condition

$$\int_0^T Mf^{2m}(t, \omega) dt < \infty, \qquad m > 1.$$

Lemma 4.12. *Let $W = (W_t, \mathscr{F}_t)$, $0 \le t \le T$, be a Wiener process, and let $f(t, \omega)$ be a nonanticipative function with*

$$\int_0^T Mf^{2m}(t, \omega) dt < \infty.$$

Then

$$M\left(\int_0^t f(s, \omega) dW_s\right)^{2m} \le [m(2m-1)]^m t^{m-1} \int_0^t Mf^{2m}(s, \omega) ds.$$

PROOF. Using the notation from the previous lemma, we find that

$$Mx_{t \wedge \tau_N}^{2m} = m(2m-1)M \int_0^{t \wedge \tau_N} x_s^{2m-2} f^2(s, \omega) ds.$$

From this formula it follows that $Mx_{t\wedge\tau_N}^{2m}$ is a nondecreasing function of t. The application of Hölder's inequality with $p = m$, $q = m/(m-1)$, provides the estimate

$$M\int_0^{t\wedge\tau_N} x_s^{2m-2}f^2(s,\omega)ds$$

$$\leq \left(M\int_0^{t\wedge\tau_N} x_s^{2m}\,ds\right)^{(m-1)/m}\left(M\int_0^{t\wedge\tau_N} f^{2m}(s,\omega)ds\right)^{1/m}$$

$$= \left(M\int_0^{t\wedge\tau_N} x_{s\wedge\tau_N}^{2m}\,ds\right)^{(m-1)/m}\left(M\int_0^{t\wedge\tau_N} f^{2m}(s,\omega)ds\right)^{1/m}$$

$$\leq \left(M\int_0^{t} x_{s\wedge\tau_N}^{2m}\,ds\right)^{(m-1)/m}\left(M\int_0^{t} f^{2m}(s,\omega)ds\right)^{1/m}$$

$$\leq t^{(m-1)/m}(Mx_{t\wedge\tau_N}^{2m})^{(m-1)/m}\left(M\int_0^{t} f^{2m}(s,\omega)ds\right)^{1/m}.$$

Hence

$$Mx_{t\wedge\tau_N}^{2m} \leq m(2m-1)^{(m-1)/m}(Mx_{t\wedge\tau_N}^{2m})^{(m-1)/m}\left(M\int_0^{t} f^{2m}(s,\omega)ds\right)^{1/m}.$$

Since $Mx_{t\wedge\tau_N}^{2m} < \infty$, the above inequality is equivalent to the following:

$$(Mx_{t\wedge\tau_N}^{2m})^{1/m} \leq m(2m-1)t^{(m-1)/m}\left(M\int_0^{t} f^{2m}(s,\omega)ds\right)^{1/m}$$

or

$$Mx_{t\wedge\tau_N}^{2m} \leq [m(2m-1)]^m t^{m-1} M\int_0^{t} f^{2m}(s,\omega)ds.$$

Applying now the Fatou lemma, we obtain the desired inequality. □

4.4 Strong and weak solutions of stochastic differential equations

4.4.1

Let $(\Omega, \mathscr{F}, P)$ be a certain probability space ($T = 1$ for simplicity), $(\mathscr{F}_t)$, $t \leq 1$, be a nondecreasing family of the sub-σ-algebras of $\mathscr{F}$, and $W = (W_t, \mathscr{F}_t)$, $t \leq 1$, be a Wiener process. Denote by $(C_1, \mathscr{B}_1)$ the measurable space of the continuous functions $x = (x_t, 0 \leq t \leq 1)$ on $[0, 1]$ with the σ-algebra $\mathscr{B}_1 = \sigma(x : x_s, s \leq 1)$. Also, set $\mathscr{B}_t = \sigma\{x : x_s, s \leq t\}$.

Let $a(t, x)$ and $b(t, x)$ be measurable nonanticipative (i.e., $\mathscr{B}_t$-measurable for each t, $0 \leq t \leq 1$) functionals.

Definition 8. We shall say that the ((P-a.s.) continuous) random process $\xi = (\xi_t)$, $0 \le t \le 1$, is a *strong solution* (or simply a *solution*) of the stochastic differential equation

$$d\xi_t = a(t, \xi)dt + b(t, \xi)dW_t \tag{4.105}$$

with the $\mathscr{F}_0$-measurable initial condition $\xi_0 = \eta$ if for each t, $0 \le t \le 1$, the variables ξ_t are $\mathscr{F}_t$-measurable

$$P\left(\int_0^1 |a(t, \xi)|dt < \infty\right) = 1, \tag{4.106}$$

$$P\left(\int_0^1 b^2(t, \xi)dt < \infty\right) = 1, \tag{4.107}$$

and with probability 1 for each t, $0 \le t \le 1$,

$$\xi_t = \eta + \int_0^t a(s, \xi)ds + \int_0^t b(s, \xi)dW_s. \tag{4.108}$$

Introduce now the concept of a weak solution of the stochastic differential equation given by (4.105).

Definition 9. We say that the stochastic differential equation given by (4.105) with the initial condition η, having the prescribed distribution function $F(x)$, has a *weak solution* (or a *solution in a weak sense*), if there are: a probability space $(\Omega, \mathscr{F}, P)$; a nondecreasing family of the sub-σ-algebras $(\mathscr{F}_t)$, $t \le 1$; a continuous random process $\xi = (\xi_t, \mathscr{F}_t)$; and a Wiener process $W = (W_t, \mathscr{F}_t)$, such that (4.106), (4.107) (4.108) are satisfied and $P\{\omega : \xi_0 \le x\} = F(x)$.

Note the principal difference between the concepts of strong and weak solutions, assuming for simplicity $\eta = 0$. When one speaks about the solution in a strong sense, then it is implied that there have been prescribed some probability space $(\Omega, \mathscr{F}, P)$, the system $(\mathscr{F}_t)$, $t \le 1$, and the Wiener process $W = (W_t, \mathscr{F}_t)$. If in this case $\mathscr{F}_t = \mathscr{F}_t^W$, then the process $\xi = (\xi_t)$, $t \le 1$, is such that with each t the variables ξ_t are $\mathscr{F}_t^W$-measurable (i.e., ξ_t is determined by the "past" values of the Wiener process). Thus, for the strong solution,

$$\mathscr{F}_t^\xi \subseteq \mathscr{F}_t^W, \qquad 0 \le t \le 1.$$

When one speaks about the weak solution of Equation (4.105) with the prescribed nonanticipative functionals $a(t, x)$ and $b(t, x)$, then it is assumed that we may construct a probability space $(\Omega, \mathscr{F}, P)$, a system $(\mathscr{F}_t)$, $0 \le t \le 1$, a process $\xi = (\xi_t, \mathscr{F}_t$ and a Wiener process $W = (W_t, \mathscr{F}_t)$, for which (4.108) is satisfied (P-a.s.). In many cases where the weak solution exists, $\mathscr{F}_t = \mathscr{F}_t^\xi$ and, consequently, the process $W = (W_t, \mathscr{F}_t^\xi)$ is a Wiener process with respect to the system of the sub-σ-algebras $(\mathscr{F}_t^{\xi})$, $t \leq 1$. Hence in the case of the weak solutions

$$\mathscr{F}_t^W \subseteq \mathscr{F}_t^{\xi}, \qquad 0 \leq t \leq 1.$$

From Definition 9 it follows that the weak solution is, actually, an aggregate of the system $\mathscr{A} = (\Omega, \mathscr{F}, \mathscr{F}_t, P, W_t, \xi_t)$, where for brevity the process $\xi = (\xi_t)$, $0 \leq t \leq 1$, will be also called a *weak solution*.

Definition 10. We shall say that the stochastic differential equation given by (4.105) has a *unique solution in a weak sense*, if for any two of its solutions $\mathscr{A} = (\Omega, \mathscr{F}, \mathscr{F}_t, P, W_t, \xi_t)$ and $\tilde{\mathscr{A}} = (\tilde{\Omega}, \tilde{\mathscr{F}}, \tilde{\mathscr{F}}_t, \tilde{P}, \tilde{W}_t, \tilde{\xi}_t)$ the distributions of the processes $\xi = (\xi_t)$ and $\tilde{\xi} = (\tilde{\xi}_t)$, $0 \leq t \leq 1$, coincide, i.e.,

$$\mu_{\xi}(A) = \tilde{\mu}_{\tilde{\xi}}(A), \qquad A \in \mathscr{B},$$

where

$$\mu_{\xi}(A) = P\{\omega : \xi \in A\}, \qquad \tilde{\mu}_{\tilde{\xi}}(A) = \tilde{P}(\tilde{\omega} : \tilde{\xi} \in A\}.$$

Definition 11. One says that the stochastic differential equation given by (4.105) has a *unique strong solution*, if for any two of its strong solutions $\xi = (\xi_t, \mathscr{F}_t)$ and $\tilde{\xi} = (\tilde{\xi}_t, \mathscr{F}_t)$, $0 \leq t \leq 1$,

$$P\left\{\sup_{0 \leq t \leq 1} |\xi_t - \tilde{\xi}_t| > 0\right\} = 0. \tag{4.109}$$

In Subsections 4.4.2–4.4.6 there will be given the main theorems on the existence and uniqueness of the strong solutions of the stochastic differential equations given by (4.105). Problems related to the weak solutions are considered in Subsection 4.4.7.

4.4.2

The simplest conditions guaranteeing the existence and uniqueness of the strong solutions of Equation (4.105) are given in the following theorem.

Theorem 4.6. *Let the nonanticipative functionals* $a(t, x)$, $b(t, x)$, $t \in [0, 1]$, $x \in C_1$, *satisfy the Lipschitz condition*

$$|a(t, x) - a(t, y)|^2 + |b(t, x) - b(t, y)|^2 \leq L_1 \int_0^t |x_s - y_s|^2 \, dK(s) + L_2 |x_t - y_t|^2 \tag{4.110}$$

and

$$a^2(t, x) + b^2(t, x) \leq L_1 \int_0^t (1 + x_s^2) dK(s) + L_2(1 + x_t^2), \tag{4.111}$$

where L_1, L_2 are constants, $K(s)$ is a nondecreasing right continuous function, $0 \le K(s) \le 1$, $x, y \in C_1$. Let $\eta = \eta(\omega)$ be an $\mathscr{F}_0$-measurable random variable $P(|\eta(\omega)| < \infty) = 1$. Then:

(1) *the equation*

$$dx_t = a(t, x)dt + b(t, x)dW_t, \qquad x_0 = \eta, \tag{4.112}$$

has, and in this case a unique, strong solution $\xi = (\xi_t, \mathscr{F}_t)$, $0 \le t \le 1$;

(2) *if $M\eta^{2m} < \infty, m \ge 1$, then there exists a constant c_m, such that*

$$M\xi_t^{2m} \le (1 + M\eta^{2m})e^{c_m t} - 1. \tag{4.113}$$

PROOF. We begin with the uniqueness. If $\xi = (\xi_t, \mathscr{F}_t)$ and $\tilde{\xi} = (\tilde{\xi}_t, \mathscr{F}_t)$ are two continuous (P-a.s.) strong solutions with $\xi_0 = \eta$, $\tilde{\xi}_0 = \eta$, then

$$\xi_t - \tilde{\xi}_t = \int_0^t [a(s, \xi) - a(s, \tilde{\xi})]ds + \int_0^t [b(s, \xi) - b(s, \tilde{\xi})]dW_s.$$

Denote

$$\chi_t^N = \chi_{\left\{\sup\limits_{s \le t}(\xi_s^2 + \tilde{\xi}_s^2) \le N\right\}}.$$

Since $\chi_t^N = \chi_t^N \cdot \chi_s^N$ for $t \ge s$, then

$$\chi_t^N[\xi_t - \tilde{\xi}_t]^2 \le 2\chi_t^N\left[\left(\int_0^t \chi_s^N(a(s, \xi) - a(s, \tilde{\xi}))ds\right)^2 + \left(\int_0^t \chi_s^N[b(s, \xi) - b(s, \tilde{\xi})]dW_s\right)^2\right]. \tag{4.114}$$

From the definition of χ_t^N it follows that the variables

$$\chi_t^N[\xi_t - \tilde{\xi}_t]^2, \qquad \chi_t^N[a(t, \xi) - a(t, \tilde{\xi})]^2, \qquad \chi_t^N[b(t, \xi) - b(t, \tilde{\xi})]^2$$

are bounded and, consequently, mathematical expectations exist for the left and the right sides of the inequality given by (4.114). Hence, using (4.110), we find that

$$\begin{aligned}
&M\chi_t^N[\xi_t - \tilde{\xi}_t]^2 \\
&\le 2\int_0^t M\chi_s^N([a(s, \xi) - a(s, \tilde{\xi})]^2 + [b(s, \xi) - b(s, \tilde{\xi})]^2)ds \\
&\le 2\left\{L_1 \int_0^t M\chi_s^N \int_0^s (\xi_u - \tilde{\xi}_u)^2\, dK(u)ds + L_2 \int_0^t M\chi_s^N(\xi_s - \tilde{\xi}_s)^2\, ds\right\} \\
&\le 2\left\{L_1 \int_0^t M\chi_s^N \int_0^s \chi_u^N(\xi_u - \tilde{\xi}_u)^2\, dK(u)ds + L_2 \int_0^t M\chi_s^N(\xi_s - \tilde{\xi}_s)^2\, ds\right\} \\
&\le 2\left\{L_1 \int_0^t \int_0^s M\chi_u^N(\xi_u - \tilde{\xi}_u)^2\, dK(u)ds + L_2 \int_0^t M\chi_s^N(\xi_s - \tilde{\xi}_s)^2\, ds\right\}.
\end{aligned} \tag{4.115}$$

For the further development of the proof it is necessary to have:

Lemma 4.13. *Let c_0, c_1, c_2, be nonnegative constants, $u(t)$ be a nonnegative bounded function, and $v(t)$ be a nonnegative integrable function, $0 \le t \le 1$, such that*

$$u(t) \le c_0 + c_1 \int_0^t v(s)u(s)ds + c_2 \int_0^t v(s)\left[\int_0^s u(s_1)dK(s_1)\right]ds, \tag{4.116}$$

where $K(s)$ is a nondecreasing right continuous function, $0 \le K(s) \le 1$. Then

$$u(t) \le c_0 \exp\left\{(c_1 + c_2)\int_0^t v(s)ds\right\}. \tag{4.117}$$

PROOF. Substitute into the right side of (4.116) the function $u(s)$ with its majorant, defined by the right side of (4.116). After n such iterations we find

$$u(t) \le c_0 \sum_{j=0}^{n} \frac{(c_1 + c_2)^j}{j!}\left(\int_0^t v_s\, ds\right)^j + \varphi_n(t), \tag{4.118}$$

where $\varphi_n(t) \to 0$, $n \to \infty$, because of the boundedness of the function $u(t)$. Passing in (4.118) to the limit over $n \to \infty$, we obtain the desired estimate given by (4.117).

Apply this lemma to (4.115), assuming $c_0 = 0$, $c_1 = 2L_1$, $c_2 = 2L_2$, $v(t) \equiv 1$ and $u(t) = M\chi_t^N[\xi_t - \tilde{\xi}_t]^2$. We find that for all t, $0 \le t \le 1$,

$$M\chi_t^N[\xi_t - \tilde{\xi}_t]^2 = 0,$$

and therefore

$$P\{|\xi_t - \tilde{\xi}_t| > 0\} \le P\left\{\sup_{0 \le s \le 1} (\xi_s^2 + \tilde{\xi}_s^2) > N\right\}.$$

But the probability $P\{\sup_{0\le s\le 1} (\xi_s^2 + \tilde{\xi}_s^2) > N\} \to 0$, $N \to \infty$, because of the continuity of the processes ξ and $\tilde{\xi}$. Hence, for any t, $0 \le t \le 1$,

$$P\{|\xi_t - \tilde{\xi}_t| > 0\} = 0,$$

and therefore, for any countable everywhere dense set S in $[0, 1]$,

$$P\left\{\sup_{t\in S} |\xi_t - \tilde{\xi}_t| > 0\right\} = 0.$$

Finally, using again the continuity of the processes ξ and $\tilde{\xi}$, we find that

$$P\left\{\sup_{0\le t\le 1} |\xi_t - \tilde{\xi}_t| > 0\right\} = P\left\{\sup_{t\in S} |\xi_t - \tilde{\xi}_t| > 0\right\},$$

which proves the uniqueness of (the continuous) strong solution.

We shall now prove the existence of such a solution, first assuming that $M\eta^2 < \infty$. Set $\xi_t^{(0)} = \eta$ (zero approximation) and

$$\xi_t^{(n)} = \eta + \int_0^t a(s, \xi^{(n-1)})ds + \int_0^t b(s, \xi^{(n-1)})dW_s. \tag{4.119}$$

Let us show that $M(\xi_t^{(n)})^2 \le d$, where the constant d depends neither on n nor on $t \le 1$.

Actually, because of (4.111),

$$\begin{aligned} M(\xi_t^{(n+1)})^2 &\le 3\left\{M\eta^2 + \int_0^t M[a^2(a, \xi^{(n)}) + b^2(s, \xi^{(n)})]ds\right\} \\ &\le 3M\eta^2 + 3L_1 \int_0^t\int_0^s [1 + M(\xi_{s_1}^{(n)})^2]dK(s_1)ds \\ &\quad + 3L_2 \int_0^t [1 + M(\xi_s^{(n)})^2]ds \\ &\le 3(M\eta^2 + L_1 + L_2) + 3L_1 \int_0^t\int_0^s M(\xi_{s_1}^{(n)})^2\, dK(s_1)ds \\ &\quad + 3L_2 \int_0^t M(\xi_s^{(n)})^2\, ds. \end{aligned}$$

From this, taking into account that $M(\xi_t^{(0)})^2 = M\eta^2 < \infty$, by induction we obtain the estimate

$$M(\xi_t^{(n+1)})^2 \le 3(L + M\eta^2)e^{3Lt} \le 3(L + M\eta^2)e^{3L} \tag{4.120}$$

with $L = L_1 + L_2$. In other words one can take $d = 3(L + M\eta^2)e^{3L}$.

Because of (4.119) and the Lipschitz condition given by (4.110),

$$\begin{aligned} M[\xi_t^{(n+1)} - \xi_t^{(n)}]^2 &\le 2\int_0^t M[(a(s, \xi^{(n)}) - a(s, \xi^{(n-1)}))^2 \\ &\quad + (b(s, \xi^{(n)}) - b(s, \xi^{(n-1)})^2)]ds \\ &\le 2\left\{L_1 \int_0^t\int_0^s M(\xi_{s_1}^{(n)} - \xi_{s_1}^{(n-1)})^2\, dK(s_1)ds \right. \\ &\quad \left. + L_2 \int_0^t M(\xi_s^{(n)} - \xi_s^{(n-1)})^2\, ds\right\}. \end{aligned}$$

Since $M \sup_{0 \le t \le 1} [\xi_t^{(1)} - \xi_t^{(0)}]^2 \le c$, where c is a certain constant, then ($L = L_1 + L_2$)

$$\begin{aligned} M[\xi_t^{(2)} - \xi_t^{(1)}]^2 &\le 2Lct, \\ M[\xi_t^{(3)} - \xi_t^{(2)}]^2 &\le 2Lc\left\{2L_1 \int_0^t\int_0^s s_1\, dK(s_1)ds + 2L_2 \int_0^t s\, ds\right\} \\ &\le 2Lc\left\{2L_1 \int_0^t sK(s)ds + 2L_2 \int_0^t s\, ds\right\} \le c\frac{(2Lt)^2}{2}. \end{aligned}$$

And, in general,

$$M[\xi_t^{(n+1)} - \xi_t^{(n)}]^2$$

$$\leq \frac{c(2L)^{n-1}}{(n-1)!}\left\{2L_1 \int_0^t \int_0^s s_1^{n-1}\, dK(s_1)ds + 2L_2 \int_0^t s^{n-1}\, ds\right\}$$

$$\leq \frac{c(2L)^{n-1}}{(n-1)!}\left\{2L_1 \int_0^t s^{n-1}K(s)ds + 2L_2 \int_0^t s^{n-1}\, ds\right\} \leq \frac{c(2Lt)^n}{n!}. \quad (4.120')$$

Further,

$$\sup_{0\leq t\leq 1} |\xi_t^{(n+1)} - \xi_t^{(n)}| \leq \int_0^1 |a(s, \xi^{(n)}) - a(s, \xi^{(n-1)})|ds$$

$$+ \sup_{0\leq t\leq 1} \left|\int_0^t [b(s, \xi^{(n)}) - b(s, \xi^{(n-1)})]dW_s\right|.$$

Make use now of (4.54) which, along with (4.120′) and the Lipschitz condition given by (4.110), lead to the inequality

$$M \sup_{0\leq t\leq 1} [\xi_t^{(n+1)} - \xi_t^{(n)}]^2$$

$$\leq 10L_1 \int_0^1 \int_0^t M[\xi_s^{(n)} - \xi_s^{(n-1)}]^2\, dK(s)dt + 10L_2 \int_0^1 M[\xi_s^{(n)} - \xi_s^{(n-1)}]^2\, ds$$

$$\leq 10L_1 c\frac{(2L)^{n-1}}{(n-1)!}\int_0^1 \int_0^t s^{n-1}\, dK(s)dt + 10L_2 c\frac{(2L)^{n-1}}{(n-1)!}\int_0^1 s^{n-1}\, ds \leq 5c\frac{(2L)^n}{n!}.$$

The series

$$\sum_{n=1}^{\infty} P\left\{\sup_{0\leq t\leq 1} |\xi_t^{(n+1)} - \xi_t^{(n)}| > \frac{1}{n^2}\right\} \leq 5c\sum_{n=1}^{\infty} \frac{(2L)^n}{n!} n^4 < \infty.$$

Hence, by the Borel–Cantelli lemma, the series $\xi_t^{(0)} + \sum_{n=0}^{\infty} |\xi_t^{(n+1)} - \xi_t^{(n)}|$ converges (P-a.s.) uniformly over t, $0 \leq t \leq 1$. Therefore, the sequence of the random processes $(\xi_t^{(n)})$, $0 \leq t \leq 1$, $n = 0, 1, 2, \ldots$, (P-a.s.) converges uniformly to the *continuous* process

$$\xi_t = \xi_t^{(0)} + \sum_{n=0}^{\infty} (\xi_t^{(n+1)} - \xi_t^{(n)}).$$

From (4.120) and Fatou's lemma it follows that

$$M\xi_t^2 \leq 3(L + M\eta^2)e^{3L}.$$

Let us next show that the constructed process $\xi = (\xi_t)$, $t \leq 1$, is the solution of Equation (4.112), i.e., that (P-a.s.) for each t, $0 \leq t \leq 1$,

$$\xi_t - \eta - \int_0^t a(s, \xi)ds - \int_0^t b(s, \xi)dW_s = 0. \quad (4.121)$$

In accordance with (4.119) the left side in (4.121) is equal to

$$[\xi_t - \xi_t^{(n+1)}] + \int_0^t [a(s, \xi^{(n)}) - a(s, \xi)]ds + \int_0^t [b(s, \xi^{(n)}) - b(s, \xi)]dW_s. \tag{4.122}$$

Because of the Lipschitz condition, (4.110),

$$\left| \int_0^t [a(s, \xi^{(n)}) - a(s, \xi)]ds \right|^2 \le L_1 \int_0^t \int_0^s |\xi_u - \xi_u^{(n)}|^2 \, dK(u)ds + L_2 \int_0^t |\xi_s - \xi_s^{(n)}|^2 \, ds \le L \sup_{0 \le s \le 1} |\xi_s - \xi_s^{(n)}|^2. \tag{4.123}$$

Similarly, according to (4.60) and (4.110), for any $\delta > 0$ and $\varepsilon > 0$,

$$\begin{aligned} P\left\{ \left| \int_0^t [b(s, \xi^{(n)}) - b(s, \xi)]dW_s \right| > \varepsilon \right\} &\le \frac{\delta}{\varepsilon^2} + P\left\{ \int_0^t [b(s, \xi^{(n)}) - b(s, \xi)]^2 \, ds > \delta \right\} \\ &\le \frac{\delta}{\varepsilon^2} + P\left\{ L_1 \int_0^t \int_0^s |\xi_u - \xi_u^{(n)}| dK(u)ds + L_2 \int_0^t |\xi_s - \xi_s^{(n)}|^2 \, ds > \delta \right\} \\ &\le \frac{\delta}{\varepsilon^2} + P\left\{ L \sup_{0 \le s \le 1} |\xi_s - \xi_s^{(n)}|^2 > \delta \right\}. \end{aligned} \tag{4.124}$$

But $P\{\sup_{0 \le s \le 1} |\xi_s - \xi_s^{(n)}|^2 > \delta\} \to 0$, $n \to \infty$; hence, from (4.123) and (4.124) it follows that (4.122) converges in probability to zero as $n \to \infty$. This proves that $\xi = (\xi_t)$, $0 \le t \le 1$, is the solution of Equation (4.121).

From the construction of the process ξ it follows that it is measurable over (t, ω) and nonanticipative, i.e., $\mathscr{F}_t^{\eta, W}$-measurable at each t. Thus with $M\eta^2 < \infty$ the existence of the strong solution of Equation (4.112) is proved.

Assume now that $M\eta^{2m} < \infty$, $m > 1$, and establish the estimate given by (4.113). Let

$$\chi_N(t) = \begin{cases} 1, & \sup_{s \le t} |\xi_s| \le |\eta| + N, \\ 0, & \sup_{s \le t} |\xi_s| > |\eta| + N, \end{cases}$$

and

$$\psi_n = \begin{cases} 1, & |\eta| \le n, \\ 0, & |\eta| > n. \end{cases}$$

By Ito's formula,

$$\xi_t^{2m} = \eta^{2m} + 2m \int_0^t \xi_s^{2m-1} a(s, \xi) ds$$

$$+ m(2m-1) \int_0^t \xi_s^{2m-2} b^2(s, \xi) ds + 2m \int_0^t \xi_s^{2m-1} b(s, \xi) dW_s.$$

From this, for $t \geq s$, taking into account the equality $\chi_N(t)\psi_n = \chi_N(t)\chi_N(s)\psi_n^2$, we find that

$$\xi_t^{2m}\chi_N(t)\psi_n = \chi_N(t)\psi_n\Bigg[\psi_n\eta^{2m} + 2m \int_0^t \psi_n\chi_N(s)\,\xi_s^{2m-1} a(s, \xi) ds$$

$$+ m(2m-1) \int_0^t \psi_n\chi_N(s)\xi_s^{2m-2} b^2(s, \xi) ds$$

$$+ 2m \int_0^t \psi_n\chi_N(s)\xi_s^{2m-1} b(s, \xi) dW_s\Bigg]$$

$$\leq \psi_n\eta^{2m} + 2m \int_0^t \psi_n\chi_N(s)\xi_s^{2m-1} a(s, \xi) ds$$

$$+ m(2m-1) \int_0^t \psi_n\chi_N(s)\xi_s^{2m-2} b^2(s, \xi) ds$$

$$+ 2m \int_0^t \psi_n\chi_N(s)\xi_s^{2m-1} b(s, \xi) dW_s. \tag{4.125}$$

Note that because of the definitions of $\chi_N(t)$ and ψ_n,

$$M \int_0^1 \psi_n\chi_N(s)\xi_s^{4m-2} b^2(s, \xi) ds < \infty.$$

Hence (see (4.125) and (4.48))

$$M\xi_t^{2m}\chi_N(t)\psi_n \leq M\eta^{2m} + 2m \int_0^t M\psi_n\chi_N(s)|\xi_s^{2m-1} a(s, \xi)| ds$$

$$+ m(2m-1) \int_0^t M\psi_n\chi_N(s)\xi_s^{2m-2} b^2(s, \xi) ds.$$

For estimating the values

$$|\xi_s^{2m-1}||a(s, \xi)| \quad \text{and} \quad \xi_s^{2m-2} b^2(s, \xi)$$

we make use of the inequality

$$a^{1/p} b^{1/q} \leq \frac{a}{p} + \frac{b}{q}, \tag{4.126}$$

valid (see [17]) for any $a \geq 0, b \geq 0, p > 1, 1/p + 1/q = 1$. In (4.126), taking $p = 2m/(2m - 1)$, $q = 2m$, we have

$$|\xi_s|^{2m-1}|a(s, \xi)| = (\xi_s^{2m})^{1/p}(a^{2m}(s, \xi))^{1/q} \leq \frac{2m-1}{2m}\xi_s^{2m} + \frac{1}{2m}a^{2m}(s, \xi).$$

Similarly, with $p = m/(m - 1)$, $q = m$,

$$\xi_s^{2m-2}b^2(s, \xi) \leq \frac{m-1}{m}\xi_s^{2m} + \frac{1}{m}b^{2m}(s, \xi).$$

Hence for each m there exists a constant a_m such that

$$M(\xi_t^{2m}\chi_N(t)\psi_n) \leq M\eta^{2m} + a_m\int_0^t M\left\{\chi_N(s)\psi_n\left(\xi_s^{2m} + \left[(1 + \xi_s^2) + \int_0^s (1 + \xi_{s_1}^2)dK(s_1)\right]^m\right)\right\}ds, \tag{4.127}$$

where

$$\left[1 + \xi_s^2 + \int_0^s (1 + \xi_{s_1}^2)dK(s_1)\right]^m \leq b_m\left[1 + \xi_s^{2m} + \int_0^s (1 + \xi_{s_1}^{2m})dK(s_1)\right] \tag{4.128}$$

for some constant b_m.

From (4.127) and (4.128) we find (c_m a constant)

$$M(\xi_t^{2m}\chi_N(t)\psi_n) \leq M\eta^{2m} + \frac{c_m}{2}\left[t + \int_0^t M(\xi_s^{2m}\chi_N(s)\psi_n)ds + \int_0^t\int_0^s M(\xi_{s_1}^{2m}\chi_N(s)\psi_n)dK(s_1)ds\right]. \quad \square \tag{4.129}$$

Before going further let us establish the following:

Lemma 4.14. *Let c, d be positive constants and $u(t)$, $t \geq 0$, be a nonnegative bounded function, such that*

$$u(t) \leq d + c + \left[t + \int_0^t u(s)ds + \int_0^t\int_0^s u(s_1)dK(s_1)ds\right] \tag{4.130}$$

where $K(s)$ is a nondecreasing right continuous function, $0 \leq K(s) \leq 1$. Then

$$u(t) \leq (1 + d)e^{2ct} - 1. \tag{4.131}$$

Proof. From (4.130) it follows that

$$1 + u(t) \leq (1 + d) + c\left[\int_0^t (1 + u(s))ds + \int_0^t\int_0^s (1 + u(s_1))dK(s_1)ds\right].$$

Applying Lemma 4.13 with $c_0 = (1 + d)$, $c_1 = c_2 = c$, $v(t) \equiv 1$, to the function $1 + u(t)$, yields the desired inequality, (4.131).

Let us make use of this lemma, in (4.129) taking $u(t) = M[\xi_t^{2m}\chi_N(t)\psi_n]$. Then, according to (4.131),

$$M[\xi_t^{2m}\chi_N(t)\psi_n] \le (1 + M\eta^{2m})e^{c_m t} - 1. \tag{4.132}$$

From this, by Fatou's lemma, it follows that

$$M\xi_t^{2m} \le \lim_{\substack{N\to\infty \\ n\to\infty}} M[\xi_t^{2m}\chi_N(t)\psi_n] \le (1 + M\eta^{2m})e^{c_m t} - 1.$$

To complete the proof of the theorem it remains to check that the solution of Equation (4.112) also exists without the assumption $M\eta^2 < \infty$. Let $\eta_n = \eta\psi_n$, where $\psi_n = \chi_{\{|\eta| \le n\}}$, and let $\xi_n = (\xi_n(t))$, $0 \le t \le 1$, be the solutions of Equation (4.112) corresponding to the initial conditions $\xi_0 = \eta_n, M\eta_n^2 \le n^2$. Let $m > n$. Then exactly in the same way as in proving the uniqueness of the solution of Equation (4.112), (assuming $M\eta^2 < \infty$), one establishes the inequality

$$\begin{aligned} M[\xi_m(t) - \xi_n(t)]^2\psi_n \le 2L_1 &\int_0^t\int_0^s M[\xi_m(u) - \xi_n(u)]^2\psi_n\, dK(u)ds \\ &+ 2L_2 \int_0^t M[\xi_m(u) - \xi_n(u)]^2\psi_n\, du, \end{aligned}$$

from which, because of Lemma 4.13, it follows that $M[\xi_m(t) - \xi_n(t)]^2\psi_n = 0$. Therefore

$$P\{|\xi_m(t) - \xi_n(t)| > 0\} \le P\{|\eta| > n\}. \tag{4.133}$$

Since by the assumption $P\{|\eta| < \infty\} = 1$, it follows from (4.133) that $P\{|\xi_m(t) - \xi_n(t)| > 0\} \to 0, m, n \to \infty$, i.e., the sequence $\{\xi_n(t), n = 1, 2, \ldots\}$ is fundamental in probability. Consequently, for each t, $0 \le t \le 1$, there exists

$$P\text{-}\lim_{n\to\infty} \xi_n(t) = \xi(t).$$

Analogous considerations show that

$$P\text{-}\lim_{n\to\infty}\left\{\int_0^1\int_0^t [\xi_s - \xi_n(s)]^2\, dK(s)dt + \int_0^1 [\xi_s - \xi_n(s)]^2\, ds\right\} = 0.$$

This equality allows us (compare with the proof of (4.121)) in the equation

$$\xi_n(t) = \eta_n + \int_0^t a(s, \xi_m)ds + \int_0^t b(s, \xi_n)dW_s$$

to pass to the limit as $n \to \infty$. This completes the proof of Theorem 4.6. ☐

Corollary. *Consider the stochastic differential equation*

$$dx_t = a(t, x_t)dt + b(t, x_t)dW_t, \tag{4.134}$$

where the functions $a(t, y)$, $b(t, y)$, $0 \le t \le 1$, $y \in \mathbb{R}^1$, *satisfy the Lipschitz condition*

$$[a(t, y) - a(t, \tilde{y})]^2 + [b(t, y) - b(t, \tilde{y})]^2 \le L[y - \tilde{y}]^2 \qquad (4.135)$$

and increase no faster than linearly:

$$a^2(t, y) + b^2(t, y) \le L(1 + y^2). \qquad (4.136)$$

Then, according to Theorem 4.6, *Equation* (4.134) *with the initial condition* $x_0 = \eta$, $P(|\eta| < \infty) = 1$, *has a unique strong solution.*

Note. Theorem 4.6 is easily generalized to the case of the vector stochastic differential equations

$$dx_t = a(t, x)dt + b(t, x)dW_t, \qquad x_0 = \eta,$$

where $\eta = (\eta_1, \ldots, \eta_n)$, $x_t = (x_1(t), \ldots, x_n(t))$, $W_t = (W_1(t), \ldots, W_n(t))$, is a Wiener process, and

$$a(t, x) = (a_1(t, x), \ldots, a_n(t, x)), b(t, x) = \|b_{ij}(t, x)\|, \qquad i, j = 1, \ldots, n, x \in C_1.$$

For the existence and uniqueness of the continuous strong solution of the equation under consideration it suffices to demand that the functionals $a_i(t, x)$, $b_{ij}(t, x)$ satisfy (4.110), (4.111) with

$$x_s^2 = \sum_{i=1}^n x_i^2(s), \qquad |x_s - y_s|^2 = \sum_{i=1}^n |x_i(s) - y_i(s)|^2,$$

$$P\left(\sum_{i=1}^n |\eta_i| < \infty\right) = 1.$$

(4.113) is generalized in the following way: if $M \sum_{i=1}^n \eta_i^{2m} < \infty$, then

$$M \sum_{i=1}^n \xi_i^{2m}(t) \le \left(1 + M \sum_{i=1}^n \eta_i^{2m}\right)e^{c_m t} - 1.$$

4.4.3

From (4.113) it is seen that the finiteness of the moments $M\eta^{2m}$ is followed by the finiteness of $M\xi_t^{2m}$ at any t, $0 \le t \le 1$ (and generally at any $t \ge 0$, if Equation (4.112) is considered on the semiline $0 \le t < \infty$). Consider now the similar problem with respect to exponential moments.

Theorem 4.7. *Let* $\xi = (\xi_t)$, $0 \le t \le T$, *be a continuous random process that is a strong solution of the stochastic differential equation*

$$dx_t = a(t, x_t)dt + b(t, x_t)dW_t, \qquad x_0 = \eta, \qquad (4.137)$$

where η *is an* $\mathscr{F}_0$*-measurable random variable with*

$$Me^{\varepsilon\eta^2} < \infty \qquad (4.138)$$

for some $\varepsilon > 0$ *and the functions* $a(t, y)$, $b(t, y)$, $y \in \mathbb{R}^1$, *such that*

$$a^2(t, y) \leq K^2(1 + y^2), \; |b(t, y)| \leq K \tag{4.139}$$

(K *is a constant*). *Then there will be a* $\delta = \delta(T) > 0$ *such that*

$$\sup_{0 \leq t \leq T} Me^{\delta \xi_t^2} < \infty. \tag{4.140}$$

PROOF. Consider first a particular case of Equation (4.137),

$$dx_t = ax_t\, dt + b\, dW_t, \qquad x_0 = \eta, \tag{4.141}$$

where $a \geq 0$ and $b \geq 0$ are constants. Let us show that the statement of the theorem is correct in this case.

It is not difficult to check that the unique (continuous) solution ξ_t of Equation (4.141) is given by the formula

$$\xi_t = e^{at}\left[\eta + b \int_0^t e^{-as}\, dW_s\right].$$

It is clear that $\gamma_t = b \int_0^t e^{-as}\, dW_s$ is a Gaussian random variable with

$$M\gamma_t = 0 \quad \text{and} \quad M\gamma_t^2 = b^2 \int_0^t e^{-2as}\, ds \leq b^2 \int_0^T e^{-2as}\, ds \, (=R).$$

Choose

$$\delta = e^{-2aT} \min\left(\frac{1}{5R}, \frac{\varepsilon}{2}\right).$$

Then, because of the independence of the variables η and γ_t,

$$\begin{aligned} Me^{\delta \xi_t^2} &\leq M \exp\{2\delta e^{2at}[\eta^2 + \gamma_t^2]\} \\ &= M \exp\{2\delta e^{2at}\eta^2\}\, M \exp\{2\delta e^{2at}\gamma_t^2\} \\ &\leq Me^{\varepsilon\eta^2} Me^{(2/5R)\gamma_t^2} \\ &\leq Me^{\varepsilon\eta^2} \sup_{0 \leq t \leq T} Me^{(2/5R)\gamma_t^2} < \infty. \end{aligned}$$

Let us now consider the general case. By Ito's formula

$$\begin{aligned} \xi_t^{2n} = \eta^{2n} &+ 2n \int_0^t \xi_s^{2n-1} a(s, \xi_s) ds + n(2n-1) \int_0^t \xi_s^{2n-2} b^2(s, \xi_s) ds \\ &+ 2n \int_0^t \xi_s^{2n-1} b(s, \xi_s) dW_s. \end{aligned}$$

Because of the assumption of (4.138), $M\eta^{2m} < \infty$ for any $m \geq 1$. Hence, according to (4.113),

$$M \int_0^t \xi_s^{4n-2} b^2(s, \xi_s) ds < \infty, \qquad 0 \leq t \leq T.$$

Consequently

$$\begin{aligned} M\xi_t^{2n} &\le M\eta^{2n} + 2n\int_0^t M|\xi_s^{2n-1}a(s, \xi_s)|ds + K^2 n(2n-1)\int_0^t M\xi_s^{2n-2}\,ds \\ &\le M\eta^{2n} + 2nK\int_0^t M(1 + 2\xi_s^{2n})ds + K^2 n(2n-1)\int_0^t M\xi_s^{2n-2}\,ds \\ &\le M\eta^2 + 2nKT + 4nK\int_0^t M\xi_s^{2n} + K^2 n(2n-1)\int_0^t M\xi_s^{2n-2}\,ds. \end{aligned} \tag{4.142}$$

Choose $r > 0$ so that

$$M(\eta^2 + r)^n \ge M\eta^{2n} + 2nKT.$$

Then from (4.142) we obtain

$$M\xi_t^{2n} \le M(\eta^2 + r)^n + 4nK\int_0^t M\xi_s^{2n} + K^2 n(2n-1)\int_0^t M\xi_s^{2n-2}\,ds. \tag{4.143}$$

Consider the linear equation

$$dy_t = 2Ky_t\,dt + K\,dW_t, \qquad y_0 = (\eta^2 + r)^{1/2}. \tag{4.144}$$

By Ito's formula,

$$My_t^{2n} = M(\eta^2 + r)^n + 4nK\int_0^t My_s^{2n}\,ds + K^2 n(2n-1)\int_0^t My_s^{2n-2}\,ds. \tag{4.145}$$

Assuming in (4.143) and (4.145) that $n - 1$, we infer that

$$M\xi_t^2 \le M(\eta^2 + r) + 4K\int_0^t M\xi_s^2\,ds + K^2 t, \tag{4.146}$$

$$My_t^2 = M(\eta^2 + r) + 4K\int_0^t My_s^2\,ds + K^2 t. \quad \square \tag{4.147}$$

Let us now prove the following lemma.

Lemma 4.15. *Let $u(t)$, $v(t)$, $t \ge 0$, be integrable functions, such that, with some $c > 0$,*

$$u(t) \le v(t) + c\int_0^t u(s)ds. \tag{4.148}$$

Then

$$u(t) \le v(t) + c\int_0^t e^{c(t-s)}v(s)ds. \tag{4.149}$$

In this case, if in (4.148) for all $t \ge 0$ there is equality, then (4.149) is satisfied also with the sign of equality.

PROOF. Denote $z(t) = \int_0^t u(s)ds$ and $g(t) = u(t) - v(t) - cz(t) \leq 0$. It is clear that

$$\frac{dz(t)}{dt} = cz(t) + v(t) + g(t), \qquad z(0) = 0.$$

From this it follows that

$$z(t) = \int_0^t e^{c(t-s)}[v(s) + g(s)]ds \leq \int_0^t e^{c(t-s)}v(s)ds,$$

and therefore,

$$u(t) \leq v(t) + cz(t) \leq v(t) + c\int_0^t e^{c(t-s)}v(s)ds,$$

which proves (4.149). The final part of the lemma follows from the fact that $g(t) \equiv 0$.

Applying this lemma to (4.146) and (4.147), we find that

$$M\xi_t^2 \leq M(\eta^2 + r) + K^2 t + 4K\int_0^t e^{4K(t-s)}[M(\eta^2 + r) + K^2 s]ds = My_t^2.$$

From this, using the same lemma, from (4.142) and (4.145) by induction we obtain the inequalities

$$M\xi_t^{2n} \leq My_t^{2n}, \qquad n \geq 1, \qquad 0 \leq t \leq T.$$

Hence, if for some $\delta > 0$, $Me^{\delta y_t^2} < \infty$, then $Me^{\delta \xi_t^2} \leq Me^{\delta y_t^2} < \infty$.

To complete the proof of Theorem 4.7 it remains only to note that if $Me^{\varepsilon\eta^2} < \infty$ for some $\varepsilon > 0$, then

$$Me^{\varepsilon y_0^2} = e^{\varepsilon r}Me^{\varepsilon\eta^2} < \infty,$$

and, hence, as it was shown above, there will be a $\delta = \delta(T) > 0$ such that $\sup_{0 \leq t \leq T} Me^{\delta y_t^2} < \infty$. □

Note. To weaken the condition $|b(t, y)| \leq K$ by replacing it with the requirement $|b(t, y)| \leq K(1 + |y|)$ is, generally speaking, impossible, as is illustrated by the following example:

$$dx_t = x_t\, dW_t, \qquad x_0 = 1.$$

In this case

$$x_t = e^{W_t - (1/2)t}, \qquad Me^{x_0^2} = e < \infty,$$

but

$$Me^{\delta x_t^2} = M\exp\{\delta e^{2W_t - t}\} = \infty$$

with any $\delta > 0$.

4.4.4

Stochastic differential equations of type different than that of Equation (4.112) will be discussed below.

Theorem 4.8. *Let $a(t, x), b(t, x), t \in [0, 1], x \in C_1$, be nonanticipative functionals satisfying (4.110) and (4.111). Let $W = (W_t, \mathscr{F}_t)$ be a Wiener process, $\varphi = (\varphi_t, \mathscr{F}_t)$ be some (P-a.s.) continuous random process, and let $\lambda_i = (\lambda_i(t), \mathscr{F}_t)$, $i = 1, 2$, be random processes with $|\lambda_i(t)| \le 1$. Then the equation*

$$x_t = \varphi_t + \int_0^t \lambda_1(s)a(s, x)ds + \int_0^t \lambda_2(s)b(s, x)dW_s \tag{4.150}$$

has a unique strong solution.

PROOF. Let us start with the uniqueness. Let $\xi = (\xi_t)$ and $\tilde{\xi} = (\tilde{\xi}_t)$, $0 \le t \le 1$, be two solutions of Equation (4.150). As in proving Theorem 4.6, we infer that

$$M\chi_t^N[\xi_t - \tilde{\xi}_t]^2 \le 2\int_0^t M\chi_s^N[(a(s, \xi) - a(s, \tilde{\xi}))^2 + (b(s, \xi) - b(s, \tilde{\xi}))^2]ds.$$

From this, because of Lemma 4.13 and the Lipschitz condition given by (4.110), we obtain $M\chi_t^N[\xi_t - \tilde{\xi}_t]^2 = 0$, leading to the relationship $P\{\sup_{t\le 1}|\xi_t - \tilde{\xi}_t| > 0\} = 0$ (compare with the corresponding proof in Theorem 4.6). This establishes uniqueness.

For proving the existence of the strong solution assume first that $M \sup_{0\le t\le 1} \varphi_t^2 < \infty$. Then, considering the sequence of continuous processes $\xi_t^{(n)}$, $n = 0, 1, 2, \ldots, 0 \le t = 1$, defined by the relationships

$$\xi_t^{(0)} = \varphi_t,$$

$$\vdots$$

$$\xi_t^{(n)} = \varphi_t + \int_0^t \lambda_1(s)a(s, \xi^{(n-1)})ds + \int_0^t \lambda_2(s)b(s, \xi^{(n-1)})dW_s,$$

as in Theorem 4.6, we find that

$$M \sup_{t\le 1} [\xi_t^{(n+1)} - \xi_t^{(n)}]^2 \le c_1 \frac{c_2^n}{n!},$$

where c_1 and c_2 are some constants.

Further, it is established that the sequence of continuous processes $\xi^{(n)} = (\xi_t^{(n)})$, $0 \le t \le 1$, $n = 0, 1, 2, \ldots$, converges (P-a.s.) uniformly (over t) to some (continuous) process $\xi = (\xi_t)$, $0 \le t \le 1$, which is the strong solution of Equation (4.150) with $\sup_{t\le 1} M\xi_t^2 < \infty$.

In the general case, where the condition $M \sup_{t\le 1} \varphi_t^2 < \infty$ ceases to be valid, in order to prove the existence of the solution consider the sequence of equations

$$\xi_m(t) = \varphi_m(t) + \int_0^t \lambda_1(s)a(s, \xi_m)ds + \int_0^t \lambda_2(s)b(s, \xi_m)dW_s, \tag{4.151}$$

where $\varphi_m(t) = \varphi_{t\wedge\tau_m}$ and $\tau_m = \inf(t \le 1: \sup_{s\le t}|\varphi_s| \ge m)$, setting $\tau_m = 1$ if $\sup_{s\le 1}|\varphi_s| < m$, $m = 1, 2, \ldots$.

Since $|\varphi_m(t)| \le m$, Equation (4.151), for each $m = 1, 2, \ldots$, has a continuous strong solution. Further, as in Theorem 4.6, it is established that for each t, $0 \le t \le 1$, $\xi_m(t)$ converges as $m \to \infty$ in probability to some process $\xi(t)$, which satisfies (P-a.s.) Equation (4.150). □

Note. The statement of Theorem 4.8 can be generalized to the case of the vector equations given by (4.150) with $x_t = (x_1(t), \ldots, x_n(t))$, $\varphi_t = (\varphi_1(t), \ldots, \varphi_n(t))$, scalar processes $\lambda_i = (\lambda_i(t), \mathscr{F}_t)$, $|\lambda_i(t)| \le 1$ $(i = 1, 2)$, and $a(t, x) = (a_1(t, x), \ldots, a_n(t, x))$, $b(t, x) = \|b_{ij}(t, x)\|$ $(i, j = 1, \ldots, n)$. It suffices to require the processes $\varphi_i = (\varphi_i(t), \mathscr{F}_t)$ to be continuous, and the functionals $a_i(t, x)$, $b_{ij}(t, x)$ to satisfy (4.110) and (4.111).

4.4.5

Consider one more type of stochastic differential equations, for which filtering problems will be discussed in detail (see Chapter 12).

Theorem 4.9. *Let the nonanticipative functionals $a_0(t, x)$, $a_1(t, x)$, $b(t, x)$, $0 \le t \le 1$, satisfy* (4.110) *and* (4.111), *and let $|a_1(t, x)| \le c < \infty$. Then, if η is an $\mathscr{F}_0$-measurable random variable with $M\eta^2 < \infty$:*

(1) *the equation*

$$dx_t = [a_0(t, x) + a_1(t, x)x_t]dt + b(t, x)dW_t, \qquad x_0 = \eta, \tag{4.152}$$

has a unique strong solution;

(2) *if $M\eta^{2m} < \infty$, $m \ge 1$, then there exists a constant $c_m > 0$ such that*

$$M\xi_t^{2m} \le (1 + M\eta^{2m})e^{c_m t} - 1. \tag{4.153}$$

Proof. If the existence of the solution in Equation (4.152) is established then the correctness of the estimate given by (4.153) will result from the proof of the corresponding inequality, (4.113), in Theorem 4.6, since in deducing it only (4.111) was used, which is obviously satisfied for the functionals $a_0(t, x)$, $a_1(t, x)x_t$, $b(t, x)$.

The Lipschitz condition, (4.110), is not satisfied for the functional $a_1(t, y)y_t$. Hence, for proving the existence and uniqueness of the solution of Equation (4.152), the immediate application of Theorem 4.6 is not possible. We shall do it in the following way.

Consider the sequence of processes $\xi^{(n)} = (\xi_t^{(n)})$, $0 \le t \le 1$, $n = 1, 2, \ldots$, which are solutions of the equations

$$d\xi_t^{(n)} = [a_0(t, \xi^{(n)}) + a_1(t, \xi^{(n)})g_n(\xi_t^{(n)})]dt + b(t, \xi^{(n)})\, dW_t, \qquad \xi_0^{(n)} = \eta, \tag{4.154}$$

with

$$g_n(z) = \begin{cases} z, & |z| \le n, \\ n, & |z| > n. \end{cases}$$

Then with each $n, n = 1, 2, \ldots$, the functional $a_1(t, y)g_n(y_t)$ satisfies, as it is not difficult to see, the Lipschitz condition given by (4.110). Consequently, for each $n, n = 1, 2, \ldots$, the strong solution $\xi_t^{(n)}$ of Equation (4.154) exists and is unique.

Analysis of the proof of the inequality given by (4.113) shows that

$$M(\xi_t^{(n)})^2 \leq (1 + M\eta^2)e^{c_1 t} - 1,$$

where the constant c_1 does not depend on n. Therefore,

$$\sup_n \sup_{0 \leq t \leq 1} M(\xi_t^{(n)})^2 \leq (1 + M\eta^2)e^{c_1} - 1 < \infty,$$

which, in conjunction with (4.54), yields the inequality

$$\sup_n M \sup_{0 \leq t \leq 1} (\xi_t^{(n)})^2 < \infty.$$

Consequently

$$P\left\{\sup_{0 \leq t \leq 1} |\xi_t^{(n)}| > n\right\} \leq \frac{1}{n^2} \sup_n M \sup_{0 \leq t \leq 1} (\xi_t^{(n)})^2 \to 0, \qquad n \to \infty. \tag{4.155}$$

Set $\tau_n = \inf(t \leq 1 : \sup_{s \leq t} |\xi_s^{(n)}| \geq n)$, and $\tau_n = 1$, if $\sup_{s \leq 1} |\xi_s^{(n)}| < n$, and, for prescribed n' and n, $n' > n$, let $\sigma = \tau_n \wedge \tau_{n'}$.

Then

$$\begin{aligned}\xi_{t \wedge \sigma}^{(n')} - \xi_{t \wedge \sigma}^{(n)} = &\int_0^{t \wedge \sigma} [a_0(s, \xi^{(n')}) - a_0(s, \xi^{(n)})]ds \\ &+ \int_0^{t \wedge \sigma} [a_1(s, \xi^{(n')})g_{n'}(\xi_s^{(n')}) - a_1(s, \xi^{(n)})g_n(\xi_s^{(n)})]ds \\ &+ \int_0^{t \wedge \sigma} [b(s, \xi^{(n')}) - b(s, \xi^{(n)})]dW_s.\end{aligned}$$

Taking into consideration the Lipschitz condition, we find that

$$\begin{aligned}M[\xi_{t \wedge \sigma}^{(n')} - \xi_{t \wedge \sigma}^{(n)}]^2 \leq c_1 &\int_0^t \int_0^s M[\xi_{u \wedge \sigma}^{(n')} - \xi_{u \wedge \sigma}^{(n)}]^2 \, dK(u)ds \\ &+ c_2 \int_0^t M[\xi_{s \wedge \sigma}^{(n')} - \xi_{s \wedge \sigma}^{(n)}]^2 \, ds,\end{aligned} \tag{4.156}$$

where c_1 and c_2 are some constants. From (4.156), according to Lemma 4.13 we obtain

$$M[\xi_{t \wedge \sigma}^{(n')} - \xi_{t \wedge \sigma}^{(n)}]^2 = 0,$$

i.e., with $t \leq \sigma = \tau_n \wedge \tau_{n'}$ the solutions $\xi_t^{(n')}$ and $\xi_t^{(n)}$ coincide (P-a.s.). Hence for any t, $0 \leq t \leq 1$,

$$\begin{aligned}P\{|\xi_t^{(n')} - \xi_t^{(n)}| > 0\} &\leq P\{\sigma < t\} = P\{\tau_n \wedge \tau_{n'} < t\} \\ &\leq P\{\tau_n < t\} + P\{\tau_{n'} < t\} \\ &\leq P\left\{\sup_{s \leq t} |\xi_s^{(n)}| > n\right\} + P\left\{\sup_{s \leq t} |\xi_s^{(n')}| > n\right\},\end{aligned}$$

which together with (4.155) leads to the relationship

$$\lim_{\substack{n\to\infty\\ n'\to\infty}} P\{|\xi_t^{(n)} - \xi_t^{(n')}| > 0\} = 0.$$

Therefore, the $\xi_t^{(n)}$ converge in probability to some limit ξ_t.

From the coincidence of the values $\xi_t^{(n)}$ and $\xi_t^{(n')}$ for $t \in [0, \sigma]$ it follows that $\tau_n \leq \tau_{n'}$ (P-a.s.) for $n' > n$. Let $n = n_1 < n_2 < \ldots$. Then $P\text{-}\lim_{k\to\infty} \xi_t^{(n_k)} = \xi_t$ and, for $t \leq \tau_{n_1}$,

$$\xi_t^{(n_1)} = \xi_t^{(n_2)} = \cdots = \xi_t \qquad (P\text{-a.s.}).$$

Hence

$$P\left\{\left|\xi_t - \eta - \int_0^t [a_0(s, \xi) + a_1(s, \xi)\xi_s]ds - \int_0^t b(s, \xi)dW_s\right| > 0\right\}$$

$$\leq P\{\tau_{n_1} < t\} = P\left\{\sup_{s\leq t}|\xi_s^{(n_1)}| > n_1\right\} \to 0, \qquad n_1 = n \to \infty.$$

Thus the existence of the strong solution of Equation (4.152) is proved.

Let now $\xi = (\xi_t)$ and $\tilde{\xi} = (\tilde{\xi}_t)$, $0 \leq t \leq 1$, be two such solutions of Equation (4.152). Then, as in Theorem 4.6, it is established (using Lemma 4.13) that $M\chi_N(t)[\xi_t - \tilde{\xi}_t]^2 = 0$, where

$$\chi_N(t) = \chi_{\left\{\sup_{s\leq t}(\xi_s^2 + \tilde{\xi}_s^2) \leq N\right\}}.$$

From this we obtain

$$P\{|\xi_t - \tilde{\xi}_t| > 0\} \leq P\left\{\sup_{s\leq t}(\xi_s^2 + \tilde{\xi}_s^2) > N\right\} \to 0, \qquad N \to \infty,$$

which, because of the continuity of the processes ξ and $\tilde{\xi}$, leads to the equality $P\{\sup_{t\leq 1}|\xi_t - \tilde{\xi}_t| > 0\} = 0$. □

4.4.6

Let us formulate one more theorem on the existence and the form of the strong solution of linear vectors of stochastic differential equations.

Theorem 4.10. *Let the elements of the vector function $a_0(t) = (a_{01}(t), \ldots, a_{0n}(t))$ and the matrices $a_1(t) = \|a_{ij}^{(1)}(t)\|$, $b(t) = \|b_{ij}(t)\|$, $i, j = 1, \ldots, n$, be measurable (deterministic) functions t, $0 \leq t \leq 1$, satisfying the conditions*

$$\int_0^1 |a_{0j}(t)|dt < \infty, \qquad \int_0^1 |a_{ij}^{(1)}(t)|dt < \infty, \qquad \int_0^1 b_{ij}^2(t)dt < \infty.$$

Then the vector stochastic differential equation

$$dx_t = [a_0(t) + a_1(t)x_t]dt + b(t)dW_t, \qquad x_0 = \eta, \tag{4.157}$$

with the Wiener (with respect to the system $(\mathscr{F}_t)$, $t \leq 1$) *process* $W_t = (W_1(t), \ldots, W_n(t))$ *has a unique strong solution, defined by the formula*

$$x_t = \Phi_t\left[\eta + \int_0^t \Phi_s^{-1} a_0(s)ds + \int_0^t \Phi_s^{-1} b(s)dW_s\right], \tag{4.158}$$

where Φ_t *is the fundamental matrix* $(n \times n)$

$$\Phi_t = E + \int_0^t a_1(s)\Phi_s \, ds \tag{4.159}$$

(E is a unit matrix of order $(n \times n)$*).*

PROOF. Let us first of all show that there exists a solution of Equation (4.159). For this purpose consider the sequence $\{\Phi_k(t), k = 0, 1, \ldots\}$ with

$$\Phi_0(t) = E, \ldots, \Phi_{k+1}(t) = E + \int_0^t a_1(s)\Phi_k(s)ds. \tag{4.160}$$

We have

$$\Phi_{k+1}(t) - \Phi_k(t) = \int_0^t a_1(s)[\Phi_k(s) - \Phi_{k-1}(s)]ds \tag{4.161}$$

and

$$\sum_{i,j=1}^n |[\Phi_{k+1}(t) - \Phi_k(t)]_{ij}| \leq \int_0^t \sum_{i,j=1}^n |a_{ij}^{(1)}(s)| \sum_{i,j=1}^n |[\Phi_k(s) - \Phi_{k-1}(s)]_{ij}| ds.$$

Since (because of (4.160))

$$\sum_{i,j=1}^n |[\Phi_1(t) - \Phi_0(t)]_{ij}| < \int_0^t \sum_{i,j=1}^n |a_{ij}^{(1)}(s)| ds < \infty, \qquad 0 \leq t \leq 1,$$

then from (4.161) we infer that

$$\sum_{i,j=1}^n |[\Phi_{k+1}(t) - \Phi_k(t)]_{ij}| \leq \frac{1}{k!}\left(\int_0^t \sum_{i,j=1}^n |a_{1}^{(1)}(s)| ds\right)^k$$
$$\leq \frac{1}{k!}\left(\int_0^1 \sum_{i,j=1}^n |a_{ij}^{(1)}(s)| ds\right)^k.$$

From this it follows that the matrix series

$$\Phi_0(t) + \sum_{k=0}^\infty [\Phi_{k+1}(t) - \Phi_k(t)]$$

converges absolutely and uniformly to the matrix Φ_t with continuous elements. Hence, after the passage to the limit with $k \to \infty$ in (4.160), we convince ourselves of the existence of the solution of Equation (4.159). The matrix Φ_t, $0 \leq t \leq 1$ is almost everywhere differentiable, and the derivative of its determinant $|\Phi_t|$,

$$\frac{d|\Phi_t|}{dt} = \operatorname{Sp} a_1(t) \cdot |\Phi_t|, \qquad |\Phi_0| = 1,$$

almost everywhere, $0 \leq t \leq 1$. From this, we find that

$$|\Phi_t| = \exp\left(\int_0^t \mathrm{Sp}\, a_1(s)ds\right), \qquad 0 \leq t \leq 1,$$

and the matrix Φ_t is nonsingular. Let us show now that the solution of Equation (4.159) is unique.

Since the matrix Φ_t does not become nonsingular, then from the identity $\Phi_t\Phi_t^{-1} = E$ we find that almost everywhere, $0 \leq t \leq 1$,

$$\frac{d\Phi_t^{-1}}{dt} = -\Phi_t^{-1}\frac{d\Phi_t}{dt}\Phi_t^{-1} = -\Phi_t^{-1}a_1(t). \tag{4.162}$$

Let $\tilde{\Phi}_t$, $\tilde{\Phi}_0 = E$, be another solution of Equation (4.159). Then, because of (4.159) and (4.162), almost everywhere, $0 \leq t \leq 1$,

$$\frac{d}{dt}(\Phi_t^{-1}\tilde{\Phi}_t) = 0,$$

proving the correspondence of the continuous matrices Φ_t and $\tilde{\Phi}_t$ for all t, $0 \leq t \leq 1$.

Now to make sure of the existence of the strong solution of the system of the stochastic differential equations, (4.157), suffice it to apply Ito's formula to the representation (4.158) for x_t.

For proving the uniqueness of the solution of the equation system given by (4.157) note that the difference $\Delta_t = x_t - \tilde{x}_t$ of any two solutions x_t, $\tilde{x}_t$ satisfies the equations

$$\Delta_t = \Delta_0 + \int_0^t a_1(s)\Delta_s$$

and

$$\sum_{i=1}^n |\Delta_t|_i \leq \sum_{i=1}^n |\Delta_0|_i + \int_0^t \sum_{i,j=1}^n |a_{ij}^{(1)}(s)| \sum_{i=1}^n |\Delta_s|_i.$$

From this, by Lemma 4.13, we have

$$\sum_{i=1}^n |\Delta_t|_i \leq \sum_{i=1}^n |\Delta_0|_i \exp\left\{\int_0^t \sum_{i,j=1}^n |a_{ij}^{(1)}(s)|ds\right\},$$

and therefore any two solutions x_t, $\tilde{x}_t$, $0 \leq t \leq 1$, with $P\{x_0 = \tilde{x}_0 = \eta\} = 1$ coincide (P-a.s.) for all t. □

Note. Along with (4.157) consider the equation

$$dx_t = [a_0(t) + a_1(t)x_t + a_2(t)\xi_t]dt + b(t)d\xi_t, \qquad x_0 = \eta, \tag{4.163}$$

where $\xi = [(\xi_1(t), \ldots, \xi_n(t)), \mathscr{F}_t]$ is an Ito process with the differential

$$d\xi_t = \alpha_t(\omega)dt + \beta_t(\omega)dW_t. \tag{4.164}$$

Here $W = ([W_1(t), \dots, W_n(t)], \mathscr{F}_t)$ is a Wiener process, $(\alpha_t(\omega), \mathscr{F}_t)$, $0 \le t \le T$, are vectors where $\alpha_t(\omega) = [\alpha_1(t, \omega), \dots, \alpha_n(t, \omega)]$ and $(\beta_t(\omega), \mathscr{F}_t)$, $0 \le t \le T$, where $\beta_t(\omega) = \|\beta_{ij}(t, \omega)\|$, and $b(t) = \|b_{ij}(t)\|$ (order $(n \times n)$) are matrices having the following properties:

$$
\begin{aligned}
&P\left\{\int_0^T |b_{ij}(t)\alpha_j(t, \omega)|\,dt < \infty\right\} = 1, \qquad i, j = 1, \dots, n;\\
&P\left\{\int_0^T (b_{ij}(t)\beta_{jk}(t, \omega))^2\,dt < \infty\right\} = 1, \qquad i, k = 1, \dots, n.
\end{aligned}
\tag{4.165}
$$

If the vector $a_0(t) = [a_{01}(t), \dots, a_{0n}(t)]$ and the matrices $a_1(t) = \|a_{ij}^{(1)}(t)\|$, $a_2(t) = \|a_{ij}^{(2)}(t)\|$ (both having order $(n \times n)$) satisfy the assumptions of Theorem 4.10, then, as in the proof of Theorem 4.10, it can be established that

$$
x_t = \Phi_t\left[\eta + \int_0^t \Phi_s^{-1}(a_0(s) + a_2(s)\xi_s)ds + \int_0^t \Phi_s^{-1}b(s)d\xi_s\right], \tag{4.166}
$$

where Φ_t satisfies (4.159) and is the unique strong solution of Equation (4.163).

4.4.7

Consider now the problem of the existence and uniqueness of the weak solution of the equation

$$
d\xi_t = a(t, \xi)dt + dW_t, \qquad \xi_0 = 0. \tag{4.167}
$$

Let $(C_1, \mathscr{B}_1)$ be a measure space of the functions $x = (x_t)$, $0 \le t \le 1$, continuous on $[0, 1]$, with $x_0 = 0$, $\mathscr{B}_t = \sigma\{x : x_s, s \le t\}$. Denote by ν a Wiener measure on $(C_1, \mathscr{B}_1)$. Then the process $\tilde{W} = (\tilde{W}_t(x))$, $0 \le t \le 1$, on the space $(C_1, \mathscr{B}_1, \nu)$ will be a Wiener process, if we define $\tilde{W}_t(x) = x_t$.

Theorem 4.11. *Let the nonanticipative functional* $a = (a(t, x))$, $0 \le t \le 1$, $x \in C_1$, *be such that*

$$
\nu\left\{x : \int_0^1 a^2(t, x)dt < \infty\right\} = 1, \tag{4.168}
$$

$$
M_\nu \exp\left\{\int_0^1 a(t, x)d\tilde{W}_t(x) - \frac{1}{2}\int_0^1 a^2(t, x)dt\right\} = 1, \tag{4.169}
$$

where M_ν *is an averaging over measure* ν. *Then Equation* (4.167) *has a weak solution.*

PROOF. For proving the existence of such a solution it suffices to construct an aggregate of the systems $\mathscr{A} = (\Omega, \mathscr{F}, \mathscr{F}_t, P, W, \xi)$ satisfying the requirements of Definition 8. Take $\Omega = C_1$, $\mathscr{F} = \mathscr{B}_1$, $\mathscr{F}_t = \mathscr{B}_t$. As the measure P consider a measure with the differential $P(d\omega) = \rho(\tilde{W}(\omega))\nu(d\omega)$, where

$$
\rho(\tilde{W}(\omega)) = \exp\left\{\int_0^1 a(t, \tilde{W}(\omega))d\tilde{W}_t(\omega) - \frac{1}{2}\int_0^1 a^2(t, \tilde{W}(\omega))dt\right\}.
$$

From (4.169) it follows that the measure P is probabilistic, since $P(\Omega) = M_\nu \rho(x) = 1$.

On the probability space $(\Omega, \mathcal{F}, P)$ consider now the process

$$W_t = \tilde{W}_t - \int_0^t a(s, \tilde{W})ds, \qquad 0 \le t \le 1. \tag{4.170}$$

According to Theorem 6.3 this process is a Wiener process (with respect to the system of the σ-algebras $\mathcal{F}_t^{\tilde{W}}$ and the measure P). Hence if we set $\xi_t = \tilde{W}_t$, then from (4.170) we find that

$$\xi_t = \int_0^t a(s, \xi)ds + W_t, \qquad 0 \le t \le 1. \tag{4.171}$$

Thus the constructed aggregate of the systems $\mathcal{A} = (\Omega, \mathcal{F}, \mathcal{F}_t, P, W, \xi)$ forms a weak solution of Equation (4.167). □

Note 1. Let μ_W and μ_ξ be measures corresponding to the processes W and ξ. Then

$$\mu_\xi(A) = P(\xi \in A) = P(\tilde{W} \in A) = \int_{\{\tilde{W} \in A\}} \rho(\tilde{W}(\omega))\nu(d\omega)$$
$$= \int_{\{W \in A\}} \rho(W(\omega))d\mu_W(\omega).$$

Hence $\mu_\xi \ll \mu_W$, and, according to Lemma 6.8, $\mu_W \ll \mu_\xi$. Therefore, $\mu_\xi \sim \mu_W$ and

$$\frac{d\mu_\xi}{d\mu_W}(W(\omega)) = \rho(W(\omega)) \qquad (P\text{-a.s.}) \tag{4.172}$$

Note 2. Because of (4.168),

$$P\left(\int_0^1 a^2(t, W)dt < \infty\right) = 1,$$

and, according to Note 1, $\mu_\xi \sim \mu_W$. Hence the weak solution constructed above is such that

$$P\left(\int_0^1 a^2(t, \xi)dt < \infty\right) = 1. \tag{4.173}$$

Theorem 4.12. *Let the conditions of Theorem* 4.11 *be satisfied. Then, in a class of solutions satisfying* (4.173) *a weak solution of Equation* (4.167) *is unique.*

PROOF. Let $\mathcal{A} = (\Omega, \mathcal{F}, \mathcal{F}_t, P, W, \xi)$ be the solution constructed above and $\mathcal{A}' = (\Omega', \mathcal{F}', \mathcal{F}'_t, P', W', \xi')$ be one more weak solution with

$$P'\left(\int_0^1 a^2(t, \xi')dt < \infty\right) = 1. \tag{4.174}$$

Then, by Theorem 7.7, $\mu_{\xi'} \sim \mu_{W'}$ and

$$\frac{d\mu_{\xi'}}{d\mu_{W'}}(W'(\omega')) = \rho(W'(\omega')),$$

which together with (4.172) yields the desired equality $\mu_{\xi'}(A) = \mu_{\xi}(A)$. □

Let us formulate, finally, one more result, being actually a corollary of Theorems 4.11 and 4.12.

Theorem 4.13. *Let the functional* $a = (a(t, x))$, $0 \le t \le 1$, $x \in C_1$, *be such that, for any* $x \in C_1$,

$$\int_0^1 a^2(t, x)dt < \infty. \tag{4.175}$$

Then (4.169) *is necessary and sufficient for the existence and uniqueness of a weak solution of Equation* (4.167).

PROOF. The sufficiency follows from Theorems 4.11 and 4.12. To prove the necessity, note that if $\mathscr{A} = (\Omega, \mathscr{F}, \mathscr{F}_t, P, W, \xi)$ is some weak solution, then from (4.175) and Theorem 7.7 it follows that $\mu_\xi \sim \mu_W$ and

$$\frac{d\mu_\xi}{d\mu_W}(W(\omega)) = \rho(W(\omega)).$$

Therefore

$$\mu_\xi(\Omega) = \int_\Omega \rho(W(\omega))d\mu_W(\omega) = 1,$$

which corresponds to Equation (4.169). □

Note. Sufficient conditions for Equation (4.169) to hold are given in Section 6.2.

4.4.8

If the nonanticipative functional $a(t, x)$ in (4.167) is such that

$$|a(t, x)| \le C \le \infty, \qquad 0 \le t \le 1, \qquad x \in C_1,$$

then, according to Theorem 4.13, Equation (4.167) has a unique weak solution. The question then arises whether (4.167) with $|a(t, x)| \le C < \infty$ also has a strong solution. It follows from Theorem 4.6 that if the functional $a(t, x)$ satisfies the integral Lipschitz condition (4.110), then (4.167) has a strong solution (on the given probability space and for the Wiener process specified thereon).

Below we give an example which shows that, in general, (4.167) may not have a strong solution, even if $a(t, x)$ is bounded.

EXAMPLE. Consider numbers t_k, $k = 0, -1, -2, \ldots$, such that

$$0 < t_{k-1} < t_k < \cdots < t_0 = 1.$$

For $x \in C_1$ let $a(0, x) = 0$ and

$$a(t, x) = \left\{ \frac{x_{t_k} - x_{t_{k-1}}}{t_k - t_{k-1}} \right\}, \qquad t_k \le t < t_{k+1}, \tag{4.176}$$

where $\{b\}$ denotes the fractional part of b.

According to (4.167), we have

$$x_{t_{k+1}} - x_{t_k} = \left\{ \frac{x_{t_k} - x_{t_{k-1}}}{t_k - t_{k-1}} \right\}(t_{k+1} - t_k) + (W_{t_{k+1}} - W_{t_k}). \tag{4.177}$$

Setting

$$\eta_k = \frac{x_{t_k} - x_{t_{k-1}}}{t_k - t_{k-1}}, \qquad \varepsilon_k = \frac{W_{t_k} - W_{t_{k-1}}}{t_k - t_{k-1}}$$

we find from (4.177) that

$$\eta_{k+1} = \{\eta_k\} + \varepsilon_{k+1}, \qquad k = 0, -1, -2, \ldots.$$

Hence

$$e^{2\pi i \eta_{k+1}} = e^{2\pi i \{\eta_k\}} e^{2\pi i \varepsilon_{k+1}} = e^{2\pi i \eta_k} e^{2\pi i \varepsilon_{k+1}} \tag{4.178}$$

Denote $Me^{2\pi i \eta_k}$ by d_k. If (4.167) has a strong solution, then (in agreement with the definition of a strong solution) η_k must be $\mathscr{F}_{t_k}^W$-measurable, and hence the variables η_k and ε_{k+1} are independent. Therefore (4.178) implies that

$$d_{k+1} = d_k M e^{2\pi i \varepsilon_{k+1}} = d_k e^{-2\pi^2/(t_{k+1} - t_k)}$$

thus

$$d_{k+1} = d_{k+1-n} \exp\left[(-2\pi^2)\left(\frac{1}{t_{k+1} - t_k} + \cdots + \frac{1}{t_{k+2-n} - t_{k+1-n}} \right) \right]$$

and consequently

$$|d_{k+1}| \le e^{-2\pi^2 n}$$

for any n. Therefore $d_k = 0$ for all $k = 0, -1, -2, \ldots$. Next, from (4.178)

$$\begin{aligned} e^{2\pi i \eta_{k+1}} &= e^{2\pi i \eta_k} e^{2\pi i \varepsilon_{k+1}} \\ &= e^{2\pi i (\varepsilon_k + \varepsilon_{k+1})} e^{2\pi i \eta_{k-1}} \\ &\vdots \\ &= e^{2\pi i (\varepsilon_{k+1-n} + \cdots + \varepsilon_k + \varepsilon_{k+1})} e^{2\pi i \eta_{k-n}}. \end{aligned}$$

If Equation (4.167) has a strong solution, then the variables η_{k-n} are $\mathscr{F}_{t_{k-n}}^W$-measurable and consequently if

$$\mathscr{G}_{t_{k-n}, t_{k+1}}^W = \sigma\{\omega : W_t - W_s, t_{k-n} \le s \le t \le t_{k+1}\},$$

the independence of σ-algebras $\mathscr{F}^w_{t_{k-n}}$ and $\mathscr{G}^w_{t_{k-n}, t_{k+1}}$ implies

$$M[e^{2\pi i\eta_{k+1}} | \mathscr{G}^w_{t_{k-n}, t_{k+1}}] = e^{2\pi i(\varepsilon_{k+1-n} + \cdots + \varepsilon_{k+1})} M e^{2\pi i \eta_{k-n}}.$$

Taking into account the identities

$$d_k = Me^{2\pi i\eta_k} = 0, \qquad k = 0, -1, -2, \ldots,$$

we conclude

$$M[e^{2\pi i\eta_{k+1}} | \mathscr{G}^w_{t_{k-n}, t_{k+1}}] = 0.$$

Since

$$\mathscr{G}^w_{t_{k-n}, t_{k+1}} \uparrow \mathscr{F}^w_{t_{k+1}}$$

as $n \uparrow \infty$, we have from Theorem 1.5 that

$$M[e^{2\pi i\eta_{k+1}} | \mathscr{F}^w_{t_{k+1}}] = 0. \tag{4.179}$$

If a strong solution exists, then the variables η_{k+1} are $\mathscr{F}^w_{t_{k+1}}$-measurable and therefore it follows from (4.179) that

$$e^{2\pi i\eta_{k+1}} = 0,$$

which is clearly impossible.

The contradiction obtained above shows that Equation (4.167) with $a(t, x)$ defined in (4.167) does not possess a strong solution.

Notes and references

4.1. The proof of Levy theorem that any Wiener process is a Brownian motion process can be found in Doob [16]. We present here another proof. The result of the continuity of (augmented) σ-algebras $\mathscr{F}^W_t$ generated by values of the Wiener process W_s, $s \leq t$, is a well-known fact.

4.2. The construction of stochastic integrals over a Wiener process from different classes of functions is due to Wiener [20], and Ito [59]. The structure of properties of stochastic integrals was discussed in recent books by Gykhman and Skorokhod [34], [36]. The integrals $\Gamma_t(f)$ have been introduced here for the first time. Lemma 4.9 is due to Yershov [52].

4.3. The Ito formula of the change of variables (see [34], [36], [47], [60]) plays a fundamental role in the theory of stochastic differential equations.

4.4. In stochastic differential equations the concepts of strong and those of weak solutions should be essentially distinguished. The weak solutions were discussed in Skorokhod [144], Yershov [52], [53], Shiryayev [166], Liptser and Shiryayev [111], Yamada and Watanabe [174]. The existence and uniqueness of the strong solution under a Lipschitz integrable condition (4.10) have been proved by Ito and Nisio [62]. The assertion of Theorem 4.7 is contained in Kallianpur, Striebel [74]. We have presented here another proof. An example showing the nonexistence of a strong solution of stochastic differential equation (4.167) was given by Cirelson [213] (the simple proof given here is due to Krylov).

5 Square integrable martingales, and structure of the functionals on a Wiener process

5.1 Doob–Meyer decomposition for square integrable martingales

5.1.1

Let $(\Omega, \mathscr{F}, P)$ be a complete probability space, and let $F = (\mathscr{F}_t)$, $t \geq 0$, be a nondecreasing (right continuous) family of sub-σ-algebras $\mathscr{F}$, each of which is augmented by sets from $\mathscr{F}$ having zero P-probability.

Denote by $\mathscr{M}_T$ the family of square integrable martingales, i.e., right continuous martingales $X = (x_t, \mathscr{F}_t)$, $0 \leq t \leq T$, with $\sup_{t \leq T} Mx_t^2 < \infty$. The martingales $X = (x_t, \mathscr{F}_t)$, $0 \leq t \leq T$, having (P-a.s.) continuous trajectories and satisfying the condition $\sup_{t \leq T} Mx_t^2 < \infty$ will be denoted by $\mathscr{M}_T^c$. Obviously $\mathscr{M}_T^c \subseteq \mathscr{M}_T$. In the case $T = \infty$ the classes $\mathscr{M}_\infty$ and $\mathscr{M}_\infty^c$ will be denoted by $\mathscr{M}$ and $\mathscr{M}^c$ respectively.

The random process $Z = (x_t^2, \mathscr{F}_t)$, $0 \leq t \leq T$, where the martingale $X = (x_t, \mathscr{F}_t) \in \mathscr{M}_T$, is a nonnegative submartingale and, by Theorem 3.7, it belongs to class DL (in the case $T < \infty$ it belongs to class D).

Applying the Doob–Meyer expansion (Theorem 3.8 and the corollary) to the submartingale $Z = (x_t^2, \mathscr{F}_t)$, $0 \leq t \leq T < \infty$, we obtain the following result.

Theorem 5.1. *For each $X \in \mathscr{M}_T$ there exists a unique (to within stochastic equivalence) natural increasing process $A_t \equiv \langle x \rangle_t$, $t \leq T$, such that for all t, $0 \leq t \leq T$,*

$$x_t^2 = m_t + \langle x \rangle_t \qquad (P\text{-a.s.}) \tag{5.1}$$

where $(m_t, \mathscr{F}_t)$, $t \leq T$, is a martingale. In the case $t \geq s$,

$$M[(x_t - x_s)^2 | \mathscr{F}_s] = M[\langle x \rangle_t - \langle x \rangle_s | \mathscr{F}_s] \qquad (P\text{-a.s.}). \tag{5.2}$$

PROOF. It suffices to establish only (5.2). But $(m_t, \mathscr{F}_t)$ and $(x_t, \mathscr{F}_t)$ are martingales, hence

$$M(m_t - m_s|\mathscr{F}_s) = 0, \qquad M[x_t^2 - x_s^2|\mathscr{F}_s] = M[(x_t - x_s)^2|\mathscr{F}_s] \qquad (P\text{-a.s.}),$$

and (5.2) follows from (5.1). □

EXAMPLE 1. Let $X = (W_t, \mathscr{F}_t)$ be a Wiener process. Then $\langle W\rangle_t = t$ (P-a.s.).

EXAMPLE 2. Let $a(t, \omega) \in \mathscr{M}_T$, and let $X = (x_t, \mathscr{F}_t)$, $t \le T$, be the continuous martingale $x_t = \int_0^t a(s, \omega)dW_s$. Then, by Ito's formula

$$x_t^2 = 2\int_0^t a(s, \omega)x_s\, dW_s + \int_0^t a^2(s, \omega)ds.$$

It is immediate that the process

$$y_t \equiv 2\int_0^t a(s, \omega)x_s\, dW_s = x_t^2 - \int_0^t a^2(s, \omega)ds$$

is a martingale, and that the process $\int_0^t a^2(s, \omega)ds$ is natural. Hence in the example considered:

$$\langle x\rangle_t = \int_0^t a^2(s, \omega)ds.$$

5.1.2

We shall need an analog of (5.1) for the product $x_t \cdot y_t$ of two square integrable martingales $X = (x_t, \mathscr{F}_t)$ and $Y = (y_t, \mathscr{F}_t)$, $0 \le t \le T$.

Theorem 5.2. *Let $X \in \mathscr{M}_T$, $Y \in \mathscr{M}_T$. Then there exists a unique (to within stochastic equivalence) process $\langle x, y\rangle_t$, which is the difference between two natural increasing processes, and the martingale $(m_t, \mathscr{F}_t)$, such that for all t, $0 \le t \le T$,*

$$x_t y_t = m_t + \langle x, y\rangle_t \qquad (P\text{-a.s.}). \tag{5.3}$$

In this case, (P-a.s.)

$$M[(x_t - x_s)(y_t - y_s)|\mathscr{F}_s] = M[\langle x, y\rangle_t - \langle x, y\rangle_s|\mathscr{F}_s]. \tag{5.4}$$

PROOF. Let us show first of all that there exist processes m_t and $\langle x, y\rangle_t$ with the above properties for which (5.3) is satisfied. According to (5.1),

$$(x_t - y_t)^2 = m_t^{x-y} + \langle x - y\rangle_t, \qquad (x_t + y_t)^2 = m_t^{x+y} + \langle x + y\rangle_t,$$

where the notation is obvious.

We define now

$$\langle x, y\rangle_t = \tfrac{1}{4}[\langle x + y\rangle_t - \langle x - y\rangle_t] \quad \text{and} \quad m_t = x_t y_t - \langle x, y\rangle_t.$$

It is clear that $\langle x, y\rangle_t$ is the difference between two natural increasing processes. Let us check that $(m_t, \mathscr{F}_t)$ is a martingale.

Because of the formula $ab = \frac{1}{4}[(a+b)^2 - (a-b)^2]$,

$$\begin{aligned} M[x_t y_t - x_s y_s | \mathscr{F}_s] &= M[(x_t - x_s)(y_t - y_s) | \mathscr{F}_s] \\ &= \tfrac{1}{4} M\{[(x_t + y_t) - (x_s + y_s)]^2 \\ &\quad - [(x_t - y_t) - (x_s - y_s)]^2 | \mathscr{F}_s\} \\ &= \tfrac{1}{4} M\{[\langle x + y\rangle_t - \langle x + y\rangle_s] \\ &\quad - [\langle x - y\rangle_t - \langle x - y\rangle_s] | \mathscr{F}_s\} \\ &= \tfrac{1}{4} M\{[\langle x + y\rangle_t - \langle x - y\rangle_t] \\ &\quad - [\langle x + y\rangle_s - \langle x - y\rangle_s] | \mathscr{F}_s\} \\ &= M[\langle x, y\rangle_t - \langle x, y\rangle_s | \mathscr{F}_s]. \end{aligned}$$

From this it follows that the process $(m_t, \mathscr{F}_t)$ is a martingale.

Let there be one more representation $x_t y_t = m_t' + A_t'$, where $(m_t', \mathscr{F}_t)$ is a martingale and A_t' is a process which is the difference between two natural increasing processes.

If the time t is discrete ($t = 0, 1, \ldots, N$), then the equalities $m_t' = m_t$, $A_t' = \langle x, y\rangle_t$ (P-a.s.) are established in the following way.

Since

$$A_{t+1}' - A_t' = (x_{t+1} y_{t+1} - x_t y_t) - (m_{t+1}' - m_t')$$

then, because of the $\mathscr{F}_t$-measurability of A_{t+1}' and Equation (5.4),

$$A_{t+1}' - A_t' = M[x_{t+1} y_{t+1} - x_t y_t | \mathscr{F}_t] = \langle x, y\rangle_{t+1} - \langle x, y\rangle_t \qquad (P\text{-a.s.}).$$

But $A_0' = \langle x, y\rangle_0 = 0$. Hence $A_t' = \langle x, y\rangle_t$ and $m_t' = m_t$ (P-a.s.) for each $t = 0, 1, \ldots, N$.

If the time t is continuous, then the uniqueness of the expansion given by (5.3) is established in the same way as in proving uniqueness in Theorem 3.8. □

Note. To avoid misunderstanding we note that, generally speaking, $\langle x + y\rangle_t \neq \langle x\rangle_t + \langle y\rangle_t$. The equality $\langle x + y\rangle_t = \langle x\rangle_t + \langle y\rangle_t$, $t \le T$, will be satisfied (P-a.s.) when the martingales $X = (x_t, \mathscr{F}_t)$ and $Y = (y_t, \mathscr{F}_t)$ are orthogonal ($X \perp Y$), i.e., $\langle x, y\rangle_t = 0$, $t \le T$. Because of the uniqueness of the expansion given by (5.3), it is not difficult to show that the condition $\langle x, y\rangle_t = 0$ is equivalent to the fact that the process $(x_t y_t, \mathscr{F}_t)$, $t \le T$, is also a martingale.

EXAMPLE 3. Let $W = (W_t, \mathscr{F}_t)$ be a Wiener process and let

$$x_t = \int_0^t a(s, \omega) dW_s, \qquad y_t = \int_0^t b(s, \omega) dW_s,$$

where

$$M \int_0^T a^2(s, \omega) ds < \infty, \qquad M \int_0^T b^2(s, \omega) ds < \infty.$$

Then $X = (x_t, \mathscr{F}_t) \in \mathscr{M}_T$, $Y = (y_t, \mathscr{F}_t) \in \mathscr{M}_T$ and, by Ito's formula,

$$x_t y_t = \int_0^s [x_s b(s, \omega) + y_s a(s, \omega)] dW_s + \int_0^t a(s, \omega) b(s, \omega) ds.$$

As in Example 2, it can be shown that the process

$$\int_0^t [x_s b(s, \omega) + y_s a(s, \omega)] dW_s$$

is a martingale, and that

$$\langle x, y \rangle_t = \int_0^t a(s, \omega) b(s, \omega) ds. \tag{5.5}$$

In particular, if $y_t \equiv W_t$, i.e., $b(s, \omega) \equiv 1$, then

$$\langle x, W \rangle_t = \int_0^t a(s, \omega) ds \qquad (P\text{-a.s.}), \qquad t \le T. \tag{5.6}$$

5.1.3

One of the crucial results of square integrable martingale theory is that the representation given by (5.6) is correct for any martingale $X = (x_t, \mathscr{F}_t) \in \mathscr{M}_T$, not only in the case where $x_t = \int_0^t a(s, \omega) dW_s$. The exact result is given in the following theorem.

Theorem 5.3. *Let the martingale $X = (x_t, \mathscr{F}_t) \in \mathscr{M}_T$, and let $W = (W_t, \mathscr{F}_t)$ be a Wiener process. Let us assume that the family of the σ-algebras $F = (\mathscr{F}_t)$, $t \le T$, is right continuous, i.e., $\mathscr{F}_t = \mathscr{F}_{t+}$ for all t, $0 \le t \le T$, where $\mathscr{F}_{T+} = \mathscr{F}_T$. Then there will be a random process $(a(t, \omega), \mathscr{F}_t)$ with $M\int_0^T a^2(t, \omega) dt < \infty$, such that for all t, $0 \le t \le T$,*

$$\langle x, W \rangle_t = \int_0^t a(s, \omega) ds \qquad (P\text{-a.s.}). \tag{5.7}$$

As a preliminary, let us prove the following lemma.

Lemma 5.1. *Let the family $F = (\mathscr{F}_t)$, $t \le T$, be right continuous, let $W = (W_t, \mathscr{F}_t)$ be a Wiener process, and let $X = (x_t, \mathscr{F}_t) \in \mathscr{M}_T$. Let the random process $(g(t, \omega), \mathscr{F}_t)$, $t \le T$, be measurable with respect to a σ-algebra on $[0, T] \times \Omega$ generated by nonanticipative processes having left continuous trajectories where $\int_0^T g^2(s, \omega)(|a\langle x, W\rangle_s| + ds) < \infty$. If $y_t = \int_0^t g(s, \omega) dW_s$, then (P-a.s.)*

$$\langle x, y \rangle_t = \int_0^t g(s, \omega) d\langle x, W \rangle_s, \tag{5.8}$$

where the integral is understood as a Lebesgue–Stieltjes integral.

If for almost all ω the function $\langle x, W \rangle_t$ is absolutely continuous then

Equation (5.8) *is satisfied for any process* $(g(t, \omega), \mathscr{F}_t)$, $t \le T$, *satisfying the condition*

$$M \int_0^T g^2(s, \omega)(|d\langle x, W\rangle_s| + ds) < \infty.$$

PROOF. Let $g^{(n)}(t, \omega)$, $n = 1, 2, \ldots$, be a sequence of simple functions,

$$g^{(n)}(t, \omega) = \sum_{k=0}^{\infty} g(t_k^{(n)}, \omega)\chi_{(t_k^{(n)}, t_{k+1}^{(n)})}(t), \qquad 0 = t_0^{(n)} < \cdots < t_n^{(n)} = T, \quad (5.9)$$

such that

$$M \int_0^T |g(t, \omega) - g^{(n)}(t, \omega)|^2(|d\langle x, W\rangle_t| + dt) \to 0, \qquad n \to \infty.$$

(The existence of such a sequence in a more general situation is proved in Lemma 5.3.) Then, because of (5.4) and (4.48), (P-a.s.)

$$\begin{aligned} M[\langle x, y\rangle_t - \langle x, y\rangle_s | \mathscr{F}_s] &= M[(x_t - x_s)(y_t - y_s) | \mathscr{F}_s] \\ &= M\left[(x_t - x_s) \int_s^t g(u, \omega) dW_u | \mathscr{F}_s\right] \\ &= M\left[x_t \int_s^t g(u, \omega) dW_u | \mathscr{F}_s\right] \\ &= \underset{n\to\infty}{\text{l.i.m.}}\, M\left[x_t \int_s^t g^{(n)}(u, \omega) dW_u | \mathscr{F}_s\right]. \end{aligned}$$

According to (5.9),

$$\int_s^t g^{(n)}(u, \omega) dW_u = \sum_{l \le k \le m} g(t_k^{(n)}, \omega)[W_{t \wedge t_{k+1}^{(n)}} - W_{s \wedge t_k^{(n)}}],$$

where l and m are found from the conditions $t_l^{(n)} \le s < t_{l+1}^{(n)}$, $t_m^{(n)} < t \le t_{m+1}^{(n)}$.

Without restricting the generality we may assume that $t_l^{(n)} = s$, $t_{m+1}^{(n)} = t$. Then

$$\begin{aligned} &M\left[x_t \int_s^t g^{(n)}(u, \omega) dW_u | \mathscr{F}_s\right] \\ &= \sum_{l \le k \le m} M\{x_t g(t_k^{(n)}, \omega)[W_{t_{k+1}^{(n)}} - W_{t_k^{(n)}}] | \mathscr{F}_s\} \\ &= \sum_{l \le k \le m} M\{M(x_t | \mathscr{F}_{t_{k+1}^{(n)}}) g(t_k^{(n)}, \omega)[W_{t_{k+1}^{(n)}} - W_{t_k^{(n)}}] | \mathscr{F}_s\} \\ &= \sum_{l \le k \le m} M\{g(t_k^{(n)}, \omega)[W_{t_{k+1}^{(n)}} - W_{t_k^{(n)}}] x_{t_{k+1}^{(n)}} | \mathscr{F}_s\}, \end{aligned} \quad (5.10)$$

where

$$
\begin{aligned}
&M\{g(t_k^{(n)}, \omega)[W_{t_{k+1}^{(n)}} - W_{t_k^{(n)}}]x_{t_{k+1}^{(n)}} | \mathscr{F}_s\} \\
&= M\{g(t_k^{(n)}, \omega)M[(W_{t_{k+1}^{(n)}} - W_{t_k^{(n)}})(x_{t_{k+1}^{(n)}} - x_{t_k^{(n)}}) | \mathscr{F}_{t_k^{(n)}}] | \mathscr{F}_s\} \\
&= M\{g(t_k^{(n)}, \omega)M[\langle x, W\rangle_{t_{k+1}^{(n)}} - \langle x, W\rangle_{t_k^{(n)}} | \mathscr{F}_{t_k^{(n)}}] | \mathscr{F}_s\} \\
&= M\{g(t_k^{(n)}, \omega)[\langle x, W\rangle_{t_{k+1}^{(n)}} - \langle x, W\rangle_{t_k^{(n)}}] | \mathscr{F}_s\}.
\end{aligned} \tag{5.11}
$$

From (5.10) and (5.11) it follows that

$$M\left[x_t \int_0^t g^{(n)}(u, \omega)dW_u | \mathscr{F}_s\right] = M\left[\int_s^t g^{(n)}(u, \omega)d\langle x, W\rangle_u | \mathscr{F}_s\right]. \tag{5.12}$$

Passing in (5.12) to the limit with $n \to \infty$ we infer that (P-a.s.)

$$\underset{n\to\infty}{\text{l.i.m.}}\, M\left[x_t \int_s^t g^{(n)}(u, \omega)dW_u | \mathscr{F}_s\right] = M\left[\int_s^t g(u, \omega)d\langle x, W\rangle_u | \mathscr{F}_s\right]. \tag{5.13}$$

Thus

$$M[\langle x, y\rangle_t - \langle x, y\rangle_s | \mathscr{F}_s] = M\left[\int_s^t g(u, \omega)d\langle x, W\rangle_u | \mathscr{F}_s\right],$$

where the process $\int_0^t g(u, \omega)d\langle x, W\rangle_u$ can be represented as the difference between two natural increasing processes. Hence, according to Theorem 5.2, $\langle x, y\rangle_t$ permits the representation given by (5.8).

The final part of the lemma follows from Lemma 5.5, which will be proved below.

PROOF OF THEOREM 5.3. Let $(g(t, \omega), \mathscr{F}_t)$, $t \le T$, be a function satisfying the conditions of Lemma 5.1 and such that $g^2(t, \omega) = g(t, \omega)$ and $\int_0^t g(t, \omega)dt = 0$ (P-a.s.). We will show that $\int_0^T g(t, \omega)d\langle x, W\rangle_t = 0$ (P-a.s.) also.

For this purpose assume $y_t = \int_0^t g(s, \omega)dW_s$. It is clear that the process $Y = (y, \mathscr{F}_t)$, $t \le T$, is a square integrable martingale and, by Lemma 5.1,

$$\langle x, y\rangle_t = \int_0^t g(s, \omega)d\langle x, W\rangle_s. \tag{5.14}$$

But $My_t^2 = M \int_0^t g^2(s, \omega)ds = 0$. Hence $y_t = 0$ (P-a.s.), $t \le T$, and, therefore, $\langle x, y\rangle_t = 0$ (P-a.s.), $t \le T$. From (5.14) it now follows that

$$\int_0^T g(s, \omega)d\langle x, W\rangle_s = 0 \qquad (P\text{-a.s.}). \tag{5.15}$$

We shall define in the measure space $([0, T] \times \Omega, \mathscr{B}_{[0,T]} \times \mathscr{F}_T)$ the measure $Q(\cdot)$ assuming it to equal

$$Q(S \times A) = \int_A \left[\int_S d\langle x, W\rangle_u\right] dP(\omega)$$

on the sets $S \times A$, $S \in \mathscr{B}_{[0,T]}$, $A \in \mathscr{F}_T$. Then it follows from (5.15) that the measure Q is absolutely continuous with respect to R where $R(S \times A) = \lambda(S)P(A)$, λ is a Lebesgue measure, $\lambda(dt) = dt$. Therefore, there exists a $\mathscr{B}_{[0,T]} \times \mathscr{F}_T$-measurable function $f(t, \omega)$ with $\int_\Omega \int_0^T |f(t, \omega)| dt\, dP(\omega) < \infty$, such that

$$Q(S \times A) = \int_A \int_S f(t, \omega) dt\, dP(\omega).$$

From this we find that

$$\int_A \langle x, W \rangle_t\, dP(\omega) = \int_A \left[\int_0^t f(s, \omega) ds \right] dP(\omega),$$

and, because of the arbitrariness of the set $A \in \mathscr{F}_T$,

$$\langle x, W \rangle_t = \int_0^t f(s, \omega) ds \qquad (P\text{-a.s.}) \tag{5.16}$$

for all t, $0 \le t \le T$.

The representation given in (5.16) thus obtained is not yet the desired representation, (5.7), since the proof so far only shows that the function $f(t, \omega)$ is $\mathscr{B}_{[0,T]} \times \mathscr{F}_T$-measurable, and it does not follow that at each fixed t it is $\mathscr{F}_t$-measurable. □

Let us show that actually there exists a version of the function $f(t, \omega)$ which is $\mathscr{F}_t$-measurable at each t, $0 \le t \le T$. Let us recall that a Radon–Nikodym derivative $f(t, \omega)$ is defined uniquely only (P-a.s.). This follows immediately from the following:

Lemma 5.2. *Let $(\Omega, \mathscr{F}, P)$ be a complete probability space and let $(\mathscr{F}_t)$, $t \ge 0$, be a right continuous family of the sub-σ-algebras $\mathscr{F}$, augmented by the sets from $\mathscr{F}$ of zero probability. Assume that the $\mathscr{B} \times \mathscr{F}$-measurable function $F(t, \omega)$ is $\mathscr{F}_t$-measurable at each $t \ge 0$ and (P-a.s.) absolutely continuous*

$$F(t, \omega) = \int_0^t f(s, \omega) ds,$$

where the $\mathscr{B} \times \mathscr{F}$-measurable function $f(s, \omega)$ is such that

$$P\left\{ \int_0^t |f(s, \omega)| ds < \infty \right\} = 1, \qquad t \ge 0.$$

Then there will be a $\mathscr{F}_t$-measurable function $\tilde{f}(t, \omega)$ for each $t \ge 0$ such that

$$F(t, \omega) = \int_0^t \tilde{f}(s, \omega) ds \qquad (P\text{-a.s.}), \qquad t \ge 0,$$

and

$$P\left\{ \int_0^t |\tilde{f}(s, \omega)| ds < \infty \right\} = 1, \qquad t \ge 0.$$

PROOF. If the function $f(t, \omega)$ is continuous (P-a.s.) (over $t \le T$), then it can be assumed that $\tilde{f}(t, \omega) = f(t, \omega)$. Actually, in this case

$$f(t, \omega) = \lim_{\Delta \downarrow 0} \frac{F(t + \Delta, \omega) - F(t, \omega)}{\Delta}, \tag{5.17}$$

and at each $t \le T$ the values $f(t, \omega)$ will be $\mathscr{F}_t$-measurable because of the right continuity of the family $F = (\mathscr{F}_t)$.

If the function $f(t, \omega)$ is not continuous, then let us consider a sequence of continuous functions $\{f_n(t, \omega) = n \int_0^t e^{-n(t-s)} f(s, \omega)ds,\ n = 1, 2, \ldots\}$. It is known that this sequence has the property that with probability one

$$\lim_{n \to \infty} \int_0^T |f(t, \omega) - f_n(t, \omega)| dt = 0. \tag{5.18}$$

Let $\tilde{f}(t, \omega)$ be the limit of this sequence over the measure $\lambda \times P$, where λ is a Lebesgue measure on $[0, T]$, and let $\{f_{n_k}(t, \omega), k = 1, 2, \ldots\}$ be a subsequence of the sequence $\{f_n(t, \omega), n = 1, 2, \ldots\}$ converging a.e. over the measure $\lambda \times P$ to $\tilde{f}(t, \omega)$.

Let us show now that for each $t \le T$ the values $f_n(t, \omega)$, $n = 1, 2, \ldots$, and therefore $f_{n_k}(t, \omega)$, $k = 1, 2, \ldots$, and $\tilde{f}(t, \omega)$ are $\mathscr{F}_t$-measurable. For this we consider a sequence of the differential equations

$$\dot{x}_t^{(n)} = -nx_t^{(n)} + nF(t, \omega), \qquad n = 1, 2, \ldots, x_0^{(n)} = 0. \tag{5.19}$$

It is clear that the variables

$$x_t^{(n)} = n \int_0^t e^{-n(t-s)} F(s, \omega) ds$$

at each $t \le T$ are $\mathscr{F}_t$-measurable. Consequently, the variables $\dot{x}_t^{(n)}$ are the same.

Let us show now that $\dot{x}_t^{(n)} = f_n(t, \omega)$. Actually, from (5.19) and the definition of $F(t, \omega)$ we find that

$$\begin{aligned}
\dot{x}_t^{(n)} &= n[F(t, \omega) - x_t^{(n)}] \\
&= n\left[\int_0^t f(s, \omega)ds - n \int_0^t e^{-n(t-s)} \int_0^s f(u, \omega)du\, ds\right] \\
&= n\left[\int_0^t f(s, \omega)ds - \int_0^t f(s, \omega)\left(n \int_s^t e^{-n(t-u)} du\right)ds\right] \\
&= n \int_0^t e^{-n(t-s)} f(s, \omega)ds,
\end{aligned}$$

which fact proves $\mathscr{F}_t$-measurability (at any $t \le T$) of the variables $f_n(t, \omega)$, $n = 1, 2, \ldots$. Finally, it follows from (5.18) that

$$\int_0^t |\tilde{f}(s, \omega)| ds = \int_0^t |f(s, \omega)| ds < \infty \qquad (P\text{-a.s.}), \qquad t \ge 0$$

thus proving the lemma.

Applying this lemma to $A_t = \langle x, W\rangle_t$, $B_t = t$, we obtain the desired representation, (5.7). It remains only to show that in this representation $M\int_0^T a^2(s, \omega)ds < \infty$. Set

$$a_n(t, \omega) = a(t, \omega)\chi\{|a(t, \omega)| < n\}. \tag{5.20}$$

The process $(y_n(t), \mathscr{F}_t)$, $t \le T$ with $y_n(t) = \int_0^t a_n(s, \omega)dW_s$ is a square integrable martingale, and hence by Lemma 5.1

$$\begin{aligned}\langle x, y_n\rangle_t &= \int_0^t a_n(s, \omega)d\langle x, W\rangle_s \\ &= \int_0^t a_n(s, \omega)\frac{d\langle x, W\rangle_s}{ds}\,ds \\ &= \int_0^t a_n(s, \omega)a(s, \omega)ds = \int_0^t a_n^2(s, \omega)ds.\end{aligned} \tag{5.21}$$

This means

$$0 \le M(x_T - y_n(T))^2 = Mx_T^2 + M\int_0^T a_n^2(t, \omega)dt - 2M\langle x, y_n\rangle_T. \tag{5.22}$$

From (5.21) and (5.22) it now follows that

$$M\int_0^T a_n^2(t, \omega)dt \le Mx_T^2.$$

Taking limits as $n \to \infty$, we obtain

$$M\int_0^T a^2(t, \omega)dt \le Mx_T^2 < \infty. \qquad \square$$

5.2 Representation of square integrable martingales

5.2.1

Let us apply Theorem 5.3 to the proof of the next crucial result, the representation of square integrable martingales in the form of a sum of two orthogonal martingales, one of which is a stochastic integral over a Wiener process.

Theorem 5.4. *Let the family $F = (\mathscr{F}_t)$, $t \le T$, be right continuous, let the martingale $X = (x_t, \mathscr{F}_t) \in \mathscr{M}_T$, and let $W = (W_t, \mathscr{F}_t)$ be a Wiener process. Then*

$$x_t = \int_0^t a(s, \omega)dW_s + z_t \qquad (P\text{-a.s.}), \qquad t \le T, \tag{5.23}$$

where the process $(a(t, \omega), \mathscr{F}_t)$, $t \le T$, is such that $\langle x, W\rangle_t = \int_0^t a(s, \omega)ds$, $M\int_0^T a^2(s, \omega)ds < \infty$, and the martingale $Z = (z_t, \mathscr{F}_t) \in \mathscr{M}_T$.

The martingales $Z = (z_t, \mathscr{F}_t)$ *and* $Y = (y_t, \mathscr{F}_t)$, *where* $y_t = \int_0^t a(s, \omega) dW_s$, *are orthogonal* ($Z \perp Y$), *i.e.,*

$$\langle z, y \rangle_t = 0, \qquad t \leq T. \tag{5.24}$$

Proof. The existence of a process $(a(t, \omega), \mathscr{F}_t)$ with $M \int_0^T a^2(t, \omega) dt < \infty$ and

$$\langle x, W \rangle_t = \int_0^t a(s, \omega) ds \tag{5.25}$$

follows from Theorem 5.3.

Let us set $y_t = \int_0^t a(s, \omega) dW_s$ and $z_t = x_t - y_t$. It is obvious that $Z = (z_t, \mathscr{F}_t) \in \mathscr{M}_T$, and, by Lemma 5.1,

$$\langle x, y \rangle_t = \int_0^t a(s, \omega) d\langle x, W \rangle_s = \int_0^t a^2(s, \omega) ds. \tag{5.26}$$

Hence

$$\langle z, y \rangle_t = \langle x - y, y \rangle_t = \langle x, y \rangle_t - \langle y \rangle_t = 0,$$

i.e., $Z \perp Y$. □

Note 1. If $M x_t^2 = M \int_0^t a^2(s, \omega) ds$, then $z_t = 0$ (P-a.s.), $t \leq T$, and

$$x_t = \int_0^t a(s, \omega) dW_s.$$

Actually,

$$M x_t^2 = M(z_t + y_t)^2 = M z_t^2 + M y_t^2.$$

But $M x_t^2 = M y_t^2 = M \int_0^t a^2(s, \omega) ds$. Hence $M z_t^2 = 0$, and, consequently, $z_t = 0$ (P-a.s.), $t \leq T$.

Note 2. If, under the conditions of Theorem 5.4, $W_t = (W_1(t), \ldots, W_n(t))$ is an n-dimensional Wiener process with respect to $(\mathscr{F}_t)$, $t \leq T$, then in the same way it is proved that for the martingale $X = (x_t, \mathscr{F}_t) \in \mathscr{M}_T$ there exist processes $(a_i(s, \omega), \mathscr{F}_s)$ with $\langle x, W_i \rangle_t = \int_0^t a_i(s, \omega) ds$, $M \int_0^T a_i^2(s, \omega) ds < \infty$, $i = 1, \ldots, n$, and a martingale $Z = (z_t, \mathscr{F}_t) \in \mathscr{M}_T$, such that

$$x_t = \sum_{i=1}^n \int_0^t a_i(s, \omega) dW_i(s) + z_t,$$

where

$$M\left(z_t \sum_{i=1}^n \int_0^t a_i(s, \omega) dW_i(s) \right) = 0, \qquad t \leq T.$$

5.2.2

Any random process $X = (x_t, \mathscr{F}_t)$, $t \leq T$, of the form

$$x_t = \int_0^t a(s, \omega) dW_s, \qquad M \int_0^T a^2(s, \omega) ds < \infty,$$

is a square integrable martingale. The reverse result is also true in a certain sense.

Theorem 5.5. *Let $W = (W_t, \mathscr{F}_t^W)$ be a Wiener process, $t \le T$, and let $\mathscr{M}_T^W$ be a class of the square integrable martingales $X = (x_t, \mathscr{F}_t^W)$ with $\sup_{t \le T} Mx_t^2 < \infty$ and with right continuous trajectories. If $X \in \mathscr{M}_T^W$ then there exists a process $(f(s, \omega), \mathscr{F}_s^W)$, $s \le T$, with $M \int_0^T f^2(s, \omega)ds < \infty$ and such that, for all t,*

$$x_t = x_0 + \int_0^t f(s, \omega)dW_s \qquad (P\text{-a.s.}). \tag{5.27}$$

PROOF. First of all we note that a (augmented) system of the σ-algebras $F^W = (\mathscr{F}_t^W)$, $t \le T$, is continuous (Theorem 4.3). By Theorem 5.3

$$\langle x, W \rangle_t = \int_0^t f(s, \omega)ds,$$

where $f(s, \omega)$ is $\mathscr{F}_s^W$-measurable, $s \le T$. Let us assume $\tilde{x} = x_t - x_0$. It is clear that $\tilde{X} = (\tilde{x}_t, \mathscr{F}_t^W) \in \mathscr{M}_T^W$ and that $\langle \tilde{x}, W \rangle_t = \int_0^t f(s, \omega)ds$. Then, by Theorem 5.4,

$$\tilde{x}_t = \int_0^t f(s, \omega)dW_s + z_t,$$

where $Mz_t \int_0^t f(s, \omega)dW_s = 0$, $t \le T$.

Let us show that in this case $z_t = 0$ (P-a.s.) for all $t \le T$. Since at each t the values z_t are $\mathscr{F}_t^W$-measurable, it suffices to establish that for any n, $n = 1, 2, \ldots,$

$$Mz_t \prod_{j=1}^n F_j(W_{t_j}) = 0, \tag{5.28}$$

where the $F_j(x)$ are bounded Borel measurable functions and $0 \le t_1 \le \cdots \le t_n \le t$.

Let us take $n = 1$, $F_1(x) = e^{i\lambda x}$, $-\infty < \lambda < \infty$, and prove that, for all $s \le t$,

$$Mz_t e^{i\lambda W_s} = 0. \tag{5.29}$$

We have

$$Mz_t e^{i\lambda W_s} = M[M(z_t | \mathscr{F}_s^W)e^{i\lambda W_s}] = Mz_s e^{i\lambda W_s}.$$

By Ito's formula,

$$e^{i\lambda W_s} = 1 + i\lambda \int_0^s e^{i\lambda W_u}\, dW_u - \frac{\lambda^2}{2} \int_0^s e^{i\lambda W_u}\, du. \tag{5.30}$$

From this we find

$$Mz_s e^{i\lambda W_s} = Mz_s + i\lambda M\left[z_s \int_0^s e^{i\lambda W_u}\, dW_u\right] - \frac{\lambda^2}{2} M\left[z_s \int_0^s e^{i\lambda W_u}\, du\right].$$

But $Mz_s = 0$,

$$M\left[z_s \int_0^s e^{i\lambda W_u}\,du\right] = M\int_0^s z_s e^{i\lambda W_u}\,du = M\left[\int_0^s M(z_s|\mathscr{F}_u^W)e^{i\lambda W_u}\,du\right],$$

and, by Lemma 5.1,

$$\begin{aligned} M\left[z_s \int_0^s e^{i\lambda W_u}\,dW_u\right] &= M\left[\tilde{x}_s - \int_0^s f(u,\omega)dW_u\right]\left[\int_0^s e^{i\lambda W_u}\,dW_u\right] \\ &= M\tilde{x}_s \int_0^s e^{i\lambda W_u}\,dW_u - M\int_0^s f(u,\omega)dW_u \int_0^s e^{i\lambda W_u}\,dW_u \\ &= M\int_0^s f(u,\omega)e^{i\lambda W_u}\,du - M\int_0^s f(u,\omega)e^{i\lambda W_u}\,du = 0. \end{aligned}$$

Hence

$$Mz_s e^{i\lambda W_s} = -\frac{\lambda^2}{2}\int_0^s M(z_u e^{i\lambda W_u})du,$$

and therefore

$$Mz_t e^{i\lambda W_s} = Mz_s e^{i\lambda W_s} = 0.$$

Because of the arbitrariness of λ, $-\infty < \lambda < \infty$, it follows that for any bounded Borel measurable function $F_1(x)$ Equation (5.29) is satisfied.

Let us now prove (5.28) by induction. For any bounded functions $F_1(x), \ldots, F_{n-1}(x)$, let

$$Mz_t \prod_{j=1}^{n-1} F_j(W_{t_j}) = 0.$$

It has to be shown that then $Mz_t \prod_{j=1}^{n} F_j(W_{t_j}) = 0$. Let us first put

$$F_n(W_{t_n}) = e^{i\lambda W_{t_n}}, \qquad -\infty < \lambda < \infty.$$

Because of (5.30),

$$e^{i\lambda W_{t_n}} = e^{i\lambda W_{t_{n-1}}} + i\lambda \int_{t_{n-1}}^{t_n} e^{i\lambda W_u}\,dW_u - \frac{\lambda^2}{2}\int_{t_{n-1}}^{t_n} e^{i\lambda W_u}\,du.$$

Hence

$$\begin{aligned} M\left[z_t e^{i\lambda W_{t_n}} \prod_{j=1}^{n-1} F_j(W_{t_j})\right] &= M\left[z_t e^{i\lambda W_{t_{n-1}}} \prod_{j=1}^{n-1} F_j(W_{t_j})\right] \\ &\quad + i\lambda M\left[z_t \prod_{j=1}^{n-1} F_j(W_{t_j}) \int_{t_{n-1}}^{t_n} e^{i\lambda W_u}\,dW_u\right] \\ &\quad - \frac{\lambda^2}{2} M\left[z_t \prod_{j=1}^{n-1} F_j(W_{t_j}) \int_{t_{n-1}}^{t_n} e^{i\lambda W_u}\,du\right]. \end{aligned} \tag{5.31}$$

For proof by induction, assume

$$Mz_t e^{i\lambda W_{t_{n-1}}} \prod_{j=1}^{n-1} F_j(W_{t_j}) = 0. \tag{5.32}$$

It is also clear that

$$M\left[z_t \prod_{j=1}^{n-1} F_j(W_{t_j}) \int_{t_{n-1}}^{t_n} e^{i\lambda W_u}\, dW_u | \mathscr{F}_{t_{n-1}}^W \right] = 0. \tag{5.33}$$

From (5.31)–(5.33) we obtain

$$\begin{aligned} Mz_t e^{i\lambda W_{t_n}} \prod_{j=1}^{n-1} F_j(W_{t_j}) &= Mz_{t_n} e^{i\lambda W_{t_n}} \prod_{j=1}^{n-1} F_j(W_{t_j}) \\ &= -\frac{\lambda^2}{2} \int_{t_{n-1}}^{t_n} Mz_s e^{i\lambda W_s} \prod_{j=0}^{n-1} F_j(W_{t_j}) ds. \end{aligned}$$

Therefore,

$$Mz_t e^{i\lambda W_{t_n}} \prod_{j=1}^{n-1} F_j(W_{t_j}) = 0.$$

Because of the arbitrariness of λ, $-\infty < \lambda < \infty$, (5.28) follows. □

Note 1. If $W_t = (W_1(t), \ldots, W_n(t))$ is an n-dimensional Wiener process and $X = (x_t, \mathscr{F}_t^W)$, $t \le T$, is a square integrable martingale with $\mathscr{F}_t^W = \sigma\{\omega: W_1(s), \ldots, W_n(s), s \le t\}$, then

$$x_t = x_0 + \sum_{i=1}^{n} \int_0^t f_i(s, \omega) dW_i(s), \tag{5.34}$$

where the variables $f_i(s, \omega)$, $i = 1, \ldots, n$, are $\mathscr{F}_s^W$-measurable and

$$\sum_{i=1}^{n} \int_0^T Mf_i^2(s, \omega) ds < \infty. \tag{5.35}$$

This can be proved in the same way as in the one-dimensional case ($n = 1$).

Note 2. From (5.27) it follows that any square integrable martingale $X = (x_t, \mathscr{F}_t^W) \in \mathscr{M}_T^W$ has continuous (P-a.s.) trajectories (more precisely, has a continuous modification).

5.3 The structure of functionals of a Wiener process

5.3.1

Let $(\Omega, \mathscr{F}, P)$ be a complete probability space and let $W = (W_t, \mathscr{F}_t^W)$, $t \le T$, be a Wiener process. We shall assume that the $\mathscr{F}_t^W$, $t \le T$, are augmented by sets from $\mathscr{F}$ having P-measure zero.

Theorem 5.6. *Let $\xi = \xi(\omega)$ be an $\mathscr{F}_T^W$-measurable random variable with $M\xi^2 < \infty$. Then there will be an F^W-adapted process $(f(t, \omega), \mathscr{F}_t^W)$, $t \le T$, with*

$$M \int_0^T f^2(t, \omega)dt < \infty \tag{5.36}$$

such that (P-a.s.)

$$\xi = M\xi + \int_0^T f(t, \omega)dW_t. \tag{5.37}$$

If, in addition, the random variable ξ and the process $W = (W_t), 0 \le t \le T$, form a Gaussian system (see Section 1.1), i.e., the mutual distribution of ξ and W is Gaussian, then there exists a deterministic measurable function $f = f(t), 0 \le t \le T$, with $\int_0^T f^2(t)dt < \infty$, such that

$$\xi = M\xi + \int_0^T f(t)dW_t. \tag{5.38}$$

Proof. Let $x_t = M(\xi | \mathscr{F}_t^W)$, where conditional expectations are selected so that the process $x_t, 0 \le t \le T$, has right continuous trajectories (this can be done because of Theorem 3.1). Then the martingale $X = (x_t, \mathscr{F}_t^W) \in \mathscr{M}_T$ and, by Theorem 5.5, we can find $f(t, \omega)$ with the above properties and such that

$$x_t = M(\xi | \mathscr{F}_0^W) + \int_0^t f(s, \omega)dW_s. \tag{5.39}$$

From this (5.37) follows since $M(\xi | \mathscr{F}_0^W) = M\xi$ (P-a.s.) and $x_T = \xi$.

Let us assume now that the mutual distribution of ξ and W is Gaussian. Put

$$\Delta = \frac{T}{2^n},$$

$$\begin{aligned}\mathscr{F}_{T,n}^W &= \sigma\{\omega: W_0, W_\Delta, \dots, W_T\} \\ &= \sigma\{\omega: W_\Delta - W_0, W_{2\Delta} - W_\Delta, \dots, W_T - W_{(T-\Delta)}\}.\end{aligned}$$

Since $\mathscr{F}_{T,n}^W \subseteq \mathscr{F}_{T,n+1}^W$ and $\mathscr{F}_T^W = \sigma(\bigcup_n \mathscr{F}_{T,n}^W)$, then, by Levy's theorem (Theorem 1.5), $\xi_n = M(\xi | \mathscr{F}_{T,n}^W) \to \xi$, as $n \to \infty$, with probability 1. The sequence of the random variables $\{(\xi_n - \xi)^2, n = 1, 2, \dots\}$ is uniformly integrable, and hence

$$\lim_{n \to \infty} M|\xi_n - \xi|^2 = 0.$$

Therefore, $\lim_{n,m \to \infty} M|\xi_n - \xi_m|^2 = 0$. But, because of Corollary 3 of the Theorem on normal correlation (Theorem 13.1),

$$\begin{aligned}\xi_n &= M(\xi | \mathscr{F}_{T,n}^W) \\ &= M\xi + \sum_{k=0}^{2^n - 1} \frac{M[(\xi - M\xi)(W_{(k+1)\Delta} - W_{k\Delta})]}{\Delta} [W_{(k+1)\Delta} - W_{k\Delta}] \\ &= M\xi + \int_0^T f_n(s)dW_s,\end{aligned}$$

where

$$f_n(s) = \frac{1}{\Delta} M[(\xi - M\xi)(W_{(k+1)\Delta} - W_{k\Delta})], \qquad k\Delta \le s < (k+1)\Delta,$$

and, obviously, $\int_0^T f_n^2(s)ds < \infty$. Consequently, by the properties of stochastic integrals,

$$\lim_{n, m \to \infty} M|\xi_n - \xi_m|^2 = \lim_{n, m \to \infty} \int_0^T [f_n(s) - f_m(s)]^2 \, ds.$$

From this it follows that there exists a function $f(s)$, $0 \le s \le T$, with $\int_0^T f^2(s)ds < \infty$, such that

$$\lim_{n \to \infty} \int_0^T [f_n(s) - f(s)]^2 \, ds = 0 \quad \text{and} \quad \text{l.i.m.}_{n \to \infty} \xi_n = M\xi + \int_0^T f(s)dW_s.$$

On the other hand, $\text{l.i.m.}_{n \to \infty} \xi_n = \xi$. Hence

$$\xi = M\xi + \int_0^T f(s)dW_s. \qquad \square$$

Note 1. Let us note that, in proving (5.38), (5.37) was not used. Actually, (5.38) is only a corollary of the Normal correlation theorem. If it is known that $\xi = M\xi + \int_0^T f(s, \omega)dW_s$, then the representation $\xi = M\xi + \int_0^T Mf(s, \omega)dW_s$ will be also valid. To make sure of this, it suffices to note that in this case we have

$$f_n(s) = \frac{1}{\Delta} \int_{k\Delta}^{(k+1)\Delta} Mf(t, \omega)dt, \qquad k\Delta \le s \le (k+1)\Delta$$

and, since $\lim_{n \to \infty} f_n(s) = Mf(s, \omega)$ for almost all s (see the proof of Lemma 4.4), the function $f(s) = Mf(s, \omega)$ can be taken as the function $f(s)$ entering (5.38).

Note 2. (5.38) becomes, generally speaking, invalid if it is assumed that the random variable ξ is normally distributed but that the joint distribution (ξ, W) is not Gaussian.

Indeed, the random process

$$\xi_t = \int_0^t S(W_s)dW_s,$$

where

$$S(x) = \begin{cases} 1, & x \ge 0, \\ -1, & x < 0, \end{cases}$$

is a Wiener process. Therefore, the random variable $\xi = \xi_T$ is Gaussian, but it cannot be represented in the form of a stochastic integral $\int_0^T f(s)dW_s$ with a deterministic function $f(s)$.

Note 3. From (5.38) it follows that

$$f(t) = \frac{d}{dt} M[(\xi - M\xi)W_t].$$

EXAMPLE 1. Let $\xi = \int_0^T W_s\, ds$. Since (ξ, W) is a Gaussian system, then ξ can be represented in the form

$$\int_0^T W_t\, dt = \int_0^T (T - t) dW_t$$

(this relationship can be also easily obtained from Ito's formula).

EXAMPLE 2. Let $\xi = W_1^4$. Then

$$W_1^4 = 3 + \int_0^1 [12(1 - t)W_t + 4W_t^3] dW_t.$$

Actually, let $x_t = M[W_1^4 | \mathscr{F}_t^W] = M[W_1^4 | W_t]$. Since the distribution $P(W_1 \le x | W_t)$ is normal, $N(W_t, 1 - t)$, then

$$\begin{aligned} x_t &= M[W_1^4 | W_t] = M[(W_1 - W_t + W_t)^4 | W_t] \\ &= M[(W_1 - W_t)^4 | W_t] + 4M[(W_1 - W_t)^3 W_t | W_t] \\ &\quad + 6M[(W_1 - W_t)^2 W_t^2 | W_t] + 4M[(W_1 - W_t) W_t^3 | W_t] + W_t^4 \\ &= 3(1 - t)^2 + 6(1 - t)W_t^2 + W_t^4. \end{aligned}$$

From this, by Ito's formula, we find that $dx_t = [12(1 - t)W_t + 4W_t^3] dW_t$, which, by virtue of the equation $MW_1^4 = 3$, leads to the desired representation, (5.37).

5.3.2

According to Theorem 5.5, any square integrable martingale $X = (x_t, \mathscr{F}_t^W) \in \mathscr{M}_T^W$ permits the representation given by (5.27), where the function $f(t, \omega)$ is such that $M \int_0^T f^2(t, \omega) dt < \infty$. Let us consider now the possibility of an analogous representation of the martingales $X = (x_t, \mathscr{F}_t^W)$ satisfying, instead of the condition $\sup_{t \le T} Mx_t^2 < \infty$, a weaker requirement, $\sup_{t \le T}$, $M|x_t| < \infty$.

Theorem 5.7. *Let* $X = (x_t, \mathscr{F}_t^W)$, $t \le T$, *be a martingale having right continuous trajectories and such that*

$$\sup_{t \le T} M|x_t| < \infty. \tag{5.40}$$

Then there will be an F^W*-adapted process* $(f(t, \omega), \mathscr{F}_t^W)$, $t \le T$, *such that*

$$P\left(\int_0^T f^2(s, \omega) ds < \infty\right) = 1 \tag{5.41}$$

and such that, for all $t \le T$,

$$x_t = x_0 + \int_0^t f(s, \omega)dW_s. \tag{5.42}$$

The representation given by (5.42) *is unique.*

PROOF. First of all let us show that the martingale under consideration, $X = (x_t, \mathscr{F}_t^W)$, has continuous trajectories.

Let $\{x_T^{(n)}, n = 1, 2, \dots\}$ be a sequence of $\mathscr{F}_T^W$-measurable functions with $M(x_T^{(n)})^2 < \infty$ such that

$$M|x_T - x_T^{(n)}| < \frac{1}{n^2}.$$

Denote by $x_t^{(n)}$ the right continuous modification $M(x_T^{(n)}|\mathscr{F}_t^W)$, existing by Theorem 3.1. Then, by Theorem 5.5,

$$x_t^{(n)} = x_0^{(n)} + \int_0^t f_n(s, \omega)dW_s, \tag{5.43}$$

where $M \int_0^T f_n^2(s, \omega)ds < \infty$. From this representation it follows that the martingale $X^{(n)} = (x_t^{(n)}, \mathscr{F}_t^W)$, $t \le T$, has a continuous modification. The process $(x_t - x_t^{(n)}, \mathscr{F}_t^W)$ has right continuous trajectories, and, by (3.6), for any $\varepsilon > 0$,

$$P\left\{\sup_{0 \le t \le T} |x_t - x_t^{(n)}| > \varepsilon\right\} \le \varepsilon^{-1}M|x_T - x_T^{(n)}| \le \varepsilon^{-1}n^{-2}.$$

Hence, by the Borel–Cantelli lemma,

$$\lim_n \sup_{0 \le t \le T} |x_t - x_t^{(n)}| = 0 \qquad (P\text{-a.s.}).$$

From this it follows that the (P-a.s.) functions $x_t, t \le T$, are continuous as uniform limits of the continuous functions $x_t^{(n)}$, $t \le T$. Let us pass now immediately to proving (5.42). Define a Markov time

$$\tau_n = \inf\{t \le T: |x_t| \ge n\},$$

assuming $\tau_n = T$ if $\sup_{t \le T}|x_t| < n$. It is clear that $\{\tau_n \le t\} \in \mathscr{F}_t^W$ and that the process $X_n = (x_n(t), \mathscr{F}_t^W)$, with $x_n(t) = x_{t \wedge \tau_n}$, forms a martingale (see (3.16)).

Because of the continuity (P-a.s.) of the process $x_t, t \le T$,

$$\sup_{t \le T} |x_{t \wedge \tau_n}| \le n.$$

Then applying Theorem 5.5 to the martingales $X_n = (x_n(t), \mathscr{F}_t^W)$ we obtain that, for each n, $n = 1, 2, \dots$,

$$x_n(t) = x^n(0) + \int_0^t f_n(s, \omega)dW_s,$$

where $M \int_0^T f_n^2(s, \omega)ds < \infty$.

Let us note that for $m \geq n$,

$$x_m(t \wedge \tau_n) = x_n(t)$$

and

$$x_m(t \wedge \tau_n) = x_n(0) + \int_0^{t \wedge \tau_n} f_m(s, \omega) dW_s$$

$$= x_n(0) + \int_0^t f_m(s, \omega) \chi_{\left\{\sup_{u \leq s} |x_u| \leq n\right\}}(s) dW_s.$$

From this, by (4.49), we find

$$\int_0^T M\{f_m(s, \omega)\chi_{\left\{\sup_{u \leq s} |x_u| \leq n\right\}}(s) - f_n(s, \omega)\}^2 \, ds = 0.$$

Consequently, on the set of those (t, ω) for which $\sup_{u \leq t}|x_u| \leq n$,

$$f_n(t, \omega) = f_{n+1}(t, \omega) = \cdots.$$

Let us assume

$$f(t, \omega) = \begin{cases} f_1(t, \omega), & \text{if} \quad \sup\limits_{u \leq t}|x_u| \leq 1, \\ f_2(t, \omega), & \text{if} \quad 1 < \sup\limits_{u \leq t}|x_u| \leq 2, \\ \dots\dots\dots\dots\dots\dots\dots\dots & \end{cases}$$

The measurable function $f = f(t, \omega)$ thus constructed is $\mathscr{F}_t^W$-measurable for each t. Further for any n, $n = 1, 2, \ldots$,

$$\left\{\omega: \int_0^T f^2(s, \omega) ds = \infty\right\} \subseteq \left\{\omega: \int_0^T [f(s, \omega) - f_n(s, \omega)]^2 \, ds > 0\right\}$$

$$\subseteq \left\{\omega: \sup_{s \leq T}|x_s| \geq n\right\}.$$

But because of the continuity of the process x_t, $t \leq T$,

$$P\left\{\sup_{s \leq T}|x_s| \geq n\right\} \to 0, \qquad n \to \infty.$$

Hence $P(\int_0^T f^2(s, \omega) ds < \infty) = 1$ and the stochastic integral $\int_0^t f(s, \omega) dW_s$, $t \leq T$, is defined. Let us assume

$$\tilde{x}_t = x_0 + \int_0^t f(s, \omega) dW_s.$$

Because of the inequality

$$P\left\{\left|\int_0^t [f(s,\omega) - f_n(s,\omega)]dW_s\right| > \varepsilon\right\} \leq P\left\{\int_0^T [f(s,\omega) - f_n(s,\omega)]^2\, ds > \delta\right\} + \frac{\delta}{\varepsilon^2}$$

(see Note 7 in Section 4.2),

$$\tilde{x}_t = P\text{-}\lim_n x_n(t).$$

On the other hand, (P-a.s.)

$$\lim_n x_n(t) = \lim_n x_{t\wedge\tau_n} = x_t, \qquad t \leq T.$$

Therefore, (P-a.s.) for all $t \leq T$, $x_t = \tilde{x}_t$ and

$$x_t = x_0 + \int_0^t f(s,\omega)dW_s.$$

It remains to establish that this representation is unique: if also $x_t = x_0 + \int_0^t f'(s,\omega)dW_s$ with a nonanticipative function $f'(s,\omega)$ such that $P(\int_0^T (f'(s,\omega))^2\, ds < \infty) = 1$, then $f(t,\omega) = f'(t,\omega)$ for almost all (t,ω).

Let $\bar{f}(t,\omega) = f(t,\omega) - f'(t,\omega)$. Then for the process $\bar{x}_t = \int_0^t \bar{f}(s,\omega)dW_s$, by Ito's formula,

$$\bar{x}_t^2 = \int_0^t \bar{f}^2(s,\omega)ds + 2\int_0^t \bar{x}_s \bar{f}(s,\omega)dW_s.$$

But $\bar{x}_t = 0$ (P-a.s.), $t \leq T$. Hence $\int_0^T \bar{f}^2(s,\omega)ds = 0$, from which it follows that $f(s,\omega) = f'(s,\omega)$ for almost all (s,ω). □

Note. Let $W_t = (W_1(t), \ldots, W_n(t))$ be an n-dimensional Wiener process, and let $\mathscr{F}_t^W = \sigma\{w: W_1(s), \ldots, W_n(s), s \leq t\}$. If $X = (x_t, \mathscr{F}_t^W)$, $t \leq T$, is a martingale and $\sup_{t\leq T} M|x_t| < \infty$, then there will be F^W-adapted processes $(f_i(t,\omega), \mathscr{F}_t^W)$, $i = 1, \ldots, n$, such that $P(\sum_{i=1}^n \int_0^T f_i^2(s,\omega)ds < \infty) = 1$ and (P-a.s.) for each $t \leq T$

$$x_t = x_0 + \sum_{i=1}^n \int_0^t f_i(s,\omega)dW_i(s).$$

The proof of this is based on (5.34) and is carried out in the same way as in the one-dimensional case.

5.3.3

From Theorem 5.7 the following useful result can be easily deduced (compare with Theorem 5.6).

Theorem 5.8. *Let $\xi = \xi(\omega)$ be an $\mathscr{F}_t^W$-measurable random variable with $M|\xi| < \infty$ and let $M(\xi|\mathscr{F}_t^W)$, $t \le T$ be a right continuous modification of conditional expectations. Then there will be a process $(f(t, \omega), \mathscr{F}_t^W)$, $0 \le t \le T$, such that $P(\int_0^T f^2(t, \omega)dt < \infty) = 1$ and such that for all t, $0 \le t \le T$,*

$$M(\xi|\mathscr{F}_t^W) = M\xi + \int_0^t f(s, \omega)dW_s \qquad (P\text{-a.s.}). \tag{5.44}$$

In particular,

$$\xi = M\xi + \int_0^T f(s, \omega)dW_s. \tag{5.45}$$

PROOF. The proof follows from Theorem 5.7, if it is assumed that $x_t = M(\xi|\mathscr{F}_t^W)$ with $x_0 = M\xi$. □

5.3.4

Theorem 5.9. *Let $\xi = \xi(\omega)$ be an $\mathscr{F}_T^W$-measurable random variable with $P(\xi > 0) = 1$ and $M\xi < \infty$. Then there will be a process $(\varphi(t, \omega), \mathscr{F}_t^W)$, $0 \le t \le T$, such that $P(\int_0^T \varphi^2(t, \omega)dt < \infty) = 1$ and for all $t \le T$ (P-a.s.)*

$$M(\xi|\mathscr{F}_t^W) = \exp\left[\int_0^t \varphi(s, \omega)dW_s - \frac{1}{2}\int_0^t \varphi^2(s, \omega)ds\right]M\xi. \tag{5.46}$$

In particular,

$$\xi = \exp\left[\int_0^T \varphi(s, \omega)dW_s - \frac{1}{2}\int_0^T \varphi^2(s, \omega)ds\right]M\xi. \tag{5.47}$$

PROOF. Let $x_t = M(\xi|\mathscr{F}_t^W)$, $t \le T$, be a right continuous modification of conditional expectations. Then, by Theorem 5.8,

$$x_t = M\xi + \int_0^t f(s, \omega)dW_s. \tag{5.48}$$

Let us show that

$$P\left(\inf_{t \le T} x_t > 0\right) = 1. \tag{5.49}$$

Indeed, the martingale $X = (x_t, \mathscr{F}_t^W)$, $t \le T$, is uniformly integrable. Henceforth, if $\tau = \tau(\omega)$ is a Markov time with $P(\tau \le T) = 1$, then, by Theorem 3.6,

$$x_\tau = M(\xi|\mathscr{F}_\tau^W). \tag{5.50}$$

Let us assume $\tau = \inf\{t \le T: x_t = 0\}$ and we will write $\tau = \infty$ if $\inf_{t \le T} x_t > 0$. On the set $\{\tau \le T\} = \{\inf_{t \le T} x_t = 0\}$ the value $x_\tau = 0$ since, according to (5.48), the process x_t, $t \le T$, is continuous (P-a.s.). Hence, because of (5.50),

$$0 = \int_{\{\tau \le T\}} x_\tau \, dP(\omega) = \int_{\{\tau \le T\}} \xi \, dP(\omega).$$

But $P(\xi > 0) = 1$. Henceforth $P(\tau \le T) = P(\inf_{t \le T} x_t = 0) = 0$.

Let us introduce a function

$$\varphi(t, \omega) = \frac{f(t, \omega)}{x_t} \left(= \frac{f(t, \omega)}{M\xi + \int_0^t f(s, \omega) dW_s} \right), \tag{5.51}$$

for which, because of the condition $P(\inf_{t \le T} x_t > 0) = 1$,

$$P\left(\int_0^T \varphi^2(t, \omega) dt < \infty \right) = 1.$$

Further on, according to (5.48) and (5.51),

$$dx_t = f(t, \omega) dW_t = \varphi(t, \omega) x_t \, dW_t.$$

A unique continuous (strong) solution of the equation

$$dx_t = \varphi(t, \omega) x_t \, dW_t, \qquad x_0 = M\xi, \tag{5.52}$$

exists and is determined by the formula

$$x_t = \exp\left[\int_0^t \varphi(s, \omega) dW_s - \frac{1}{2} \int_0^t \varphi^2(s, \omega) ds \right] M\xi. \tag{5.53}$$

Indeed, the fact that (5.53) provides a solution of Equation (5.52) follows from the Ito formula (see Example 3, Section 4.3). Let y_t, $t \le T$, be another solution of this equation. Then it is not difficult to check, making use of the Ito formula again, that $d(y_t/x_t) = 0$. From this we find $P\{\sup_{t \le T} |x_t - y_t| > 0\} = 0$. □

5.4 Stochastic integrals over square integrable martingales

5.4.1

In Chapter 4 the stochastic integral $I_t(f) = \int_0^t f(s, \omega) dW_s$ over a Wiener process $W = (W_t, \mathscr{F}_t)$, $t \ge 0$, for nonanticipative functions $f = f(t, \omega)$ satisfying the condition $M \int_0^\infty f^2(t, \omega) dt < \infty$ was defined. Among nontrivial properties of this integral the following two properties are most important:

$$M \int_0^t f(s, \omega) dW_s = 0, \tag{5.54}$$

$$M\left[\int_0^t f(s, \omega) dW_s \right]^2 = M \int_0^t f^2(s, \omega) ds. \tag{5.55}$$

A Wiener process is a square integrable martingale, $M(W_t - W_s | \mathscr{F}_s) = 0$, $t \ge s$, having the property that

$$M[(W_t - W_s)^2 | \mathscr{F}_s] = t - s. \tag{5.56}$$

The comparison of (5.56) with (5.2) shows that for a Wiener process the corresponding natural increasing process is $A_t \equiv \langle W \rangle_t = t$.

The analysis of the integral structure $I_t(f)$ implies that an analogous integral $\int_0^t f(s, \omega)dx_s$ can be defined over square integrable martingales $X = (x_t, \mathscr{F}_t) \in \mathscr{M}$. Indeed, they satisfy an equality,

$$M[(x_t - x_s)^2 | \mathscr{F}_s] = M[\langle x \rangle_t - \langle x \rangle_s | \mathscr{F}_s], \tag{5.57}$$

which is analogous to Equation (5.56), playing the key role in defining the stochastic integrals $I_t(f)$ over a Wiener process.

Denote $A_t = \langle x \rangle_t$, $t \geq 0$. One could expect that the natural class of functions $f = f(t, \omega)$, for which the stochastic integrals $\int_0^t f(s, \omega)dx_s$, $t \geq 0$, are to be defined, is a class of nonanticipative functions, satisfying the condition

$$M \int_0^\infty f^2(t, \omega)dA_t < \infty. \tag{5.58}$$

(5.58) is necessary if the stochastic integral is to have properties analogous to (5.54) and (5.55).

However, while considering arbitrary martingales $X = (x_t, \mathscr{F}_t) \in \mathscr{M}$ there emerges the additional fact that the class of functions for which the stochastic integral $\int_0^t f(s, \omega)dx_s$ can be defined depends essentially on properties of the natural processes $A_t = \langle x \rangle_t$ corresponding to the martingale $X = (x_t, \mathscr{F}_t) \in \mathscr{M}$.

We shall introduce three classes of functions: Φ_1, Φ_2, Φ_3, $(\Phi_1 \supseteq \Phi_2 \supseteq \Phi_3)$, for which stochastic integrals will be defined according to properties of the natural processes A_t, $t \geq 0$.

Let $(\Omega, \mathscr{F}, P)$ be a complete probability space and let $(\mathscr{F}_t)$, $t \geq 0$, be a right continuous nondecreasing family of the sub-σ-algebra $\mathscr{F}$ augmented by sets of $\mathscr{F}$ probability zero.

Definition 1. The measurable function $f = f(t, \omega)$, $t \geq 0$, $\omega \in \Omega$, *belongs to class* Φ_1 if it is nonanticipative, i.e.,

$$f(t, \omega) \text{ is } \mathscr{F}_t\text{-measurable} \tag{5.59}$$

for each $t \geq 0$.

Definition 2. The measurable function $f = f(t, \omega)$ *belongs to class* Φ_2 if it is strongly nonanticipative, i.e.,

$$f(\tau, \omega) \text{ is } \mathscr{F}_\tau\text{-measurable} \tag{5.60}$$

for each bounded Markov time τ (with respect to $(\mathscr{F}_t)$, $t \geq 0$).

Definition 3. The measurable function $f = f(t, \omega)$ *belongs to class* Φ_3 it if is nonanticipative and measurable with respect to the smallest σ-algebra on $\mathbb{R}^+ \times \Omega$ generated by nonanticipative processes having left continuous trajectories.

Note 1. Any function $f \in \Phi_3$ is strongly nonanticipative (corollary of Lemma 1.8).

Note 2. The left continuous functions $f = f(t, \omega)$ are *predictable* in the sense that $f(t, \omega) = \lim_{s \uparrow t} f(s, \omega)$ for each $t \geq 0$. Hence the functions of class Φ_3 are called nonanticipative and predictable.

By $L_A^2(\Phi_i)$ we shall denote the functions from class Φ_i satisfying the condition

$$M \int_0^\infty f^2(s, \omega) dA_s < \infty.$$

Definition 4. The function $f \in L_A^2(\Phi_1)$ is called a *simple function* if there exists a finite decomposition $0 = t_0 < \cdots < t_n < \infty$, such that

$$f(t, \omega) = \sum_{k=0}^{n-1} f(t_k, \omega) \chi_{(t_k, t_{k+1}]}(t). \tag{5.61}$$

Definition 5. The function $f \in L_A^2(\Phi_2)$ is called a *simple stochastic function* if there exists a sequence $\tau_0, \tau_1, \ldots, \tau_n$ of Markov times such that $0 = \tau_0 < \tau_1 < \cdots < \tau_n < \infty$ (P-a.s.) and

$$f(t, \omega) = \sum_{k=0}^{n-1} f(\tau_k, \omega) \chi_{(\tau_k, \tau_{k+1}]}(t). \tag{5.62}$$

The classes of simple functions and simple stochastic functions we shall denote by $\mathscr{E}$ and $\mathscr{E}_s$, respectively.

5.4.2

Let $X = (x_t, \mathscr{F}_t) \in \mathscr{M}$, $x_0 = 0$ (for simplicity) and $A_t = \langle x \rangle_t$, $t \geq 0$. We shall define the stochastic integral $I(f)$ (denoted by $\int_0^\infty f(s, \omega) dx_s$ as well) over a simple stochastic function $f = f(t, \omega)$, as follows:

$$I(f) = \sum_{k=0}^{n-1} f(\tau_k, \omega)[x_{\tau_{k+1}} - x_{\tau_k}]. \tag{5.63}$$

In particular, if $f = f(t, \omega)$ is the simple function defined in (5.61), then, by definition,

$$I(f) = \sum_{k=0}^{n-1} f(t_k, \omega)[x_{t_{k+1}} - x_{t_k}]. \tag{5.64}$$

If $f \in \mathscr{E}_s$, then under the stochastic integral $I_\tau(f) = \int_0^\tau f(s, \omega) dx_s$ an integral $I(g)$ over the function

$$g(s, \omega) = f(s, \omega) \chi_{\{s \leq \tau\}}(s) \tag{5.65}$$

will be understood.

Similarly, under the integral $I_{\sigma,\tau}(f) = \int_\sigma^\tau f(s, \omega) dx_s$, where $P(\sigma \leq \tau) = 1$, an integral over the function

$$g(s, \omega) = f(s, \omega) \chi_{\{\sigma < s \leq \tau\}}(s)$$

will be understood.

The stochastic integrals thus defined have the following properties (f, f_1 and f_2 are simple stochastic functions):

$$I_t(af_1 + bf_2) = aI_t(f_1) + bI_t(f_2) \quad (P\text{-a.s.}), \quad a, b = \text{const.}, t \geq 0; \tag{5.66}$$

$$\int_0^t f(s, \omega)dx_s = \int_0^u f(s, \omega)dx_s + \int_u^t f(s, \omega)dx_s \quad (P\text{-a.s.}); \tag{5.67}$$

$I_t(f)$ is a right continuous function over $t \geq 0$ (P-a.s.); (5.68)

$$M\left[\int_0^t f(u, \omega)dx_u | \mathscr{F}_s\right] = \int_0^s f(u, \omega)dx_u \quad (P\text{-a.s.}); \tag{5.69}$$

$$M\left[\int_0^t f_1(u, \omega)dx_u \int_0^t f_2(u, \omega)dx_u\right] = M\int_0^t f_1(u, \omega)f_2(u, \omega)dA_u. \tag{5.70}$$

In particular, from (5.69) and (5.70) it follows that:

$$M\int_0^t f(u, \omega)dx_u = 0; \tag{5.71}$$

$$M\left[\int_0^t f(u, \omega)dx_u\right]^2 = M\int_0^t f^2(u, \omega)dA_u. \tag{5.72}$$

As in the case of a Wiener process, the stochastic integral $\int_0^\infty f(s, \omega)dx_s$ for a measurable function $f = f(s, \omega)$ satisfying the condition $M\int_0^\infty f^2(s, \omega)dA_s < \infty$ will be constructed as the limit of the integrals $\int_0^\infty f_n(s, \omega)dx_s$ over the simple functions approximating (in a certain sense) $f(s, \omega)$.

In the lemmas given below the classes of functions permitting approximation by simple functions according to properties of the processes $A_t, t \geq 0$, are described.

5.4.3

Lemma 5.3. *Let $X = (x_t, \mathscr{F}_t) \in \mathscr{M}$, and let $A_t = \langle x \rangle_t$, $t \geq 0$, be a natural increasing process corresponding to a martingale X. Then the space $\mathscr{E}$ of simple functions is dense in $L_A^2(\Phi_3)$.*

Note 1. In a general case without additional restrictions on the martingale $X \in \mathscr{M}$, the closure of $\mathscr{E}$ (in $L_A^2(\Phi_3)$), does not necessarily include nonanticipative functions having right continuous trajectories.

Note 2. If $\tilde{A} = (\tilde{A}_t, \mathscr{F}_t)$, $t \geq 0$, is a modification of the process $A = (A_t, \mathscr{F}_t)$, then it is not difficult to show that $L_{\tilde{A}}^2(\Phi_3) = L_A^2(\Phi_3)$.

Lemma 5.4. *Let $X = (x_t, \mathscr{F}_t) \in \mathscr{M}$, with the corresponding natural process $A_t = \langle x \rangle_t$, $t \geq 0$, being continuous with probability one. Then the space $\mathscr{E}$ of simple functions is dense in $L_A^2(\Phi_2)$.*

Note 3. If the martingale $X = (x_t, \mathscr{F}_t) \in \mathscr{M}$ is quasi-continuous to the left (i.e., with probability one $x_{\tau_n} \to x_\tau$, if the sequence of Markov times $\tau_n \uparrow \tau$,

$P(\tau < \infty) = 1$), then the process $A_t = \langle x \rangle_t$, $t \geq 0$, is continuous (P-a.s.) (Theorem 3.11 and its corollary).

Lemma 5.5. *Let the martingale $X = (x_t, \mathscr{F}_t) \in \mathscr{M}$, with the corresponding natural process $A_t = \langle x \rangle_t$, $t \geq 0$, being absolutely continuous with probability one. Then the space $\mathscr{E}$ of simple functions is dense in $L_A^2(\Phi_1)$.*

We proceed now to proving these lemmas.

PROOF OF LEMMA 5.3. First we note that the σ-algebra Σ on $\mathbb{R}_+ \times \Omega$, generated by nonanticipative processes having left continuous trajectories, coincides with the σ-algebra generated by sets of the form $(a, b] \times B$, where $B \in \mathscr{F}_a$. Indeed, if the function $f = f(t, \omega)$ is nonanticipative, has left continuous trajectories and is bounded, then it is the limit of the sequence of the functions

$$f_n(t, \omega) = \sum_{k=0}^{n-1} f(t_k^{(n)}, \omega)\chi_{(t_k^{(n)}, t_{k+1}^{(n)}]}(t),$$

where

$$0 = t_0^{(n)} < t_1^{(n)} < \cdots < t_{k_n}^{(n)} = T \quad \text{and} \quad \max_{0 \leq k < k_n - 1} |t_{k+1}^{(n)} - t_k^{(n)}| \to 0, \quad n \to \infty.$$

From this it follows that it suffices to prove the lemma for the function $\chi = \chi_M(t, \omega)$, which is a characteristic function of a set $M \in \Sigma$ such that $M \subseteq [a, b] \times \Omega$.

Denote by $\nu = \nu(\cdot)$ the measure on $(\mathbb{R}_+ \times \Omega, \Sigma)$, defined on sets of the form $S \times B$ by the equality

$$\nu(S \times B) = M\left[\int_S dA_t; B\right] = \int_B \left[\int_S dA_t(\omega)\right] P(d\omega).$$

According to the definition of σ-algebra Σ, for the set $M \in \Sigma$ under consideration there will be a sequence of the sets $\{M_n, n = 1, 2, \ldots\}$ of the form $\bigcup_{i=0}^{n-1} (t_i, t_{i+1}] \times B_i$, where $a = t_0 < t_1 < \cdots < t_n = b$ and the sets B_i are $\mathscr{F}_{t_i}$-measurable, such that $M \supseteq M_n$ and $\nu(M - M_n) \leq 1/n$, i.e.,

$$\int |\chi_M(t, \omega) - \chi_{M_n}(t, \omega)|^2 \, d\nu(t, \omega) \leq \frac{1}{n}.$$

In other words,

$$M \int_0^\infty |\chi_M(t, \omega) - \chi_{M_n}(t, \omega)|^2 \, dA_t \leq \frac{1}{n},$$

which proves the lemma. □

Lemmas 5.5 and 5.6 (see below) will be used in the proof of Lemma 5.4.

PROOF OF LEMMA 5.5. In the case $A_t \equiv t$ the statement of the lemma is established in Chapter 4 (Lemma 4.4), where it was shown that there exist decompositions $0 = t_0^{(n)} < t_1^{(n)} < \cdots < t_n^{(n)} < \infty$ such that, for $f \in \mathscr{M}_\infty$,

$$M \int_0^\infty |f(t, \omega) - f_n(t, \omega)|^2 \, dt \to 0, \qquad n \to \infty, \tag{5.73}$$

$$f_n(t, \omega) = \sum_{k=0}^{\infty} f(t_k^{(n)}, \omega) \chi_{(t_k^{(n)} < t \leq t_{k+1}^{(n)})}(t). \tag{5.74}$$

Therefore, for some subsequence $n_i \uparrow \infty$, $i \to \infty$,

$$|f(t, \omega) - f_{n_i}(t, \omega)|^2 \to 0, \qquad i \to \infty, \tag{5.75}$$

for almost all (t, ω) (over measure $dt\, dP$). Hence, it also holds that $|f(t, \omega) - f_{n_i}(t, \omega)|^2 a(t, \omega) \to 0$, $i \to \infty$, for almost all (t, ω). Without restricting generality, the function f can be considered to be finite and such that $|f(t, \omega)| \leq K$. Then $|f(t, \omega) - f_n(t, \omega)|^2 a(t, \omega) \leq 4K^2 a(t, \omega)$, where

$$M \int_0^\infty a(t, \omega) dt = M A_\infty < \infty.$$

Consequently,

$$\lim_{i \to \infty} M \int_0^\infty |f(t, \omega) - f_{n_i}(t, \omega)|^2 \, dA_t$$

$$= \lim_{i \to \infty} M \int_0^\infty |f(t, \omega) - f_{n_i}(t, \omega)|^2 a(t, \omega) dt = 0, \tag{5.76}$$

proving the lemma for bounded functions $f = f(t, \omega)$ equal to zero beyond some finite interval. The general case is thus reduced to one which has already been proved (compare with the proof of Lemma 4.4). □

Lemma 5.6. *Let $0 \leq a < b < \infty$ and let $\alpha = \alpha(t)$, $t \in [a, b]$, be a continuous nondecreasing function. For each $u \subset \mathbb{R}$ we set*

$$\beta(u) = \begin{cases} \inf\{a \leq t \leq b : \alpha(t) > u\}, & \text{if } \alpha(b) > u, \\ b, & \text{if } \alpha(\beta) \leq u. \end{cases}$$

Then the function $\beta = \beta(u)$, $u \in \mathbb{R}$, has the following properties:

(1) *it does not decrease and is right continuous;*
(2) *if $\alpha(a) \leq u \leq \alpha(b)$, then $\alpha(\beta(u)) = u$;*
(3) *if $a < t \leq b$, then $\beta(u) < t \Leftrightarrow u < \alpha(t)$;*
(4) *if $\varphi = \varphi(t)$, $a \leq t \leq b$, is a measurable (Borel) bounded function, then*

$$\int \chi_{(a, b]}(t) \varphi(t) d\alpha(t) = \int_{\alpha(a)}^{\alpha(b)} \varphi(\beta(u)) du. \tag{5.77}$$

The proof of (1)–(3) is elementary, while (4) was noted in Section 1.1.

PROOF OF LEMMA 5.4. Let the function $f(t, \omega) \in L_A^2(\Phi_2)$ be bounded and equal to zero beyond some finite interval $[a, b]$, and let the process $A_t = A_t(\omega)$, $t \geq 0$, (P-a.s.) be continuous.

$$\beta_\omega(u) = \begin{cases} \inf\{a \leq t \leq b: A_t(\omega) > u\}, & \text{if} \quad A_b(\omega) > u, \\ b, & \text{if} \quad A_b(\omega) \leq u. \end{cases}$$

For each $u \in [0, \infty)$ the random variable $\beta_\omega(u)$ is a Markov time with values in $[a, b]$. Indeed, according to (3) of Lemma 5.6,

$$\{\omega: \beta_\omega(u) < t\} = \{\omega: u < A_t\}$$

for any $a \leq t \leq b$. Hence the Markov property of time $\beta_\omega(u)$ follows from Lemma 1.2.

We set $\tilde{\mathscr{F}}_u = \mathscr{F}_{\beta_\omega(u)}$ and $\tilde{f}(u, \omega) = f(\beta_\omega(u), \omega)$. Since the process $\beta_\omega(u)$, $u \geq 0$, has (P-a.s.) left continuous trajectories, then it is measurable (even progressively measurable). Hence from a measurability of the process $f = f(u, \omega)$ it follows that the process $\tilde{f} = \tilde{f}(u, \omega)$ will be also measurable.

According to the assumption made under the condition of the lemma, the function $f = f(t, \omega)$ is strongly nonanticipative and, therefore, with each $u \geq 0$ the random variables $\tilde{f}(u, \omega) = f(\beta_\omega(u), \omega)$ are $\tilde{\mathscr{F}}_u = \mathscr{F}_{\beta_\omega(u)}$-measurable. Because of the definition of the function $\beta_\omega(u)$,

$$u > A_b(\omega) \Rightarrow \beta_\omega(u) = b, \qquad u < A_a(\omega) \Rightarrow \beta_\omega(u) = a.$$

Hence, if $c = \sup_{t,\omega} |f(t, \omega)|$ and $f(t, \omega) = 0$ for $t \notin [a, b]$, then

$$M \int_0^\infty |\tilde{f}(u, \omega)|^2 \, du = M \int_{A_a}^{A_b} |\tilde{f}(u, \omega)|^2 \, du \leq c^2 M |A_b - A_a| < \infty.$$

Consequently, Lemma 5.5. is applicable to the function $\tilde{f} = \tilde{f}(u, \omega)$, $u \geq 0$; according to this lemma, for given $\varepsilon > 0$ a finite decomposition $0 = u_0 < u_1 < \cdots < u_n < \infty$ can be found such that

$$M \int_0^\infty |\tilde{f}(u, \omega) - \tilde{f}_n(u, \omega)|^2 \, du < \varepsilon,$$

where

$$\tilde{f}_n(u, \omega) = \sum_{k=0}^{n-1} \tilde{f}(u_k, \omega) \chi_{(u_k, u_{k+1}]}(u) = \sum_{k=0}^{n-1} f(\beta_\omega(u_k), \omega) \chi_{(u_k, u_{k+1}]}(u).$$

We shall show that the function

$$\varphi_n(t, \omega) = \chi_{(a, b]}(t) \tilde{f}_n(A_t, \omega) \tag{5.78}$$

is an ε-approximation of the function $f \in L_A^2(\Phi_2)$ under consideration, i.e.,

$$M \int_0^\infty |f(t, \omega) - \varphi_n(t, \omega)|^2 \, dA_t \leq \varepsilon.$$

For this purpose we shall note that, according to Lemma 5.6 (3), for any t, $a < t \leq b$, and k, $k = 0, 1, \ldots, n-1$,

$$\{\omega : u_k < A_t \leq u_{k+1}\} = \{\omega : \beta_\omega(u_k) < t \leq \beta_\omega(u_{k+1})\}.$$

Hence, taking into account the fact that $\beta_\omega(u_k) \in [a, b]$ for all $\omega \in \Omega$ and all k, $k = 0, 1, \ldots, n-1$, we conclude that the function $\varphi_n = \varphi_n(t, \omega)$, defined in (5.78), can be written in the following form:

$$\begin{aligned}
\varphi_n(t, \omega) &= \chi_{(a, b]}(t)\tilde{f}_n(A_t, \omega) \\
&= \chi_{(a, b]}(t)\left[\sum_{k=0}^{n-1} f(\beta_\omega(u_k), \omega)\chi_{(u_k, u_{k+1}]}(A_t)\right] \\
&= \chi_{(a, b]}(t) \sum_{k=0}^{n-1} f(\beta_\omega(u_k), \omega)\chi_{\{\beta_\omega(u_k) < t \leq \beta_\omega(u_{k+1})\}}(t) \\
&= \sum_{k=0}^{n-1} f(\beta_\omega(u_k), \omega)\chi_{\{\beta_\omega(u_k) < t \leq \beta_\omega(u_{k+1})\}}(t).
\end{aligned} \tag{5.79}$$

By assumption, the process $A_t = A_t(\omega)$, $t \geq 0$, is continuous (P-a.s.). Hence, from Lemma 5.6 (2), it follows that if $A_a(\omega) \leq u \leq A_b(\omega)$, then

$$A_{\beta_\omega(u)} = u \tag{5.80}$$

and $\beta_\omega(u) \in (a, b]$. Consequently, if $A_a(\omega) \leq u \leq A(\omega)$, then

$$\varphi_n(\beta_\omega(u), \omega) = \chi_{(a, b]}(\beta_\omega(u))\tilde{f}_n(A_{\beta_\omega(u)}, \omega) = \tilde{f}_n(u, \omega).$$

Then, according to Lemma 5.6 (4),

$$\begin{aligned}
M \int_0^\infty |f(t, \omega) - \varphi_n(t, \omega)|^2 \, dA_t &= M \int_{A_a}^{A_b} |f(\beta_\omega(u), \omega) - \varphi_n(\beta_\omega(u), \omega)|^2 \, du \\
&= M \int_{A_a}^{A_b} |f(\beta_\omega(u), \omega) - \tilde{f}_n(u, \omega)|^2 \, du \\
&\leq M \int_0^\infty |f(\beta_\omega(u), \omega) - \tilde{f}_n(u, \omega)|^2 \, du \\
&= M \int_0^\infty |\tilde{f}(u, \omega) - \tilde{f}_n(u, \omega)|^2 \, du < \varepsilon.
\end{aligned}$$

Thus, the simple stochastic function

$$\varphi_n(t, \omega) = \sum_{k=0}^{n-1} f(\tau_k, \omega)\chi_{(\tau_k < t \leq \tau_{k+1}]}(t),$$

where $\tau_k = \beta_\omega(u_k)$, is an ε-approximation of the function $f(t, \omega)$ in $L_A^2(\Phi_2)$. Hence, if it is established that the simple stochastic function

$$\chi(t, \omega) = \chi_{(0 < t \leq \tau]}(t) \in L_A^2(\Phi_2)$$

($P(\tau \leq K < \infty) = 1$) can be approximized arbitrarily closely by simple functions, then Lemma 5.4 will have been proved.

Let $\chi_n(t, \omega)$ be a simple function defined for $k/2^n < t \le (k+1)/2^n$ as follows:

$$\chi_n(t, \omega) = \begin{cases} 1, & \text{if} \quad \tau(\omega) \ge k/2^n, \\ 0, & \text{if} \quad \tau(\omega) < k/2^n. \end{cases}$$

Then

$$M \int_0^\infty [\chi(t, \omega) - \chi_n(t, \omega)]^2 \, dA_t \le M[A_{\tau + 2^{-n}} - A_\tau] \to 0, \qquad n \to \infty.$$

Hence Lemma 5.4 is proved for the bounded functions $f(t, \omega) \in L_A^2(\Phi_2)$ equal to zero beyond some finite interval. The general case of the functions $f(t, \omega) \in L_A^2(\Phi_2)$ can easily be reduced to the case considered. □

5.4.4

Lemmas 5.3–5.5 enable us to define the stochastic integrals $I(f) = \int_0^\infty f(t, \omega) dx_t$ over the martingale $X = (x_t, \mathscr{F}_t) \in \mathscr{M}$ for some classes of functions $f = f(t, \omega)$ satisfying the condition $M \int_0^\infty f^2(t, \omega) dA_t < \infty$ as the limits in the mean square of the integrals $I(f_n) = \int_0^\infty f_n(t, \omega) dx_t$, where the $f_n = f_n(t, \omega)$ are simple functions approximating $f = f(t, \omega)$ in terms of

$$M \int_0^\infty |f(t, \omega) - f_n(t, \omega)|^2 \, dA_t \to 0, \qquad n \to \infty$$

(compare with the corresponding construction, Section 4.2, for a Wiener process).

The precise result is formulated in the following way.

Theorem 5.10. *Let $X = (x_t, \mathscr{F}_t)$, $t \ge 0$, be a square integrable martingale from $\mathscr{M}$, and let $A_t = \langle x \rangle_t$, $t \ge 0$, be a corresponding natural increasing process.*

Let one of the three conditions be satisfied:

(1) *the function $f \in L_A^2(\Phi_3)$;*
(2) *the function $f \in L_A^2(\Phi_2)$ and the process A_t, $t \ge 0$, is (P-a.s.) continuous;*
(3) *the function $f \in L_A^2(\Phi_1)$, and the process A_t, $t \ge 0$, is absolutely continuous (P-a.s.).*

Then there is a uniquely defined (within stochastic equivalence) random variable $I(f)$ corresponding in the case of simple functions f to the stochastic integral introduced above such that

$$MI(f) = 0, \tag{5.81}$$

$$M[I(f)]^2 = M \int_0^\infty f^2(t, \omega) dA_t. \tag{5.82}$$

The value of the random variable $I(f)$ does not depend (P-a.s.) on the choice of the approximizing sequence of simple functions.

(The random variable $I(f)$ is also denoted by $\int_0^\infty f(t, \omega) dx_t$ and called a stochastic integral of the function $f = f(t, \omega)$ over the martingale $X = (x_t, \mathscr{F}_t) \in \mathscr{M}$.)

PROOF. The existence of $I(f)$ follows immediately from Lemmas 5.3–5.5. (5.81) and (5.82) follow from the respective properties for the integrals of simple functions $f_n = f_n(t, \omega)$ and the fact that $I(f) = \text{l.i.m.}\ I(f_n)$. □

5.4.5

Under the integral

$$I_\tau(f) = \int_0^\tau f(s, \omega)dx_s$$

the integral

$$\int_0^\infty f(s, \omega)\chi_{(s \le \tau)}(s)dx_s$$

will be understood.

Theorem 5.11. *If the martingale $X = (x_t, \mathscr{F}_t) \in \mathscr{M}^c$ (has continuous trajectories (P-a.s.)) and $f \in L_A^2(\Phi_2)$, then the integrals $I_t(f) = \int_0^t f(s, \omega)dx_s$ have a continuous modification.*

PROOF. If the function $f \in L_A^2(\Phi_2)$ is simple, then the continuity of $I_t(f)$ is obvious. In the general case it is proved in the same way as it is for a Wiener process (see Section 4.2). □

5.4.6

If $X = (x_t, \mathscr{F}_t) \in \mathscr{M}$, and $f \in L_A^2(\Phi_3)$, then the process $(I_t(f), \mathscr{F}_t)$ will be a square integrable martingale. According to Theorem 3.1, $I_t(f)$ has a right continuous modification.

5.4.7

In the case where the natural process $A_t = \langle x \rangle_t, t \ge 0$, corresponding to the martingale $X = (x_t, \mathscr{F}_t) \in \mathscr{M}$, is continuous, one can uniquely define the stochastic integral $I(f) = \int_0^\infty f(t, \omega)dx_t$ for functions $f \in \Phi_2$ satisfying only the assumption

$$P\left(\int_0^\infty f^2(t, \omega)dA_t < \infty\right) = 1.$$

5.4.8

We make use of Theorem 5.10 for proving the following result, a generalization of Levy's theorem (Theorem 4.1).

Theorem 5.12 (Doob). *Let the martingale $X = (x_t, \mathscr{F}_t) \in \mathscr{M}_T^c$(have continuous trajectories) and*

$$A_t \equiv \langle x \rangle_t = \int_0^t a^2(s, \omega)ds,$$

where the nonanticipative function $a^2(s, \omega) > 0$ *almost everywhere with respect to the measure* $dt\,dP$ *on* $([0, T] \times \Omega, \mathscr{B}_{[0,t]} \times \mathscr{F})$. *Then on the space* $(\Omega, \mathscr{F}, P)$ *there exists a Wiener process* $W = (W_t, \mathscr{F}_t)$, $t \le T$, *such that with probability one*

$$x_t = x_0 + \int_0^t a(s, \omega) dW_s. \tag{5.83}$$

PROOF. Define the process

$$W_t = \int_0^t \frac{dx_s}{a(s, \omega)}, \tag{5.84}$$

where $a^{-1}(s, \omega) = 0$ if $a(s, \omega) = 0$. The integral given by (5.84) is defined because of Theorem 5.10 (3) since the process A_t, $t \ge 0$, is absolutely continuous (P-a.s.) and

$$M \int_0^T a^{-2}(s, \omega) dA_s = T < \infty.$$

According to Theorem 5.11 the process W_t, $t \le T$, has a continuous (P-a.s.) modification.

Further, because of (5.81) and (5.82),

$$M[W_t | \mathscr{F}_s] = W_s,$$
$$M[(W_t - W_s)^2 | \mathscr{F}_s] = t - s, \qquad (P\text{-a.s.}) \qquad t \ge s.$$

Hence, by Theorem 4.1, the process $W = (W_t, \mathscr{F}_t)$, $t \le T$, is a Wiener process.

We now note that for any nonanticipative functions $\varphi = \varphi(t, \omega)$ with $M \int_0^T \varphi^2(t, \omega) ds < \infty$,

$$\int_0^t \varphi(s, \omega) dW_s = \int_0^t \frac{\varphi(s, \omega)}{a(s, \omega)} dx_s,$$

since this equality holds for simple functions. In particular, assuming $\varphi(s, \omega) = a(s, \omega)$, we obtain the equality

$$\int_0^t a(s, \omega) dW_s = x_t - x_0 \qquad (P\text{-a.s.}), \qquad t \le T,$$

from which (5.83) follows. □

5.5 Integral representations of the martingales which are conditional expectations, and the Fubini theorem for stochastic integrals

5.5.1

Let $(\mathscr{F}_t)$, $0 \le t \le T$, be a nondecreasing family of the continuous sub-σ-algebras $\mathscr{F}$, let $X = (x_t, \mathscr{F}_t)$ be a martingale with right continuous trajectories, and let $W = (W_t, \mathscr{F}_t)$ be a Wiener process. In this section we study the representations of conditional expectations $y_t = M(x_t | \mathscr{F}_t^W)$ in the form of stochastic integrals over a Wiener process.

Lemma 5.7. *The process $Y = (y_t, \mathscr{F}_t^W)$, $0 \le t \le T$, is a martingale.*

PROOF. Because of the Jensen inequality,

$$M|y_t| = M|M(x_t|\mathscr{F}_t^W)| \le M|x_t|, \qquad t \le T.$$

Further, if $s \le t$, then (P-a.s.)

$$\begin{aligned} M(y_t|\mathscr{F}_s^W) &= M[M(x_t|\mathscr{F}_t^W)|\mathscr{F}_s^W] = M(x_t|\mathscr{F}_s^W) \\ &= M[M(x_t|\mathscr{F}_s)|\mathscr{F}_s^W] = M(x_s|\mathscr{F}_s^W) = y_s, \end{aligned}$$

proving the lemma. □

Note. If $X = (x_t, \mathscr{F}_t)$ is a square integrable martingale, then the martingale $Y = (y_t, \mathscr{F}_t^W)$ is also square integrable.

Theorem 5.13. *If $X = (x_t, \mathscr{F}_t)$ is a square integrable martingale, then the martingale $Y = (y_t, \mathscr{F}_t^W)$, $y_t = M(x_t|\mathscr{F}_t^W)$, permits the representation*

$$y_t = Mx_0 + \int_0^t M(a_s|\mathscr{F}_s^W)dW_s, \qquad 0 \le t \le T, \tag{5.85}$$

where the process $a = (a_s, \mathscr{F}_s)$, $s \le t$, is such that

$$\langle x, W\rangle_t = \int_0^t a_s\, ds \tag{5.86}$$

and

$$\int_0^T Ma_s^2\, ds < \infty. \tag{5.87}$$

PROOF. First of all we note that $y_0 = M(x_0|\mathscr{F}_0^W) = Mx_0$ (P-a.s.), since the σ-algebra $\mathscr{F}_0^W$ is trivial ($\mathscr{F}_0^W = \{\Omega, \varnothing\}$). Further, because of the remark to Lemma 5.7, the process $Y = (y_t, \mathscr{F}_t^W)$ is a square integrable martingale; and by Theorem 5.5,

$$y_t = Mx_0 + \int_0^t f_s(\omega)dW_s, \tag{5.88}$$

where the process $f = (f_s(\omega), \mathscr{F}_s^W)$ is such that

$$\int_0^T Mf_s^2(\omega)ds < \infty.$$

By Theorem 5.3 there exists a random process $a = (a_s, \mathscr{F}_s)$, $s \le t$, such that (P-a.s.)

$$\langle x, W\rangle_t = \int_0^t a_s\, ds, \qquad 0 \le t \le T,$$

and $\int_0^T Ma_s^2\, ds < \infty$. We will show that in (5.88) $f_s(\omega) = M(a_s|\mathscr{F}_s^W)$ (P-a.s.) for almost each s, $0 \le s \le T$.

Let $g = (g_t(\omega), \mathscr{F}_t^W)$ be a bounded random process satisfying the conditions of Lemma 5.1, and let $z_t = \int_0^t g_s(\omega)dW_s$. Then

$$My_t z_t = M\{M(x_t|\mathscr{F}_t^W)z_t\} = Mx_t z_t. \tag{5.89}$$

From (5.88) and the properties of stochastic integrals we obtain

$$My_t z_t = \int_0^t M[f_s(\omega)g_s(\omega)]ds. \tag{5.90}$$

By Theorem 5.2 and Lemma 5.1,

$$Mx_t z_t = M\langle x, z\rangle_t = M\int_0^t g_s(\omega)a_s\,ds = \int_0^t M[M(a_s|\mathscr{F}_s^W)g_s(\omega)]ds. \tag{5.91}$$

From (5.89)–(5.91) we find that

$$\int_0^t M\{[f_s(\omega) - M(a_s|\mathscr{F}_s^W)]g_s(\omega)\}ds = 0.$$

From this, because of the arbitrariness of the function $g_s(\omega)$, we infer that (P-a.s.) for almost all s, $0 \le s \le T$,

$$f_s(\omega) = M(a_s|\mathscr{F}_s^W). \qquad \square$$

Corollary 1. *Let $X = (x_t, \mathscr{F}_t)$ be a square integrable martingale,*

$$x_t = \int_0^t a_s\,dW_s, \tag{5.92}$$

and let $M\int_0^T a_s^2\,ds < \infty$. Then ($P$-a.s.) for all t, $0 \le t \le T$,

$$M\left[\int_0^t a_s\,dW_s|\mathscr{F}_t^W\right] = \int_0^t M(a_s|\mathscr{F}_s^W)dW_s. \tag{5.93}$$

PROOF. Indeed, from (5.92) and (5.6) we obtain

$$\langle x, W\rangle_t = \int_0^t a_s\,ds.$$

Hence (5.93) follows from (5.85). $\square$

Corollary 2. *Let $W = (W_t, \mathscr{F}_t)$, $\tilde{W} = (\tilde{W}_t, \mathscr{F}_t)$ be two independent Wiener processes and let $X = (x_t, \mathscr{F}_t)$ be a martingale where*

$$x_t = \int_0^t a_s\,d\tilde{W}_s, \qquad \int_0^T Ma_s^2\,ds < \infty.$$

Then, (P-a.s.)

$$M\left[\int_0^t a_s\,d\tilde{W}_s|\mathscr{F}_t^W\right] = 0. \tag{5.94}$$

PROOF. To prove (5.94) it suffices to establish that $\langle x, W\rangle_t = 0$ (P-a.s.) for all $t, 0 \le t \le T$.

We have

$$x_t + W_t = \int_0^t a_s\, d\tilde{W}_s + W_t, \qquad x_t - W_t = \int_0^t a_s\, d\tilde{W}_s - W_t.$$

From this it is not difficult to infer that $\langle x + W\rangle_t = \int_0^t (a_s^2 + 1)ds$, $\langle x - W\rangle_t = \int_0^t (a_s^2 + 1)ds$, and, therefore, that

$$\langle x, W\rangle_t = \tfrac{1}{4}\{\langle x + W\rangle_t - \langle x - W\rangle_t\} = 0. \qquad \square$$

5.5.2

In the following theorem, Equation (5.93) is generalized to a wider class of martingales.

Theorem 5.14. *Let* $X = (x_t, \mathscr{F}_t)$, $0 \le t \le T$, *be a martingale;*

$$x_t = \int_0^t a_s\, dW_s, \qquad P\left(\int_0^T a_s^2\, ds < \infty\right) = 1. \tag{5.95}$$

If $M|a_s| < \infty$, $0 \le s \le T$, *and*

$$P\left(\int_0^T [M(|a_s||\mathscr{F}_s^W)]^2\, ds < \infty\right) = 1, \tag{5.96}$$

then (P-a.s.) *for all* $t, 0 \le t \le T$,

$$M\left(\int_0^t a_s\, dW_s|\mathscr{F}_t^W\right) = \int_0^t M(a_s|\mathscr{F}_s^W)dW_s. \tag{5.97}$$

PROOF. (5.97) can be reformulated to state that the martingale $Y = (y_t, \mathscr{F}_t^W)$, with $y_t = M(x_t|\mathscr{F}_t^W)$, permits the representation

$$y_t = \int_0^t M(a_s|\mathscr{F}_s^W)dW_s \qquad (P\text{-a.s.}), \qquad 0 \le t \le T.$$

For proving (5.98) we introduce Markov times

$$\tau_n = \begin{cases} \inf\left\{t \le T: \int_0^t a_s^2\, ds \ge n\right\}, \\ T, \quad \text{if } \int_0^T a_s^2\, ds < n. \end{cases}$$

Then the martingale $X^{(n)} = (x_t^{(n)}, \mathscr{F}_t)$ with

$$x_t^{(n)} = \int_0^t \chi_{(\tau_n \ge s)} a_s\, dW_s \tag{5.98}$$

is square integrable,

$$\int_0^T M\chi_{(\tau_n \geq s)} a_s^2 \, ds < \infty,$$

and, by Corollary 1 of Theorem 5.13, for the martingale $Y^{(n)} = (y_t^{(n)}, \mathscr{F}_t^W)$, with $y_t^{(n)} = M(x_t^{(n)} | \mathscr{F}_t^W)$, there is the representation

$$y_t^{(n)} = \int_0^t M\{\chi_{(\tau_n \geq s)} a_s | \mathscr{F}_s^W\} dW_s. \tag{5.99}$$

We shall show that $y_t^{(n)} \xrightarrow{P} y_t$ (in probability) with $n \to \infty$ for each t, $0 \leq t \leq T$. For this purpose we note that, because of (5.98), the process $x_t^{(n)}$ has a continuous modification, and hence

$$x_t^{(n)} = x_{t \wedge \tau_n} = M(x_T | \mathscr{F}_{t \wedge \tau_n}).$$

From this it follows that the sequence of the random variables $\{x_t^{(n)}, n = 1, 2, \ldots\}$ is uniformly integrable (see Theorem 2.7). But $x_t^{(n)} \to x_t$ (in probability), $n \to \infty$. Hence, from these two facts and Note 1 to Theorem 1.3, it follows that

$$\lim_{n \to \infty} M|x_t - x_t^{(n)}| = 0.$$

But $M|y_t - y_t^{(n)}| \leq M|x_t - x_t^{(n)}|$. Consequently, $y_t^{(n)} \xrightarrow{P} y_t$ for each t, $0 \leq t \leq T$.

To complete the proof of the theorem it remains to show that, with $n \to \infty$,

$$\int_0^t M\{\chi_{(\tau_n \geq s)} a_s | \mathscr{F}_s^W\} dW_s \xrightarrow{P} \int_0^t M(a_s | \mathscr{F}_s^W) dW_s.$$

According to (4.60), for this purpose it suffices to establish that

$$\int_0^T [M\{\chi_{(\tau_n < s)} a_s | \mathscr{F}_s^W\}]^2 \, ds \xrightarrow{P} 0, \qquad n \to \infty. \tag{5.100}$$

First of all we note that $M\{\chi_{(\tau_n < s)} a_s | \mathscr{F}_s^W\} \xrightarrow{P} 0$, $n \to \infty$, since $M|a_s| < \infty$, $\chi_{(\tau_n < s)} \xrightarrow{P} 0$, $n \to \infty$, and

$$M|M\{\chi_{(\tau_n < s)} a_s | \mathscr{F}_s^W\}| \leq M\chi_{(\tau_n < s)} |a_s| \to 0, \qquad n \to \infty.$$

Let us denote

$$\sigma_N = \begin{cases} \inf\left\{t \leq T: \int_0^t \{M(|a_s| | \mathscr{F}_s^W)\}^2 \, ds \geq N\right\}, \\ T, \quad \text{if } \int_0^T \{M(|a_s| | \mathscr{F}_s^W)\}^2 \, ds < N. \end{cases}$$

Then, for $\varepsilon > 0$,

$$P\left\{\int_0^T [M(\chi_{(\tau_n < s)}a_s | \mathscr{F}_s^W)]^2\, ds > \varepsilon\right\}$$

$$= P\left\{\int_0^T [M(\chi_{(\tau_n < s)}a_s | \mathscr{F}_s^W)]^2\, ds > \varepsilon; \sigma_N = T\right\}$$

$$+ P\left\{\int_0^T [M(\chi_{(\tau_n < s)}a_s | \mathscr{F}_s^W)]^2\, ds > \varepsilon; \sigma_N < T\right\}$$

$$\leq P\left\{\int_0^{T \wedge \sigma_N} [M(\chi_{(\tau_n < s)}a_s | \mathscr{F}_s^W)]^2\, ds > \varepsilon; \sigma_N = T\right\} + P\{\sigma_N < T\}$$

$$\leq P\left\{\int_0^T \chi_{(\sigma_N \geq s)} [M(\chi_{(\tau_n > s)}a_s | \mathscr{F}_s^W)]^2\, ds > \varepsilon\right\} + P(\sigma_N < T). \tag{5.101}$$

Here $P\{\sigma_N < T\} \to 0$, $N \to \infty$, because of (5.96). Further since

$$M(\chi_{(\tau_n < s)}a_s | \mathscr{F}_s^W) \xrightarrow{P} 0,$$

then, by the Lebesgue dominated convergence theorem,

$$\lim_{n \to \infty} M \int_0^T \chi_{(\sigma_N \geq s)}[M(\chi_{(\tau_n < s)}a_s | \mathscr{F}_s^W]^2\, ds = 0.$$

Hence passing in (5.101) to the limit (at first over $n \to \infty$ and then over $N \to \infty$), we obtain the required relationship, (5.100). □

5.5.3

Equation (5.97), established in Theorem 5.14, enables us to prove for stochastic integrals a statement (Theorem 5.15), similar to the Fubini theorem.

Let $(\Omega, \mathscr{F}, P)$, $(\tilde{\Omega}, \tilde{\mathscr{F}}, \tilde{P})$ be two probability spaces, let $(\bar{\Omega}, \bar{\mathscr{F}}, \bar{P}) = (\Omega \times \tilde{\Omega}, \mathscr{F} \times \tilde{\mathscr{F}}, P \times \tilde{P})$, and let $(\mathscr{F}_t)$ and $(\tilde{\mathscr{F}}_t)$, $0 \leq t \leq 1$, be nondecreasing families of the sub-σ-algebras $\mathscr{F}$ and $\tilde{\mathscr{F}}$. Let $W = (W_t(\omega), \mathscr{F}_t)$, $0 \leq t \leq 1$, be a Wiener process, $\mathscr{F}_t^W = \sigma\{\omega: W_s, s \leq t\}$.

Theorem 5.15. *Consider a random process $(g_t(\omega, \tilde{\omega}), \mathscr{F}_t^W \times \tilde{\mathscr{F}}_t)$, $t \leq 1$. If*

$$M \times \tilde{M} \int_0^1 g_t^2(\omega, \tilde{\omega})dt < \infty \tag{5.102}$$

($M \times \tilde{M}$ is an averaging over the measure $P \times \tilde{P}$), then for each t, $0 \leq t \leq 1$, (P-a.s.)

$$\int_{\tilde{\Omega}}\left[\int_0^t g_s(\omega, \tilde{\omega})dW_s(\omega)\right]d\tilde{P}(\tilde{\omega}) = \int_0^t\left[\int_{\tilde{\Omega}} g_s(\omega, \tilde{\omega})d\tilde{P}(\tilde{\omega})\right]dW_s(\omega). \tag{5.103}$$

Proof. Let us denote

$$x_t(\omega, \tilde{\omega}) = \int_0^t g_s(\omega, \tilde{\omega})dW_s(\omega)$$

and set $\hat{W}_s(\omega, \tilde{\omega}) = W_s(\omega)$. Then using the construction of the stochastic integral described in Chapter 4, one can define an integral $\int_0^t g_s(\omega, \tilde{\omega})dW_s(\omega)$ so that it coincides ($P \times \tilde{P}$-a.s.) with the integral

$$\hat{x}_t(\omega, \tilde{\omega}) = \int_0^t g_s(\omega, \tilde{\omega})d\hat{W}_s(\omega, \tilde{\omega}),$$

which is $\mathscr{F}_t^W \times \tilde{\mathscr{F}}_t$ measurable.

It is not difficult to show that ($P \times \tilde{P}$-a.s.) $\int_{\tilde{\Omega}} x_t(\omega, \tilde{\omega})d\tilde{P}(\tilde{\omega})$ is one of the versions of the conditional expectation $M \times \tilde{M}[x_t(\omega, \tilde{\omega})|\mathscr{F}_t^{\hat{W}}]$, i.e.,

$$M \times \tilde{M}[x_t(\omega, \tilde{\omega})|\mathscr{F}_t^{\hat{W}}] = \int_{\tilde{\Omega}} x_t(\omega, \tilde{\omega})d\tilde{P}(\tilde{\omega}) \qquad (P \times \tilde{P}\text{-a.s.}).$$

Similarly,

$$M \times \tilde{M}[g_t(\omega, \tilde{\omega})|\mathscr{F}_t^{\hat{W}}] = \int_{\tilde{\Omega}} g_t(\omega, \tilde{\omega})d\tilde{P}(\tilde{\omega}) \qquad (P \times P\text{-a.s.}).$$

Hence, taking into account (5.97), we find ($P \times \tilde{P}$-a.s.)

$$\begin{aligned}\int_{\tilde{\Omega}} x_t(\omega, \tilde{\omega})d\tilde{P}(\tilde{\omega}) &= M \times \tilde{M}[x_t(\omega, \tilde{\omega})|\mathscr{F}_t^{\hat{W}}] \\ &= M \times \tilde{M}\left[\int_0^t g_s(\omega, \tilde{\omega})dW_s(\omega)|\mathscr{F}_t^{\hat{W}}\right] \\ &= M \times \tilde{M}\left[\int_0^t g_s(\omega, \tilde{\omega})d\hat{W}_s(\omega, \tilde{\omega})|\mathscr{F}_t^{\hat{W}}\right] \\ &= \int_0^t M \times \tilde{M}[g_s(\omega, \tilde{\omega})|\mathscr{F}_s^{\hat{W}}]d\hat{W}_s(\omega, \tilde{\omega}) \\ &= \int_0^t\left[\int_{\tilde{\Omega}} g_s(\omega, \tilde{\omega})d\tilde{P}(\tilde{\omega})\right]d\hat{W}_s(\omega, \tilde{\omega}) \\ &= \int_0^t\left[\int_{\tilde{\Omega}} g_s(\omega, \tilde{\omega})d\tilde{P}(\tilde{\omega})\right]dW_s(\omega).\end{aligned}$$

This proves (5.103), if only it is noted that $\mathscr{F}_t^{\hat{W}} = \mathscr{F}_t^W \times (\tilde{\Omega}, \varnothing)$. □

5.6 The structure of the functionals from processes of the diffusion type

5.6.1

From Theorem 5.5 it follows that any square integrable martingale $X = (x_t, \mathscr{F}_t^W)$, $t \leq T$, where $\mathscr{F}_t^W$ is the σ-algebra generated by the Wiener process $W_s, s \leq t$, permits the representation

$$x_t = x_0 + \int_0^t f_s(\omega)dW_s,$$

where the process $f = (f_s(\omega), \mathscr{F}_s^W)$ is such that $\int_0^T Mf_s^2(\omega)ds < \infty$.

In this section this result as well as Theorems 5.7 and 5.8 will be extended to the martingales $X = (x_t, \mathcal{F}_t^\xi)$, where $\xi = (\xi_t, \mathcal{F}_t)$, $t \leq T$, is a process of the diffusion type with the differential

$$d\xi_t = a_t(\xi)dt + b_t(\xi)dW_t. \tag{5.104}$$

It will be shown, in particular, that (subject to the assumptions formulated further on) any square integrable martingale $X = (x_t, \mathcal{F}_t^\xi)$ permits the representation $x_t = x_0 + \int_0^t f_s(\omega)dW_s$ (P-a.s.), $0 \leq t \leq T$, where the process $f = (f_s(\omega), \mathcal{F}_s^\xi)$, $s \leq T$, is such that

$$M \int_0^t f_s^2(\omega)ds < \infty. \tag{5.105}$$

5.6.2

We begin with the consideration of a particular case of Equation (5.104).

Theorem 5.16. *Let the process $\xi = (\xi_t, \mathcal{F}_t)$ be a (strong) solution of the equation*

$$\xi_t = \xi_0 + \int_0^t b_s(\xi)dW_s, \tag{5.106}$$

where the nonanticipative functional[1] $b = (b_t(x), \mathcal{B}_t)$, $t \leq T$, *is assumed to be such that* $P(\int_0^T b_t^2(\xi)ds < \infty) = 1$ *and*

$$b_t^2(x) \geq c > 0. \tag{5.107}$$

Then any martingale $X = (x_t, \mathcal{F}_t^\xi)$, $0 \leq t \leq T$, has a continuous modification, which permits (P-a.s.) the representation

$$x_t = x_0 + \int_0^t f_s(\omega)dW_s, \qquad 0 \leq t \leq T, \tag{5.108}$$

where the process $f = (f_s(\omega), \mathcal{F}_s^\xi)$ is such that

$$P\left(\int_0^T f_s^2(\omega)ds < \infty\right) = 1. \tag{5.109}$$

If the martingale $X = (x_t, \mathcal{F}_t^\xi)$ is square integrable, then

$$M \int_0^T f_s^2(\omega)ds < \infty. \tag{5.110}$$

PROOF. We shall show first that the family of (augmented) σ-algebras $(\mathcal{F}_t^\xi)$, $0 \leq t \leq T$, is continuous. Let $\mathcal{F}_t^{\xi_0, W} = \mathcal{F}_0^\xi \vee \mathcal{F}_t^W$ where $\mathcal{F}_0^\xi = \sigma\{\omega: \xi_0(\omega)\}$.

Since ξ is the strong solution of Equation (5.106), then

$$\mathcal{F}_t^{\xi_0, W} \supseteq \mathcal{F}_t^\xi. \tag{5.111}$$

[1] $\mathcal{B}_t = \sigma\{x: x_s, s \leq t\}$ where x belongs to the space of continuous (on $[0, T]$) functions.

On the other hand, by virtue of (5.107) for each t, $0 \le t \le T$,

$$W_t = \int_0^t \frac{d\xi_s}{b_s(\xi)} \qquad (P\text{-a.s.}) \tag{5.112}$$

(see Theorem 5.12). Hence $\mathscr{F}_t^{\xi_0, W} \subseteq \mathscr{F}_t^{\xi}$, which together with (5.111) leads to the equality[2])

$$\mathscr{F}_t^{\xi_0, W} = \mathscr{F}_t^{\xi}. \tag{5.113}$$

According to Theorem 4.3 the family of the (augmented) σ-algebras $(\mathscr{F}_t^W)$, $0 \le t \le T$, is continuous. As it is not difficult to show, the family $(\mathscr{F}_t^{\xi_0, W})$, $0 \le t \le T$, as well as $(\mathscr{F}_t^{\xi})$ have the same property.

Because of Theorem 3.1, it follows that any martingale $Y = (y_t, \mathscr{F}_t^{\xi_0, W})$ has a right continuous modification (which will be assumed from now on).

We assume now that $X = (x_t, \mathscr{F}_t^{\xi})$ is a square integrable martingale. If $W = (W_t, \mathscr{F}_t)$ is a Wiener process, then, as it is not difficult to check, the process $(W_t, \mathscr{F}_t^{\xi})$ will also be a Wiener process. Hence, according to Theorem 5.3, there exists a process $f = (f_t(\omega)\mathscr{F}_t^{\xi})$ such that

$$M \int_0^T f_t^2(\omega)dt < \infty$$

and

$$\langle x, W \rangle_t = \int_0^t f_s(\omega)ds. \tag{5.114}$$

We set

$$\tilde{x}_t = x_0 + \int_0^t f_s(\omega)dW_s$$

and show that $P(\tilde{x}_t = x_t) = 1$, $0 \le t \le T$.

We first consider the decomposition $0 = t_0 < t_1 < \cdots < t_n = t$ of the interval $[0, t]$. If it is shown that

$$M(\tilde{x}_t - x_t)\exp\left\{i\left(z_0 \xi_0 + \sum_{k=1}^n z_k W_{t_k}\right)\right\} = 0 \tag{5.115}$$

for any z_i with $|z_i| < \infty$, $i = 0, 1, \dots, n$, then from this the required equality $P(\tilde{x}_t = x_t) = 1$ will follow, since the random variables

$$\exp\left\{i\left(z_0 \xi_0 + \sum_{k=1}^n z_k W_{t_k}\right)\right\}$$

can be used for the approximation of any bounded $\mathscr{F}_t^{\xi_0, W}$ $(= \mathscr{F}_t^{\xi})$-measurable random variable.

[2] If $\xi_0 \equiv 0$ the assertion of the theorem can be easily deduced from Theorem 5.5 and the fact that by (5.113) $\mathscr{F}_t^W = \mathscr{F}_t^{\xi}$, $0 \le t \le T$.

We start with the case $n = 1$. Set $y_t = x_t - \tilde{x}_t$. It is clear that $Y = (y_t, \mathscr{F}_t^\xi)$ is also a square integrable martingale and, according to (5.6) and (5.114),

$$\langle y, W\rangle_s = 0 \qquad (P\text{-a.s.}), \qquad 0 \le s \le T.$$

Because of Lemma 5.1, it follows that

$$M\left[y_t \int_0^t \exp(iz_1 W_u) dW_u | \mathscr{F}_0^{\xi_0, W}\right] = M\left[\int_0^t \exp(iz_1 W_u) d\langle y, W\rangle_u | \mathscr{F}_0^\xi\right] = 0. \tag{5.116}$$

Further, by Ito's formula

$$\exp\{i(z_0\xi_0 + z_1 W_t)\} = \exp\{iz_0\xi_0\} + iz_1 \exp\{iz_0\xi_0\} \int_0^t \exp(iz_1 W_u) dW_u - \frac{z_1^2}{2} \exp\{iz_0\xi_0\} \int_0^t \exp(iz_1 W_u) du.$$

Hence, taking into account (5.116) and the fact that $y_0 = 0$, we find

$$\begin{aligned} My_t \exp\{i(z_0\xi_0 + z_1 W_t)\} &= My_t \exp\{iz_0\xi_0\} \\ &\quad + iz_1 M\left\{y_t \exp(iz_0\xi_0) \int_0^t \exp(iz_1 W_u) dW_u\right\} \\ &\quad - \frac{z_1^2}{2} M\left\{y_t \exp(iz_0\xi_0) \int_0^t \exp(iz_1 W_u) du\right\} \\ &= M\{M(y_t | \mathscr{F}_0^{\xi_0, W}) \exp(iz_0\xi_0)\} + iz_1 \\ &\quad \times M\left\{\exp(iz_0\xi_0) M\left[y_t \int_0^t \exp(iz_1 W_u) dW_u | \mathscr{F}_0^{\xi_0, W}\right]\right\} \\ &\quad - \frac{z_1^2}{2} \int_0^t M\{M(y_t | \mathscr{F}_u^{\xi_0, W}) \exp[i(z_0\xi_0 + z_1 W_u)]\} du \\ &= -\frac{z_1^2}{2} \int_0^t M(y_u \exp[i(z_0\xi_0 + z_1 W_u)]\} du. \end{aligned}$$

Consequently, $u_t = My_t \exp[i(z_0\xi_0 + z_1 W_t)]$ satisfies the linear equation

$$\dot{u}_t = -\frac{z_1^2}{2} u_t, \qquad u_0 = 0,$$

the solution of which is identically equal to zero.

Thus we have proved Equation (5.115) with $n = 1$.

Let now $n > 1$, and for $n - 1$, let Equation (5.115) be proven. By the Ito formula,

$$\exp\left\{i\left(z_0\xi_0 + \sum_{k=1}^{n} z_k W_{t_k}\right)\right\} = \exp\left\{i\left(z_0\xi_0 + \sum_{k=1}^{n-1} z_k W_{t_k}\right)\right\}$$
$$+ iz_n \int_{t_{n-1}}^{t_n} \exp\left\{i\left(z_0\xi_0 + \sum_{k=1}^{n-1} z_k W_{t_k} + z_n W_u\right)\right\} dW_u$$
$$- \frac{z_n^2}{2} \int_{t_{n-1}}^{t_n} \exp\left\{i\left(z_0\xi_0 + \sum_{k=1}^{n-1} z_k W_{t_k} + z_n W_u\right)\right\} du. \tag{5.117}$$

Noting now that, by the induction assumption,

$$M y_t \exp\left\{i\left(z_0\xi_0 + \sum_{k=1}^{n-1} z_k W_{t_k}\right)\right\}$$
$$= M\left\{M(y_t \mid \mathscr{F}_{t_{n-1}}^{\xi_0, W}) \exp\left[i\left(z_0\xi_0 + \sum_{k=1}^{n-1} z_k W_{t_k}\right)\right]\right\}$$
$$= M y_{t_{n-1}} \exp\left\{i\left(z_0\xi_0 + \sum_{k=1}^{n-1} z_k W_{t_k}\right)\right\} = 0,$$

from (5.117), just as in the case $n = 1$, it is easily deduced that

$$M\left[y_t \exp\left\{i\left(z_0\xi_0 + \sum_{k=1}^{n} z_k W_{t_k}\right)\right\}\right]$$
$$= -\frac{z_n^2}{2} \int_{t_{n-1}}^{t_n} M\left[y_u \exp\left\{i\left(z_0\xi_0 + \sum_{k=1}^{n-1} z_k W_{t_k} + z_n W_u\right)\right\}\right] du.$$

From this we obtain

$$M y_t \exp\left\{i\left(z_0\xi_0 + \sum_{k=1}^{n} z_k W_{t_k}\right)\right\} = 0.$$

Thus, (5.115) for the case of a square integrable martingale $X = (x_t, \mathscr{F}_t^{\xi_0, W})$ is proved.

When the martingale $X = (x_t, \mathscr{F}_t^{\xi})$ is not square integrable, the proof of the representation given by (5.108) corresponds almost word for word to the proof of Theorem 5.7. □

Corollary. *Let the functional* $b = (b_t(x), \mathscr{B}_t)$ *satisfy* (4.110), (4.111), *where* $b_t^2(x) \geq c > 0$. *Then according to Theorem* 4.6 *a strong solution of Equation* (5.106) *exists and any martingale* $X = (x_t, \mathscr{F}_t^{\xi})$ *permits the representation given by* (5.108).

5.6.3

We pass now to the consideration of the general case.

Theorem 5.17. *Let $\xi = (\xi_t, \mathcal{F}_t)$, $0 \le t \le T$, be a process of the diffusion type with the differential*

$$d\xi_t = a_t(\xi)dt + b_t(\xi)dW_t, \tag{5.118}$$

where $a = (a_t(x), \mathcal{B}_t)$ and $b = (b_t(x), \mathcal{B}_t)$ are nonanticipative functionals. We shall assume that the coefficient $b_t(x)$ satisfies (4.110) *and* (4.111), *and that for almost all $t \le T$,*

$$b_t^2(x) \ge c > 0 \tag{5.119}$$

Suppose

$$P\left(\int_0^T a_t^2(\xi)dt < \infty\right) = P\left(\int_0^T a_t^2(\eta)dt < \infty\right) = 1, \tag{5.120}$$

where η is a (strong) solution of the equation

$$d\eta_t = b_t(\eta)dW_t, \qquad \eta_0 = \xi_0. \tag{5.121}$$

Then any martingale $X = (x_t, \mathcal{F}_t^\xi)$ has a continuous modification with the representation

$$x_t = x_0 + \int_0^t f_s(\omega)dW_s \tag{5.122}$$

with an $\mathcal{F}^\xi$-adapted process $(f_t(\omega), \mathcal{F}_t^\xi)$ such that

$$P\left(\int_0^T f_s^2(\omega)ds < \infty\right) = 1.$$

If $X = (x_t, \mathcal{F}_t^\xi)$, $0 \le t \le T$, is a square integrable martingale, then also

$$\int_0^T Mf_t^2(\omega)dt < \infty. \tag{5.123}$$

PROOF. According to the assumptions made and Theorem 7.19 the measures μ_ξ and μ_η are equivalent. Here the density

$$\varkappa_t(\xi) = \frac{d\mu_\eta}{d\mu_\xi}(t, \xi)$$

is given by the formula (see (7.124))

$$\begin{aligned}\varkappa_t(\xi) &= \exp\left(-\int_0^t a_s(\xi)(b_s^2(\xi))^{-1}\,d\xi_s + \frac{1}{2}(a_s(\xi)b_s^{-1}(\xi))^2\,ds\right)\\ &= \exp\left(-\int_0^t a_s(\xi)(b_s^2(\xi))^{-1}\,dW_s - \frac{1}{2}\int_0^t(a_s(\xi)b_s^{-1}(\xi))^2\,ds\right).\end{aligned} \tag{5.124}$$

We consider a new probability space $(\Omega, \mathscr{F}, \tilde{P})$ with the measure

$$\tilde{P}(d\omega) = \varkappa_T(\xi(\omega))P(d\omega)$$

(it is clear that $\tilde{P} \ll P$ and, because of Lemma 6.8, $P \ll \tilde{P}$; therefore, $P \sim \tilde{P}$).

We have

$$\tilde{P}(\xi \in \Gamma\} = \int_{\{\omega:\xi\in\Gamma\}} \varkappa_T(\xi(\omega))P(d\omega) = \int_\Gamma \varkappa_T(x)d\mu_\xi(x) = \mu_\eta(\Gamma).$$

Thus, the random process $\xi = (\xi_t)$, $0 \le t \le T$, on a new probability space $(\Omega, \mathscr{F}, \tilde{P})$, has the same distribution that the process $\eta = (\eta_t)$, $0 \le t \le T$, has on the space $(\Omega, \mathscr{F}, P)$.

Further, by Theorem 6.2, the process $(\tilde{W}_t, \mathscr{F}_t)$, where

$$\tilde{W}_t = W_t + \int_0^t a_s(\xi)b_s^{-1}(\xi)ds, \tag{5.125}$$

is a Wiener process over measure $\tilde{P}$.

From (5.125) and (4.80) it follows that ($\tilde{P}$-a.s.) and (P-a.s.)

$$\xi_0 + \int_0^t b_s(\xi)d\tilde{W}_s = \xi_0 + \int_0^t a_s(\xi)ds + \int_0^t b_s(\xi)dW_s = \xi_t.$$

Therefore, the process $\xi = (\xi_t)$, $0 \le t \le T$, on $(\Omega, \mathscr{F}, \tilde{P})$, is a solution of the equation

$$\xi_t = \xi_0 + \int_0^t b_s(\xi)d\tilde{W}_s \tag{5.126}$$

(compare with Equation (5.121)).

According to the assumptions on the coefficient $b_s(x)$ made under the conditions of the theorem, a (strong) solution of Equation (5.126), as well as of Equation (5.121), exists and is unique. Then, by Theorem 5.16, any martingale $Y = (y_t, \mathscr{F}_t^\xi)$, $0 \le t \le T$, defined on a probability space $(\Omega, \mathscr{F}, \tilde{P})$ has a continuous modification which permits ($\tilde{P}$-a.s.) the representation

$$y_t = y_0 + \int_0^t g_s(\omega)d\tilde{W}_s, \qquad 0 \le t \le T, \tag{5.127}$$

where $\tilde{P}(\int_0^T g_s^2(\omega)ds < \infty) = 1$.

Let $X = (x_t, \mathscr{F}_t^\xi)$ be a martingale. We show that the process $Y = (y_t, \mathscr{F}_t^\xi)$, with $y_t = x_t/\varkappa_t(\xi)$, on $(\Omega, \mathscr{F}, \tilde{P})$ is also a martingale.

Indeed,

$$\tilde{M}|y_t| = \int_\Omega |y_t| d\tilde{P} = \int_\Omega |y_t| \varkappa_T(\xi)dP = \int_\Omega \frac{|x_t|}{\varkappa_t(\xi)} \varkappa_T(\xi)dP$$
$$= \int_\Omega \frac{|x_t|}{\varkappa_t(\xi)} M(\varkappa_T(\xi)|\mathscr{F}_t^\xi)dP = \int_\Omega |x_t|dP = M|x_t| < \infty;$$

and with $t \geq s$, according to Lemma 6.6, ($\tilde{P}$-a.s.)

$$\tilde{M}(y_t | \mathscr{F}_s^\xi) = \varkappa_s^{-1}(\xi) M(y_t \varkappa_t(\xi) | \mathscr{F}_s^\xi) = \varkappa_s^{-1}(\xi) M(x_t | \mathscr{F}_s^\xi) = \frac{x_s}{\varkappa_s(\xi)} = y_s.$$

Consequently, to the martingale $Y = (y_t, \mathscr{F}_t^\xi)$ with $y_t = x_t/\varkappa_t(\xi)$ we apply the result (5.127), according to which ($\tilde{P}$-a.s.) and (P-a.s.) for each t, $0 \leq t \leq T$,

$$\begin{aligned} \frac{x_t}{\varkappa_t(\xi)} &= x_0 + \int_0^t g_s(\omega) d\tilde{W}_s \\ &= x_0 + \int_0^t g_s(\omega) W_s + \int_0^t g_s(\omega) a_s(\xi) b_s^{-1}(\xi), \end{aligned}$$

or

$$x_t = \varkappa_t(\xi) z_t, \tag{5.128}$$

where

$$z_t = x_0 + \int_0^t g_s(\omega) dW_s + \int_0^t g_s(\omega) a_s(\xi) b_s^{-1}(\xi) ds. \tag{5.129}$$

Applying the Ito formula, we find from (5.128), (5.129) and (5.124) that

$$\begin{aligned} dx_t &= \varkappa_t(\xi) dz_t + z_t\, d\varkappa_t(\xi) - \varkappa_t(\xi) g_t(\omega) a_t(\xi) b_t^{-1}(\xi) dt \\ &= \varkappa_t(\xi) g_t(\omega) dW_t + \varkappa_t(\xi) g_t(\omega) a_t(\xi) b_t^{-1}(\xi) dt - z_t \varkappa_t(\xi) a_t(\xi) b_t^{-1}(\xi) dW_t \\ &\quad - \varkappa_t(\xi) g_t(\omega) a_t(\xi) b_t^{-1}(\xi) dt \\ &= f_t(\omega) dW_t, \end{aligned}$$

where

$$f_t(\omega) = \varkappa_t(\xi) g_t(\omega) - x_t a_t(\xi) b_t^{-1}(\xi). \tag{5.130}$$

In other words, (P-a.s.)

$$x_t = x_0 + \int_0^t f_s(\omega) dW_s$$

where $P(\int_0^T f_s^2(\omega) ds < \infty) = 1$, which fact follows from (5.130), because of the equivalence of measures P and $\tilde{P}$ (Lemma 6.8), the continuity (P-a.s.) of the processes $\varkappa_t(\xi)$ and $x_t = \varkappa_t(\xi) z_t$ as well as the conditions

$$P\left(\int_0^t g_t^2(\omega) dt < \infty\right) = P\left(\int_0^T (a_t(\xi) b_t^{-1}(\xi))^2\, dt < \infty\right) = 1.$$

To complete proving the theorem it only remains to check that in the case of square integrable martingales $X = (x_t, \mathscr{F}_t^\xi)$ the functional $f_s(\omega)$, $s \leq T$, satisfies (5.110). This follows from the following general result.

Lemma 5.8. *Let $F = (\mathscr{F}_t)$, $0 \le t \le T$, be a nondecreasing family of the sub-σ-algebras of $\mathscr{F}$, and let $f = (f_t(\omega), \mathscr{F}_t)$ be a process with*

$$P\left(\int_0^T f_t^2(\omega)dt < \infty\right) = 1.$$

In order that the (continuous) martingale $X = (x_t, \mathscr{F}_t)$ $t \le T$, with

$$x_t = \int_0^t f_s(\omega)dW_s$$

be square integrable, it is necessary and sufficient that

$$\int_0^T Mf_s^2(\omega)ds < \infty. \tag{5.131}$$

PROOF. The sufficiency of (5.131) follows from the property of stochastic integrals (over a Wiener process; see (4.49)). To prove necessity we assume that for n, $n = 1, 2, \ldots,$

$$\tau_n = \begin{cases} \inf\left(t \le T: \int_0^t f_s^2\, ds \ge n\right), \\ T, \quad \text{if } \int_0^T f_s^2\, ds < n. \end{cases}$$

Because of Theorem 3.6 and the continuity of the trajectories of the martingale $X = (x_t, \mathscr{F}_t)$, (P-a.s.)

$$\int_0^{t \wedge \tau_n} f_s(\omega)dW_s = x_{t \wedge \tau_n} = M[x_T | \mathscr{F}_{t \wedge \tau_n}].$$

Since in addition to this the martingale $X = (x_t, \mathscr{F}_t)$ is square integrable, we have, because of the Jensen inequality,

$$Mx_{t \wedge \tau_n}^2 = M[M(x_T | \mathscr{F}_{t \wedge \tau_n})]^2 \le Mx_T^2 < \infty.$$

On the other hand, since $M \int_0^{T \wedge \tau_n} f_s^2(\omega)ds \le n < \infty$, then

$$Mx_{T \wedge \tau_n}^2 = M\left(\int_0^{T \wedge \tau_n} f_s(\omega)dW_s\right)^2 = M \int_0^{T \wedge \tau_n} f_s^2(\omega)ds.$$

Consequently, for any n, $n = 1, 2, \ldots,$

$$M \int_0^{T \wedge \tau_n} f_s^2(\omega)ds \le Mx_T^2,$$

and, therefore,

$$M \int_0^T f_s^2(\omega)ds = \lim_{n \to \infty} M \int_0^{T \wedge \tau_n} f_s^2(\omega)ds \le Mx_T^2 < \infty,$$

which proves the lemma. □

Note. If $X = (x_t, \mathscr{F}_t)$, $0 \le t \le T$, is a square integrable martingale with

$$x_t = x_0 + \int_0^t f_s(\omega)dW_s,$$

where $P(\int_0^T f_s^2(\omega)ds < \infty) = 1$, then

$$M\int_0^T f_s^2(\omega)ds \le M[x_T - x_0]^2 = Mx_T^2 - Mx_0^2 \le Mx_T^2 < \infty.$$

5.6.4

In the next theorem the condition (5.120) appearing in the formulation of the previous theorem is weakened.

Theorem 5.18. *Let the assumptions of Theorem* 5.17 *be fulfilled with the exception of* (5.120), *which is replaced by the weaker condition that*

$$P\left(\int_0^T a_t^2(\xi)dt < \infty\right) = 1. \tag{5.132}$$

Then the conclusions of Theorem 5.17 *remain true.*

Proof. (5.120) provided the equivalence $\mu_\xi \sim \mu_\eta$. By (5.132), according to Theorem 7.20 we have only $\mu_\xi \ll \mu_\eta$.

Let $n = 1, 2, \ldots$, and let $\xi^{(n)} = (\xi_t^{(n)}, \mathscr{F}_t)$ be a process which is a (strong) solution of the equation

$$\xi_t^{(n)} = \xi_{\int_0^t \chi_s^{(n)} ds} + \int_0^t [1 - \chi_s^{(n)}]b_s(\xi^{(n)})dW_s, \tag{5.133}$$

where

$$\chi_s^{(n)} = \chi_{[\int_0^s (a_u(\xi)b_u^{-1}(\xi))^2 ds < n]}.$$

Because of the assumptions, the coefficient $b_s(x)$ satisfies the conditions (4.110) and (4.111). Hence, from Theorem 4.8 it follows that a strong solution of Equation (5.133) actually exists.

As it is shown in proving Theorem 7.19, the process $\xi^{(n)} = (\xi_t^{(n)}, \mathscr{F}_t)$ permits the differential

$$d\xi_t^{(n)} = a_t^{(n)}(\xi^{(n)})dt + b_t(\xi^{(n)})dW_t, \tag{5.134}$$

where

$$a_t^{(n)}(x) = a_t(x)\chi_{[\int_0^t (a_s(x)b_s^{-1}(x))^2 ds < n]}.$$

Since

$$P\left(\int_0^T (a_t^{(n)}(\xi^{(n)})b_t^{-1}(\xi^{(n)}))^2\, dt \le n\right) = 1,$$

by Theorem 7.18 $\mu_{\xi^{(n)}} \sim \mu_\eta$.

We assume now $x_t^{(n)} = M[x_T|\mathscr{F}_t^{\xi^{(n)}}]$. Then, because of Theorem 5.17, for the martingale $X^{(n)} = (x_t^{(n)}, \mathscr{F}_t^{\xi(n)})$ we have the representation

$$x_t^{(n)} = x_0^{(n)} + \int_0^t f_s^{(n)}(\omega^{(n)})dW_s, \tag{5.135}$$

where the process $(f_s^{(n)}(\omega), \mathscr{F}_s^{\xi(n)})$ is such that

$$P\left(\int_0^T [f_s^{(n)}(\omega)]^2\, ds < \infty\right) = 1.$$

We note that $x_0^{(n)} = x_0$ (P-a.s.). Indeed, since $\xi_0^{(n)} = \xi_0$ (P-a.s.),

$$x_0^{(n)} = M[x_T|\mathscr{F}_0^{\xi(n)}] = M[x_T|\xi_0^{(n)}] = M[x_t|\xi_0] = x_0.$$

Let

$$\tau_n(x) = \begin{cases} \inf\left[t: \int_0^t (a_s(x)b_s^{-1}(x))^2\, ds \geq n\right], \\ T, \quad \text{if } \int_0^T (a_s(x)b_s^{-1}(x))^2\, ds < n. \end{cases}$$

From the construction of the process $\xi^{(n)}$ it follows that $\xi_t^{(n)} = \xi_t$ for $t < \tau_n(\xi)$. Hence $\tau_n(\xi) = \tau_n(\xi^{(n)})$ (P-a.s.). From this it is not difficult to deduce that for any t, $0 \leq t \leq T$,

$$\mathscr{F}^{\xi}_{t \wedge \tau_n(\xi)} = \mathscr{F}^{\xi(n)}_{t \wedge \tau_n(\xi^{(n)})}. \tag{5.136}$$

From (5.135) it follows that the martingale $X^{(n)} = (x_t^{(n)}, \mathscr{F}_t^{\xi(n)})$ has continuous trajectories. Hence, by Theorem 3.6,

$$\begin{aligned} M(x_t^{(n)}|\mathscr{F}^{\xi(n)}_{t \wedge \tau_n(\xi^{(n)})}) &= x^{(n)}_{t \wedge \tau_n(\xi^{(n)})} \\ &= x_0 + \int_0^{t \wedge \tau_n(\xi^{(n)})} f_s^{(n)}(\omega)dW_s \\ &= x_0 + \int_0^{t \wedge \tau_n(\xi)} f_s^{(n)}(\omega)dW_s, \end{aligned} \tag{5.137}$$

since $\tau_n(\xi) = \tau_n(\xi^{(n)})$ (P-a.s.) and, with $s \leq \tau_n(\xi)$, $\xi_s = \xi_s^{(n)}$ (P-a.s.).

We note now that $x_t^{(n)} = M(x_T|\mathscr{F}_t^{\xi(n)})$. Then, according to Theorem 3.6 and (5.136),

$$x^{(n)}_{t \wedge \tau_n(\xi^{(n)})} = M(x_T|\mathscr{F}^{\xi(n)}_{t \wedge \tau_n(\xi^{(n)})}) = M(x_T|\mathscr{F}^{\xi}_{t \wedge \tau_n(\xi)}),$$

which together with (5.137) yields the equality

$$M(x_T|\mathscr{F}^{\xi}_{t \wedge \tau_n(\xi)}) = x_0 + \int_0^{t \wedge \tau_n(\xi)} f_s^{(n)}(\omega)dW_s, \qquad 0 \leq t \leq T. \tag{5.138}$$

Denote for brevity the right side in (5.138) by $\tilde{x}_t^{(n)}$ and set $\tilde{\mathscr{F}}_t^{(n)} = \mathscr{F}_{t \wedge \tau_n(\xi)}^{\xi}$. The process $\tilde{X}^{(n)} = (\tilde{x}_t^{(n)}, \tilde{\mathscr{F}}_t^{(n)})$ is a martingale, since $M|\tilde{x}_t^{(n)}| \le M|x_T| < \infty$ and, for $t \le s$,

$$M(\tilde{x}_t^{(n)} | \tilde{\mathscr{F}}_s^{(n)}) = M[M(x_T | \mathscr{F}_{t \wedge \tau_n(\xi)}^{\xi} | \mathscr{F}_{s \wedge \tau_n(\xi)}^{\xi})]$$

$$= M(x_T | \mathscr{F}_{s \wedge \tau_n(\xi)}^{\xi}) = x_0 + \int_0^{s \wedge \tau_n(\xi)} f_s^{(n)}(\omega) dW_s$$

$$= \tilde{x}_s^{(n)} \qquad (P\text{-a.s.}).$$

Let $m \le n$. Then $\tau_m(\xi) \le \tau_n(\xi)$ and, by Theorem 3.6, (P-a.s.)

$$M(\tilde{x}_T^{(n)} | \tilde{\mathscr{F}}_{t \wedge \tau_m(\xi)}^{(n)}) = \tilde{x}_{t \wedge \tau_m(\xi)}^{(n)} = x_0 + \int_0^{t \wedge \tau_m(\xi)} f_s^{(n)}(\omega) dW_s. \tag{5.139}$$

On the other hand, since

$$\tilde{\mathscr{F}}_{t \wedge \tau_m(\xi)}^{(n)} = \mathscr{F}_{t \wedge \tau_n(\xi) \wedge \tau_m(\xi)}^{\xi} = \mathscr{F}_{t \wedge \tau_m(\xi)}^{\xi},$$

then (P-a.s.)

$$M(\tilde{x}_T^{(n)} | \tilde{\mathscr{F}}_{t \wedge \tau_m(\xi)}) = M[M(x_T | \mathscr{F}_{t \wedge \tau_n(\xi)}^{\xi}) | \mathscr{F}_{t \wedge \tau_m(\xi)}^{\xi}]$$

$$= M(x_T | \mathscr{F}_{t \wedge \tau_m(\xi)}^{\xi}) = x_0 + \int_0^{t \wedge \tau_m(\xi)} f_s^{(m)}(\omega) dW_s. \tag{5.140}$$

Comparing (5.139) and (5.140), we see that

$$\int_0^{t \wedge \tau_m(\xi)} [f_s^{(n)}(\omega) - f_s^{(m)}(\omega)] dW_s = 0, \qquad 0 \le t \le T. \tag{5.141}$$

From (5.141), with the help of the Ito formula, we find that

$$0 = \left(\int_0^{T \wedge \tau_m(\xi)} [f_s^{(n)}(\omega) - f_s^{(m)}(\omega)] dW_s \right)^2$$

$$= \int_0^{T \wedge \tau_m(\xi)} [f_s^{(n)}(\omega) - f_s^{(m)}(\omega)]^2 ds$$

$$+ 2 \int_0^{T \wedge \tau_m(\xi)} \left\{ \int_0^t [f_s^{(n)}(\omega) - f_s^{(m)}(\omega)] dW_s \right\}$$

$$\times \{f_t^{(n)}(\omega) - f_t^{(m)}(\omega)\} dW_t$$

$$= \int_0^{T \wedge \tau_m(\xi)} [f_s^{(n)}(\omega) - f_s^{(m)}(\omega)]^2 ds,$$

since on the set $\{\omega : t \le \tau_m(\xi)\}$

$$\int_0^t [f_s^{(n)}(\omega) - f_s^{(m)}(\omega)] dW_s = 0.$$

Thus, for $n \geq m$, on the set $\{\tau_m(\xi) = T\}$

$$\int_0^T [f_s^{(n)}(\omega) - f_s^{(m)}(\omega)]^2 ds = 0. \tag{5.142}$$

From this it follows that for almost all t, $0 \leq t \leq T$,

$$f_t^{(n)}(\omega) = f_t^{(n)}(\omega) \qquad \{\tau_m(\xi) = T\}\ (P\text{-a.s.}).$$

We define now a function $f_t(\omega)$:

$$f_t(\omega) = \begin{cases} f_t^{(1)}(\omega), & \text{if } \displaystyle\int_0^t a_s^2(\xi)ds < 1, \\ f_t^{(2)}(\omega), & \text{if } 1 \leq \displaystyle\int_0^t a_s^2(\xi)ds < 2, \\ \dots\dots\dots\dots\dots\dots & \\ f_t^{(n)}(\omega), & \text{if } n - 1 \leq \displaystyle\int_0^t a_s^2(\xi)ds < n, \\ \dots\dots\dots\dots\dots\dots & \end{cases} \tag{5.143}$$

From the definition it is clear that the function $f_t(\omega)$, $0 \leq t \leq T$, is $\mathscr{B}_{[0, T]} \times \mathscr{F}_T^\xi$-measurable and $\mathscr{F}_t^\xi$-measurable for each fixed t, $0 \leq t \leq T$.

Further,

$$\begin{aligned} \int_0^T f_t^2(\omega)dt &= \sum_{n=0}^{\infty} \int_{\tau_n(\xi)}^{\tau_{n+1}(\xi)} f_t^2(\omega)dt \\ &= \sum_{n=0}^{\infty} \int_{\tau_n(\xi)}^{\tau_{n+1}(\xi)} [f_t^{(n+1)}(\omega)]^2 dt \\ &= \sum_{n=0}^{N} \int_{\tau_n(\xi)}^{\tau_{n+1}(\xi)} [f_t^{(n+1)}(\omega)]^2 dt \\ &\quad + \sum_{n=N+1}^{\infty} \int_{\tau_n(\xi)}^{\tau_{n+1}(\xi)} [f_t^{(n+1)}(\omega)]^2 dt. \end{aligned}$$

On the set $\{\omega : \tau_{N+1}(\xi) = T\}$,

$$\int_0^T f_t^2(\omega)dt = \sum_{n=0}^{N} \int_{\tau_n(\xi)}^{\tau_{n+1}(\xi)} [f_t^{(n+1)}(\omega)]^2 dt < \infty.$$

Therefore, for any N,

$$\left\{\omega : \int_0^T f_t^2(\omega)dt = \infty\right\} \subseteq \{\omega : \tau_{N+1}(\xi) < T\}.$$

But $\tau_N(\xi) \uparrow T$ (P-a.s.) with $N \to \infty$. Hence

$$P\left\{\int_0^T f_t^2(\omega)dt < \infty\right\} = 1.$$

In a similar way one easily establishes also the inclusion

$$\left\{\omega: \int_0^T [f_t(\omega) - f_t^{(n)}(\omega)]^2\,dt > 0\right\} \subseteq \{\omega: \tau_n(\xi) < T\}.$$

Hence, with $n \to \infty$,

$$\int_0^T [f_t(\omega) - f_t^{(n)}(\omega)]^2\,dt \xrightarrow{P} 0,$$

and consequently,

$$P\text{-}\lim_{n\to\infty} \int_0^t f_s^{(n)}(\omega)\,dW_s = \int_0^t f_s(\omega)\,dW_s. \tag{5.144}$$

It is also clear that

$$\int_0^{t\wedge\tau_n(\xi)} f_s^{(n)}(\omega)\,dW_s = \int_0^t \chi_{\{\tau_n(\xi)>s\}} f_s^{(n)}(\omega)\,dW_s$$

and for any t, $0 \le t \le T$,

$$P\text{-}\lim_n \int_0^t \chi_{\{\tau_n(\xi)>s\}} f_s^{(n)}(\omega)\,dW_s = \int_0^t f_s(\omega)\,dW_s, \tag{5.145}$$

since

$$\begin{aligned}
&P\left\{\int_0^T [f_s(\omega) - f_s^{(n)}(\omega)\chi_{\{\tau_n(\xi)>s\}}]^2\,ds > \varepsilon\right\}\\
&= P\left\{\int_0^T [f_s(\omega) - f_s^{(n)}(\omega)\chi_{\{\tau_n(\xi)>s\}}]^2\,ds > \varepsilon;\ \tau_n(\xi) = T\right\}\\
&\quad + P\left\{\int_0^T [f_s(\omega) - f_s^{(n)}(\omega)\chi_{\{\tau_n(\xi)>s\}}]^2\,ds > \varepsilon,\ \tau_n(\xi) < T\right\}\\
&\le P\left\{\int_0^T [f_s(\omega) - f_s^{(n)}(\omega)]^2\,ds > \varepsilon\right\} + P\{\tau_n(\xi) < T\} \to 0,\ n \to \infty.
\end{aligned}$$

Let $t < T$. We take the limit in (5.138) as $n \to \infty$. The left side of the equality, because of Theorem 1.5, tends to $M(x_T|\mathscr{F}_t^\xi) = x_t$, and the right side, according to (5.145), converges in probability to

$$x_0 + \int_0^t f_s(\omega)\,dW_s.$$

Thus, with $t < T$,

$$x_t = x_0 + \int_0^t f_s(\omega)\,dW_s \quad (P\text{-a.s.}). \tag{5.146}$$

But if $t = T$ then $\mathscr{F}_{T\wedge\tau_n(\xi)}^\xi \uparrow \mathscr{F}_{T-}^\xi$, and therefore,

$$M(x_T|\mathscr{F}_{T-}^\xi) = x_0 + \int_0^T f_s(\omega)\,dW_s.$$

But the process $\xi = (\xi_t)$, $0 \le t \le T$, has (P-a.s.) continuous trajectories, and hence $\mathscr{F}_{t-}^{\xi} = \mathscr{F}_t^{\xi}$ (compare with the proof of Theorem 4.3), which proves Theorem 5.18. □

5.6.5

In Theorem 4.3 it was shown that the (augmented) σ-algebras $\mathscr{F}_t^W$, generated by the Wiener process W_s, $s \le t$, are continuous, i.e., $\mathscr{F}_{t-}^W = \mathscr{F}_t^W = \mathscr{F}_{t+}^W$. We establish a similar result for processes of the diffusion type as well.

Theorem 5.19. *Let the conditions of Theorem 5.18 be fulfilled. Then the (augmented) σ-algebras $\mathscr{F}_t^{\xi}$ are continuous:*

$$\mathscr{F}_{t-}^{\xi} = \mathscr{F}_t^{\xi} = \mathscr{F}_{t+}^{\xi}.$$

PROOF. As it was noted above, the relationship $\mathscr{F}_{t-}^{\xi} = \mathscr{F}_t^{\xi}$ is proved in a similar way as in the case of a Wiener process (see Theorem 4.3). We establish the right continuity of the σ-algebras $\mathscr{F}_t^{\xi}$.

Let η be a bounded random variable, $|\eta| \le c$. Then by Theorem 5.18, the martingale $X = (x_t, \mathscr{F}_t^{\xi})$ with $x_t = M(\eta | \mathscr{F}_t^{\xi})$ has a continuous modification. We show that

$$M(\eta | \mathscr{F}_t^{\xi}) = M(\eta | \mathscr{F}_{t+}^{\xi}) \qquad (P\text{-a.s.}). \tag{5.147}$$

For any $\varepsilon > 0$

$$M(\eta | \mathscr{F}_{t+}^{\xi}) = M(M(\eta | \mathscr{F}_{t+\varepsilon}^{\xi}) | \mathscr{F}_{t+}^{\xi}) = M(x_{t+\varepsilon} | \mathscr{F}_{t+}^{\xi}). \tag{5.148}$$

But the random variables $x_t = M(\eta | \mathscr{F}_t^{\xi})$ are bounded, $|x_t| \le c$, and because of the continuity of the process x_t from (5.148), passing to the limit with $\varepsilon \downarrow 0$, we find that (P-a.s.)

$$M(\eta | \mathscr{F}_{t+}^{\xi}) = M(x_t | \mathscr{F}_{t+}^{\xi}) = x_t = M(\eta | \mathscr{F}_t^{\xi}).$$

From this relationship (5.147) follows.

We take now an $\mathscr{F}_{t+}^{\xi}$-measurable bounded random variable for η. Then $M(\eta | \mathscr{F}_{t+}^{\xi}) = \eta$, and, therefore, $\eta = M(\eta | \mathscr{F}_t^{\xi})$.

Consequently, the random variable η is $\mathscr{F}_t^{\xi}$-measurable which proves the inclusion $\mathscr{F}_{t+}^{\xi} \subseteq \mathscr{F}_t^{\xi}$. The inverse inclusion $\mathscr{F}_t^{\xi} \subseteq \mathscr{F}_{t+}^{\xi}$ is obvious. □

5.6.6

A particular case of Theorems 5.17 and 5.18 deserving special attention is when the coefficient $b_t(x) \equiv 1$ (or $b_t(x) \equiv c \ne 0$).

Theorem 5.20. *Let $\xi = (\xi_t, \mathscr{F}_t)$, $0 \le t \le T$, be a process of the diffusion type with the differential*

$$d\xi_t = a_t(\xi)dt + dW_t, \tag{5.149}$$

where $a = (a_t(x), \mathscr{B}_t)$ is a nonanticipative functional with

$$P\left(\int_0^T a_t^2(\xi)dt < \infty\right) = 1.$$

Then any martingale $X = (x_t, \mathscr{F}_t^\xi)$ has a continuous modification for which there is the representation

$$x_t = x_0 + \int_0^t f_s(\omega) dW_s,$$

where the process $(f_s(\omega), \mathscr{F}_s^\xi)$ is such that

$$P\left(\int_0^T f_s^2(\omega) ds < \infty\right) = 1.$$

If $X = (x_t, \mathscr{F}_t^\xi)$ is a square integrable martingale, then, in addition

$$M \int_0^T f_s^2(\omega) ds < \infty.$$

5.6.7

We consider now the structure of the functionals on processes of the diffusion type in the Gaussian case. We shall assume that the random process $\xi = (\xi_t, \mathscr{F}_t)$, $0 \le t \le T$, has the differential

$$d\xi_t = a_t(\xi) dt + b(t) dW_t, \qquad \xi_0 = 0, \tag{5.150}$$

where $W = (W_t, \mathscr{F}_t)$ is a Wiener process, and $b(t)$, $0 \le t \le T$, is a deterministic function where $b^2(t) \ge c \ge 0$ ($\int_0^T b^2(t) dt < \infty$).

Theorem 5.21. *Let $X = (x_t, \mathscr{F}_t^\xi)$, $0 \le t \le T$, be a Gaussian martingale. If the process $(W, \xi, X) = (W_t, \xi_t, x_t)$, $0 \le t \le T$, forms a Gaussian system and*

$$P\left(\int_0^T a_t^2(\xi) dt < \infty\right) = 1, \tag{5.151}$$

then the martingale $X = (x_t, \mathscr{F}_t^\xi)$ has a continuous modification and

$$x_t = x_0 + \int_0^t f(s) dW_s, \qquad 0 \le t \le T, \tag{5.152}$$

where the measurable deterministic function $f = f(t)$ is such that

$$\int_0^T f^2(t) dt < \infty. \tag{5.153}$$

PROOF. The Gaussian martingale X is square integrable. Hence, according to Theorem 5.18, we can find a process $g = (g_t(\omega), \mathscr{F}_t^\xi)$, $0 \le t \le T$, with $\int_0^T M g_t^2(\omega) dt < \infty$, such that

$$x_t = x_0 + \int_0^t g_s(\omega) dW_s. \tag{5.154}$$

From (5.150) it follows that for each t the random variables W_t are $\mathscr{F}_t^\xi$-measurable. Consequently, not only the process $(W_t, \mathscr{F}_t)$ but also the process $(W_t, \mathscr{F}_t^\xi)$ are martingales, and, therefore, $M(W_t|\mathscr{F}_s^\xi) = W_s$ (P-a.s.), $t \geq s$. From this it follows that the expression

$$M|(x_t - x_s)(W_t - W_s)|\mathscr{F}_s^\xi| = M[(x_t - M(x_t|\mathscr{F}_s^\xi))((W_t - M(W_t|\mathscr{F}_s^\xi))|\mathscr{F}_s^\xi] \tag{5.155}$$

is the conditional covariance $\operatorname{cov}(x_t, W_t|\mathscr{F}_s^\xi)$ (see notations in Section 13.1).

We show that because of the Gaussian behavior of the process (W, ξ, X),

$$\begin{aligned}\operatorname{cov}(x_t, W_t|\mathscr{F}_s^\xi) &= M[(x_t - x_s)(W_t - W_s)|\mathscr{F}_s^\xi] \\ &= M[(x_t - x_s)(W_t - W_s)] \qquad (P\text{-a.s.}).\end{aligned} \tag{5.156}$$

To prove this we note first that

$$M[(x_t - x_s)(W_t - W_s)|\mathscr{F}_s^\xi] = M[x_t W_t|\mathscr{F}_s^\xi] - x_s W_s.$$

Now let $\mathscr{F}_{s,n}^\xi = \sigma\{\omega: \xi_0, \xi_{(s/2^n)}, \xi_{2(s/2^n)}, \ldots, \xi_s\}$. Then $\mathscr{F}_{s,n}^\xi \uparrow \mathscr{F}_s^\xi$ and, therefore, by Theorem 1.5 (P-a.s.)

$$M[x_t W_t|\mathscr{F}_{s,n}^\xi] \to M[x_t W_t|\mathscr{F}_s^\xi], \qquad M[x_s W_s|\mathscr{F}_{s,n}^\xi] \to x_s W_s.$$

Therefore,

$$\begin{aligned}M[(x_t - x_s)(W_t - W_s)|\mathscr{F}_s^\xi] &= \lim_n M[x_t W_t|\mathscr{F}_{s,n}^\xi] - x_s W_s \\ &= \lim_n \{M[x_t W_t|\mathscr{F}_{s,n}^\xi] \\ &\quad - M[x_t|\mathscr{F}_{s,n}^\xi] M[W_t|\mathscr{F}_{s,n}^\xi]\} \\ &\quad + \lim_n \{M[x_t|\mathscr{F}_{s,n}^\xi] M[W_t|\mathscr{F}_{s,n}^\xi]\} - x_s W_s.\end{aligned}$$

Since $\mathscr{F}_s^\xi \supseteq \mathscr{F}_{s,n}^\xi$, $M[x_t|\mathscr{F}_{s,n}^\xi] = M[M(x_t|\mathscr{F}_s^\xi)|\mathscr{F}_{s,n}^\xi] = M(x_s|\mathscr{F}_{s,n}^\xi) \to x_s$ and, similarly, $M[W_t|\mathscr{F}_{s,n}^\xi] = M[W_s|\mathscr{F}_{s,n}^\xi] \to W_s$. Consequently

$$\begin{aligned}&M[(x_t - x_s)(W_t - W_s)|\mathscr{F}_s^\xi] \\ &= \lim_n \{M[x_t W_t|\mathscr{F}_{s,n}^\xi] - M[x_t|\mathscr{F}_{s,n}^\xi] M[W_t|\mathscr{F}_{s,n}^\xi]\} \\ &= \lim_n M\{[x_t - M(x_t|\mathscr{F}_{s,n}^\xi)] W_t - M(W_t|\mathscr{F}_{s,n}^\xi)] | \mathscr{F}_{s,n}^\xi\}.\end{aligned}$$

But, by the theorem on normal correlation (Theorem 13.1),

$$\begin{aligned}&M\{[x_t - M(x_t|\mathscr{F}_{s,n}^\xi)][W_t - M(W_t|\mathscr{F}_{s,n}^\xi)]|\mathscr{F}_{s,n}^\xi\} \\ &= M\{[x_t - M(x_t|\mathscr{F}_{s,n}^\xi)][W_t - M(W_t|\mathscr{F}_{s,n}^\xi)]\} \qquad (P\text{-a.s.}).\end{aligned}$$

Thus,

$$\begin{aligned}&M[(x_t - x_s)(W_t - W_s)|\mathscr{F}_s^\xi] \\ &= \lim_n M\{[x_t - M(x_t|\mathscr{F}_{s,n}^\xi)][W_t - M(W_t|\mathscr{F}_{s,n}^\xi)]\},\end{aligned}$$

which, because of the uniform integrability of the variables $\{M(x_t|\mathscr{F}^{\xi}_{s,n}),$ $n = 1, 2, \ldots\}$ and $\{M(W_t|\mathscr{F}^{\xi}_{s,n}), n = 1, 2, \ldots\}$, leads to (5.156).

From (5.156) and (5.154) we infer that

$$\int_s^t M[g_u(\omega)|\mathscr{F}^{\xi}_s]du = \int_s^t Mg_u(\omega)du. \tag{5.157}$$

We now consider for fixed t, $0 \le t \le T$, the decomposition

$$0 \equiv t_0^{(n)} < t_1^{(n)} < \cdots < t_n^{(n)} \equiv t$$

with $\max_j [t_{j+1}^{(n)} - t_j^{(n)}] \to 0$, $n \to \infty$, and

$$g_n(u) = M[g_u(\omega)|\mathscr{F}^{\xi}_{t_j^{(n)}}], \qquad t_j^{(n)} \le u < t_j^{(n+1)}.$$

Then, according to (5.157),

$$\begin{aligned}\int_0^{t} Mg_u(\omega)du &= \sum_{j=0}^{n-1} \int_{t_j^{(n)}}^{t_{j+1}^{(n)}} Mg_u(\omega)du \\ &= \sum_{j=0}^{n-1} \int_{t_j^{(n)}}^{t_{j+1}^{(n)}} g_n(u)du = \int_0^t g_n(u)du.\end{aligned}$$

By Theorem 5.19 the σ-algebras $\mathscr{F}^{\xi}_t$, $0 \le t \le T$, are continuous. Hence for each u, $0 \le u \le t$, with probability one as $n \to \infty$

$$g_n(u) \to M[g_u(\omega)|\mathscr{F}^{\xi}_u] = g_u(\omega). \tag{5.158}$$

By the Jensen inequality $Mg_n^2(u) \le Mg_u^2(\omega)$, and, therefore,

$$\int_0^t Mg_n^2(u)du \le \int_0^t Mg_u^2(\omega)du < \infty.$$

Thus by Theorem 1.8 the family of the random functions $\{g_n(u), n = 1, 2, \ldots\}$ is uniformly integrable (over measure $P(d\omega \times du)$ and, because of (5.158),

$$\begin{aligned}&M\left|\int_0^t [Mg_u(\omega) \quad g_u(\omega)]du\right. \\ &\le M\left|\int_0^t [Mg_u(\omega) - g_n(u)]du\right| \\ &\quad + M\left|\int_0^t [g_u(\omega) - g_n(u)]\, du\right| \le \int_0^t M|g_u(\omega) \\ &\quad - g_n(u)|\, du \to 0, \qquad n \to \infty.\end{aligned}$$

From this, for each t, $0 \le t \le T$, we obtain

$$\int_0^t Mg_u(\omega)du = \int_0^t g_u(\omega)du \qquad (P\text{-a.s.}) \tag{5.159}$$

and, therefore, for almost all t, $0 \leq t \leq T$, (P-a.s.)

$$g_t(\omega) = Mg_t(\omega).$$

Together with (5.154) this proves the correctness of the representation (5.152) with $f(t) = Mg_t(\omega)$. □

Corollary 1. *The function $f(t)$, $0 \leq t \leq T$, entering in* (5.152) *can be defined from the equality.*

$$f(t) = \frac{dM[x_t W_t]}{dt}.$$

Corollary 2. *Let $\eta = \eta(\omega)$ be an $\mathscr{F}_T^{\xi}$-measurable Gaussian random variable. We assume that (η, W, ξ) forms a Gaussian system. Then there will be a deterministic function $f(s)$, $0 \leq s \leq T$, such that* (P-a.s.)

$$\eta(\omega) = M\eta(\omega) + \int_0^T f(s)dW_s, \tag{5.160}$$

where $\int_0^T f^2(s)ds < \infty$.

By the theorem on normal correlation a martingale $x_t = M(\eta | \mathscr{F}_t^{\xi})$ will be Gaussian, as will be the system (W, ξ, X) with $X = (x_t, \mathscr{F}_t^{\xi})$, $0 \leq t \leq T$. Hence (5.160) follows from (5.152) by taking into account that $x_T = \eta$ and that $x_0 = M\eta$ (P-a.s.).

Notes and references

5.1,5.2. For the proofs of Theorems 5.1–5.4, see also Meyer [126], Kunita and Watanabe [95], and Wentzel [18]. Theorem 5.5 was proved in a different way by Clark. [85]. The proof of Theorem 5.5 is similar to that of Wentzel [18].

5.3. The assertions of Theorem 5.6 were partially presented in Clark [85]. The proof of the representation for Gaussian random variables is due to the authors. Theorem 5.7 was proved by Clark [85]. The assertions of the type of Theorems 5.8 and 5.9 can also be found in Wentzel [18].

5.4. The structure of a stochastic integral over square integrable martingales given here is due to Courrege [96].

5.5. Theorems 5.13 and 5.14 appear to be new. A Fubini theorem for stochastic integrals was first presented in Kallianpur and Striebel [75]. For its extensions see also Yershov [51]. The proof given here is based on a result related to Theorem 5.14.

5.6. The structure of the functionals on diffusion-type processes in the case $b_t(\xi) \equiv 1$ was discussed in Fujisaki, Kallianpur, Kunita [156]. The general case presented here as well as the proof of continuity of the σ-algebras $\mathscr{F}_t^{\xi}$ (Theorem 5.19) are new. Theorem 5.21 is also new.

6 Nonnegative supermartingales and martingales, and the Girsanov theorem

6.1 Nonnegative supermartingales

6.1.1

Let $(\Omega, \mathscr{F}, P)$ be a complete probability space, and let $(\mathscr{F}_t)$, $0 \leq t \leq T$, be a nondecreasing family of the sub-σ-algebras $\mathscr{F}$, augmented by sets of $\mathscr{F}$ probability zero. Let $W = (W_t, \mathscr{F}_t)$ be a Wiener process and let $\gamma = (\gamma_t, \mathscr{F}_t)$ be a random process with

$$P\left(\int_0^T \gamma_s^2\, ds < \infty\right) = 1. \tag{6.1}$$

In investigating questions about the absolute continuity of measures corresponding to the Ito processes with respect to a Wiener measure (see next chapter) an essential role is played by nonnegative continuous (P-a.s.) random processes $\varkappa = (\varkappa_t, \mathscr{F}_t)$, $0 \leq t \leq T$, permitting the representation

$$\varkappa_t = 1 + \int_0^t \gamma_s\, dW_s. \tag{6.2}$$

In the following lemma it is shown that processes of this type are necessarily supermartingales.

Lemma 6.1. *Let the process* $\gamma = (\gamma_t, \mathscr{F}_t)$, $t \leq T$, *satisfy* (6.1) *and let* $\varkappa_t \geq 0$ (P-a.s.), $0 \leq t \leq T$. *Then the random process* $\varkappa = (\varkappa_t, \mathscr{F}_t)$ *is a (nonnegative) supermartingale,*

$$M(\varkappa_t | \mathscr{F}_s) \leq \varkappa_s \qquad (P\text{-a.s.}), \qquad t \geq s, \tag{6.3}$$

and, in particular,

$$M\varkappa_t \leq 1. \tag{6.4}$$

PROOF. We set[1], for $n \geq 1$,

$$\tau_n = \inf\left\{t \leq T: \int_0^t \gamma_s^2 \, ds \geq n\right\},$$

and $\tau_n = T$ if $\int_0^T \gamma_s^2 \, ds < n$. Then, according to (4.63), for $t > s$,

$$\varkappa_{t \wedge \tau_n} = 1 + \int_0^{t \wedge \tau_n} \gamma_u \, dW_u = \varkappa_{s \wedge \tau_n} + \int_{s \wedge \tau_n}^{t \wedge \tau_n} \gamma_u \, dW_u.$$

Since

$$M\left[\int_{s \wedge \tau_n}^{t \wedge \tau_n} \gamma_n \, dW_u | \mathscr{F}_s\right] = 0 \qquad (P\text{-a.s.}),$$

therefore

$$M[\varkappa_{t \wedge \tau_n} | \mathscr{F}_s] = \varkappa_{s \wedge \tau_n} \qquad (P\text{-a.s.}).$$

But $\tau_t \rightarrow T$ with probability one as $n \rightarrow \infty$; hence, because of the nonnegativity and continuity of the process $\varkappa_t$, $0 \leq t \leq T$, by the Fatou lemma $M(\varkappa_t | \mathscr{F}_s) \leq \varkappa_s$. □

6.1.2

Lemma 6.2 *The nonnegative supermartingale* $\varkappa = (\varkappa_t, \mathscr{F}_t)$, $0 \leq t \leq T$, *with* $\varkappa_t = 1 + \int_0^t \gamma_t \, dW_s$, $P(\int_0^T \gamma_s^2 \, ds < \infty) = 1$, *permits the representation*

$$\varkappa_t = \exp\left(\Gamma_t(\beta) - \frac{1}{2}\int_0^t \beta_s^2 \, ds\right), \tag{6.5}$$

where

$$\beta_s = \varkappa_s^+ \gamma_s, \qquad \varkappa_s^+ = \begin{cases} \varkappa_s^{-1}, & \varkappa_s > 0, \\ 0, & \varkappa_s = 0, \end{cases} \tag{6.6}$$

and[2]

$$\Gamma_t(\beta) = P\text{-}\lim_n \chi_{(\int_0^t \beta_s^2 \, ds < \infty)} \int_0^t \beta_s^{(n)} \, dW_s, \qquad \beta_s^{(n)} = \beta_s \chi_{(\int_0^s \beta_u^2 \, du \leq n)}.$$

PROOF. Let

$$\sigma_n = \inf\left\{t \leq T: \varkappa_t = \frac{1}{n}\right\}$$

[1] According to the note to Lemma 4.4, the process $\int_0^t \gamma_s^2 \, ds, t \leq T$, has a progressively measurable modification which will be considered in this and other similar cases. Then time τ_n will be Markov with respect to a system $(\mathscr{F}_t)$, $0 \leq t \leq T$.

[2] The random variables $\Gamma_t(\beta)$ were discussed in detail in Subsection 4.2.9.

($\sigma_n = \infty$ if $\inf_{t \le T} \varkappa_t > 1/n$). Also, let

$$\sigma = \inf\{t \le T : \varkappa_t = 0\}$$

($\sigma = \infty$, if $\inf_{t \le T} \varkappa_t > 0$). It is clear that ($P$-a.s.) $\sigma_n \uparrow \sigma$, $n \to \infty$. According to Note 2 to Theorem 3.5,

$$\varkappa_t = 0, \qquad (P\text{-a.s.}) \qquad T \ge t \ge \sigma.$$

Hence for all t, $0 \le t \le T$,

$$\varkappa_t = \varkappa_{t \wedge \sigma} \qquad (P\text{-a.s.}) \tag{6.7}$$

and

$$\varkappa_t \varkappa_t^+ = \begin{cases} 1, & t < \sigma, \\ 0, & t \ge \sigma. \end{cases} \tag{6.8}$$

From (6.7) and (6.8) we infer that (P-a.s.)

$$\varkappa_t = \varkappa_{t \wedge \sigma} = 1 + \int_0^{t \wedge \sigma} \gamma_s \, dW_s = 1 + \int_0^t \varkappa_s \varkappa_s^+ \gamma_s \, dW_s,$$

i.e.,

$$\varkappa_t = 1 + \int_0^t \varkappa_s \beta_s \, dW_s \tag{6.9}$$

with $\beta_s = \varkappa_s^+ \gamma_s$.

It is clear that

$$P\left(\int_0^T (\varkappa_s \beta_s)^2 \, ds < \infty\right) = P\left(\int_0^T \gamma_s^2 \, ds < \infty\right) = 1. \tag{6.10}$$

Hence

$$\left(\frac{1}{n}\right)^2 \int_0^{\sigma_n \wedge T} \beta_s^2 \, ds \le \int_0^{\sigma_n \wedge T} (\varkappa_s \beta_s)^2 \, ds < \infty.$$

From this we obtain $P(\int_0^{\sigma_n \wedge T} \beta_s^2 \, ds < \infty) = 1$, and, applying the Ito formula to $\ln \varkappa_{t \wedge \sigma_n}$, from (6.9) we find that

$$\varkappa_{t \wedge \sigma_n} = \exp\left(\int_0^{t \wedge \sigma_n} \beta_s \, dW_s - \frac{1}{2} \int_0^{t \wedge \sigma_n} \beta_s^2 \, ds\right). \tag{6.11}$$

We note now, for each $t \le T$, that on the set $\{\omega : t < \sigma \le T\}$

$$\int_0^t \beta_s^2 \, ds < \infty \qquad (P\text{-a.s.}),$$

and that on the set $\{\omega : T \ge t \ge \sigma\}$

$$\varkappa_t = 0 \qquad (P\text{-a.s.}).$$

Hence

$$\{\omega : \varkappa_t > 0\} \subseteq \left\{\omega : \int_0^t \beta_s^2 \, ds < \infty\right\},$$

and, denoting

$$\chi_t = \chi_{(\int_0^t \beta_s^2\, ds < \infty)},$$

we obtain

$$\begin{aligned}
z_t = z_t \chi_t = z_{t\wedge\sigma}\chi_t &= P\text{-}\lim_n \chi_t z_{t\wedge\sigma_n} \\
&= P\text{-}\lim_n \chi_t \exp\left(\int_0^{t\wedge\sigma_n} \beta_s\, dW_s - \frac{1}{2}\int_0^{t\wedge\sigma_n} \beta_s^2\, ds\right) \\
&= P\text{-}\lim_n \chi_t \exp\left(\chi_t \int_0^{t\wedge\sigma_n} \beta_s\, dW_s - \frac{1}{2}\int_0^{t\wedge\sigma_n} \beta_s^2\, ds\right) \\
&= \chi_t \exp\left(P\text{-}\lim_n \chi_t \int_0^{t\wedge\sigma_n} \beta_s\, dW_s - \frac{1}{2}\int_0^{t\wedge\sigma} \beta_s^2\, ds\right). \qquad (6.12)
\end{aligned}$$

Since

$$P\text{-}\lim_n \chi_t \int_{t\wedge\sigma_n}^{t\wedge\sigma} \beta_s^2\, ds = 0,$$

then, according to Subsection 4.2.9, there exists

$$\Gamma_{t\wedge\sigma}(\beta) = P\text{-}\lim_n \chi_t \int_0^{t\wedge\sigma_n} \beta_s\, dW_s.$$

Consequently, (P-a.s.) for each t, $0 \le t \le T$,

$$z_t = \chi_t \exp\left(\Gamma_{t\wedge\sigma}(\beta) - \frac{1}{2}\int_0^{t\wedge\sigma} \beta_s^2\, ds\right). \qquad (6.13)$$

Hence, on the set $\{\sigma \le T\}$, (P-a.s.)

$$\chi_\sigma \exp\left(\Gamma_\sigma(\beta) - \frac{1}{2}\int_0^{\sigma} \beta_s^2\, ds\right) = 0. \qquad (6.14)$$

We deduce from this that, on the set $\{\sigma \le T\}$,

$$\int_0^{\sigma} \beta_s^2\, ds = \infty \qquad (P\text{-a.s.}).$$

Indeed, we assume the opposite, i.e., that

$$P\left\{(\sigma \le T) \cap \left(\int_0^{\sigma} \beta_s^2\, ds < \infty\right)\right\} > 0.$$

Then, on the basis of Lemma 4.7,

$$P\left\{(\sigma \le T) \cap \left(\int_0^{\sigma} \beta_s^2\, ds < \infty\right) \cap \left(\sup_n \left| \int_0^{\sigma_n} \beta_s\, dW_s \right| = \infty\right)\right\} = 0,$$

and, consequently, on the set $(\sigma \leq T) \cap (\int_0^\sigma \beta_s^2\, ds < \infty)$ of positive probability,

$$\varkappa_{\sigma_n} = \exp\left(\int_0^{\sigma_n} \beta_s\, dW_s - \frac{1}{2}\int_0^{\sigma_n} \beta_s^2\, ds\right) \nrightarrow 0, \qquad n \to \infty,$$

which contradicts the fact that $\varkappa_{\sigma_n} \to \varkappa_\sigma = 0$ (P-a.s.) on the set $\{\sigma \leq T\}$.

Thus

$$\{\omega:\sigma \leq T\} \cap \left\{\omega: \int_0^\sigma \beta_s^2\, ds = \infty\right\} = \{\omega:\sigma \leq T\}. \tag{6.15}$$

We show now that for each $t \leq T$ (P-a.s.) the right-hand side in (6.13) is equal to

$$\chi_t \exp\left(\Gamma_{t\wedge\sigma}(\beta) - \frac{1}{2}\int_0^{t\wedge\sigma} \beta_s^2\, ds\right) = \exp\left(\Gamma_t(\beta) - \frac{1}{2}\int_0^t \beta_s^2\, ds\right). \tag{6.16}$$

We fix t, $0 \leq t \leq T$. Then, if ω is such that $t < \sigma$, (6.16) is satisfied in an obvious way, since in this case $\chi_t = 1$, and $t \wedge \sigma = t$. Let now $T \geq t \geq \sigma$. Then the left-hand side in (6.16) is equal to zero. The right-hand side is also equal to zero, since on the set $\{\sigma \leq T\}$,

$$\int_0^\sigma \beta_s^2\, ds = \infty \qquad \text{and} \qquad \Gamma_\sigma(\beta) = 0 \qquad (P\text{-a.s.})$$

(compare with Subsection 4.2.9). □

6.1.3

An important particular case of nonnegative continuous (P-a.s.) supermartingales permitting the representation given by (6.2) is represented by processes $\varphi = (\varphi_t, \mathscr{F}_t)$, $t \leq T$, with

$$\varphi_t = \exp\left(\int_0^t \beta_s\, dW_s - \frac{1}{2}\int_0^t \beta_s^2\, ds\right), \tag{6.17}$$

where the process $\beta = (\beta_t, \mathscr{F}_t)$, $t \leq T$, is such that $P(\int_0^T \beta_s^2\, ds < \infty) = 1$.

The fact that such processes permit the representation given by (6.2) follows immediately from Ito's formula, leading to the equation

$$\varphi_t = 1 + \int_0^t \varphi_s \beta_s\, dW_s. \tag{6.18}$$

In this way (6.2) is obtained with $\gamma_s = \varphi_s \beta_s$, where $P(\int_0^T \gamma_s^2\, ds < \infty) = 1$.

6.1.4

We shall investigate now, in detail, questions of the existence and uniqueness of continuous solutions of equations of the type given by (6.18), and we shall consider also the feasibility of representation of these solutions in the form given by (6.17) or (6.5).

Thus we seek nonnegative continuous (P-a.s.) solutions of the equation

$$dx_t = x_t \alpha_t \, dW_t, \qquad x_0 = 1, \qquad t \le T, \tag{6.19}$$

satisfying the assumption $P(\int_0^T x_t^2 \alpha_t^2 \, dt < \infty) = 1$.

If the random process $\alpha = (\alpha_t, \mathcal{F}_t), t \le T$, is such that $P(\int_0^T \alpha_t^2 \, dt < \infty) = 1$, then there exists a unique nonnegative solution of such an equation given by the formula

$$x_t = \exp\left(\int_0^t \alpha_s \, dW_s - \frac{1}{2}\int_0^t \alpha_s^2 \, ds\right). \tag{6.20}$$

(If y_t, $t \le T$, is another continuous solution, then by the Ito formula we find that $d(y_t/x_t) \equiv 0$, so that $y_t = x_t$, (P-a.s.), $t \le T$.)

If it is known that the process $\alpha = (\alpha_t, \mathcal{F}_t)$, $t \le T$, is such that Equation (6.19) has a continuous nonnegative solution, then from the proof of Lemma 6.2 it follows that such a solution can be represented in the form

$$x_t = \exp\left(\Gamma_t(\alpha) - \frac{1}{2}\int_0^t \alpha_s^2 \, ds\right), \tag{6.21}$$

where this solution is unique.

Naturally the question arises: under what assumptions on the process $\alpha = (\alpha_t, \mathcal{F}_t)$, $t \le T$, does Equation (6.19) have a nonnegative continuous solution? The answer to this question is contained in the lemma given below for the formulation of which we introduce the following notation.

Let

$$\tau_n = \begin{cases} \inf\left\{t \le T : \int_0^t \alpha_s^2 \, ds \ge n^2\right\}, \\ \infty, \quad \text{if} \quad \int_0^T \alpha_s^2 \, ds < n^2, \end{cases} \tag{6.22}$$

and let $\tau = \lim_n \tau_n$.

It is clear that $\int_0^\tau \alpha_s^2 \, ds = \infty$ on the set $\{\omega : \tau \le T\}$.

Lemma 6.3. *In order for Equation* (6.19) *to have a nonnegative continuous* (P-a.s.) *solution, it is necessary and sufficient that* $P(\tau_1 > 0) = 1$, *and that, on the set*[3] $\{\omega : \tau \le T\}$,

$$\lim_n \int_0^{\tau_n} \alpha_s^2 \, ds = \infty. \tag{6.23}$$

This solution is unique and is given by (6.21).

[3] (6.23) implies that on the set $\{\omega : \tau \le T\}$ "the departure" of the integral $\int_0^t \alpha_s^2(\omega) ds$ for infinity with $t \to \tau(\omega)$ occurs in a continuous manner.

PROOF. Necessity: Let the equation

$$x_t = 1 + \int_0^t x_s \alpha_s \, dW_s \tag{6.24}$$

have a solution $\varkappa_t$, $0 \le t \le T$, with

$$P\left(\int_0^T \varkappa_s^2 \alpha_s^2 \, ds < \infty\right) = 1. \tag{6.25}$$

According to Lemma 6.2,

$$\varkappa_t = \exp\left(\Gamma_t(\alpha) - \frac{1}{2}\int_0^t \alpha_s^2 \, ds\right). \tag{6.26}$$

Hence, if for some n ($n = 1, 2, \ldots$), $P(\tau_n = 0) > 0$, then it would imply that $\int_0^t \alpha_s^2 \, ds = \infty$ with positive probability for any $t > 0$. But then from (6.26) it would follow that with positive probability $\varkappa_0 = 0$. This fact, however, contradicts the assumption $P(\varkappa_0 = 1) = 1$.

Further, $\int_0^\tau \alpha_s^2 \, ds = \infty$ on the set $\{\omega : \tau \le T\}$, and, therefore, $\varkappa_\tau = 0$. Hence, on the set $\{\tau \le T\}$, (P-a.s.)

$$0 = \varkappa_\tau = P\text{-}\lim_n \varkappa_{\tau_n} = P\text{-}\lim_n \exp\left(\Gamma_{\tau_n}(\alpha) - \frac{1}{2}\int_0^{\tau_n} \alpha_s^2 \, ds\right).$$

From this, with the help of Lemma 4.7, it is not difficult now to deduce that (6.23) is satisfied.

Sufficiency: Let the process $\alpha = (\alpha_t, \mathscr{F}_t)$, $0 \le t \le T$, satisfy the conditions of the lemma. We show that then

$$\varkappa_t = \exp\left(\Gamma_t(\alpha) - \frac{1}{2}\int_0^t \alpha_s^2 \, ds\right) \tag{6.27}$$

is a solution of Equation (6.19). For this purpose it has to be checked that: first $\varkappa_0 = 1$; second $P(\int_0^t (\varkappa_s \alpha_s)^2 \, ds < \infty) = 1$; third, $\varkappa_t$, $t \le T$, is continuous (P-a.s.); and, finally, $d\varkappa_t = \varkappa_t \alpha_t \, dW_t$.

The condition $\varkappa_0 = 1$ follows from the fact that $P(\tau_n > 0) = 1$, $n = 1, 2, \ldots$

Let us check the continuity (P-a.s.): of $\varkappa_t$, $t \le T$, and the condition $P(\int_0^T (\varkappa_s \alpha_s)^2 \, ds < \infty) = 1$.

From (6.27) and Subsection 4.2.9 it follows that on $\{\omega : \tau_n \le T\}$

$$\varkappa_{\tau_n} = \exp\left(\int_0^{\tau_n} \alpha_s \, dW_s - \frac{1}{2}\int_0^{\tau_n} \alpha_s^2 \, ds\right), \tag{6.28}$$

and, consequently, by the Ito formula

$$\varkappa_{\tau_n \wedge T} = 1 + \int_0^{\tau_n \wedge T} \varkappa_s \alpha_s \, dW_s.$$

As in Lemma 6.1, from this it is not difficult to deduce that the sequence $(\varkappa_{\tau_n \wedge T}, \mathscr{F}_{\tau_n \wedge T})$, $n = 1, 2, \ldots$, is a (nonnegative) supermartingale with

$M\varkappa_{\tau_n \wedge T} \leq 1$. Hence, according to Theorem 2.6, there exists (P-a.s.) $\lim_{n\to\infty} \varkappa_{\tau_n \wedge T}$ $(=\varkappa^*)$, where $M\varkappa^* \leq 1$. From this it follows that

$$P(\varkappa^* < \infty) = 1.$$

We show that the process $\varkappa_t$, $0 \leq t \leq T$, defined in (6.27) is (P-a.s.) continuous. Since

$$\varkappa_{t\wedge\tau_n} = \exp\left(\int_0^{t\wedge\tau_n} \alpha_s \, dW_s - \frac{1}{2}\int_0^{t\wedge\tau_n} \alpha_s^2 \, ds\right), \tag{6.29}$$

then $\varkappa_t$ is a (P-a.s.) continuous function for $t \leq \tau_n$. For $\tau \leq t \leq T$ $\varkappa_t = 0$ (P-a.s.), since, on the set $\{\omega : \tau \leq t \leq T\}$, $\int_0^\tau \alpha_s^2 \, ds = \infty$. Hence, $\varkappa_t$, $t \leq T$, will be a (P-a.s.) continuous function, if it is shown that $P(\varkappa^* = 0) = 1$.

From (6.28), by Ito's formula,

$$\exp(-\varkappa_{\tau_n\wedge T}) = e^{-1} - \int_0^{\tau_n\wedge T} e^{-\varkappa_s}\varkappa_s\alpha_s \, dW_s + \frac{1}{2}\int_0^{\tau_n\wedge T} e^{-\varkappa_s}\varkappa_s^2\alpha_s^2 \, ds. \tag{6.30}$$

Here

$$M\int_0^{\tau_n\wedge T} e^{-\varkappa_s}\varkappa_s\alpha_s \, dW_s = 0,$$

because

$$M\int_0^{\tau_n\wedge T} e^{-2\varkappa_s}\varkappa_s^2\alpha_s^2 \, ds \leq \sup_{0\leq z\leq\infty} e^{-2z}z^2n^2 < \infty.$$

Hence from (6.30) it follows that

$$M\int_0^{\tau_n\wedge T} e^{-\varkappa_s}\varkappa_s^2\alpha_s^2 \, ds = 2M(\exp(-\varkappa_{\tau_n\wedge T}) - e^{-1}] \leq 2.$$

Passing in this inequality to a limit with $n \to \infty$ we find

$$M\int_0^{\tau\wedge T} e^{-\varkappa_s}\varkappa_s^2\alpha_s^2 \, ds \leq 2. \tag{6.31}$$

From (6.31) it follows that (P-a.s.)

$$\infty > \int_0^{\tau\wedge T} e^{-\varkappa_s}\varkappa_s^2\alpha_s^2 \, ds \geq \int_{\tau_n\wedge T}^{\tau_{n+1}\wedge T} e^{-\varkappa_s}\varkappa_s^2\alpha_s^2 \, ds$$

$$\geq \inf_{\tau_n\wedge T\leq s\leq\tau_{n+1}\wedge T} [e^{-\varkappa_s}\varkappa_s^2] \int_{\tau_n\wedge T}^{\tau_{n+1}\wedge T} \alpha_s^2 \, ds. \tag{6.32}$$

On the set $\{\omega : \tau \leq T\}$, because of (6.23) and (6.22),

$$\int_{\tau_n\wedge T}^{\tau_{n+1}\wedge T} \alpha_s^2 \, ds = \int_{\tau_n}^{\tau_{n+1}} \alpha_s^2 \, ds = 2n + 1.$$

Hence from (6.32) it follows that on $\{\tau \leq T\}$

$$\inf_{\tau_n \leq s \leq \tau_{n+1}} [e^{-\varkappa_s}\varkappa_s^2] \leq \frac{\int_0^\tau e^{-\varkappa_s}\varkappa_s^2\alpha_s^2\, ds}{2n+1} \to 0, \qquad n \to \infty.$$

Therefore, on $\{\tau \leq T\}$, (P-a.s.)

$$e^{-\varkappa^*}(\varkappa^*)^2 = 0.$$

But $P(\varkappa^* < \infty) = 1$, hence $P(\varkappa^* = 0) = 1$.

Thus the continuity (P-a.s.) of the trajectories of the process $\varkappa_t$, $0 \leq t \leq T$, is proved.

Further, as in (6.32) we find that

$$\infty > \int_0^{\tau \wedge T} e^{-\varkappa_s}\varkappa_s^2\alpha_s^2\, ds \geq \inf_{0 \leq s \leq \tau \wedge T} e^{-\varkappa_s} \int_0^{\tau \wedge T} \varkappa_s^2\alpha_s^2\, ds. \qquad (6.33)$$

Since $\varkappa_s$, $s \leq T$, is a continuous process, then

$$P\Big(\inf_{0 \leq s \leq \tau \wedge T} e^{-\varkappa_s} > 0\Big) = 1,$$

which together with (6.33) yields

$$\int_0^T \varkappa_s^2\alpha_s^2\, ds = \int_0^{\tau \wedge T} \varkappa_s^2\alpha_s^2\, ds \leq \frac{\int_0^{\tau \wedge T} e^{-\varkappa_s}\varkappa_s^2\alpha_s^2\, ds}{\inf_{0 \leq s \leq \tau \wedge T} e^{-\varkappa_s}} < \infty \qquad (P\text{-a.s.}), \qquad (6.34)$$

i.e., $P(\int_0^T \varkappa_s^2\alpha_s^2\, ds < \infty) = 1$. From this condition it follows that the stochastic integrals $\int_0^t \varkappa_s\alpha_s\, dW_s$ are defined for all $t \leq T$.

Denote

$$y_t = 1 + \int_0^t \varkappa_s\alpha_s\, dW_s, \qquad t < T. \qquad (6.35)$$

Because of (6.28),

$$\varkappa_{t \wedge \tau_n} = 1 + \int_0^{t \wedge \tau_n} \varkappa_s\alpha_s\, dW_s, \qquad t \leq T.$$

Hence $y_t = \varkappa_t$ (P-a.s.) for all $t \leq \tau_n \leq T$, and because of the continuity of the trajectories of these processes $y_t = \varkappa_t$ (P-a.s.) for $t \leq \tau \leq T$.

Thus if $\tau \geq T$, then $y_t = \varkappa_t$ (P-a.s.) for all $t \leq T$. But if $\tau < T$, then $y_\tau = \varkappa_\tau = 0$ and for $t > \tau$, $y_t = y_\tau + \int_\tau^t \varkappa_s\alpha_s\, dW_s = y_\tau = 0$, since $\varkappa_s = 0$ for $s \geq \tau$. Consequently, $y_t = \varkappa_t$ (P-a.s.) for all $t \leq T$, and, therefore, according to (6.35)

$$\varkappa_t = 1 + \int_0^t \varkappa_s\alpha_s\, dW_s.$$

We show now that the solution of Equation (6.19) given by (6.27) is unique up to stochastic equivalence. Let $\tilde{\varkappa}_t$, $t \le T$, be another nonnegative continuous solution to Equation (6.19). Then, $d(\tilde{\varkappa}_t/\varkappa_t) \equiv 0$ with $t < \tau \equiv \lim_n \tau_n$ (compare with Subsection 5.3.4). Hence $\varkappa_t = \tilde{\varkappa}_t$ (P-a.s.) with $t < \tau \wedge T$ and over the continuity $\varkappa_\tau = \tilde{\varkappa}_\tau$. Consequently, on the set $\{\omega:\tau > T\}$ $\varkappa_t = \tilde{\varkappa}_t$, $t \le T$. We consider now the set $\{\omega:\tau \le T\}$. Since both the processes $\varkappa_t$ and $\tilde{\varkappa}_t$ are (as solutions to Equation (6.19)) supermartingales, then $\varkappa_t = \tilde{\varkappa}_t = 0$ (P-a.s.) on the set $\{\omega:\tau \le t \le T\}$.

Thus, $\varkappa_t = \tilde{\varkappa}_t$ (P-a.s.) for each t, $0 \le t \le T$. From the continuity of these processes it follows that their trajectories coincide (P-a.s.), i.e., $P\{\sup_{t \le T} |\varkappa_t - \tilde{\varkappa}_t| > 0\} = 0$. □

6.2 Nonnegative martingales

6.2.1

Under some simple assumptions, the supermartingale $\varphi = (\varphi_t, \mathscr{F}_t)$, $t \ge 0$, introduced in (6.17) turns out to be a martingale. This section will be concerned with the investigation of this question.

We begin by proving the following:

Lemma 6.4. *If* $\xi = (\xi_t, \mathscr{F}_t)$, $t \le T$, *is a supermartingale and*

$$M\xi_0 = M\xi_T, \tag{6.36}$$

then it is a martingale.

PROOF. Because of the supermartingale property of ξ,

$$M\xi_T \le M\xi_t \le M\xi_0.$$

Hence, according to (6.36), $M\xi_t = \text{const.}$, $t \le T$.

Denote $A = \{\omega: M(\xi_t|\mathscr{F}_s) < \xi_s\}$, where $0 \le s < t \le T$, and assume that $P(A) > 0$. Then

$$\begin{aligned} M\xi_T &= M\xi_t = MM(\xi_t|\mathscr{F}_s) \\ &= M\{\chi_A M(\xi_t|\mathscr{F}_s)\} + M\{(1-\chi_A)M(\xi_t|\mathscr{F}_s)\} \\ &< M\chi_A\xi_s + M(1-\chi_A)\xi_s = M\xi_s, \end{aligned}$$

which contradicts the equality $M\xi_T = M\xi_s$. Hence $P(A) = 0$, and, therefore, the process $\xi = (\xi_t, \mathscr{F}_t)$, $t \le T$, is a martingale.

6.2.2

Theorem 6.1. *Let* $\beta = (\beta_t, \mathscr{F}_t)$, $t \le T$, *be a random process with*

$$P\left(\int_0^T \beta_s^2\, ds < \infty\right) = 1.$$

Then, if

$$M \exp\left(\frac{1}{2}\int_0^T \beta_s^2\, ds\right) < \infty, \tag{6.37}$$

the supermartingale $\varphi(\beta) = (\varphi_t(\beta), \mathscr{F}_t)$, $t \le T$, *with*

$$\varphi_t(\beta) = \exp\left(\int_0^t \beta_s\, dW_s - \frac{1}{2}\int_0^t \beta_s^2\, ds\right)$$

is a martingale, and, in particular, $M\varphi_t(\beta) = 1$, $t \le T$.

PROOF. Let $a > 0$ and let

$$\sigma_a = \begin{cases} \inf\left\{t \le T: \int_0^t \beta_s\, dW_s - \int_0^t \beta_s^2\, ds = -a\right\}, \\ T, \qquad \text{if } \inf\limits_{0 \le t \le T}\left[\int_0^t \beta_s\, dW_s - \int_0^t \beta_s^2\, ds\right] > -a. \end{cases}$$

We set $\lambda \le 0$ and show first that

$$M\varphi_{\sigma_a}(\lambda\beta) = 1. \tag{6.38}$$

For this purpose we note that

$$\varphi_{\sigma_a}(\lambda\beta) = 1 + \lambda \int_0^{\sigma_a} \varphi_s(\lambda\beta)\beta_s\, dW_s.$$

Hence, for proving Equation (6.38), it suffices to show that

$$M\int_0^{\sigma_a} \varphi_s^2(\lambda\beta)\beta_s^2\, ds < \infty. \tag{6.39}$$

Because of the assumption given in (6.37),

$$M\int_0^{\sigma_a} \beta_s^2\, ds \le 2M \exp\left(\frac{1}{2}\int_0^{\sigma_a} \beta_s^2\, ds\right) \le 2M \exp\left(\frac{1}{2}\int_0^T \beta_s^2\, ds\right) < \infty. \tag{6.40}$$

On the other hand, with $\lambda \le 0$ and $0 \le s \le \sigma_a$,

$$\begin{aligned} \varphi_s(\lambda\beta) &= \exp\left(\lambda \int_0^s \beta_u\, dW_u - \frac{\lambda^2}{2}\int_0^s \beta_u^2\, du\right) \\ &= \exp\left\{\lambda\left[\int_0^s \beta_u\, dW_u - \int_0^s \beta_u^2\, du\right]\right\}\exp\left\{\left(\lambda - \frac{\lambda^2}{2}\right)\int_0^s \beta_u^2\, du\right\} \\ &\le \exp\left\{\lambda\left[\int_0^s \beta_u\, dW_u - \int_0^s \beta_u^2\, du\right]\right\} \le \exp\{|\lambda| a\}. \end{aligned}$$

Consequently, $\varphi_s^2(\lambda\beta) \le \exp\{2a|\lambda|\}$, with $s \le \sigma_a$, and (6.39) follows from (6.40).

We prove now that Equation (6.38) also remains valid with $\lambda \le 1$. For this purpose, we denote

$$\rho_{\sigma_a}(\lambda\beta) = e^{\lambda a}\varphi_{\sigma_a}(\lambda\beta).$$

If $\lambda \le 0$ then, according to (6.38),

$$M\rho_{\sigma_a}(\lambda\beta) = e^{\lambda a}. \tag{6.41}$$

Denote

$$A(\omega) = \int_0^{\sigma_a} \beta_t^2\, dt, \qquad B(\omega) = \int_0^{\sigma_a} \beta_t\, dW_t - \int_0^{\sigma_a} \beta_t^2\, dt + a \ge 0,$$

and let $u(z) = \rho_{\sigma_a}(\lambda\beta)$, where $\lambda = 1 - \sqrt{1 - z}$. It is clear that $0 \le z \le 1$ implies that $0 \le \lambda \le 1$.

Because of the definition of the function $\lambda_{\sigma_a}(\lambda\beta)$,

$$u(z) = \exp\left\{\frac{z}{2} A(\omega) + (1 - \sqrt{1 - z})B(\omega)\right\}.$$

With $z < 1$ the function $u(z)$ is representable (P-a.s.) in the form of the series

$$u(z) = \sum_{k=0}^{\infty} \frac{z^k}{k!} p_k(\omega),$$

where, as it is not difficult to check, $p_k(\omega) \ge 0$ (P-a.s.) for all k, $k = 0, 1, \ldots$.

If $z \le 1$, then, because of Lemma 6.1,

$$Mu(z) \le \exp(a(1 - \sqrt{1 - z}))$$

and, in particular, for any $0 \le z_0 < 1$,

$$Mu(z_0) < \infty.$$

Hence, for $|z| \le z_0$,

$$M \sum_{k=0}^{\infty} \frac{|z|^k}{k!} p_k(\omega) \le M \sum_{k=0}^{\infty} \frac{z_0^k}{k!} p_k(\omega) = Mu(z_0) < \infty.$$

From this, because of the Fubini theorem, it follows that for any $|z| < 1$,

$$Mu(z) = M\sum_{k=0}^{\infty} \frac{z^k}{k!} p_k(\omega) = \sum_{k=0}^{\infty} \frac{z^k}{k!} Mp_k(\omega). \tag{6.42}$$

With $z < 1$,

$$\exp(a(1 - \sqrt{1 - z})) = \sum_{k=0}^{\infty} \frac{z^k}{k!} c_k.$$

where $c_k \ge 0$, $k = 0, 1, \ldots$.

Because of this equality and the formulas (6.41) and (6.42), for $-1 < z \le 0$,

$$\sum_{k=0}^{\infty} \frac{z^k}{k!} Mp_k(\omega) = \sum_{k=0}^{\infty} \frac{z^k}{k!} c_k.$$

Hence $Mp_k(\omega) = c_k$, $k = 0, 1, \ldots$, and, therefore, (see (6.42)) for $0 \leq z < 1$,

$$Mu(z) = \sum_{k=0}^{\infty} \frac{z^k}{k!} c_k = \exp(a(1 - \sqrt{1 - z}))$$

which proves the validity of Equation (6.41) for all $\lambda < 1$. Since $B(\omega) \geq 0$, $A(\omega) \geq 0$ (P-a.s.),

$$\rho_{\sigma_a}(\lambda\beta) = \exp\left\{\lambda B(\omega) + \left(\lambda - \frac{\lambda^2}{2}\right)A(\omega)\right\} \uparrow \rho_{\sigma_a}(\beta)$$

with $\lambda \uparrow 1$. Hence, by Theorem 1.1 (on a monotone convergence),

$$\lim_{\lambda \uparrow 1} M\rho_{\sigma_u}(\lambda\beta) = M\rho_{\sigma_a}(\beta),$$

and, therefore, because of (6.41),

$$M\rho_{\sigma_a}(\beta) = \lim_{\lambda \uparrow 1} M\rho_{\sigma_a}(\lambda\beta) = e^a,$$

and, therefore,

$$M\varphi_{\sigma_a}(\beta) = 1.$$

From this

$$\begin{aligned}1 = M\varphi_{\sigma_a}(\beta) - M[\varphi_{\sigma_a}(\beta)\chi_{(\sigma_a < T)}] + M[\varphi_{\sigma_a}(\beta)\chi_{(\sigma_a = T)}] \\ = M[\varphi_{\sigma_a}(\beta)\chi_{(\sigma_a < T)}] + M[\varphi_T(\beta)\chi_{(\sigma_a = T)}]\end{aligned}$$

and

$$M\varphi_T(\beta) = 1 - M[\varphi_{\sigma_a}(\beta)\chi_{(\sigma_a < T)}] + M[\varphi_T(\beta)\chi_{(\sigma_a < T)}]. \tag{6.43}$$

But $P\text{-}\lim_{a\to\infty} \chi_{(\sigma_a < T)} = 0$ and $M\varphi_T(\beta) \leq 1$. Hence

$$\lim_{a\to\infty} M\chi_{(\sigma_a < T)}\varphi_T(\beta) = 0. \tag{6.44}$$

Further, on the set $(\sigma_a < T)$

$$\varphi_{\sigma_a}(\beta) = \exp\left\{-a + \frac{1}{2}\int_0^{\sigma_a} \beta_s^2\, ds\right\} \leq \exp\left\{-a + \frac{1}{2}\int_0^T \beta_s^2\, ds\right\},$$

and, therefore,

$$M\chi_{(\sigma_a < T)}\varphi_{\sigma_a}(\beta) \leq e^{-a} M \exp\left(\frac{1}{2}\int_0^T \beta_s^2\, ds\right) \to 0, \qquad a \to \infty. \tag{6.45}$$

From (6.43)–(6.45) comes the required result: $M\varphi_T(\beta) = 1$, from which, according to Lemma 6.4, it follows that $\varphi(\beta) = (\varphi_t(\beta), \mathscr{F}_t)$, $t \leq T$, is a martingale. □

Note. Theorem 6.1 is valid with the substitution of T for any Markov time τ (with respect to $(\mathscr{F}_t)$, $t \geq 0$). In particular, the statement of the theorem holds with $T = \infty$.

Corollary. *Let $W = (W_t, \mathscr{F}_t)$, $t \geq 0$, be a Wiener process and let $\tau = \tau(\omega)$ be a Markov time (with respect to $(\mathscr{F}_t)$, $t \geq 0$) with*

$$Me^{(1/2)\tau} < \infty.$$

Then

$$Me^{W_\tau - (1/2)\tau} = 1.$$

6.2.3

We give a number of examples in which the supermartingale $\varphi(\beta) = (\varphi_t(\beta), \mathscr{F}_t)$, $t \leq T$, with

$$\varphi_t(\beta) = \exp\left(\int_0^t \beta_s \, dW_s - \frac{1}{2}\int_0^t \beta_s^2 \, ds\right),$$

is a martingale and, in particular, $M\varphi_t(\beta) = 1$, $t \leq T$.

EXAMPLE 1. If $|\beta_t| \leq K < \infty$ (P-a.s.), $t \leq T$, then $M\varphi_t(\beta) = 1, t \leq T$, because of the fact that

$$M \exp\left(\frac{1}{2}\int_0^T \beta_t^2 \, dt\right) \leq \exp\left(\frac{T}{2} K^2\right) < \infty.$$

EXAMPLE 2. Let $\tau_n = \inf\{t \leq T : \int_0^t \beta_s^2 \, ds = n\}$, with $\tau_n = T$ if $\int_0^T \beta_s^2 \, ds < n$. Then

$$M \exp\left(\frac{1}{2}\int_0^{\tau_n} \beta_s^2 \, ds\right) \leq e^{n/2} < \infty$$

and $M\varphi_{\tau_n}(\beta) = 1$ according to the note to Theorem 6.1.

EXAMPLE 3. For some $\delta > 0$, let

$$\sup_{t \leq T} M \exp(\delta\beta_t^2) < \infty. \tag{6.46}$$

Then $M\varphi_t(\beta) = 1$, $t \leq T$. Indeed, by the Jensen inequality,

$$\exp\left(\frac{1}{2}\int_0^T \beta_t^2 \, dt\right) = \exp\left(\frac{1}{T}\int_0^T \frac{T\beta_t^2}{2} \, dt\right) \leq \frac{1}{T}\int_0^T \exp\left(\frac{T\beta_t^2}{2}\right) dt,$$

Hence, if $T \leq 2\delta$, then

$$M \exp\left(\frac{1}{2}\int_0^T \beta_t^2 \, dt\right) \leq \sup_{0 \leq t \leq T} M \exp(\delta\beta_t^2) < \infty,$$

and, by Theorem 6.1, $M\varphi_T(\beta) = 1$.

Let now $T > 2\delta$. Let us represent $\varphi_T(\beta)$ as a product

$$\varphi_T(\beta) = \prod_{j=0}^{n-1} \varphi_{t_j}^{t_{j+1}}(\beta),$$

where $0 = t_0 < t_1 < \cdots < t_n = T$,

$$\varphi_{t_j}^{t_{j+1}}(\beta) = \exp\left(\int_{t_j}^{t_{j+1}} \beta_t\, dW_t - \frac{1}{2}\int_{t_j}^{t_{j+1}} \beta_t^2\, dt\right),$$

and $\max_j [t_{j+1} - t_j] \le 2\delta$. Then $M\varphi_{t_j}^{t_{j+1}}(\beta) = 1$ and $M[\varphi_{t_j}^{t_{j+1}}(\beta)|\mathscr{F}_{t_j}] = 1$ (P-a.s.). Therefore,

$$M\varphi_T(\beta) = M[M(\varphi_T(\beta)|\mathscr{F}_{t_{n-1}})] = M\varphi_{t_{n-1}}(\beta) = \cdots = M\varphi_{t_1} = 1.$$

A condition of the type given by (6.46) can be easily checked in the following two cases.

(a) Let $\beta = (\beta_t, \mathscr{F}_t)$, $t \le T$, be a Gaussian process with

$$\sup_{t \le T} M|\beta_t| < \infty, \qquad \sup_{t \le T} D\beta_t < \infty.$$

Then, making the choice of

$$\delta < \frac{1}{2 \sup_{t \le T} D\beta_t},$$

we find that

$$\sup_{t \le T} M \exp(\delta\beta_t^2) = \sup_{t \le T} \frac{\exp\left[\dfrac{\delta(M\beta_t)^2}{1 - 2\delta D\beta_t}\right]}{\sqrt{1 - 2\delta D\beta_t}}.$$

(b) Let $y = (y_t, \mathscr{F}_t)$, $t \le T$, be a random process permitting the differential

$$dy_t = a(t, y_t)dt + b(t, y_t)dW_t, \qquad y_0 = \eta,$$

where

$$|a(t, y)| \le K(1 + |y|), \qquad |b(t, y)| \le K < \infty,$$

and $M \exp(\varepsilon\eta^2) < \infty$ for some $\varepsilon > 0$.

By Theorem 4.7, there will be $\delta_1 > 0$, such that $\sup_{t \le T} M \exp(\delta_1 y_t^2) < \infty$, and, therefore, for some $\delta > 0$,

$$\sup_{t \le T} M \exp(\delta a^2(t, y_t)) < \infty,$$

and, therefore, $M\varphi_t(\beta) = 1$, where $\beta_t = a(t, y_t)$.

Example 4. Let the processes $\beta = (\beta_t, \mathscr{F}_t)$ and $W = (W_t, \mathscr{F}_t)$, $t \le T$, be independent and let $P(\int_0^T \beta_t^2\, dt < \infty) = 1$. Then $M\varphi_t(\beta) = 1$, $t \le T$.

For proving this, along with $(\Omega, \mathscr{F}, P)$ let us consider the identical space $(\bar{\Omega}, \bar{F}, \bar{P})$ and on a probability space $(\Omega \times \bar{\Omega}, \mathscr{F} \times \bar{\mathscr{F}}, P \times \bar{P})$ define a random variable

$$\varphi_T(\omega, \bar{\omega}) = \exp\left(\int_0^T \beta_t(\omega)dW_t(\bar{\omega}) - \frac{1}{2}\int_0^T \beta_t^2(\omega)dt\right).$$

Due to the independence of the processes β and W,

$$M\varphi_T(\beta) = \int_{\Omega\times\bar{\Omega}} \varphi_T(\omega, \bar{\omega}) d(P \times \bar{P})(\omega, \bar{\omega}).$$

By the Fubini theorem,

$$\int_{\Omega\times\bar{\Omega}} \varphi_T(\omega, \bar{\omega}) d(P \times \bar{P})(\omega, \bar{\omega}) = \int_{\Omega}\left[\int_{\bar{\Omega}} \varphi_T(\omega, \bar{\omega}) d\bar{P}(\bar{\omega})\right] dP(\omega).$$

But (P-a.s.)

$$\int_{\bar{\Omega}} \exp\left(\frac{1}{2}\int_0^T \beta_t^2(\omega)dt\right) d\bar{P}(\bar{\omega}) = \exp\left(\frac{1}{2}\int_0^T \beta_t^2(\omega)dt\right) < \infty,$$

and hence, because of Theorem 6.1.

$$\int_{\bar{\Omega}} \varphi_T(\omega, \bar{\omega}) d\bar{P}(\bar{\omega}) = 1 \qquad (P\text{-a.s.})$$

and, therefore, also $M\varphi_T(\beta) = 1$.

EXAMPLE 5. The condition of the independence of the processes β and W formulated in the preceding example can be weakened by substituting for it the independence of the σ-algebras

$$\mathscr{F}^{\beta}_{u+\varepsilon} = \sigma\{\omega: \beta_v,\ v \le u+\varepsilon\} \quad \text{and} \quad \mathscr{F}^{W}_{s,t} = \sigma\{\omega: W_v - W_s,\ s \le v \le t\}$$

for $0 \le u \le s < t \le T$, $\varepsilon > 0$.

Indeed, let $0 = t_0 < t_1 < \cdots < t_n = T$ and let $\max_j [t_{j+1} - t_j] \le \varepsilon$. Then, according to Example 4,

$$M\varphi_{t_j}^{t_{j+1}}(\beta) = 1 \quad \text{and} \quad M(\varphi_{t_j}^{t_{j+1}}(\beta) | \mathscr{F}_{t_j}) = 1 \qquad (P\text{-a.s.}).$$

Applying the technique used in Example 3, we find that

$$M\varphi_T(\beta) = M\prod_{j=0}^{n-1} \varphi_{t_j}^{t_{j+1}}(\beta) = M\varphi_0^{t_1}(\beta) = 1.$$

6.2.4

We show now that in Theorem 6.1 the condition

$$M \exp\left(\frac{1}{2}\int_0^T \beta_s^2\, ds\right) < \infty$$

cannot, generally speaking, be improved in the sense that the fulfillment for any $\varepsilon > 0$ of the condition

$$M \exp\left(\left(\frac{1}{2} - \varepsilon\right)\int_0^T \beta_s^2\, ds\right) < \infty$$

does NOT imply the equality $M\varphi_T(\beta) = 1$.

EXAMPLE 6. Let $\tau_\varepsilon = \inf\{t : W_t - (1 - \varepsilon)t = -a\}$, where $0 < \varepsilon < \frac{1}{2}$, $a > 0$. We show first that

$$M \exp((\tfrac{1}{2} - \varepsilon)\tau_\varepsilon) = \exp((1 - 2\varepsilon)a), \tag{6.47}$$

and then establish that $M\varphi_{\tau_\varepsilon} < 1$, where $\varphi_{\tau_\varepsilon} = \exp(W_{\tau_\varepsilon} - \tau_\varepsilon/2)$.

Let us define Markov time

$$\tau_\varepsilon^{(n)} = \inf\{t : n \le W_t \le -a + (1 - \varepsilon)t\}$$

and establish that

$$M \exp[(\tfrac{1}{2} - \varepsilon)\tau_\varepsilon^{(n)}] = V_n(0), \tag{6.48}$$

where

$$V_n(x) = \frac{e^{-2\varepsilon n} - e^{-(1-2\varepsilon)a - n}}{e^{-(a+2\varepsilon n)} - e^{-(1-2\varepsilon)a}} e^x + \frac{e^{-(a+n)} - 1}{e^{-(a+2\varepsilon n)} - e^{-(1-2\varepsilon)a}} e^{(1-2\varepsilon)x}$$

is a solution of the differential equation

$$V_n''(x) - 2(1 - \varepsilon)V_n'(x) + (1 - 2\varepsilon)V_n(x) = 0$$

with

$$V_n(-a) = V_n(n) = 1.$$

For proving (6.48) we consider the function $V_n(x)e^{(1/2-\varepsilon)t}$. By the Ito formula,

$$V_n(x_{\tau_\varepsilon^{(n)}})\exp[(\tfrac{1}{2} - \varepsilon)\tau_\varepsilon^{(n)}] = V_n(0) + \int_0^{\tau_\varepsilon^{(n)}} V_n'(x_s)\exp[(\tfrac{1}{2} - \varepsilon)s]dW_s,$$

where $x_t = W_t - (1 - \varepsilon)t$. It is also clear that for any $N \ge 0$,

$$V_n(x_{\tau_\varepsilon^{(n)} \wedge N})\exp[(\tfrac{1}{2} - \varepsilon)\tau_\varepsilon^{(n)} \wedge N] = V_n(0) + \int_0^{\tau_\varepsilon^{(n)} \wedge N} V_n'(x_s)\exp[(\tfrac{1}{2} - \varepsilon)s]dW_s.$$

Hence, since for $-a \le x \le n$ the function $V_n'(x)$ is bounded,

$$M \int_0^{\tau_\varepsilon^{(n)} \wedge N} V_n'(x_s)\exp[(\tfrac{1}{2} - \varepsilon)s]dW_s = 0$$

and, therefore,

$$MV_n(x_{\tau_\varepsilon^{(n)} \wedge N})\exp[(\tfrac{1}{2} - \varepsilon)\tau_\varepsilon^{(n)} \wedge N] = V_n(0). \tag{6.49}$$

It can be easily checked that

$$0 < \inf_{-a \le x \le n} V_n(x) < \sup_{-a \le x \le n} V_n(x) < \infty,$$

and, therefore,

$$M \exp[(\tfrac{1}{2} - \varepsilon)\tau_\varepsilon^{(n)} \wedge N] \le \frac{V_n(0)}{\inf_{-a \le x \le n} V_n(x)} < \infty.$$

From this, after the passage to the limit ($N \to \infty$), we obtain

$$M \exp[(\tfrac{1}{2} - \varepsilon)\tau_\varepsilon^{(n)}] \le \frac{V_n(0)}{\inf_{-a \le x \le n} V_n(x)}.$$

From this inequality and the inequality

$$V_n(x_{\tau_\varepsilon^{(n)} \wedge N})\exp[(\tfrac{1}{2} - \varepsilon)\tau_\varepsilon^{(n)} \wedge N] \le \sup_{-a \le x \le n} V_n(x)\exp[(\tfrac{1}{2} - \varepsilon)\tau_\varepsilon^{(n)}]$$

it follows that in (6.49) the passage to the limit under the sign of mathematical expectation with $N \to \infty$ is feasible; taking into account the equality

$$V_n(x_{\tau_\varepsilon^{(n)}}) = 1 \qquad (P\text{-a.s.}),$$

this leads to (6.48). Taking limit as $n \to \infty$, we obtain the required relation, (6.47).

Finally, let us note that

$$\begin{aligned}\varphi_{\tau_\varepsilon} &= \exp\left(W_{\tau_\varepsilon} - \frac{\tau_\varepsilon}{2}\right) = \exp(W_{\tau_\varepsilon} - (1 - \varepsilon)\tau_\varepsilon)\exp((\tfrac{1}{2} - \varepsilon)\tau_\varepsilon) \\ &= \exp[-a + (\tfrac{1}{2} - \varepsilon)\tau_\varepsilon].\end{aligned}$$

From this, because of (6.47), it follows that

$$M\varphi_{\tau_\varepsilon} = e^{-2\varepsilon a} < 1.$$

6.2.5

We give two more examples in which the equality $M\varphi_t(\beta) = 1$ is violated, and therefore, (6.37) is not satisfied. In the first example $T = \infty$, in the second example $T = 1$.

EXAMPLE 7. Let $\varphi_t = \exp(W_t - t/2)$ and $\tau = \inf\{t: W_t = -1\}$. Then $P(\tau < \infty) = 1$ (see Section 1.3) and

$$\varphi_\tau = \exp\left(-1 - \frac{\tau}{2}\right) < e^{-1}.$$

Consequently, $M\varphi_\tau < e^{-1} < 1$.

EXAMPLE 8. For $0 \le t \le 1$, let

$$\beta_t = -\frac{2W_t}{(1 - t)^2}\chi_{(\tau \ge t)},$$

where

$$\tau = \inf\{t \le 1: W_t^2 = 1 - t\}.$$

Then, since $P(0 < \tau < 1) = 1$,

$$\int_0^1 \beta_t^2\,dt = 4\int_0^1 \frac{W_t^2}{(1 - t)^4}\chi_{(\tau \ge t)}\,dt = 4\int_0^\tau \frac{W_t^2}{(1 - t)^4}\,dt < \infty \qquad (P\text{-a.s.}).$$

By the Ito formula for $t < 1$,

$$d\left(\frac{W_t^2}{(1-t)^2}\right) = \frac{2W_t^2}{(1-t)^3}\,dt + \frac{2W_t}{(1-t)^2}\,dW_t + \frac{1}{(1-t)^2}\,dt,$$

from which

$$-\int_0^1 \frac{2W_t}{(1-t)^2}\chi_{(\tau \ge t)}\,dW_t - \frac{1}{2}\int_0^1 \frac{4W_t^2}{(1-t)^4}\chi_{(\tau \ge t)}\,dt$$

$$= -\frac{W_\tau^2}{(1-\tau)^2} + \int_0^\tau \frac{2W_t^2}{(1-t)^3}\,dt + \int_0^\tau \frac{dt}{(1-t)^2} - \int_0^\tau \frac{2W_t^2}{(1-t)^4}\,dt$$

$$= -\frac{1}{1-\tau} + \int_0^\tau \left\{2W_t^2\left[\frac{1}{(1-t)^3} - \frac{1}{(1-t)^4}\right] + \frac{1}{(1-t)^2}\right\}dt$$

$$\le -\frac{1}{1-\tau} + \int_0^\tau \frac{1}{(1-t)^2}\,dt = -1.$$

Hence, $M\varphi_1(\beta) \le e^{-1} < 1$.

6.3 The Girsanov theorem and its generalization

6.3.1

We consider on a probability space $(\Omega, \mathscr{F}, P)$ a Wiener process $W = (W_t, \mathscr{F}_t)$, $t \le T$, and a random process $\gamma = (\gamma_t, \mathscr{F}_t)$, $t \le T$, with $P(\int_0^T \gamma_t^2\,dt < \infty) = 1$. Let $\varkappa = (\varkappa_t, \mathscr{F}_t)$, $t \le T$, be a nonnegative continuous supermartingale with

$$\varkappa_t = 1 + \int_0^t \gamma_s\,dW_s. \tag{6.50}$$

If $M\varkappa_T = 1$, then the process $\varkappa = (\varkappa_t, \mathscr{F}_t)$, $t \le T$, will be a nonnegative martingale (Lemma 6.4), and on the measurable space $(\Omega, \mathscr{F}_T)$ there will be defined a probability measure $\tilde{P}$ with $d\tilde{P} = \varkappa_T(\omega)dP$.

Theorem 6.2. *If $M\varkappa_T(\omega) = 1$, then on the probability space $(\Omega, \mathscr{F}, \tilde{P})$ the random process $\tilde{W} = (\tilde{W}_t, \mathscr{F})$, $t \le T$, with*[4]

$$\tilde{W}_t = W_t - \int_0^t \varkappa_s^+ \gamma_s\,ds \tag{6.51}$$

is a Wiener process (with respect to the system $(\mathscr{F}_t)$, $0 \le t \le T$, and measure $\tilde{P}$).

[4] $\varkappa_s^+ = \varkappa_s^{-1}$ if $\varkappa_s > 0$, and $\varkappa_s^+ = 0$ with $\varkappa_s = 0$. From Lemma 6.5, given below, it follows that $\varkappa_s^+ = \varkappa_s^{-1}$ ($\tilde{P}$-a.s.).

6.3.2

Before proving this theorem we shall make a few additional assertions.

Lemma 6.5. *Let* $M\varkappa_T = 1$. *Then*

$$\tilde{P}\left(\inf_{0 \le t \le T} \varkappa_t = 0\right) = 0.$$

PROOF. By definition of measure $\tilde{P}$,

$$\tilde{P}\left(\inf_{0 \le t \le T} \varkappa_t = 0\right) = \int_{\{\omega:\ \inf_{0 \le t \le T} \varkappa_t = 0\}} \varkappa_T \, dP(\omega).$$

Let $\tau = \inf\{t \le T : \varkappa_t = 0\}$ and $\tau = \infty$ if $\inf_{0 \le t \le T} \varkappa_t > 0$. Then

$$\left\{\omega:\ \inf_{0 \le t \le T} \varkappa_t = 0\right\} = \{\omega : \tau \le T\}$$

and, therefore, by Theorem 3.6

$$\tilde{P}\left(\inf_{0 \le t \le T} \varkappa_t = 0\right) = \int_{\{\omega : \tau \le T\}} \varkappa_T \, dP = \int_{\{\omega : \tau \le T\}} \varkappa_\tau \, dP = 0. \qquad \square$$

Lemma 6.6. *Let* $M\varkappa_T = 1$, *and let* $\eta = \eta(\omega)$ *be an* $\mathscr{F}_t$*-measurable random variable with*[5] $\tilde{M}|\eta(\omega)| < \infty$ *and* $0 \le t \le T$. *Let* $\tilde{M}(\eta | \mathscr{F}_s)$ *be one of the versions of the conditional expectation* $0 \le s \le T$. *Then, if* $s \le t$,

$$\tilde{M}(\eta | \mathscr{F}_s) = \varkappa_s^+ M(\eta \varkappa_t | \mathscr{F}_s) \qquad (\tilde{P}\text{-a.s.}). \tag{6.52}$$

PROOF. Let $\lambda = \lambda(\omega)$ be a bounded $\mathscr{F}_s$-measurable random variable and let $s \le t$. Then

$$\begin{aligned} \tilde{M}(\eta\lambda) &= \tilde{M}[\lambda \tilde{M}(\eta | \mathscr{F}_s)] = M[\lambda \tilde{M}(\eta | \mathscr{F}_s) \varkappa_T] \\ &= M[\lambda \tilde{M}(\eta | \mathscr{F}_s) M(\varkappa_T | \mathscr{F}_s)] = M[\lambda \varkappa_s \tilde{M}(\eta | \mathscr{F}_s)]. \end{aligned} \tag{6.53}$$

On the other hand,

$$\tilde{M}(\eta\lambda) = M(\lambda\eta\varkappa_T) = M(\lambda\eta M(\varkappa_T | \mathscr{F}_t)) = M(\lambda\eta\varkappa_t) = M[\lambda M(\eta\varkappa_t | \mathscr{F}_s)]. \tag{6.54}$$

From (6.53) and (6.54) it follows that P- and ($\tilde{P}$-a.s.)

$$\varkappa_s \tilde{M}(\eta | \mathscr{F}_s) = M(\eta \varkappa_t | \mathscr{F}_s). \tag{6.55}$$

But $\tilde{P}(\varkappa_s > 0) = 1$. Hence $\tilde{P}(\varkappa_s^{-1} = \varkappa_s^+) = 1$, and (6.52) in the case $s \le t$ follows from (6.55). $\square$

[5] $\tilde{M}$ denotes averaging over measure $\tilde{P}$.

Note. If $\eta \equiv 1$, then from (6.52) it follows that

$$\varkappa_s \varkappa_s^+ = 1 \qquad (\tilde{P}\text{-a.s.}),$$

so that $\tilde{M}\varkappa_s \varkappa_s^+ = 1$. But

$$M\varkappa_s \varkappa_s^+ = P(\varkappa_s > 0).$$

Lemma 6.7. *Let $\{\xi_n \geq 0, n = 1, 2, \ldots\}$ be a sequence of random variables such that $\xi_n \to \xi$ (in probability), $n \to \infty$. If $M\xi_n = M\xi = C$, then*

$$\lim_{n \to \infty} M|\xi - \xi_n| = 0. \tag{6.56}$$

PROOF. We have

$$\begin{aligned} M|\xi - \xi_n| &= M(\xi - \xi_n)\chi_{(\xi \geq \xi_n)} + M(\xi_n - \xi)\chi_{(\xi < \xi_n)} \\ &= M(\xi - \xi_n)\chi_{(\xi \geq \xi_n)} + M(\xi_n - \xi) - M(\xi_n - \xi)\chi_{(\xi \geq \xi_n)}. \end{aligned}$$

But $M(\xi_n - \xi) = 0$. Hence

$$M|\xi - \xi_n| = 2M(\xi - \xi_n)\chi_{(\xi \geq \xi_n)}, \tag{6.57}$$

where $0 \leq (\xi - \xi_n)\chi_{(\xi \geq \xi_n)} \leq \xi$. Hence by the Lebesgue bounded convergence theorem, $\lim_{n\to\infty} M(\xi - \xi_n)\chi_{(\xi \geq \xi_n)} = 0$, which together with (6.57) proves (6.56). □

Lemma 6.8. *Let two nonnegative measures ν and $\tilde{\nu}$ be given on some measurable space $(X, \mathscr{X})$ with $\tilde{\nu} \ll \nu$ and*

$$g(x) = \frac{d\tilde{\nu}}{d\nu}(x).$$

If $\nu\{x : g(x) = 0\} = 0$, then $\nu \ll \tilde{\nu}$ and

$$\frac{d\nu}{d\tilde{\nu}}(x) = g^{-1}(x) \qquad (\tilde{\nu}\text{-a.s.}).$$

PROOF. Let $A \in \mathscr{X}$. Then

$$\int_A g^+(x) d\tilde{\nu}(x) = \int_A g^+(x) g(x) d\nu(x).$$

But

$$g^+(x)g(x) = \begin{cases} 1, & g(x) > 0, \\ 0, & g(x) = 0. \end{cases}$$

Hence

$$\int_A g^+(x) d\tilde{\nu}(x) = \nu[A \cap \{x : g(x) > 0\}] = \nu(A) - \nu[A \cap \{x : g(x) = 0\}],$$

where, by the assumption of the lemma,

$$\nu[A \cap \{x:g(x) = 0\}] \le \nu\{x:g(x) = 0\} = 0.$$

Consequently,

$$\nu(A) = \int_A g^+(x)d\tilde{\nu}(x),$$

which proves the lemma, since $g^+(x)$ coincides ν- and $\tilde{\nu}$-a.s. with $g^{-1}(x)$. □

6.3.3

PROOF OF THEOREM 6.2. Since $\varkappa_s^+ = \varkappa_s^{-1}$ ($\tilde{P}$-a.s.), $0 \le s \le t$, and

$$\tilde{P}\left(\inf_{s \le T} \varkappa_s = 0\right) = 0,$$

the process $\varkappa^+ = (\varkappa_s^+)$, $0 \le s \le T$, has ($\tilde{P}$-a.s.) continuous trajectories and, therefore, $\tilde{P}(\sup_{s \le T} \varkappa_s^+ < \infty) = 1$.

Further, measure $\tilde{P}$ is absolutely continuous with respect to measure P ($\tilde{P} \ll P$) and

$$\tilde{P}\left(\int_0^T \gamma_s^2\, ds < \infty\right) = P\left(\int_0^T \gamma_s^2\, ds < \infty\right) = 1.$$

We note also that

$$\int_0^T (\varkappa_t^+ \gamma_t)^2\, dt \le \sup_{t \le T} (\varkappa_t^+)^2 \int_0^T \gamma_t^2\, dt.$$

Hence

$$\tilde{P}\left(\int_0^T (\varkappa_t^+ \gamma_t)^2\, dt < \infty\right) = 1,$$

and, therefore, the integral $\int_0^t \varkappa_s^+ \gamma_s\, ds$ entering into (6.51) is defined.

For proving the theorem it suffices to establish that ($\tilde{P}$-a.s.)

$$\tilde{M}\{\exp[iz(\tilde{W}_t - \tilde{W}_s)]|\mathscr{F}_s\} = \exp\left(-\frac{z^2}{2}(t - s)\right) \tag{6.58}$$

for any z, $-\infty < z < \infty$, and s, t, $0 \le s \le t \le T$.

It will be assumed first that

$$P\left\{0 < c_1 \le \inf_{t \le T} \varkappa_t \le \sup_{t \le T} \varkappa_t \le c_2 < \infty\right\} = 1, \tag{6.59}$$

$$M \int_0^T \gamma_t^2\, dt < \infty, \tag{6.60}$$

where c_1 and c_2 are constants. Denote $\eta(t, s) = \exp[iz(\tilde{W}_t - \tilde{W}_s)]$. Then, by Lemma 6.6, ($\tilde{P}$-a.s.)

$$\tilde{M}(\eta(t, s)|\mathscr{F}_s) = \varkappa_s^+ M(\eta(t, s)\varkappa_t|\mathscr{F}_s). \tag{6.61}$$

By the Ito formula

$$\eta(t, s)\varkappa_t = \varkappa_s + \int_s^t \eta(u, s)\varkappa_u \varkappa_u^+ \gamma_u \, dW_u + iz \int_s^t \eta(u, s)\varkappa_u \, dW_u - \frac{z^2}{2} \int_s^t \eta(u, s)\varkappa_u \, du.$$

The assumptions in (6.59) and (6.60) guarantee that

$$M\left[\int_s^t \eta(u, s)\varkappa_u \varkappa_u^+ \gamma_u \, dW_u | \mathscr{F}_s\right] = 0 \qquad (P\text{-a.s.})$$

and that

$$M\left[\int_s^t \eta(u, s)\varkappa_u \, dW_u | \mathscr{F}_s\right] = 0 \qquad (P\text{-a.s.}).$$

Hence P- and ($\tilde{P}$-a.s.)

$$\varkappa_s^+ M(\eta(t, s)\varkappa_t | \mathscr{F}_s) = \varkappa_s^+ \varkappa_s - \frac{z^2}{2} \int_s^t \varkappa_s^+ M(\eta(u, s)\varkappa_u | \mathscr{F}_s) du. \tag{6.62}$$

Denote

$$f(t, s) = \varkappa_s^+ M(\eta(t, s)\varkappa_t | \mathscr{F}_s).$$

Then, because of (6.62), ($\tilde{P}$-a.s.)

$$f(t, s) = 1 - \frac{z^2}{2} \int_s^t f(u, s) du,$$

and, hence,

$$f(t, s) = \exp\left(-\frac{z^2}{2}(t - s)\right). \tag{6.63}$$

But according to (6.61), $\tilde{M}(\eta(t, s) | \mathscr{F}_s) = f(t, s)$ ($\tilde{P}$-a.s.), which, together with (6.63), proves the statement of the theorem under the assumptions (6.59) and (6.60).

Let these assumptions not be satisfied. We shall introduce Markov times τ_n, $n = 1, 2, \ldots,$ assuming that

$$\tau_n = \begin{cases} \inf\left\{t \le T: \left[\int_0^t \gamma_s^2 \, ds + \sup_{s \le t} \varkappa_s + \left(\inf_{s \le t} \varkappa_s\right)^{-1}\right] \ge n\right\}, \\ T, \quad \text{if} \quad \left[\int_0^T \gamma_s^2 \, ds + \sup_{s \le T} \varkappa_s + \left(\inf_{s \le T} \varkappa_s\right)^{-1}\right] < n. \end{cases}$$

Since

$$P\left(\int_0^T \gamma_s^2 \, ds + \sup_{s \le T} \varkappa_s < \infty\right) = 1, \qquad \tilde{P}\left(\inf_{s \le T} \varkappa_s > 0\right) = 1$$

and $\tilde{P} \ll P$, then ($\tilde{P}$-a.s.) $\tau_n \uparrow T$, $n \to \infty$. We set

$$\varkappa_t^{(n)} = \varkappa_{t \wedge \tau_n}, \qquad \gamma_t^{(n)} = \gamma_t \chi_{(\tau_n \ge t)}$$

and

$$\tilde{W}_t^{(n)} = W_t - \int_0^{t \wedge \tau_{(n)}} (z_s^{(n)})^+ \gamma_s \, ds.$$

Then

$$z_t^{(n)} = 1 + \int_0^t \gamma_s^{(n)} \, dW_s \quad \text{and} \quad \tilde{W}_t^{(n)} = W_t - \int_0^t (z_s^{(n)})^+ \gamma_s^{(n)} \, ds.$$

Let measure $\tilde{P}^{(n)}$ be defined by the equation $d\tilde{P}^{(n)} = z_T^{(n)}(\omega)dP$. The process $z^{(n)} = (z_t^{(n)}, \mathscr{F}_t)$, $0 \le t \le T$, is a martingale with $Mz_T^{(n)} = 1$, and for this process (6.59) is satisfied with $c_2 = n$, $c_1 = n^{-1}$. In addition, $M \int_0^T (\gamma_t^{(n)})^2 \, dt \le n < \infty$, and, therefore, by what was proved, ($\tilde{P}$-a.s.)

$$\tilde{M}^{(n)}\{\exp[iz(\tilde{W}_t^{(n)} - \tilde{W}_s^{(n)})] | \mathscr{F}_s\} = \exp\left\{-\frac{z^2}{2}(t - s)\right\}, \tag{6.64}$$

where $\tilde{M}^{(n)}$ is an averaging over measure $\tilde{P}^{(n)}$.

To complete the proof it remains only to show that with $n \to \infty$,

$$\tilde{M}^{(n)}\{\exp[iz(\tilde{W}_t^{(n)} - \tilde{W}_s^{(n)})] | \mathscr{F}_s\} \xrightarrow{\tilde{P}} \tilde{M}\{\exp[iz(\tilde{W}_t - \tilde{W}_s)] | \mathscr{F}_s\}. \tag{6.65}$$

Since with $n \to \infty$

$$\tilde{M}\{\exp[iz(\tilde{W}_t^{(n)} - \tilde{W}_s^{(n)})] | \mathscr{F}_s\} \xrightarrow{\tilde{P}} \tilde{M}\{\exp[iz(\tilde{W}_t - \tilde{W}_s)] | \mathscr{F}_s\},$$

in order to prove (6.65) it suffices to check that

$$\overline{\lim_{n \to \infty}} \tilde{M} | \tilde{M}^{(n)}\{\exp[iz(\tilde{W}_t^{(n)} - \tilde{W}_s^{(n)}) | \mathscr{F}_s\} - \tilde{M}\{\exp[iz(\tilde{W}_t^{(n)} - \tilde{W}_s^{(n)})] | \mathscr{F}_s\} | = 0. \tag{6.66}$$

Because of Lemma 6.8, for each n, $n = 1, 2, \ldots$, measure $\tilde{P}^{(n)}$ is equivalent to measure P, and, therefore,

$$\tilde{P} \ll \tilde{P}^{(n)}. \tag{6.67}$$

According to Lemma 6.6, ($\tilde{P}^{(n)}$-a.s.) and (because of (6.67)) ($\tilde{P}$-a.s.)

$$\tilde{M}^{(n)}\{\exp[iz(\tilde{W}_t^{(n)} - \tilde{W}_s^{(n)})] | \mathscr{F}_s\} = M\left\{\exp[iz(\tilde{W}_t^{(n)} - \tilde{W}_s^{(n)})] \frac{z_t^{(n)}}{z_s^{(n)}} \middle| \mathscr{F}_s\right\} \tag{6.68}$$

and

$$\tilde{M}\{\exp[iz(\tilde{W}_t^{(n)} - \tilde{W}_s^{(n)})] | \mathscr{F}_s\} = M[\exp[iz(\tilde{W}_t^{(n)} - \tilde{W}_s^{(n)})] z_s^+ z_t | \mathscr{F}_s\}. \tag{6.69}$$

Hence

$$\begin{aligned}
&\tilde{M} | \tilde{M}^{(n)}\{\exp[iz(\tilde{W}_t^{(n)} - \tilde{W}_s^{(n)})] | \mathscr{F}_s\} - \tilde{M}\{\exp[iz(\tilde{W}_t^{(n)} - \tilde{W}_s^{(n)})] | \mathscr{F}_s\} | \\
&= \tilde{M} \left| M\left\{\exp[iz(\tilde{W}_t^{(n)} - \tilde{W}_s^{(n)})] \left[\frac{z_t^{(n)}}{z_s^{(n)}} - z_s^+ z_t\right] \middle| \mathscr{F}_s\right\}\right| \\
&\le \tilde{M}M\left(\left|\frac{z_t^{(n)}}{z_s^{(n)}} - z_s^+ z_t\right| \middle| \mathscr{F}_s\right) = M\left|\frac{z_s z_t^{(n)}}{z_s^{(n)}} - z_s z_s^+ z_t\right| \\
&= M | z_s z_{s \wedge \tau_n}^+ z_{t \wedge \tau_n} - z_s z_s^+ z_t |.
\end{aligned} \tag{6.70}$$

Let us show now that $\varkappa_s \varkappa^+_{s \wedge \tau_n} \varkappa_{t \wedge \tau_n} \to \varkappa_s \varkappa_s^+ \varkappa_t$ (P-a.s.) with $n \to \infty$. We shall introduce time $\tau = \inf_{s \leq T} (t \leq T : \varkappa_t = 0)$, assuming $\tau = T$ if $\inf_{s \leq T} \varkappa_s > 0$. Then, since

$$P\left(\int_0^T \gamma_s^2 \, ds < \infty\right) = 1, \qquad P\left(\sup_{s \leq T} \varkappa_s < \infty\right) = 1,$$

the Markov time τ_n, $n = 1, 2, \ldots$, introduced above, has the property that (P-a.s.) $\tau_n \uparrow \tau$, $n \to \infty$. According to Note 2 to Theorem 3.5, $\varkappa_t = 0$ ($\{t \geq \tau\}$; (P-a.s.)). From this, for all $0 \leq t \leq T$ we obtain $\varkappa_t = \varkappa_{t \wedge \tau}$ (P-a.s.).

Hence it suffices to show that

$$\lim_{n \to \infty} \varkappa_{s \wedge \tau} \varkappa^+_{s \wedge \tau_n} \varkappa_{t \wedge \tau_n} = \varkappa_{s \wedge \tau} \varkappa^+_{s \wedge \tau} \varkappa_{t \wedge \tau}. \tag{6.71}$$

Because of the continuity of $\varkappa_t$, $0 \leq t \leq T$, (6.71) will be available if $\varkappa_{s \wedge \tau} \varkappa^+_{s \wedge \tau_n} \to \varkappa_{s \wedge \tau} \varkappa^+_{s \wedge \tau}$ (P-a.s.), $n \to \infty$. But $\varkappa_{s \wedge \tau} \varkappa^+_{s \wedge \tau} = 0$ on the set $\{s \geq \tau\}$ and for all n, $n = 1, 2, \ldots$, $\varkappa_{s \wedge \tau} \varkappa^+_{s \wedge \tau_n} = 0$, and, on the set $\{\tau > s\}$, $\inf_n \varkappa_{s \wedge \tau_n} > 0$; therefore,

$$\varkappa_{s \wedge \tau} \varkappa^+_{s \wedge \tau_n} \to \varkappa_{s \wedge \tau} \varkappa^+_{s \wedge \tau}, \qquad (P\text{-a.s.}), \qquad n \to \infty.$$

Thus, (P-a.s.)

$$\frac{\varkappa_s \varkappa_t^{(n)}}{\varkappa_s^{(n)}} \to \varkappa_s \varkappa_s^+ \varkappa_t. \tag{6.72}$$

Further

$$M \varkappa_s \varkappa_s^+ \varkappa_t = M[\varkappa_s \varkappa_s^+ M(\varkappa_t | \mathscr{F}_s)] = M[\varkappa_s \varkappa_s^+ \varkappa_s] = M \varkappa_s = 1$$

and

$$M \varkappa_s \frac{\varkappa_t^{(n)}}{\varkappa_s^{(n)}} = M\left[\frac{\varkappa_s}{\varkappa_s^{(n)}} M(\varkappa_t^{(n)} | \mathscr{F}_s)\right] = M\left[\frac{\varkappa_s}{\varkappa_s^{(n)}} \varkappa_s^{(n)}\right] = M \varkappa_s = 1.$$

Hence, by Lemma 6.7,

$$\lim_{n \to \infty} M \left| \frac{\varkappa_s \varkappa_t^{(n)}}{\varkappa_s^{(n)}} - \varkappa_s \varkappa_s^+ \varkappa_t \right| = 0,$$

from which there follows (6.66), and thus Theorem 6.2. □

6.3.4

Let $\varkappa = (\varkappa_t, \mathscr{F}_t)$, $t \leq T$, be a supermartingale of a special form with

$$\varkappa_t = \exp\left(\int_0^t \beta_s \, dW_s - \frac{1}{2} \int_0^t \beta_s^2 \, ds\right), \tag{6.73}$$

where $P(\int_0^T \beta_s^2 \, ds < \infty) = 1$. Then

$$\varkappa_t = 1 + \int_0^t \gamma_s \, dW_s$$

with $\gamma_s = \varkappa_s \beta_s$.

From Theorem 6.2 for the case under consideration we obtain the following result.

Theorem 6.3 (I. V. Girsanov). *If $M\varkappa_T = 1$, then the random process*

$$\tilde{W}_t = W_t - \int_0^t \beta_s\, ds$$

is a Wiener process with respect to the system $(\mathscr{F}_t)$, $0 \le t \le T$, and the probability measure $\tilde{P}(d\tilde{P} = \varkappa_T(\omega)dP)$.

Note. As an example of a nonnegative martingale $\varkappa_t = 1 + \int_0^t \gamma_s\, dW_s$, $0 \le t \le T$, with $P(\int_0^T \gamma_s^2\, ds < \infty) = 1$, which cannot be represented in the form given by (6.73) one can take the martingale

$$\varkappa_t = 1 + W_{t \wedge \tau}, \qquad 0 \le t \le T,$$

where

$$\tau = \inf\{t : W_t = -1\}.$$

6.3.5

We shall also give a multi-dimensional version of Theorem 6.2. Let $\gamma_i = (\gamma_i(t), \mathscr{F}_t)$, $0 \le t \le T$, $i = 1, \ldots, n$, be random processes with $P(\int_0^T \gamma_i^2(t)dt < \infty) = 1$, $i = 1, \ldots, n$, and let $W = (W_1(t), \ldots, W_n(t), \mathscr{F}_t)$, $0 \le t \le T$, be an n-dimensional Wiener process.

We shall introduce the random process

$$\varkappa_t = 1 + \int_0^t \sum_{i=1}^n \gamma_i(s) dW_i(s), \tag{6.74}$$

which from now on will play the same role as the process defined in (6.49).

Lemma 6.9. *There exists a Wiener process $\hat{W} = (\hat{W}_t, \mathscr{F}_t)$, $0 \le t \le T$, such that for each t, $0 \le t \le T$, (P-a.s.)*

$$\varkappa_t = 1 + \int_0^t \hat{\gamma}_s\, d\hat{W}_s \tag{6.75}$$

with

$$\hat{\gamma}_s = \sqrt{\sum_{i=1}^n \gamma_i^2(s)}.$$

PROOF. If

$$P(\hat{\gamma}_s > 0, 0 \le s \le T\} = 1, \tag{6.76}$$

then we set

$$\hat{W}_t = \int_0^t \hat{\gamma}_s^{-1} \sum_{i=1}^n \gamma_i(s) d W_i(s).$$

Then from Theorem 4.1 it follows that the process $(\hat{W}_t, \mathscr{F}_t)$, $0 \le t \le T$, is a Wiener process.

In the general case let us define

$$\hat{W}_t = \int_0^t \hat{\gamma}_s^+ \sum_{i=1}^n \gamma_i(s) dW_i(s) dW_i(s) + \int_0^t (1 - \hat{\gamma}_s^+ \hat{\gamma}_s) dz_s,$$

where $(z_t, \mathscr{F}_t)$, $0 \le t \le T$, is a Wiener process independent of the process W. (By this we assume that the initial probability space $(\Omega, \mathscr{F}, P)$ is sufficiently "rich"; otherwise, instead of $(\Omega, \mathscr{F}, P)$ a space $(\Omega \times \Omega, \mathscr{F} \times \mathscr{F}, P \times P)$, for example, should be taken.)

The process $\hat{W} = (\hat{W}_t, \mathscr{F}_t)$ is a continuous square integrable martingale. We will show that

$$M[(\hat{W}_t - \hat{W}_s)^2 | \mathscr{F}_s] = t - s \qquad (P\text{-a.s.}). \tag{6.77}$$

By Ito's formula,

$$\hat{W}_t^2 - \hat{W}_s^2 = 2 \int_s^t \hat{W}_u \left[\hat{\gamma}_u^+ \sum_{i=1}^n \gamma_j(u) dW_i(u) + (1 - \hat{\gamma}_u^+ \hat{\gamma}_u) dz_u \right]$$

$$+ \int_s^t [(\hat{\gamma}_u^+ \hat{\gamma}_u)^2 + (1 - \hat{\gamma}_u^+ \hat{\gamma}_u)^2] du.$$

But (P-a.s.)

$$(\hat{\gamma}_u^+ \hat{\gamma}_u)^2 + (1 - \hat{\gamma}_u^+ \hat{\gamma}_u)^2 = \hat{\gamma}_u^+ \hat{\gamma}_u + (1 - \hat{\gamma}_u^+ \hat{\gamma}_u) = 1.$$

Consequently, (P-a.s.)

$$M[(\hat{W}_t - \hat{W}_s)^2 | \mathscr{F}_s] = M[\hat{W}_t^2 - \hat{W}_s^2 | \mathscr{F}_s] = t - s.$$

From Theorem 4.1 it follows that the process $\hat{W} = (\hat{W}_t, \mathscr{F}_t)$ is a Wiener process. It remains to check the validity (P-a.s.) of Equation (6.75).

We have

$$1 + \int_0^t \hat{\gamma}_s \, d\hat{W}_s = 1 + \int_0^t \hat{\gamma}_s \hat{\gamma}_s^+ \sum_{i=1}^n \gamma_i(s) dW_i(s) + \int_0^t \hat{\gamma}_s (1 - \hat{\gamma}_s^+ \hat{\gamma}_s) dz_s$$

$$= 1 + \int_0^t \hat{\gamma}_s \hat{\gamma}_s^+ \sum_{i=1}^n \gamma_i(s) dW_i(s),$$

since (P-a.s.) for any s, $0 \le s \le T$,

$$\hat{\gamma}_s (1 - \hat{\gamma}_s^+ \hat{\gamma}_s) = \hat{\gamma}_s - \hat{\gamma}_s \hat{\gamma}_s^+ \hat{\gamma}_s = \hat{\gamma}_s - \hat{\gamma}_s = 0.$$

Therefore,

$$1 + \int_0^t \hat{\gamma}_s \, d\hat{W}_s = \varkappa_t - \int_0^t (1 - \hat{\gamma}_s \hat{\gamma}_s^+) \sum_{i=1}^n \gamma_i(s) dW_i(s). \tag{6.78}$$

But

$$M\left(\int_0^t (1 - \hat{\gamma}_s\hat{\gamma}_s^+) \sum_{i=1}^n \gamma_i(s)dW_i(s)\right)^2 = M\int_0^t (1 - \hat{\gamma}_s\hat{\gamma}_s^+)^2\hat{\gamma}_s\,ds$$

$$= M\int_0^t (1 - \hat{\gamma}_s\hat{\gamma}_s^+)\hat{\gamma}_s\,ds = 0,$$

which together with (6.78) proves (6.75). □

From the above lemma the following properties of the process $\varkappa = (\varkappa_t, \mathscr{F}_t)$, $0 \le t \le T$, can be easily deduced.

Property 1. If $\varkappa_t \ge 0$ (P-a.s.), then the process $\varkappa = (\varkappa_t, \mathscr{F}_t)$ is a supermartingale, $M(\varkappa_t | \mathscr{F}_s) \le \varkappa_s$ (P-a.s.), $t \ge s$, and, in particular, $M\varkappa_t \le 1$.

Property 2. If $P(\inf_{0 \le t \le T} \varkappa_t > 0) = 1$, then $\varkappa_t$ has the representation

$$\varkappa_t = \exp\left(\int_0^t \sum_{i=1}^n \beta_i(s)dW_i(s) - \frac{1}{2}\int_0^t \sum_{i=1}^2 \beta_i^2(s)ds,\right),$$

where $\beta_i(t) = \varkappa_t^{-1}\gamma_i(t)$.

Let now $W = (W_t, \mathscr{F}_t)$, $0 \le t \le T$, be an n-dimensional Wiener process where (a vector-column)

$$W_t = [W_1(t), \ldots, W_n(t)].$$

Let $\gamma = (\gamma_t, \mathscr{F}_t)$, $0 \le t \le T$, also be an n-dimensional process with (a vector-column)

$$\gamma_t = [\gamma_1(t), \ldots, \gamma_n(t)] \quad \text{and} \quad P\left(\sum_{i=1}^n \int_0^T \gamma_i^2(t)dt < \infty\right) = 1.$$

Set

$$\varkappa_t = 1 + \int_0^t \gamma_s^*\,dW_s, \tag{6.79}$$

where γ_s^* is a vector-row transposed to γ_s.

As in the one-dimensional case ($n = 1$), the following (multi-dimensional) analog of Theorem 6.2 is proved.

Theorem 6.4. *Let $M\varkappa_T = 1$. Then the n-dimensional random process*

$$\tilde{W}_t = W_t - \int_0^t \varkappa_s^+ \gamma_s\,ds$$

is (with respect to the system $(\mathscr{F}_t)$, $t \le T$, and measure $\tilde{P}$ with $d\tilde{P} = \varkappa_T(\omega)dP$) a Wiener process.

Notes and references

6.1. The results related to this section are due to the authors.

6.2. Theorem 6.1 was proved by Novikov [133]. This theorem was proved by Gykhman and Skorokhod [36] and by Liptser and Shiryayev [118], in the case of the substitution of a multiplier $\frac{1}{2}$ for $1 + \varepsilon$ and $\frac{1}{2} + \varepsilon$ respectively.

6.3. Theorem 6.2 generalizes the important result obtained by Girsanov [31], formulated in Theorem 6.3.

7 Absolute continuity of measures corresponding to the Ito processes and processes of the diffusion type

7.1 The Ito processes, and the absolute continuity of their measures with respect to Wiener measure

7.1.1

Let $(\Omega, \mathscr{F}, P)$ be a complete probability space, let $F = (\mathscr{F}_t)$, $t \geq 0$, be a nondecreasing family of sub-σ-algebras, and let $W = (W_t, \mathscr{F}_t)$, $t \geq 0$, be a Wiener process.

We shall consider the random Ito process[1] $\xi = (\xi_t, \mathscr{F}_t)$, $0 \leq t \leq T$, with the differential[2]

$$d\xi_t = \beta_t(\omega)dt + dW_t, \qquad \xi_0 = 0, \tag{7.1}$$

where the process $\beta = (\beta_t(\omega), \mathscr{F}_t)$, $0 \leq t \leq T$, is such that

$$P\left(\int_0^T |\beta_t(\omega)|\,dt < \infty\right) = 1.$$

Denote by $(C_T, \mathscr{B}_T)$ a measurable space of the continuous functions $x = (x_s)$, $s \leq T$, with $x_0 = 0$, and let μ_ξ, μ_W be measures in $(C_T, \mathscr{B}_T)$, corresponding to the processes $\xi = (\xi_s)$, $s \leq T$, and $W = (W_s)$, $s \leq T$:

$$\mu_\xi(B) = P\{\omega : \xi \in B\}, \qquad \mu_W(B) = P\{\omega : W \in B\}. \tag{7.2}$$

In this section we shall discuss the problem of the absolute continuity and equivalence of measures μ_ξ and μ_W for the case where ξ is an Ito process.

[1] In case $T = \infty$ it is assumed that $0 \leq t < \infty$.

[2] See Definition 6 in Section 4.2.

Let us agree on some notation we shall use from now on. Let $\mu_{t,\xi}$ and $\mu_{t,W}$ be restrictions of the measures μ_ξ and μ_W on $\mathscr{B}_t = \sigma\{x:x_s, s \le t\}$. By

$$\frac{d\mu_\xi}{d\mu_W}(t, x) \quad \text{and} \quad \frac{d\mu_W}{d\mu_\xi}(t, x)$$

we denote Radon–Nykodim derivatives of the measures $\mu_{t,\xi}$ w.r.t. $\mu_{t,W}$ and $\mu_{t,W}$ w.r.t. $\mu_{t,\xi}$. In the case $t = T$ the T-index will be omitted:

$$\frac{d\mu_W}{d\mu_\xi}(x) = \frac{d\mu_W}{d\mu_\xi}(T, x), \qquad \frac{d\mu_\xi}{d\mu_W}(x) = \frac{d\mu_\xi}{d\mu_W}(T, x).$$

By

$$\frac{d\mu_\xi}{d\mu_W}(\xi), \qquad \frac{d\mu_\xi}{d\mu_W}(t, \xi)$$

we denote $\mathscr{F}_T^\xi$-measurable and $\mathscr{F}_t^\xi$-measurable random variables, respectively, obtained as a result of the substitution of x for the function $\xi = (\xi_s(\omega))$, $s \le T$, in

$$\frac{d\mu_\xi}{d\mu_W}(x), \qquad \frac{d\mu_\xi}{d\mu_W}(t, x).$$

In a similar way,

$$\frac{d\mu_\xi}{d\mu_W}(t, W), \qquad \frac{d\mu_\xi}{d\mu_W}(W), \ldots,$$

are defined.

7.1.2

Theorem 7.1. *Let $\xi = (\xi_t, \mathscr{F}_t)$, $t \le T$, be an Ito process with the differential given in* (7.1). *If*

$$P\left(\int_0^T \beta_t^2 \, dt < \infty\right) = 1, \tag{7.3}$$

$$M \exp\left\{-\int_0^T \beta_t \, dW_t - \frac{1}{2}\int_0^T \beta_t^2 \, dt\right\} = 1, \tag{7.4}$$

then $\mu_\xi \sim \mu_W$ and (P-a.s.)[3]

$$\frac{d\mu_W}{d\mu_\xi}(\xi) = M\left[\exp\left\{-\int_0^T \beta_t \, d\xi_t + \frac{1}{2}\int_0^T \beta_t^2 \, dt\right\} \middle| \mathscr{F}_T^\xi\right]. \tag{7.5}$$

PROOF. Denote

$$\varkappa_t = \exp\left(-\int_0^t \beta_s \, dW_s - \frac{1}{2}\int_0^t \beta_s^2 \, ds\right).$$

[3] Regarding the definition of the stochastic integral $\int_0^t \beta_s d\xi_s$, see Section 4.2.

Since by the assumption in (7.4) $M\varkappa_T = 1$, then (Lemma 6.4) $\varkappa = (\varkappa_t, \mathscr{F}_t)$, $t \le T$, is a martingale. Let $\tilde{P}$ be a measure on $(\Omega, \mathscr{F})$ with $d\tilde{P} = \varkappa_T(\omega)dP$. By Theorem 6.3 the process $\xi = (\xi_t, \mathscr{F}_t)$, $t \le T$, is a Wiener process (over measure $\tilde{P}$) and, therefore, for $A \in \mathscr{B}_T$,

$$\mu_W(A) = \tilde{P}(\xi \in A) = \int_{\{\omega:\xi\in A\}} \varkappa_T(\omega)dP = \int_{\{\omega:\xi\in A\}} M(\varkappa_T(\omega)|\mathscr{F}_T^\xi)dP. \quad (7.6)$$

The random variable $M(\varkappa_T(\omega)|\mathscr{F}_T^\xi)$ is $\mathscr{F}_T^\xi$-measurable and therefore[4] there will be a $\mathscr{B}_T$-measurable nonnegative function $\Phi(x)$, such that (P-a.s.)

$$M(\varkappa_T(\omega)|\mathscr{F}_T^\xi) = \Phi(\xi(\omega)). \quad (7.7)$$

(For the sake of clarity this function $\Phi(x)$ will be denoted also by $M(\varkappa_T(\omega)|\mathscr{F}_T^\xi)_{\xi=x}$. Similar notations are used in other cases.)

Then the formula (7.6) can be rewritten in the following form:

$$\mu_W(A) = \int_{\{\omega:\xi\in A\}} \Phi(\xi(\omega))dP(\omega) = \int_A \Phi(x)d\mu_\xi(x).$$

From this we obtain $\mu_W \ll \mu_\xi$ and

$$\frac{d\mu_W}{d\mu_\xi}(x) = \Phi(x) \qquad (\mu_\xi\text{-a.s.}).$$

Hence, because of (7.7),

$$\frac{d\mu_W}{d\mu_\xi}(\xi) = M(\varkappa_T(\omega)|\mathscr{F}_T^\xi) \qquad (P\text{-a.s.})$$

which, together with (7.1), proves the representation given in (7.5).

It remains only to show that $\mu_\xi \ll \mu_W$. For proving this we note that

$$\frac{d\tilde{P}}{dP}(\omega) = \varkappa_T(\omega),$$

with $P(\varkappa_T(\omega) = 0) = 0$ since because of (7.3) we have

$$P\left(\left|\int_0^T \beta_t\, dW_t\right| < \infty\right) = 1.$$

Hence, by Lemma 6.8, $P \ll \tilde{P}$ and

$$\frac{dP}{d\tilde{P}}(\omega) = \varkappa_T^{-1}(\omega).$$

Further,

$$\mu_\xi(A) = P\{\omega:\xi\in A\} = \int_{\{\omega:\xi\in A\}} \varkappa_T^{-1}(\omega)d\tilde{P}(\omega)$$

$$= \int_{\{\omega:\xi\in A\}} \tilde{M}[\varkappa_T^{-1}(\omega)|\mathscr{F}_T^\xi]d\tilde{P}(\omega) = \int_A \tilde{M}[\varkappa_T^{-1}(\omega)|\mathscr{F}_T^\xi]_{\xi=x}\, d\mu_W(x),$$

[4] See Section 1.2.

since $\tilde{P}\{\omega: \xi \in A\} = \mu_W(A)$. Consequently, $\mu_\xi \ll \mu_W$ and (P-a.s.)

$$\frac{d\mu_\xi}{d\mu_W}(\xi) = \tilde{M}[\varkappa_T^{-1}(\omega) | \mathscr{F}_T^\xi]. \quad \square \tag{7.8}$$

Note. Theorem 7.1 holds true if, in place of T, a Markov time σ (with respect to the system $(\mathscr{F}_t)$, $t \geq 0$) is considered. If

$$P\left(\int_0^\sigma \beta_t^2 \, dt < \infty\right) = 1,$$

$$M \exp\left\{-\int_0^\sigma \beta_t \, dW_t - \frac{1}{2}\int_0^\sigma \beta_t^2 \, dt\right\} = 1,$$

then the restrictions of measures μ_ξ and μ_W, on the σ-algebra $\mathscr{B}_\sigma$, are equivalent.

Corollary. *For each t, $0 \leq t \leq T$, let the random variables $\beta_t = \beta_t(\omega)$ be $\mathscr{F}_t^\xi$-measurable. Without introducing new notations, we shall immediately assume that $\beta_t = \beta_t(\xi(\omega))$. Let also (7.3), (7.4) be satisfied. Then (P-a.s.)*

$$\frac{d\mu_W}{d\mu_\xi}(\xi) = \exp\left(-\int_0^T \beta_t(\xi) d\xi_t + \frac{1}{2}\int_0^T \beta_t^2(\xi) dt\right). \tag{7.9}$$

Since $\mu_\xi \sim \mu_W$, then

$$\frac{d\mu_\xi}{d\mu_W}(x) = \left[\frac{d\mu_W}{d\mu_\xi}(x)\right]^{-1}.$$

From (7.9) and Lemma 4.10 it is not difficult to deduce that the derivative $d\mu_\xi/d\mu_W(W)$ can be represented in the following form:

$$\frac{d\mu_\xi}{d\mu_W}(W) = \exp\left(\int_0^T \beta_t(W) dW_t - \frac{1}{2}\int_0^T \beta_t^2(W) dt\right) \quad (P\text{-a.s.}). \tag{7.10}$$

EXAMPLE 1. Let $\xi_t = \theta \cdot t + W_t$, $t \leq 1$, where $\theta = \theta(\omega)$ is an $\mathscr{F}_0$-measurable normally distributed random variable, $N(0, 1)$, independent of a Wiener process W. According to Example 4, Section 6.2, $M \exp(-\theta W_1 - \theta^2/2) = 1$, and by Theorem 7.1, $\mu_\xi \sim \mu_W$,

$$\frac{d\mu_W}{d\mu_\xi}(\xi) = M\left[\exp\left(-\theta\xi_1 + \frac{\theta^2}{2}\right) \middle| \mathscr{F}_1^\xi\right].$$

The conditional distribution $P(\theta \leq y | \mathscr{F}_1^\xi)$ is normal, $N(\xi_1/2, 1/2)$. Hence,

$$M\left[\exp\left(-\theta\xi_1 + \frac{\theta^2}{2}\right) \middle| \mathscr{F}_1^\xi\right] = \sqrt{2}\exp\left(-\frac{\xi_1^2}{4}\right).$$

Consequently,

$$\frac{d\mu_W}{d\mu_\xi}(x) = \sqrt{2}\exp\left(-\frac{x_1^2}{4}\right), \qquad \frac{d\mu_\xi}{d\mu_W}(x) = \frac{1}{\sqrt{2}}\exp\left(\frac{x_1^2}{4}\right). \tag{7.11}$$

Looking ahead, we note that for these derivatives other expressions can be given (see Section 7.4). Thus, according to Theorem 7.13, (P-a.s.)

$$\frac{d\mu_W}{d\mu_\xi}(\xi) = \exp\left[-\int_0^1 \frac{\xi_s}{1+s}\,d\xi_s + \frac{1}{2}\int_0^1 \left(\frac{\xi_s}{1+s}\right)^2 ds\right]. \tag{7.12}$$

7.1.3

Theorem 7.2. *Let $\xi = (\xi_t, \mathscr{F}_t)$, $t \le T$, be an Ito process with the differential given by* (7.1). *If $P(\int_0^T \beta_t^2\,dt < \infty) = 1$, then $\mu_\xi \ll \mu_W$.*

PROOF. For $n = 1, 2, \ldots$, set

$$\tau_n = \begin{cases} \inf\left\{t \le T: \int_0^t \beta_s^2\,ds \ge n\right\}, \\ T, \quad \text{if } \int_0^T \beta_s^2\,ds < n, \end{cases}$$

and

$$\chi_t^{(n)} = \chi_{\{\int_0^t \beta_s^2\,ds \le n\}}, \qquad \beta_t^{(n)} = \chi_t^{(n)}\beta_t.$$

Let

$$\xi_t^{(n)} = \int_0^t \beta_s^{(n)}\,ds + W_t, \qquad 0 \le t \le T.$$

Then, since $P(\int_0^T (\beta_s^{(n)})^2\,ds \le n) = 1$, by Theorem 6.1

$$M\exp\left(-\int_0^T \beta_s^{(n)}\,dW_s - \frac{1}{2}\int_0^T (\beta_s^{(n)})^2\,ds\right) = 1.$$

Consequently, according to Theorem 7.1, $\mu_{\xi^{(n)}} \sim \mu_W$ for each n, $n = 1, 2, \ldots$.

It will be noted now that on the set $\{\tau_n = T\}$, $\xi_t^{(n)} = \xi_t$ (P-a.s.), $0 \le t \le T$, and hence, for any $\Gamma \in \mathscr{B}_T$,

$$\begin{aligned}\mu_\xi(\Gamma) &= P\{\omega : \xi(\omega) \in \Gamma\} \\ &= P\{\xi(\omega) \in \Gamma, \tau_n = T\} + P\{\xi(\omega) \in \Gamma, \tau_n < T\} \\ &= P\{\xi^{(n)}(\omega) \in \Gamma, \tau_n = T\} + P\{\xi(\omega) \in \Gamma, \tau_n < T\}.\end{aligned}$$

Let $\mu_W(\Gamma) = 0$. Then, since $\mu_{\xi^{(n)}} \sim \mu_W$, $\mu_{\xi^{(n)}}(\Gamma) = 0$ and

$$P\{\xi^{(n)} \in \Gamma, \tau_n = T\} \le P\{\xi^{(n)} \in \Gamma\} = \mu_{\xi^{(n)}}(\Gamma) = 0.$$

Consequently,

$$\mu_\xi(\Gamma) = P\{\xi \in \Gamma, \tau_n < T\} \le P\{\tau_n \le T\} = P\left\{\int_0^T \beta_t^2\,dt > n\right\} \to 0, \qquad n \to \infty.$$

From this it follows that $\mu_\xi(\Gamma) = 0$, and, therefore, $\mu_\xi \ll \mu_W$. □

7.1.4

Theorems 7.1 and 7.2 permit extensions to the multi-dimensional case. We shall give the corresponding results restricting ourselves to the statements alone, since their proofs are similar to those in the one-dimensional case.

Let $W = (W_t, \mathscr{F}_t)$, $0 \le t \le T$, be an n-dimensional[5] Wiener process, $W_t = (W_1(t), \ldots, W_n(t))$, and let $\beta = (\beta_t, \mathscr{F}_t), 0 \le t \le T, \beta_t = (\beta_1(t), \ldots, \beta_n(t))$.

Theorem 7.3. *Let $\xi = (\xi_t, \mathscr{F}_t)$, $0 \le t \le T$, be an n-dimensional Ito process, $\xi_t = (\xi_1(t), \ldots, \xi_n(t))$, with the differential*

$$d\xi_t = \beta_t\, dt + dW_t, \qquad \xi_0 = 0. \tag{7.13}$$

If

$$P\left(\int_0^T \beta_t^* \beta_t\, dt < \infty\right) = 1, \tag{7.14}$$

$$M \exp\left(-\int_0^T \beta_t^*\, dW_t - \frac{1}{2}\int_0^T \beta_t^* \beta_t\, dt\right) = 1, \tag{7.15}$$

then $\mu_\xi \sim \mu_W$ and (P-a.s.)

$$\frac{d\mu_W}{d\mu_\xi}(\xi) = M\left[\exp\left(-\int_0^T \beta_t^*\, d\xi_t + \frac{1}{2}\int_0^T \beta_t^* \beta_t\, dt\right)\middle| \mathscr{F}_T^\xi\right]. \tag{7.16}$$

Theorem 7.4. *Let $\xi = (\xi_t, \mathscr{F}_t)$, $0 \le t \le T$, be an n-dimensional Ito process, $\xi_t = (\xi_1(t), \ldots, \xi_n(t))$, with the differential*

$$d\xi_t = \beta_t\, dt + dW_t, \qquad \xi_0 = 0,$$

let

$$P\left(\int_0^T \beta_t^* \beta_t\, dt < \infty\right) = 1.$$

Then $\mu_\xi \ll \mu_W$.

7.2 Processes of the diffusion type: the absolute continuity of their measures with respect to Wiener measure

7.2.1

Let $W = (W_t, \mathscr{F}_t)$, $0 \le t \le T$, be a Wiener process, prescribed on a probability space $(\Omega, \mathscr{F}, P)$ with a distinguished family of sub-σ-algebras $(\mathscr{F}_t)$, $0 \le t \le T$.

[5] Here and further on vectors are taken to be column vectors.

Let us consider a random process $\xi = (\xi_t, \mathscr{F}_t)$, $0 \le t \le T$, of the diffusion type[6] with the differential

$$d\xi_t = \alpha_t(\xi)dt + dW_t, \qquad \xi_0 = 0, \tag{7.17}$$

where the nonanticipative process $\alpha = (\alpha_t(x), \mathscr{B}_{t+})$, prescribed on $(C_T, \mathscr{B}_T)$ is such that

$$P\left(\int_0^T |\alpha_t(\xi)|dt < \infty\right) = 1. \tag{7.18}$$

According to Theorem 7.2, the condition $P(\int_0^T \alpha_t^2(\xi)dt < \infty) = 1$ provides the absolute continuity of the measure μ_ξ over the measure μ_W. It turns out that for a process of the diffusion type this condition is not only sufficient but also necessary.

Theorem 7.5. *Let $\xi = (\xi_t, \mathscr{F}_t)$, $0 \le t \le T$, be a process of the diffusion type with the differential given by* (7.17). *Then*

$$P\left(\int_0^T \alpha_t^2(\xi)dt < \infty\right) = 1 \Leftrightarrow \mu_\xi \ll \mu_W. \tag{7.19}$$

PROOF. Sufficiency follows from Theorem 7.2. To prove necessity let us denote

$$\varkappa_t(x) = \frac{d\mu_\xi}{d\mu_W}(t, x), \qquad 0 \le t \le T.$$

We shall show that the process $\varkappa = (\varkappa_t(W), \mathscr{F}_T^W)$, $0 \le t \le T$, is a martingale.

Let $s < t$ and let $\lambda(W)$ be a bounded $\mathscr{F}_t^W$-measurable random variable. Then

$$M\lambda(W)\varkappa_t(W) = \int \lambda(x)\frac{d\mu_\xi}{d\mu_W}(t, x)d\mu_W(x)$$

$$= \int \lambda(x)d\mu_{t,\xi}(x) = \int \lambda(x)d\mu_{s,\xi}(x) = \int \lambda(x)\varkappa_s(x)d\mu_{s,W}(x),$$

from which we obtain $M(\varkappa_t(W)|\mathscr{F}_s^W) = \varkappa_s(W)$ (P-a.s.), $t > s$.

Let us apply Theorem 5.7 to the martingale $\varkappa_t = (\varkappa_t(W), \mathscr{F}_t^W)$, $0 \le t \le T$. According to this theorem there will be a process

$$\gamma = (\gamma_t(\omega), \mathscr{F}_t^W), 0 \le t \le T, \quad \text{with} \quad P\left(\int_0^T \gamma_t^2(\omega)dt < \infty\right) = 1,$$

such that for each t, $0 \le t \le T$, (P-a.s.)

$$\varkappa_t(W) = 1 + \int_0^t \gamma_s(\omega)dW_s. \tag{7.20}$$

Here the process $\varkappa_t(W)$, $0 \le t \le T$, is continuous (P-a.s.).

[6] See Definition 7 in Section 4.2.

We shall consider now on a probability space $(\Omega, \mathscr{F}, \tilde{P})$ with $d\tilde{P}(\omega) = \varkappa_T(\omega)dP(\omega)$ the random process $\tilde{W} = (\tilde{W}_t, \mathscr{F}_t^W)$, $0 \le t \le T$, with

$$\tilde{W}_t = W_t - \int_0^t B_s(\omega)ds, \tag{7.21}$$

where $B_s(\omega) = \varkappa_s^+(W)\gamma_s(\omega)$. By Theorem 6.2 this process is a Wiener process. In proving this theorem it was also shown that

$$\tilde{P}\left(\int_0^T B_s^2(\omega)ds < \infty\right) = 1. \tag{7.22}$$

According to Lemma 4.9, there will be a functional $\beta = (\beta_s(x), \mathscr{B}_{s+})$, such that for almost all $0 \le s \le T$ (P-a.s.)

$$B_s(\omega) = \beta_s(W(\omega)),$$

and, consequently,

$$\tilde{W}_t = W_t - \int_0^t \beta_s(W)ds \qquad (P\text{-a.s.}).$$

Because of (7.22)

$$\tilde{P}\left(\int_0^T \beta_s^2(W)ds < \infty\right) = 1.$$

From this equality and the assumption $\mu_\xi \ll \mu_W$, it follows that

$$P\left(\int_0^T \beta_s^2(\xi)ds < \infty\right) = 1. \tag{7.23}$$

Indeed,

$$P\left(\int_0^T \beta_s^2(\xi)ds < \infty\right) = \mu_\xi\left\{x: \int_0^T \beta_t^2(x)dt < \infty\right\}$$

$$= \int \chi_{\{\int_0^T \beta_t^2(x)dt < \infty\}}(x)d\mu_\xi(x) = \int \chi_{\{\int_0^T \beta_t^2(x)dt < \infty\}}(x)\varkappa_T(x)d\mu_W(x)$$

$$= M\chi_{\{\int_0^T \beta_t^2(W)dt < \infty\}}(W)\varkappa_T(W)$$

$$= \tilde{P}\left(\int_0^T \beta_s^2(W)ds < \infty\right) = 1.$$

We define now on a probability space $(\Omega, \mathscr{F}, P)$ a process $\hat{W} = (\hat{W}_t(\xi), \mathscr{F}_t^\xi)$, $0 \le t \le T$, assuming that

$$\hat{W}_t(x) = x_t - \int_0^t \beta_s(x)ds, \qquad x \in C_T. \tag{7.24}$$

This process with $x = \xi$ is a Wiener process. Actually, let $\lambda = \lambda(\xi)$ be a bounded $\mathscr{F}_s^\xi$-measurable random variable. Then

$$M\lambda(\xi)e^{iz[\hat{W}_t(\xi) - \hat{W}_s(\xi)]} = \int \lambda(x)e^{iz[\hat{W}_t(x) - \hat{W}_s(x)]}\,d\mu_\xi(x)$$

$$= \int \lambda(x)e^{iz[\hat{W}_t(x) - \hat{W}_s(x)]}\varkappa_T(x)d\mu_W(x)$$

$$= \int \lambda(W)e^{iz[\tilde{W}_t - \tilde{W}_s]}\varkappa_T(W)dP = \tilde{M}\lambda(W)e^{iz[\tilde{W}_t - \tilde{W}_s]}$$

$$= \tilde{M}\{\lambda(W)\tilde{M}[e^{iz(\tilde{W}_t - \tilde{W}_s)}|\mathscr{F}_s]\} = \tilde{M}\lambda(W)e^{-(z^2/2)(t-s)}$$

$$= e^{-(z^2/2)(t-s)}\int \lambda(W)\varkappa_T(W)dP = e^{-(z^2/2)(t-s)}M\lambda(\xi).$$

On the other hand,

$$M\lambda(\xi)e^{iz[\hat{W}_t - \hat{W}_s]} = M\{\lambda(\xi)M[e^{iz(\hat{W}_t - \hat{W}_s)}|\mathscr{F}_s^\xi]\},$$

and, therefore,

$$M[e^{iz(\hat{W}_t - \hat{W}_s)}|\mathscr{F}_s^\xi] = e^{-(z^2/2)(t-s)}.$$

From (7.18) and (7.24) we obtain

$$\hat{W}_t(\xi) - W_t = \int_0^t [\alpha_s(\xi) - \beta_s(\xi)]ds, \tag{7.25}$$

where $(\hat{W}_t, \mathscr{F}_t^\xi)$ and $(W_t, \mathscr{F}_t^\xi)$, $0 \le t \le T$, are two Wiener processes. Therefore, on one hand the process $(\hat{W}_t - W_t, \mathscr{F}_t^\xi)$, $0 \le t \le T$, is a square integrable martingale, and on the other hand it has a special form given by (7.25). From the lemma given below it follows that in such a case $\hat{W}_t - W_t = 0$ (P-a.s.) for all t, $0 \le t \le T$.

Lemma 7.1. *Let $\eta = (\eta_t, \mathscr{F}_t)$, $0 \le t \le T$, be a square integrable martingale permitting the representation*

$$\eta_t = \int_0^t f_s\,ds \qquad (P\text{-a.s.}), \tag{7.26}$$

where the nonanticipative process $f = (f_s, \mathscr{F}_s)$, $0 \le s \le T$, is such that $P(\int_0^T |f_s|ds < \infty) = 1$. Then with probability one $f_t = 0$ for almost all t, $0 \le t \le T$.

Proof. Let $\tau_N = \inf(t \le T: \int_0^t |f_s|ds \ge N)$, and let $\tau_N = T$ if $\int_0^T |f_s|ds < N$. Denote $\chi_t^{(N)} = \chi_{\{\tau_N \ge t\}}$ and $\eta_t^{(N)} = \int_0^t \chi_s^{(N)} f_s\,ds$.

The process $(\eta_t^{(N)}, \mathscr{F}_t^{(N)})$, $0 \le t \le T$, with $\mathscr{F}_t^{(N)} = \mathscr{F}_{t \wedge \tau_N}$ will be a square integrable martingale (Theorem 3.6) and, hence,

$$M(\eta_t^{(N)})^2 = \lim_{n\to\infty} \sum_{i=0}^{n-1} M[\eta_{t_{j+1}^{(N)}} - \eta_{t_j^{(N)}}]^2,$$

where $0 = t_0 < \cdots < t_n = t$ and $\max_j |t_{j+1} - t_j| \to 0$, $n \to \infty$.

Since

$$\eta_{t_{j+1}^{(N)}} - \eta_{t_j^{(N)}} = \int_{t_j}^{t_{j+1}} \chi_s^{(N)} f_s\, ds,$$

then

$$\begin{aligned} M(\eta_t^{(N)})^2 &= \lim_{n\to\infty} \sum_{j=0}^{n-1} M\left(\int_{t_j}^{t_{j+1}} \chi_s^{(N)} f_s\, ds\right)^2 \\ &\le \varliminf_{n\to\infty} M\left\{\max_{j\le n-1} \int_{t_j}^{t_{j+1}} \chi_s^{(N)} |f_s| ds \sum_{j=0}^{n-1} \int_{t_j}^{t_{j+1}} \chi_s^{(N)} |f_s| ds\right\} \\ &= \varliminf_{n\to\infty} M\left\{\max_{j\le n-1} \int_{t_j}^{t_{j+1}} \chi_s^{(N)} |f_s| ds \int_0^T \chi_s^{(N)} |f_s| ds\right\} \\ &\le N \varliminf_{n\to\infty} M\left\{\max_{j\le n-1} \int_{t_j}^{t_{j+1}} \chi_s^{(N)} |f_s| ds\right\}. \end{aligned}$$

But

$$\max_{j\le n-1} \int_{t_j}^{t_{j+1}} \chi_s^{(N)} |f_s| ds \le N$$

and with $n \to \infty$ with probability one tends to zero. Consequently, $M(\eta_t^{(N)})^2 = 0$ and by the Fatou lemma

$$M\eta_t^2 = M\left(\lim_{N\to\infty} \eta_t^{(N)}\right)^2 \le \varliminf_{N\to\infty} M(\eta_t^{(N)})^2 = 0. \qquad \square$$

We return now to the proof of Theorem 7.5. Since $\hat{W}_t - W_t = 0$ (P-a.s.) for all t, $0 \le t \le T$, then from (7.25) and Lemma 7.1 it follows that $\alpha_s(\xi) = \beta_s(\xi)$ (P-a.s.) for almost all s, $0 \le s \le T$. But according to (7.23),

$$P\left(\int_0^T \beta_s^2(\xi) ds < \infty\right) = 1.$$

Hence, $P(\int_0^T \alpha_s^2(\xi) ds < \infty) = 1$, completing proof of Theorem 7.5. $\square$

7.2.2

According to Theorem 7.5, for processes of the diffusion type the condition $P(\int_0^T \alpha_t^2(\xi) dt < \infty) = 1$ is necessary and sufficient for the absolute continuity of the measure μ_ξ w.r.t. the measure μ_W. Let us investigate now the processes

$$\varkappa_t(\xi) = \frac{d\mu_\xi}{d\mu_W}(t, \xi) \quad \text{and} \quad \varkappa_t(W) = \frac{d\mu_\xi}{d\mu_W}(t, W).$$

Theorem 7.6. *Let $\xi = (\xi_t, \mathscr{F}_t)$, $0 \le t \le T$, be a process of the diffusion type with*

$$d\xi_t = \alpha_t(\xi)dt + dW_t, \qquad \xi_0 = 0. \tag{7.27}$$

If $P(\int_0^T \alpha_t^2(\xi)dt < \infty) = 1$, then the process $\varkappa_t(W)$, $0 \le t \le T$, is the unique solution to the equation

$$\varkappa_t(W) = 1 + \int_0^t \varkappa_s(W)\alpha_s(W)dW_s; \tag{7.28}$$

$$\frac{d\mu_\xi}{d\mu_W}(t, W) = \exp\left(\Gamma_t(\alpha(W)) - \frac{1}{2}\int_0^t \alpha_s^2(W)ds\right) \qquad (P\text{-a.s.}), \tag{7.29}$$

$$\frac{d\mu_\xi}{d\mu_W}(t, \xi) = \exp\left(\int_0^T \alpha_s(\xi)d\xi_s - \frac{1}{2}\int_0^t \alpha_s^2(\xi)ds\right) \qquad (P\text{-a.s.}), \tag{7.30}$$

$$P\left(\int_0^t \alpha_s^2(W)ds < \infty\right) = M\exp\left(-\int_0^t \alpha_s(\xi)dW_s - \frac{1}{2}\int_0^t \alpha_s^2(\xi)ds\right). \tag{7.31}$$

PROOF. To prove the first statement we shall show first that the process $\varkappa_t(W)$, $t \le T$, is such that $P(\int_0^T (\varkappa_s(W)\alpha_s(W))^2\,ds < \infty) = 1$. For this purpose, making use of the notations for proving Theorem 7.5, we establish first that for almost each s, $0 \le s \le T$, (P-a.s.)

$$\varkappa_s(W)\alpha_s(W) = \varkappa_s(W)\beta_s(W). \tag{7.32}$$

It is shown in Theorem 7.5 that $P(\alpha_s(\xi) \ne \beta_s(\xi)) = 0$ for almost all $s \le T$. Second, $P(\varkappa_s(\xi) = 0) = 0$, $s \le T$, since

$$P(\varkappa_s(\xi) = 0) = \mu_\xi(x : \varkappa_s(x) = 0) = M\varkappa_s(W)\chi_{\{\varkappa_s(W)=0\}} = 0.$$

Consequently,

$$0 = \mu_\xi(\varkappa_s(x)[\alpha_s(x) - \beta_s(x)] \ne 0) = M\varkappa_s(W)\chi_{\{\varkappa_s(W)[\alpha_s(W)-\beta_s(W)]\ne 0\}},$$

which proves (7.32).

Further, by definition $\beta_s(W) = \varkappa_s^+(W)\gamma_s(W)$. Hence $\varkappa_s(W)\alpha_s = \varkappa_s(W)\varkappa_s^+(W)\gamma_s(W)$ (P-a.s.) for almost all $s \le T$ and

$$P\left(\int_0^T (\varkappa_s(W)\alpha_s(W))^2\,ds < \infty\right) = P\left(\int_0^T (\varkappa_s(W)\varkappa_s^+(W)\gamma_s(W))^2\,ds < \infty\right)$$
$$\ge P\left(\int_0^T \gamma_s^2(W)ds < \infty\right) = 1.$$

Thus $P(\int_0^T (\varkappa_s(W)\alpha_s(W))^2\,ds < \infty) = 1$ and, therefore, the stochastic integral $\int_0^t \varkappa_t(W)\alpha_s(W)dW_s$ is defined.

Let us show that (P-a.s.)

$$1 + \int_0^t \varkappa_s(W)\alpha_s(W)dW_s = 1 + \int_0^t \gamma_s(W)dW_s. \tag{7.33}$$

According to (7.20), $1 + \int_0^t \gamma_s(W)dW_s = \varkappa_t(W)$. Since the process $(\varkappa_t(W), \mathscr{F}_t^{(W)})$, $0 \le t \le T$, is a nonnegative martingale, then (P-a.s.) $\varkappa_t(W) \equiv 0$ with $T \ge t \ge \tau$, where

$$\tau = \begin{cases} \inf(t \le T : \varkappa_t = 0), \\ \infty, \qquad \inf_{t \le T} \varkappa_t > 0. \end{cases}$$

By definition,

$$1 + \int_0^t \varkappa_s(W)\alpha_s(W)dW_s = 1 + \int_0^t \varkappa_s(W)\varkappa_s^+(W)\gamma_s(W)dW_s,$$

and, therefore, with $T \ge \tau \ge t$ Equation (7.33) is satisfied (P-a.s.), and, in particular, for $\tau \le T$,

$$1 + \int_0^\tau \varkappa_s(W)\alpha_s(W)dW_s = 0.$$

With $T \ge t \ge \tau$ both sides of (7.33) are equal to zero. Equation (7.28) now follows from (7.33) and (7.20).

To prove (7.29) and (7.30) we shall make use of Lemma 6.2, according to which the process $\varkappa_t(W)$, $t \le T$, considered as a solution to Equation (7.28) can be represented by (7.29). (7.30) follows from (7.29) and Lemma 4.10 if it is noted that

$$P\left(\int_0^T \alpha_s^2(\xi)ds < \infty\right) = 1.$$

To prove (7.31) it will be noted that, because of (7.27),

$$M \exp\left(-\int_0^t \alpha_s(\xi)dW_s - \frac{1}{2}\int_0^t \alpha_s^2(\xi)ds\right)$$

$$= M \exp\left(-\int_0^t \alpha_s(\xi)d\xi_s + \frac{1}{2}\int_0^t \alpha_s^2(\xi)ds\right) = M\varkappa_t^+(\xi). \tag{7.34}$$

On the other hand,

$$M\varkappa_t^+(\xi) = \int \varkappa_t^+(x)\varkappa_t(x)d\mu_W(x) = \mu_W\{x : \varkappa_t(x) > 0\} = P(\varkappa_t(W) > 0). \tag{7.35}$$

But according to (7.29),

$$P(\varkappa_t(W) > 0) = P\left(\int_0^t \alpha_s^2(W)ds < \infty\right),$$

which together with (7.34) and (7.35) leads to the proof of Equation (7.31). □

7.2.3

Theorem 7.7. *Let $\xi = (\xi_t, \mathscr{F}_t)$, $0 \le t \le T$, be a process of diffusion type with the differential*

$$d\xi_t = \alpha_t(\xi)dt + dW_t, \qquad \xi_0 = 0, \qquad 0 \le t \le T.$$

Then

$$\left.\begin{aligned} &P\left(\int_0^T \alpha_t^2(\xi)dt < \infty\right) = 1, \\ &P\left(\int_0^T \alpha_s^2(W)dt < \infty\right) = 1 \end{aligned}\right\} \Leftrightarrow \mu_\xi \sim \mu_W. \tag{7.36}$$

Here (P-a.s.)

$$\frac{d\mu_\xi}{d\mu_W}(t, W) = \exp\left(\int_0^t \alpha_s(W)dW_s - \frac{1}{2}\int_0^t \alpha_s^2(W)ds\right), \tag{7.37}$$

$$\frac{d\mu_W}{d\mu_\xi}(t, \xi) = \exp\left(-\int_0^t \alpha_s(\xi)d\xi_s + \frac{1}{2}\int_0^t \alpha_s^2(\xi)ds\right). \tag{7.38}$$

PROOF.

Sufficiency: By Theorem 7.5, from the condition $P(\int_0^T \alpha_t^2(\xi)dt < \infty) = 1$ we obtain $\mu_\xi \ll \mu_W$. From Theorem 7.6 follows (7.37) since $P(\int_0^T \alpha_t^2(W)dt < \infty) = 1$ and, therefore, $\Gamma_t(\alpha(W)) = \int_0^t \alpha_s(W)dW_s$.

Due to the condition $P(\int_0^T \alpha_s^2(W)ds < \infty) = 1$,

$$P\left(\left|\int_0^T \alpha_s(W)dW_s\right| < \infty\right) = 1$$

(see Note 7 in Subsection 4.2.6). Hence, from (7.37) it follows that

$$\mu_W\left\{x: \frac{d\mu_\xi}{d\mu_W}(x) = 0\right\} = 0.$$

Then, by Lemma 6.8, $\mu_W \ll \mu_\xi$ and

$$\frac{d\mu_W}{d\mu_\xi}(x) = \left[\frac{d\mu_\xi}{d\mu_W}(x)\right]^{-1},$$

which together with (7.37) and Lemma 4.6 yields (7.38).

Necessity: If $\mu_\xi \ll \mu_W$, then by Theorem 7.5 $P(\int_0^T \alpha_t^2(\xi)dt < \infty) = 1$. But since $\mu_\xi \sim \mu_W$ then obviously also $P(\int_0^T \alpha_t^2(W)dt < \infty) = 1$. □

7.2.4

In this subsection the conditions providing the absolute continuity of measure μ_W over measure μ_ξ will be discussed.

As a preliminary let us introduce the following notation. Let $\alpha = (\alpha_t(x), \mathscr{B}_t)$, $0 \le t \le T$, be a nonanticipative process and for each n, $n = 1, 2, \ldots$, let

$$\tau_n(x) = \begin{cases} \inf\left\{t \le T: \int_0^T \alpha_s^2(x)ds \ge n\right\}, \\ \infty, \quad \text{if} \int_0^T \alpha_s^2(x) < n, \end{cases} \qquad \tau(x) = \lim_{n\to\infty} \tau_n(x).$$

Theorem 7.8. *Let $\xi = (\xi_t, \mathscr{F}_t)$, $t \le T$, be a process of the diffusion type with the differential*

$$d\xi_t = \alpha_t(\xi)dt + dW_t, \qquad \xi_0 = 0,$$

where

$$P\left(\int_0^T |\alpha_t(\xi)|dt < \infty\right) = 1 \quad \text{and} \quad P(\tau_n(\xi) > 0) = 1, \qquad n = 1, 2, \ldots,$$

and, on the set $(\tau(\xi) \le T)$,

$$\lim_n \int_0^{\tau_n(\xi)} \alpha_t^2(\xi)dt = \infty.$$

Then

$$P\left(\int_0^T \alpha_t^2(W)dt < \infty\right) = 1 \Rightarrow \mu_W \ll \mu_\xi; \tag{7.39}$$

and if $\mathscr{F}_t^\xi = \mathscr{F}_t^W$, $0 \le t \le T$, then

$$P\left(\int_0^T \alpha_t^2(W)dt < \infty\right) = 1 \Leftarrow \mu_W \ll \mu_\xi. \tag{7.40}$$

PROOF. Because of the condition $P(\int_0^T \alpha_s^2(W)ds < \infty) = 1$,

$$P(\tau(W) = \infty) = 1.$$

But the condition $P(\int_0^T \alpha_s^2(\xi)ds < \infty) = 1$, generally speaking, is not satisfied, and hence $P(\tau(\xi) = \infty) \le 1$. From the condition $P(\tau_n(\xi) > 0) = 1$, $n = 1, 2, \ldots$, it only follows that $P(\tau(\xi) > 0) = 1$.

Denote

$$\chi_t^{(n)}(x) = \chi_{\{x: \int_0^t \alpha_s^2(x)ds < n\}}$$

and

$$\alpha_t^{(n)}(x) = \alpha_t(x)\chi_t^{(n)}(x).$$

Also set

$$\xi_t^{(n)} = \int_0^t \alpha_s^{(n)}(\xi)ds + W_t.$$

Since $\xi_t^{(n)} = \xi_t$ (P-a.s.) with $0 \le t \le \tau_n(\xi)$,

$$P\left(\int_0^t \alpha_s^{(n)}(\xi)ds = \int_0^t \alpha_s^{(n)}(\xi^{(n)})ds, 0 \le t \le T\right) = 1,$$

and, therefore,

$$d\xi_t^{(n)} = \alpha_t^{(n)}(\xi^{(n)})dt + dW_t, \qquad \xi_0^{(n)} = 0.$$

It is clear that

$$P\left(\int_0^T (\alpha_s^{(n)}(W))^2\, ds < \infty\right) = 1,$$

$$P\left(\int_0^T (\alpha_s^{(n)}(\xi^{(n)}))^2\, ds < \infty\right) = 1.$$

Hence, by Theorem 7.7, $\mu_{\xi^{(n)}} \sim \mu_W$ and

$$\frac{d\mu_W}{d\mu_{\xi^{(n)}}}(t, \xi^{(n)}) = \exp\left\{-\int_0^t \alpha_s^{(n)}(\xi^{(n)})d\xi_s^{(n)} + \frac{1}{2}\int_0^t (\alpha_s^{(n)}(\xi^{(n)}))^2\, ds\right\}. \tag{7.41}$$

Denote

$$\rho_t^{(n)}(x) = \frac{d\mu_W}{d\mu_{\xi^{(n)}}}(t, x).$$

Then, if $A \in \mathscr{B}_T$,

$$\begin{aligned}\mu_W(A) &= \lim_n \mu_W\{A \cap (\tau^{(n)}(x) = \infty)\}\\ &= \lim_n \int_{A \cap (\tau_n(x) = \infty)} \rho_T^{(n)}(x)d\mu_{\xi^{(n)}}(x) = \lim_n \int_{A \cap (\tau_n(x) = \infty)} \rho_T^{(n)}(x)d\mu_\xi(x)\\ &= \lim_n \int_A \rho_T^{(n)}(x)\chi_T^{(n)}(x)d\mu_\xi(x).\end{aligned}$$

It will be shown that the condition $P(\int_0^T \alpha_s^2(W)ds < \infty) = 1$ provides the uniform integrability of the family of the values $\{\rho_T^{(n)}(\xi)\chi_T^{(n)}(\xi), n = 1, 2, \ldots\}$.

For any $N > 1$, we have

$$\int_{\{\omega:\rho_T^{(n)}(\xi)\chi_T^{(n)}(\xi)>N\}} \rho_T^{(n)}(\xi)\chi_T^{(n)}(\xi)dP(\omega) = \int_{\{x:\rho_T^{(n)}(x)\chi_T^{(n)}(x)>N\}} \rho_T^{(n)}(x)\chi_T^{(n)}(x)d\mu_{\xi^{(n)}}(x)$$

$$\leq \mu_W\{x:\rho_T^{(n)}(x)\chi_T^{(n)}(x) > N\}$$

$$\leq \mu_W\{x:\rho_T^{(n)}(x) > N\}$$

$$= P\left\{-\int_0^T \alpha_s^{(n)}(W)dW_s + \frac{1}{2}\int_0^T (\alpha_s^{(n)}(W))^2\, ds > \ln N\right\}$$

$$\leq P\left\{\left|\int_0^T \alpha_s^{(n)}(W)dW_s\right| > \frac{\ln N}{2}\right\} + P\left\{\int_0^T (\alpha_s^{(n)}(W))^2\, ds > \ln N\right\}$$

$$\leq \frac{4}{\ln N} + 2P\left\{\int_0^T (\alpha_s^{(n)}(W))^2\, ds > \ln N\right\}$$

$$\leq \frac{4}{\ln N} + 2P\left\{\int_0^T \alpha_s^2(W)ds > \ln N\right\} \tag{7.42}$$

where the estimate

$$P\left\{\left|\int_0^T \alpha_s^{(n)}(W)dW_s\right| > \frac{\ln N}{2}\right\} \leq \frac{4}{\ln N} + P\left\{\int_0^T (\alpha_s^{(n)}(W))^2\, ds > \ln N\right\}$$

is being used (see Lemma 4.6).

Since $P\{\int_0^T \alpha_s^2(W)ds < \infty\} = 1$, from (7.42) it follows that the sequence of the values $\{\rho_T^{(n)}(\xi)\chi_T^{(n)}(\xi), n = 1, 2, \ldots\}$ is uniformly integrable.

Consider the variables

$$\rho_T^{(n)}(\xi)\chi_T^{(n)}(\xi) = \chi_{\{\int_0^T \alpha_s^2(\xi)ds<n\}} \exp\left\{-\int_0^T \alpha_s^{(n)}(\xi)dW_s - \frac{1}{2}\int_0^T (\alpha_s^{(n)}(\xi))^2\, d\xi\right\}.$$

From the results of Subsection 4.2.9, it follows that there exists

$$\Gamma_T(\alpha(\xi)) = P\text{-}\lim_n \chi_{\{\int_0^T \alpha_s^2(\xi)ds<n\}} \int_0^T \alpha_s^{(n)}(\xi)dW_s.$$

Hence, according to Note 1 to Theorem 1.3,

$$\lim_n \int_A \rho_T^{(n)}(x)\chi_T^{(n)}(x)d\mu_\xi(x) = \lim_n \int_{\{\omega:\xi(\omega)\in A\}} \rho_T^{(n)}(\xi)\chi_T^{(n)}(\xi)dP(\omega)$$

$$= \int_{\{\omega:\xi(\omega)\in A\}} \rho_T(\alpha(\xi))dP(\omega),$$

where $\rho_T(\xi) = \exp[-\Gamma_T(\alpha(\xi)) - \frac{1}{2}\int_0^T \alpha_s^2(\xi)ds]$. Consequently, $\mu_W \ll \mu_\xi$ and

$$\frac{d\mu_W}{d\mu_\xi}(\xi) = \exp\left[-\Gamma_T(\alpha(\xi)) - \frac{1}{2}\int_0^T \alpha_s^2(\xi)ds\right]. \tag{7.43}$$

Let us now prove (7.40). Let $\mu_W \ll \mu_\xi$ and $\mathscr{F}_t^\xi = \mathscr{F}_t^W$, $t \leq T$. Consider the derivative

$$\rho_t(\xi) = \frac{d\mu_W}{d\mu_\xi}(t, \xi), \qquad t \leq T.$$

Since the σ-algebras $\mathscr{F}_t^W$ and $\mathscr{F}_t^\xi$ coincide, there exists an $\mathscr{F}_t^W$-measurable function $\tilde{\rho}_t(W)$ such that $\tilde{\rho}_t(W) = \rho_t(\xi)$ (P-a.s.), $t \leq T$.

The process $(\rho_t(\xi), \mathscr{F}_t^\xi)$, $t \leq T$, is a nonnegative martingale. Consequently, the process $(\tilde{\rho}_t(W), \mathscr{F}_t^W)$, $t \leq T$, has the same property. By Theorem 5.7 there will be a process $\tilde{\gamma} = (\tilde{\gamma}_t(W), \mathscr{F}_t^W)$, $t \leq T$, with $P(\int_0^T \tilde{\gamma}_t^2(W)dt < \infty) = 1$, such that ($P$-a.s.)

$$\tilde{\rho}_t(W) = 1 + \int_0^t \tilde{\gamma}_s(W)dW_s. \tag{7.44}$$

According to Theorem 6.2 the process $\tilde{W} = (\tilde{W}_t, \mathscr{F}_t^W)$, $t \leq T$, with

$$\tilde{W}_t = W_t - \int_0^t \tilde{\beta}_s(W)ds, \qquad \tilde{\beta}_s(W) = \tilde{\rho}_s^+(W)\tilde{\gamma}_s(W), \tag{7.45}$$

considered on $(\Omega, \tilde{P})$, $\tilde{P}(d\omega) = \tilde{\rho}_T(W(\omega))P(d\omega)$, is a Wiener process. Here

$$\tilde{P}\left(\int_0^T \beta_s^2(W)ds < \infty\right) = 1. \tag{7.46}$$

Let us set $\gamma_s(\xi) = \tilde{\gamma}_s(W)$, $\beta_s(\xi) = \rho_s^+(\xi)\gamma_s(\xi)$. Then ($P$-a.s.) $\beta_s(\xi) = \tilde{\beta}_s(W)$, $s \leq T$. Hence from (7.45) and the equation

$$\xi_t = \int_0^t \alpha_s(\xi)ds + W_t, \tag{7.47}$$

it follows that

$$\tilde{W}_t - \xi_t = -\int_0^t [\alpha_s(\xi) + \beta_s(\xi)]ds. \tag{7.48}$$

The process $(\xi_t, \mathcal{F}_t^W)$, $t \le T$, considered on $(\Omega, \tilde{P})$ is also a Wiener process, since $\mathcal{F}_t^\xi = \mathcal{F}_t^W$ and

$$\tilde{P}(\xi \in \Gamma) = \int_{\{\omega:\xi(\omega)\in\Gamma\}} \rho_T(\xi(\omega))dP(\omega) = \int_\Gamma \frac{d\mu_W}{d\mu_\xi}(T, x)d\mu_\xi(x) = \mu_W(\Gamma).$$

Consequently, the process $(\tilde{W}_t - \xi_t, \mathcal{F}_t^W)$, $t \le T$, is a square integrable martingale and, because of (7.48) and Lemma 7.1, $\alpha(\xi) = \beta_s(\xi)$ (P-a.s.) with almost all $s \le T$. Hence

$$P\left(\int_0^T \alpha_t^2(W)dt < \infty\right) = \tilde{P}\left(\int_0^T \alpha_s^2(\xi)dt < \infty\right)$$
$$= \tilde{P}\left(\int_0^T \beta_t^2(\xi)dt < \infty\right) = \tilde{P}\left(\int_0^T \tilde{\beta}_t^2(W)dt < \infty\right) = 1,$$

which proves (7.40). □

7.2.5

Theorem 7.9. *Let $\xi = (\xi_t, \mathcal{F}_t)$, $t \le T$, be a process of the diffusion type with*

$$d\xi_t = \alpha_t(\xi)dt + dW_t, \qquad \xi_0 = 0, \tag{7.49}$$

where $P(\int_0^T |\alpha_t(\xi)|dt < \infty) = 1$, $P(\int_0^T \alpha_t^2(W)dt < \infty) = 1$, and the assumptions of Theorem 7.8 are fulfilled. Then the process

$$\rho_t(\xi) = \frac{d\mu_W}{d\mu_\xi}(t, \xi), \qquad t \le T,$$

is a unique solution of the equations

$$\rho_t(\xi) = 1 - \int_0^t \rho_s(\xi)\alpha_s(\xi)dW_s \qquad (P\text{-a.s.}), \tag{7.50}$$

$$\frac{d\mu_W}{d\mu_\xi}(t, W) = \exp\left(-\int_0^t \alpha_s(W)dW_s + \frac{1}{2}\int_0^t \alpha_s^2(W)ds\right) \qquad (P\text{-a.s.}), \tag{7.51}$$

$$\frac{d\mu_W}{d\mu_\xi}(t, \xi) = \exp\left(-\Gamma_t(\alpha(\xi)) - \frac{1}{2}\int_0^T \alpha_s^2(\xi)ds\right) \qquad (P\text{-a.s.}), \tag{7.52}$$

$$P\left(\int_0^T \alpha_s^2(\xi)ds < \infty\right) = M\exp\left(\int_0^T \alpha_s(W)dW_s - \frac{1}{2}\int_0^T \alpha_s^2(W)ds\right). \tag{7.53}$$

PROOF. The representation given by (7.52) was proved in the preceding theorem (see (7.43)). (7.51) follows from (7.52) and Lemma 4.10, if only it is noted that

$$P\text{-}\lim_n \chi_{\{\int_0^t \alpha_s^2(\xi)ds < \infty\}} \exp\left(-\int_0^t \alpha_s^{(n)}(\xi)dW_s - \frac{1}{2}\int_0^t (\alpha_s^{(n)}(\xi))^2\, ds\right)$$
$$= P\text{-}\lim_n \chi_{\{\int_0^t \alpha_s^2(\xi)ds < \infty\}} \exp\left(-\int_0^t \alpha_s^{(n)}(\xi)d\xi_s + \frac{1}{2}\int_0^t (\alpha_s^{(n)}(\xi))^2\, ds\right)$$

and that in the assumption $P(\int_0^T \alpha_s^2(W)ds < \infty) = 1$,

$$\Gamma_t(\alpha(W)) = \int_0^t \alpha_s(W)dW_s.$$

Equation (7.53) can be established in the same way as (7.30) in Theorem 7.6. The validity of Equation (7.50) can be proved in the same way as in Lemma 6.3. □

7.2.6

In the problems of sequential estimation (Sections 17.5–6) to be considered further on, there arises a question of the absolute continuity of measures corresponding to processes of the diffusion type where the time of the observation (T) is a random variable.

Let $(C, \mathscr{B})$ be a space of functions $x = (x_t)$ continuous on $[0, \infty)$, $t \geq 0$, $x_0 = 0$, $\mathscr{B}_t = \sigma\{x:x_s, s \leq t\}$, and let $\sigma = \sigma_x$ be a Markov time with respect to the system $(\mathscr{B}_t)$, $t \geq 0$.

It will be assumed that the process $\xi = (\xi_t)$, $t \geq 0$, has the differential

$$d\xi_t = \alpha_t(\xi)dt + dW_t, \qquad \xi_0 = 0, \tag{7.54}$$

where $P(\int_0^\infty |\alpha_t(\xi)|dt < \infty) = 1$. By $\mu_{\sigma,\xi}$ and $\mu_{\sigma,W}$ we denote the restrictions of the measures μ_ξ and μ_W on the algebra $\mathscr{B}_\sigma$.

Theorem 7.10

(1) *If* $P(\int_0^{\sigma_\xi} \alpha_s^2(\xi)ds < \infty) = 1$, *then* $\mu_{\sigma,\xi} \ll \mu_{\sigma,W}$ *and*

$$P\left(\int_0^{\sigma_W} \alpha_t^2(W)dt < \infty\right) = M \exp\left\{-\int_0^{\sigma_\xi} \alpha_t(\xi)dW_t - \frac{1}{2}\int_0^{\sigma_\xi} \alpha_t^2(\xi)dt\right\}, \tag{7.55}$$

where $\sigma_W = \sigma_{W(\omega)}$, $\sigma_\xi = \sigma_{\xi(\omega)}$.

(2) *If*

$$P\left(\int_0^{\sigma_\xi} \alpha_t^2(\xi)dt < \infty\right) = P\left(\int_0^{\sigma_W} \alpha_t^2(W)dt < \infty\right) = 1,$$

then $\mu_{\sigma,\xi} \sim \mu_{\sigma,W}$ *and* $(\mu_{\sigma,W})$ (P-a.s.)

$$\frac{d\mu_{\sigma,\xi}}{d\mu_{\sigma,W}}(\sigma, W) = \exp\left(\int_0^{\sigma_W} \alpha_s(W)dW_s - \frac{1}{2}\int_0^{\sigma_W} \alpha_s^2(W)ds\right). \tag{7.56}$$

PROOF. Let

$$\tilde{\alpha}_t(x) = \alpha_t(x)\chi_{\{t<\sigma_x\}}, \tag{7.57}$$

and let

$$\tilde{\xi}_t = \begin{cases} \xi_t, & t < \sigma_\xi, \\ \xi_{\sigma_\xi} + [W_t - W_{\sigma_\xi}], & t \geq \sigma_\xi, \end{cases} \tag{7.58}$$

i.e., $\tilde{\xi}_t = \int_0^{t \wedge \sigma_\xi} \alpha_s(\xi) ds + W_t$. It is not difficult to see that

$$d\tilde{\xi}_t = \tilde{\alpha}_t(\tilde{\xi})dt + dW_t. \tag{7.59}$$

According to the above assumption,

$$P\left(\int_0^{\sigma_\xi} \alpha_s^2(\xi)ds < \infty\right) = 1,$$

and, consequently,

$$P\left(\int_0^{\infty} \tilde{\alpha}_s^2(\tilde{\xi})ds < \infty\right) = 1. \tag{7.60}$$

Hence, by Theorem 7.5 (with $T = \infty$), $\mu_{\tilde{\xi}} \ll \mu_W$ and

$$P\left(\int_0^{\infty} \tilde{\alpha}_t^2(W)dt < \infty\right) = M \exp\left(-\int_0^{\infty} \tilde{\alpha}_t(\xi)dW_t - \frac{1}{2}\int_0^{\infty} \tilde{\alpha}_t^2(\xi)dt\right). \tag{7.61}$$

But $\mu_{\sigma,\xi}(A) = \mu_{\tilde{\xi}}(A)$ and $\mu_{\sigma,W}(A) = \mu_W(A)$ on the sets $A \in \mathscr{B}_{\sigma_x}$. Therefore $\mu_{\sigma,\xi} \ll \mu_{\sigma,W}$ and (7.55) follows from (7.61) and (7.57). □

Similarly, from Theorem 7.7 one can deduce a statement of the equivalence of measures $\mu_{\sigma,\xi}$ and $\mu_{\sigma,W}$, and also (7.56).

7.2.7

Let $W = (W_t, \mathscr{F}_t)$, $0 \le t \le T$, be an n-dimensional Wiener process, $W_t = (W_1(t), \ldots, W_n(t))$, and let $\xi_t = (\xi_1(t), \ldots, \xi_n(t))$ be a process with the differential

$$d\xi_t = \alpha_t(\xi)dt + dW_t, \qquad \xi_0 = 0,$$

where $\alpha_t(x) = (\alpha_1(t, x), \ldots, \alpha_n(t, x))$ is a vector formed of nonanticipative functionals.

Theorems 7.5–7.10 permit generalization to the multi-dimensional case under consideration. All the formulations remain the same, except for replacing $\alpha_t^2(x)$ by $\alpha_t^*(x)\alpha_t(x)$. Thus, for example, a multi-dimensional analog of (7.19) (Theorem 7.5) can be formulated as follows:

$$P\left(\int_0^{T} \alpha_t^*(\xi)\alpha_t(\xi)dt < \infty\right) = 1 \Leftrightarrow \mu_\xi \ll \mu_W. \tag{7.62}$$

7.3 The structure of processes whose measure is absolutely continuous with respect to Wiener measure

If $\xi = (\xi_t, \mathscr{F}_t)$, $0 \le t \le T$, is a process of the diffusion type with the differential

$$d\xi_t = \alpha_t(\xi)dt + dW_t, \qquad \xi_0 = 0, \tag{7.63}$$

then according to Theorem 7.5 the condition $P(\int_0^T \alpha_t^2(\xi)dt < \infty) = 1$ is necessary and sufficient for $\mu_\xi \ll \mu_W$. In this section it will be established that if some random process $\xi = (\xi_t, \mathscr{F}_t)$, $0 \le t \le T$, is such that its measure μ_ξ is absolutely continuous with respect to a Wiener measure μ_W, then this process is a process of the diffusion type. More precisely, we have:

Theorem 7.11. *On a complete probability space $(\Omega, \mathscr{F}, P)$ let there be prescribed a nondecreasing family of sub-σ-algebras $(\mathscr{F}_t)$, $0 \le t \le T$, a random process $\xi = (\xi_t, \mathscr{F}_t)$, and a Wiener process $W = (W_t, \mathscr{F}_t)$, $0 \le t \le T$, such that $\mu_\xi \ll \mu_W$. Then there will be a Wiener process $\hat{W} = (\hat{W}_t, \mathscr{F}_t^\xi)$ and a nonanticipative process $\alpha = (\alpha_t(x), \mathscr{B}_{t+})$, $0 \le t \le T$, such that*

$$\xi_t = \int_0^t \alpha_s(\xi)ds + \hat{W}_t \qquad (P\text{-a.s.}), \tag{7.64}$$

$$P\left(\int_0^T \alpha_s^2(\xi)ds < \infty\right) = 1. \tag{7.65}$$

If, in addition, $\mu_\xi \sim \mu_W$, then

$$P\left(\int_0^T \alpha_s^2(W)ds < \infty\right) = 1. \tag{7.66}$$

PROOF. By assumption $\mu_\xi \ll \mu_W$. Let

$$\varkappa_t(x) = \frac{d\mu_\xi}{d\mu_W}(t, x).$$

The process $\varkappa = (\varkappa_t(W), \mathscr{F}_t^W)$ is a nonnegative martingale with $M\varkappa_t(W) = 1$, and, according to Theorem 5.7, there exists a process $\gamma = (\gamma_t(\omega), \mathscr{F}_t^W)$ with $P(\int_0^T \gamma_t^2(\omega)dt < \infty) = 1$, such that ($P$-a.s.)

$$\varkappa_t(W) = 1 + \int_0^t \gamma_s(\omega)dW_s, \qquad 0 \le t \le T. \tag{7.67}$$

We shall consider a new probability space $(\Omega, \mathscr{F}_T^W, \tilde{P})$ with $d\tilde{P}(\omega) = \varkappa_T(W(\omega))dP(\omega)$ and define on it a random process $\tilde{W} = (\tilde{W}_t, \mathscr{F}_t^W)$ with

$$\tilde{W}_t = W_t - \int_0^t \alpha_s(W)ds,$$

where the functional $\alpha = (\alpha_s(x), \mathscr{B}_{s+})$ is such[7] that, (P-a.s.) for almost all $0 \le s \le T$, $\alpha_s(W) = \varkappa_s^+(W)\gamma_s(\omega)$. According to Theorem 6.2, the process $\tilde{W} = (\tilde{W}_t, \mathscr{F}_t^W)$, $0 \le t \le T$, is a Wiener process, where $\tilde{P}(\int_0^T \alpha_s^2(W)ds < \infty) = 1$ (see Subsection 6.3.3).

[7] The existence of such a functional follows from Lemma 4.9.

It will now be noted that $\mu_\xi(A) = \tilde{P}(W \in A)$, since

$$\tilde{P}(W \in A) = \int_{\{\omega: W \in A\}} \varkappa_T(W(\omega))dP(\omega) = \int_A \varkappa_T(x)d\mu_W(x) = \mu_\xi(A).$$

Hence

$$P\left(\int_0^T \alpha_s^2(\xi)ds < \infty\right) = \mu_\xi\left(\int_0^T \alpha_s^2(x)ds < \infty\right) = \tilde{P}\left(\int_0^T \alpha_s^2(W)ds < \infty\right) = 1, \tag{7.68}$$

which enables us to define a process

$$\hat{W}_t = \xi_t - \int_0^t \alpha_s(\xi)ds, \qquad 0 \le t \le T.$$

The process $\hat{W} = (\hat{W}_t, \mathscr{F}_t^\xi)$ on $(\Omega, \mathscr{F}, P)$ is a Wiener process; this can be shown in the same way as in Theorem 7.5.

Thus (7.64) and (7.65) of Theorem 7.11 are proved. (7.66) follows from the equivalence of the measures μ_ξ and μ_W and Equation (7.65). □

Note 1. From the theorem just proved it follows that, if the process $\xi = (\xi_t, \mathscr{F}_t)$ is such that its measure μ_ξ is absolutely continuous with respect to Wiener measure, then this process is necessarily a weak solution to an equation of the type given by (7.63).

Note 2. If $\mu_\xi \sim \mu_W$, then from Theorems 7.7 and 7.11 it follows that there exists a nonanticipative functional $\alpha = (\alpha_t(x), \mathscr{B}_{t+})$, $0 \le t \le T$, such that the densities

$$\frac{d\mu_\xi}{d\mu_W}(t, W) \quad \text{and} \quad \frac{d\mu_\xi}{d\mu_W}(t, \xi)$$

can be found by (7.37) and (7.38).

7.4 Representation of the Ito processes as processes of the diffusion type, innovation processes, and the structure of functionals on the Ito process

7.4.1

As indicated in Section 7.1 (Theorem 7.2) for the Ito processes $\xi = (\xi_t, \mathscr{F}_t)$, $0 \le t \le T$, with the differential

$$d\xi_t = \beta_t(\omega)dt + dW_t, \qquad \xi_0 = 0, \tag{7.69}$$

the condition $P(\int_0^T \beta_t^2(\omega)dt < \infty) = 1$ provides the absolute continuity of the measure μ_ξ w.r.t. a Wiener measure μ_W. However, it is not, generally speaking, feasible to obtain an explicit formula for the density $d\mu_\xi/d\mu_W$.

On the other hand, if the process ξ is a process of the diffusion type ($\beta_t(\omega) = \alpha_t(\xi(\omega))$), then, according to Theorem 7.6, for the densities $d\mu_\xi/d\mu_W$ one can give simple expressions (see (7.29) and (7.30)). In the same manner the structure of functionals on processes of the diffusion type may be adequately investigated (see Section 5.6). In studying functionals on Ito processes there arises immediately the question as to whether an Ito process can be represented as a process of the diffusion type (perhaps with respect to another Wiener process). The positive answer to this question is contained in the following theorem, in which the structure of functionals on the Ito process is described as well.

Theorem 7.12. *Let $\xi = (\xi_t, \mathscr{F}_t)$, $t \le T$, be an Ito process with the differential given by* (7.69), *where*

$$\int_0^T M|\beta_t(\omega)|dt < \infty. \tag{7.70}$$

Let $\alpha = (\alpha_t(x), \mathscr{B}_{t+})$, $0 \le t \le T$, be a functional such[8] *that for almost all t, $0 \le t \le T$,* (P-a.s.)

$$\alpha_t(\xi) = M(\beta_t | \mathscr{F}_t^\xi). \tag{7.71}$$

(1) *The random process $\overline{W} = (\overline{W}_t, \mathscr{F}_t^\xi)$, $0 \le t \le T$, with*

$$\overline{W}_t = \xi_t - \int_0^t \alpha_s(\xi)ds \tag{7.72}$$

is a Wiener process, and the process ξ is a process of the diffusion type with respect to the process $\overline{W}$:

$$d\xi_t = \alpha_t(\xi)dt + d\overline{W}_t. \tag{7.73}$$

(2) *If*

$$P\left(\int_0^T \beta_t^2(\omega)dt < \infty\right) = 1,$$

then any martingale $X = (x_t, \mathscr{F}_t^\xi)$, $0 \le t \le T$, permits a continuous modification for which we have the representation

$$x_t = x_0 + \int_0^t f_s(\omega)d\overline{W}_s, \tag{7.74}$$

where the process $f = (f_s(\omega), \mathscr{F}_s^\xi)$, $0 \le s \le T$, is such that

$$P\left(\int_0^T f_s^2(\omega)ds < \infty\right) = 1.$$

[8] The existence of such a functional follows from Lemma 4.9.

If, in addition, the martingale $X = (x_t, \mathscr{F}_t^\xi)$ is square integrable, then

$$\int_0^T Mf_s^2(\omega)ds < \infty.$$

PROOF. Because of (7.69) and (7.72),

$$\overline{W}_t = W_t + \int_0^t [\beta_s(\omega) - \alpha_s(\xi)]ds.$$

From this, by the Ito formula with $0 \le s \le t \le T$, we find

$$e^{iz(\overline{W}_t - \overline{W}_s)} = 1 + iz \int_0^t e^{iz(\overline{W}_u - \overline{W}_s)}\, dW_u$$
$$+ iz \int_s^t e^{iz(\overline{W}_u - \overline{W}_s)}[\beta_u(\omega) - \alpha_u(\xi)]du - \frac{z^2}{2}\int_s^t e^{iz(\overline{W}_u - \overline{W}_s)}\, du. \quad (7.75)$$

But

$$M\left[\int_s^t e^{iz(\overline{W}_u - \overline{W}_s)}\, dW_u | \mathscr{F}_s^\xi\right] = 0$$

and

$$M\left[\int_s^t e^{iz(\overline{W}_u - \overline{W}_s)}(\beta_u(\omega) - \alpha_u(\xi))du | \mathscr{F}_s^\xi\right]$$
$$= M\left[\int_s^t e^{iz(\overline{W}_u - \overline{W}_s)} M(\beta_u(\omega) - \alpha_u(\xi) | \mathscr{F}_u^\xi)du | \mathscr{F}_s^\xi\right] = 0.$$

Hence, taking in (7.75) the conditional expectation $M(\cdot | \mathscr{F}_s^\xi)$ from the left-hand and the right-hand sides, we find

$$M(e^{iz(\overline{W}_t - \overline{W}_s)} | \mathscr{F}_s^\xi) = 1 - \frac{z^2}{2}\int_s^t M(e^{iz(\overline{W}_u - \overline{W}_s)} | \mathscr{F}_s^\xi)du.$$

From this we obtain

$$M(e^{iz(\overline{W}_t - \overline{W}_s)} | \mathscr{F}_s^\xi) = e^{-(z^2/2)(t-s)} \qquad (P\text{-a.s.}), \qquad 0 \le s \le T.$$

Consequently, the process $\overline{W} = (\overline{W}_t, \mathscr{F}_t^\xi)$ is a Wiener process. To complete the proof it only remains to note that the representation given by (7.74) follows immediately from (7.73) and Theorems 5.20, 7.2 and 7.5. □

Corollary. *Let $\eta = \eta(\omega)$ be an $\mathscr{F}_T^\xi$-measurable random variable with $M|\eta| < \infty$, and let the second condition of Theorem 7.12 be satisfied. Then there will be a process*

$$f = (f_t(\omega), \mathscr{F}_t^\xi), \qquad 0 \le t \le T,$$

$$P\left(\int_0^T f_t^2(\omega)dt < \infty\right) = 1,$$

such that

$$\eta = M\eta + \int_0^T f_t(\omega)d\overline{W}_t.$$

If, in addition, $M\eta^2 < \infty$, *then* $\int_0^T Mf_t^2(\omega)dt < \infty$.

PROOF. It suffices to note that $x_t = M(\eta|\mathscr{F}_t^\xi)$ is a martingale, and that $x_0 = M\eta$, $x_T = \eta$ (P-a.s.). □

7.4.2

The feasibility of representation of Ito processes $\xi = (\xi_t, \mathscr{F}_t)$, $0 \le t \le T$, with the differential given by (7.69) as processes of the diffusion type (see (7.73)) with the Wiener process $\overline{W} = (\overline{W}_t, \mathscr{F}_t^\xi)$ is of great importance in deducing the general equations of optimal nonlinear filtering, interpolation, and extrapolation, (Chapter 8), and other scattered results (see, for example, Chapter 10). According to the definition of the process $\overline{W} = (\overline{W}_t, \mathscr{F}_t^\xi)$,

$$\mathscr{F}_t^{\overline{W}} \subseteq \mathscr{F}_t^\xi$$

for all t, $0 \le t \le T$.

In many cases the inverse inclusion

$$\mathscr{F}_t^{\overline{W}} \supseteq \mathscr{F}_t^\xi, \qquad 0 \le t \le T,$$

may be valid and, consequently,

$$\mathscr{F}_t^{\overline{W}} = \mathscr{F}_t^\xi, \qquad 0 \le t \le T.$$

The correspondence of the σ-algebras $\mathscr{F}_t^{\overline{W}}$ and $\mathscr{F}_t^\xi$, $0 \le t \le T$, indicates that the process $\overline{W}$ carries the same "information" as the process ξ does. This property of the process $\overline{W}$ justifies the following:

Definition. A Wiener process $\overline{W} = (\overline{W}_t, \mathscr{F}_t^\xi)$ is called an *innovation process* (*with respect to the process* $\xi = (\xi_t, \mathscr{F}_t)$, $0 \le t \le T$) if for each t, $0 \le t \le T$,

$$\mathscr{F}_t^{\overline{W}} = \mathscr{F}_t^\xi.$$

The investigation of the question when the Wiener process in (7.73) is an innovation process presents a crucial and difficult problem. If Equation (7.73) has a unique strong solution, then undoubtedly the process $\overline{W}$ will be an innovation process. However, to determine when this equation has a strong solution is as a rule rather difficult. One sufficiently general case of the correspondence of the σ-algebras $\mathscr{F}_t^{\overline{W}}$ and $\mathscr{F}_t^\xi$ will be treated in the next section (Theorem 7.16). As to the correspondence of these σ-algebras in other cases see Theorems 12.5 and 13.5.

EXAMPLE 2. Let $\xi = (\xi_t, \mathscr{F}_t)$, $0 \le t \le T$, have the differential

$$d\xi_t = \theta dt + dW_t, \qquad \xi_0 = 0,$$

where θ is an $\mathscr{F}_0$-measurable normal random variable, $N(m, \gamma)$, independent of the Wiener process $W = (W_t, \mathscr{F}_t)$. Then

$$M(\theta | \mathscr{F}_t^\xi) = \frac{m + \gamma\xi_t}{1 + \gamma_t}$$

(see, for example, Chapter 12, Theorem 12.2) and, consequently, the process ξ is a process of the diffusion type with the differential

$$d\xi_t = \frac{m + \gamma\xi_t}{1 + \gamma_t} dt + d\overline{W}_t. \tag{7.76}$$

One can immediately be convinced that in this example $\mathscr{F}_t^\xi = \mathscr{F}_t^{\overline{W}}$, $0 \le t \le T$.

7.4.3

Thus the conditions given in (7.70) guarantee that any Ito process is at the same time a process of the diffusion type (with respect to a Wiener process $\overline{W}$).

Let us make use of this fact for deducing formulas of the densities $d\mu_\xi/d\mu_W$ and $d\mu_W/d\mu_\xi$.

Theorem 7.13. *Let $\xi = (\xi_t, \mathscr{F}_t)$, $t \le T$, be an Ito process with the differential*

$$d\xi_t = \beta_t(\omega)dt + dW_t, \tag{7.77}$$

where

$$\int_0^T M|\beta_t(\omega)|dt < \infty, \tag{7.78}$$

$$P\left(\int_0^T \beta_t^2(\omega)dt < \infty\right) = 1. \tag{7.79}$$

If, in addition,

$$M \exp\left(-\int_0^T \beta_t\, dW_t - \frac{1}{2}\int_0^T \beta_t^2\, dt\right) = 1, \tag{7.80}$$

then

$$\mu_\xi \sim \mu_W, \qquad P\left(\int_0^T \alpha_s^2(\xi)ds < \infty\right) = P\left(\int_0^T \alpha_s^2(W)ds < \infty\right) = 1$$

and

$$\frac{d\mu_\xi}{d\mu_W}(t, W) = \exp\left(\int_0^t \alpha_s(W)dW_s - \frac{1}{2}\int_0^t \alpha_s^2(W)ds\right), \tag{7.81}$$

$$\frac{d\mu_\xi}{d\mu_W}(t, \xi) = \exp\left(\int_0^t \alpha_s(\xi)d\xi_s - \frac{1}{2}\int_0^t \alpha_s^2(\xi)ds\right), \tag{7.82}$$

where the functional $\alpha = (\alpha_t(x), \mathscr{B}_{t+})$, $t \le T$ is such that (P-a.s.) $\alpha_t(\xi) = M[\beta_t(\omega)|\mathscr{F}_t^\xi]$ for almost all $t \le T$.

PROOF. From (7.79), (7.80) and Theorem 7.1, it follows that $\mu_\xi \sim \mu_W$. Because of (7.77), (7.78) and Theorem 7.12, the process ξ is at the same time a process of the diffusion type with the differential (7.74) where $\overline{W}$ is a Wiener process. But the measures μ_W and $\mu_{\overline{W}}$ coincide, hence $\mu_\xi \sim \mu_{\overline{W}}$ and, by Theorem 7.7,

$$P\left(\int_0^T \alpha_s^2(\xi)ds < \infty\right) = P\left(\int_0^T \alpha_s^2(\overline{W})ds < \infty\right) = P\left(\int_0^T \alpha_s^2(W)ds < \infty\right) = 1,$$

where the functional $\alpha = (\alpha_t(x), \mathscr{B}_{t+})$, $t \le T$, is such that (P-a.s.) $\alpha_t(\xi) = M[\beta_t(\omega)|\mathscr{F}_t^\xi]$, $t \le T$, for almost all $t \le T$. (For the sake of avoiding misunderstanding, we note that, generally speaking, (P-a.s.) $\alpha_t(\overline{W}) \neq M(\beta_t(\omega)|\mathscr{F}_t^{\overline{W}})$. Indeed, because $\mathscr{F}_t^{\overline{W}} \subseteq \mathscr{F}_t^\xi$, then $M(\beta_t(\omega)|\mathscr{F}_t^{\overline{W}}) = M[M(\beta_t(\omega)|\mathscr{F}_t^\xi)|\mathscr{F}_t^{\overline{W}}] = M[\alpha_t(\xi)|\mathscr{F}_t^{\overline{W}})$, which might not be equal to $\alpha_t(\overline{W})$.) Equations (7.81), (7.82) follow from (7.37), (7.38) and the fact that

$$\frac{d\mu_\xi}{d\mu_W}(t, \xi) = \frac{d\mu_\xi}{d\mu_{\overline{W}}}(t, \xi) \qquad (P\text{-a.s.}), \qquad t \le T. \qquad \square$$

Note 1. Comparing (7.5) and (7.82), one can see that

$$M\left[\exp\left\{-\int_0^t \beta_s\, d\xi_s + \frac{1}{2}\int_0^t \beta_s^2\, ds\right\}\bigg|\mathscr{F}_t^\xi\right]$$
$$= \exp\left[-\int_0^t M(\beta_s|\mathscr{F}_s^\xi)d\xi_s + \frac{1}{2}\int_0^t (M(\beta_s|\mathscr{F}_s^\xi))^2\, ds\right]. \quad (7.83)$$

In other words, in (7.76)–(7.80) the conditional expectation entering the left-hand side of (7.83) can be "carried over" under the sign of an exponent.

Note 2. If $M \exp\{\frac{1}{2}\int_0^T \beta_s^2\, ds < \infty\} < \infty$, then (7.81) and (7.82) hold true.

For proving this it suffices to refer to Theorem 6.1 and note that from the condition $\int_0^T M|\beta_t(\omega)|dt < \infty$ it follows that $M|\beta_t(\omega)| < \infty$ for almost all $t \in [0, T]$. Without any restriction of generality one can consider $M|\beta_t(\omega)| < \infty$ for all $t \in [0, T]$ since otherwise, without changing the process ξ, one could pass to a new function $\tilde{\beta}_t(\omega)$ which, for almost all $t \in [0, T]$, coincides with $\beta_t(\omega)$ and, at other points t, is, for example, equal to zero.

Note 3. If the processes $\beta = (\beta_t(\omega), \mathscr{F}_t)$ and $W = (W_t, \mathscr{F}_t)$, $t \le T$, are independent, $P(\int_0^T \beta_t^2(\omega)dt < \infty) = 1$ and $M|\beta_t(\omega)| < \infty$, $\int_0^T M|\beta_t(\omega)|dt < \infty$, then the measures μ_ξ and μ_W are equivalent and (7.81) and (7.82) are valid.

For proof it suffices to note that, according to Example 4 in Section 6.2, (7.80) is satisfied.

EXAMPLE 3. Let us consider further the example from the previous subsection. (7.77) (7.80) are satisfied and hence (compare also with (7.12)), (P-a.s.)

$$\frac{d\mu_\xi}{d\mu_W}(t, \xi) = \exp\left(\int_0^t \frac{m + \gamma\xi_s}{1 + \gamma s}\, d\xi_s - \frac{1}{2}\int_0^t \left(\frac{m + \gamma\xi_s}{1 + \gamma s}\right)^2 ds\right),$$

$$\frac{d\mu_\xi}{d\mu_W}(t, W) = \exp\left(\int_0^t \frac{m + \gamma^{W_s}}{1 + \gamma s}\, dW_s - \frac{1}{2}\int_0^t \left(\frac{m + \gamma^{W_s}}{1 + \gamma s}\right)^2 ds\right).$$

7.4.4

Theorem 7.14. *Let $\xi = (\xi_t, \mathscr{F}_t)$, $t \le T$, be an Ito process with the differential given by (7.77) and let (7.78) and (7.79) be satisfied. Then*

$$P\left(\int_0^T \alpha_t^2(\xi)dt < \infty\right) = 1, \qquad \mu_\xi \ll \mu_W$$

and

$$\begin{aligned}\frac{d\mu_\xi}{d\mu_W}(t, W) &= \exp\left(\Gamma_t(W) - \frac{1}{2}\int_0^t \alpha_s^2(W)ds\right),\\ \frac{d\mu_\xi}{d\mu_W}(t, \xi) &= \exp\left(\int_0^t \alpha_s(\xi)d\xi_s - \frac{1}{2}\int_0^t \alpha_s^2(\xi)ds\right).\end{aligned} \tag{7.84}$$

PROOF. From (7.79) and Theorem 7.2 it follows that $\mu_\xi \ll \mu_W$. According to Theorem (7.12), ξ is a process of the diffusion type with the differential given by (7.74), where $\overline{W}$ is a Wiener process. But the measures μ_W and $\mu_{\overline{W}}$ coincide; hence, $\mu_\xi \ll \mu_{\overline{W}}$ and, by Theorem 7.5, $P(\int_0^T \alpha_t^2(\xi)dt < \infty) = 1$. (7.84) follows from Theorem 7.6. □

7.5 The case of Gaussian processes

7.5.1

In this section we shall consider the Ito processes $\xi = (\xi_t, \mathscr{F}_t)$, $0 \le t \le T$, with the differential

$$d\xi_t = \beta_t(\omega)dt + dW_t, \qquad \xi_0 = 0, \tag{7.85}$$

on the assumption that the process $\beta = (\beta_t(\omega), \mathscr{F}_t)$, $0 \le t \le T$, is Gaussian.

Theorem 7.15. *Let $\beta = (\beta_t(\omega), \mathscr{F}_t)$, $0 \le t \le T$, be a continuous (in the mean square) Gaussian process. Then $\mu_\xi \sim \mu_W$ and*

$$P\left(\int_0^T \alpha_t^2(\xi)dt < \infty\right) = P\left(\int_0^T \alpha_t^2(W)dt < \infty\right) = 1, \tag{7.86}$$

$$\frac{d\mu_\xi}{d\mu_W}(t, W) = \exp\left(\int_0^t \alpha_s(W)dW_s - \frac{1}{2}\int_0^t \alpha_s^2(W)ds\right), \tag{7.87}$$

$$\frac{d\mu_W}{d\mu_\xi}(t, \xi) = \exp\left(-\int_0^t \alpha_s(\xi)d\xi_s + \frac{1}{2}\int_0^t \alpha_s^2(\xi)ds\right), \tag{7.88}$$

where the functional $\alpha = (\alpha_t(x), \mathscr{B}_{t+})$ is such that (P-a.s.) $\alpha_t(\xi) = M[\beta_t(\omega)|\mathscr{F}_t^\xi]$ for almost all t, $0 \le t \le T$.

PROOF. By assumption the process $\beta_t = \beta_t(\omega)$, $0 \le t \le T$, is continuous in the mean square, hence $M\beta_t$ and $M\beta_t^2$ are continuous over t and

$$\int_0^T M\beta_t^2\, dt < \infty. \tag{7.89}$$

Consequently, $P(\int_0^T \beta_t^2\,dt < \infty) = 1$, and, by Theorem 7.2, $\mu_\xi \ll \mu_W$.
Further, in Section 6.2 it was shown (see Example 3(a)) that

$$M \exp\left(- \int_0^T \beta_s\,dW_s - \frac{1}{2}\int_0^T \beta_s^2\,ds \right) = 1.$$

Hence, because of Theorem 7.1, $\mu_\xi \sim \mu_W$.
Since

$$\int_0^T M\alpha_t^2(\xi)dt = \int_0^T M[M(\beta_t | \mathscr{F}_t^\xi)]^2\,dt \le \int_0^T M\beta_t^2\,dt < \infty,$$

(7.86) is satisfied and, therefore, the densities

$$\frac{d\mu_\xi}{d\mu_W}(t, W) \quad \text{and} \quad \frac{d\mu_W}{d\mu_\xi}(t, \xi)$$

are given by (7.87) and (7.88) according to Theorem 7.7. □

7.5.2

Let us remove now the assumption of continuity in the mean square of the Gaussian process $\beta_t(\omega)$, $t \le T$.

Theorem 7.16. *Let $\beta = (\beta_t(\omega), \mathscr{F}_t)$, $0 \le t \le T$, be a Gaussian process with*

$$P\left(\int_0^T \beta_t^2(\omega)dt < \infty \right) = 1. \tag{7.90}$$

(1) *Then $\mu_\xi \ll \mu_W$ and (P-a.s.)*

$$\frac{d\mu_\xi}{d\mu_W}(t, W) = \exp\left(\Gamma_t(\alpha_t(W)) - \frac{1}{2}\int_0^t \alpha_s^2(W)ds \right), \tag{7.91}$$

$$\frac{d\mu_\xi}{d\mu_W}(t, \xi) = \exp\left(\int_0^t \alpha_s(\xi)d\xi_s - \frac{1}{2}\int_0^t \alpha_s^2(\xi)ds \right). \tag{7.92}$$

(2) *If, in addition, the system $(\beta, W) = (\beta_t, W_t)$, $0 \le t \le T$, is Gaussian, then for all t, $0 \le t \le T$,*

$$\mathscr{F}_t^\xi = \mathscr{F}_t^{\overline{W}},$$

where $\overline{W} = (\overline{W}_t, \mathscr{F}_t^\xi)$ is a (innovation) process with

$$\overline{W}_t = \xi_t - \int_0^t \alpha_s(\xi)ds, \qquad \alpha_s(\xi) = M(\beta_s(\omega) | \mathscr{F}_s^\xi).$$

Here any martingale $X = (x_t, \mathscr{F}_t^\xi)$, $0 \le t \le T$, forming together with (β, W) a Gaussian system, can be represented in the form

$$x_t = x_0 + \int_0^t f(s)d\overline{W}_s, \tag{7.93}$$

where the deterministic function $f(s)$, $0 \le s \le T$, is such that $\int_0^T f^2(s)ds < \infty$.

Before proving this theorem let us assume the following lemma, which is of interest by itself.

Lemma 7.2. *Let $\beta_t = \beta_t(\omega)$, $0 \le t \le T$, be a measurable Gaussian process. Then*

$$P\left(\int_0^T \beta_s^2\, ds < \infty\right) = 1 \Leftrightarrow \int_0^T M\beta_s^2\, ds < \infty. \tag{7.94}$$

PROOF. Necessity is obvious. In proving sufficiency, it can be assumed that $M\beta_t \equiv 0$. Indeed, let us assume that it has been established that

$$P\left(\int_0^T \tilde{\beta}_s^2\, ds < \infty\right) = 1 \Rightarrow \int_0^T M\tilde{\beta}_s^2\, ds < \infty$$

for the Gaussian processes $\tilde{\beta}_t = \tilde{\beta}_t(\omega)$, $0 \le t \le T$, with $M\tilde{\beta}_t \equiv 0$. Then, along with the initial process β_t, we shall consider an independent Gaussian process $\bar{\beta}_t$, having the same distributions as the process β_t.

The process $\tilde{\beta}_t = \beta_t - \bar{\beta}_t$, $0 \le t \le T$, has the zero mean and therefore, from the condition $P(\int_0^T \beta_t^2\, dt < \infty) = P(\int_0^T \bar{\beta}_t^2\, dt < \infty) = 1$, it follows that

$$\int_0^T M(\beta_t - \bar{\beta}_t)^2\, dt < \infty.$$

But

$$\int_0^T M(\beta_t - \bar{\beta}_t)^2\, dt = 2\int_0^T [M\beta_t^2 - (M\beta_t)^2]dt = 2\int_0^T M(\beta_t - M\beta_t)^2\, dt.$$

Consequently

$$P\left(\int_0^T (\beta_t - M\beta_t)^2\, dt < \infty\right) = 1.$$

Since $M\beta_t = \beta_t - (\beta_t - M\beta_t)$,

$$\int_0^T (M\beta_t)^2\, dt \le 2\int_0^T \beta_t^2\, dt + 2\int_0^T (\beta_t - M\beta_t)^2\, dt.$$

The right side of this inequality is finite with probability one and, therefore, $\int_0^T (M\beta_t)^2\, dt < \infty$. Hence, if sufficiency is proved for processes with the zero mean, then from the condition $P(\int_0^T \beta_t^2\, dt < \infty) = 1$ it will follow that $\int_0^T (M\beta_t)^2\, dt < \infty$ and $\int_0^T M\tilde{\beta}_t^2\, dt < \infty$, where $\tilde{\beta}_t = \beta_t - M\beta_t$. Thus

$$\int_0^T M\beta_t^2\, dt \le 2\int_0^T M\tilde{\beta}_t^2\, dt + 2\int_0^T (M\beta_t)^2\, dt < \infty.$$

Thus we shall assume that $M\beta_t = 0$, $0 \le t \le T$.

Assume now also that the process β_t, $0 \le t \le T$, is continuous in the mean square. Let us show that

$$M \int_0^T \beta_t^2 \, dt \le \left[M \exp\left(- \int_0^T \beta_t^2 \, dt \right) \right]^{-2}. \tag{7.95}$$

Indeed, according to the Karhunen expansion (see [34], Chapter 5, Section 2), with $0 \le t \le T$, (P-a.s.)

$$\beta_t = \sum_{i=1}^{\infty} \eta_i \varphi_i(t),$$

where the $\{\varphi_i(t), i = 1, 2, \dots\}$ are orthonormal eigenfunctions of the kernel $M\beta_t \beta_s$:

$$\int_0^T M\beta_t \beta_s \varphi_i(s) ds = \lambda_i \varphi_i(t), \qquad \int_0^T \varphi_i(t) \varphi_j(t) dt = \delta(i - j),$$

and

$$\eta_i = \int_0^T \beta_t \varphi_i(t) dt$$

are independent Gaussian random variables with $M\eta_i = 0$ and $M\eta_i^2 = \lambda_i$. Then

$$M \int_0^T \beta_t^2 \, dt = M \int_0^T \left(\sum_{i=1}^{\infty} \eta_i \varphi_i(t) \right)^2 dt = \sum_{i=1}^{\infty} M\eta_i^2 = \sum_{i=1}^{\infty} \lambda_i. \tag{7.96}$$

It is easy to calculate that

$$\begin{aligned} 0 < M \exp\left(- \int_0^T \beta_t^2 \, dt \right) &= M \exp\left(- \int_0^T \left(\sum_{i=1}^{\infty} \eta_i \varphi_i(t) \right)^2 dt \right) \\ &= M \exp\left(- \sum_{i=1}^{\infty} \eta_i^2 \right) \\ &= \prod_{i=1}^{\infty} M \exp(-\eta_i^2) = \prod_{i=1}^{\infty} (1 + 2\lambda_i)^{-1/2}. \end{aligned} \tag{7.97}$$

Comparing the right-hand sides in (7.96) and (7.97) we arrive at (7.95).

Let now $\beta_t = \beta_t(\omega)$, $0 \le t \le T$, be an arbitrary Gaussian process (not necessarily continuous in the mean square) with $M\beta_t = 0$, $0 \le t \le T$, and $P(\int_0^T \beta_t^2 \, dt < \infty) = 1$. Denote by $f = (f_i(t), i = 1, 2, \dots, 0 \le t \le T)$ some complete orthonormal (in $L_2[0, T]$) system of continuous functions, and for $n = 1, 2, \dots$, set

$$\beta_t^{(n)} = \sum_{i=1}^{n} \alpha_i f_i(t),$$

where[9]

$$\alpha_i = \int_0^T \beta_t f_i(t) dt.$$

[9] As to the Gaussian behavior of the variables α_i and the variables η_i considered above, see a note to this lemma, below.

It is easy to check that for each n, $n = 1, 2, \ldots$, the processes $\beta_t^{(n)}, 0 \le t \le T$, are continuous in the mean square, and that

$$\lim_n \int_0^T [\beta_t - \beta_t^{(n)}]^2\,dt = 0, \qquad \int_0^T \beta_t^2\,dt = \lim_n \int_0^T (\beta_t^{(n)})^2\,dt \qquad (P\text{-a.s.}). \tag{7.98}$$

Then, from the inequality proved above,

$$M\int_0^T (\beta_t^{(n)})^2\,dt \le \left[M\exp\left(-\int_0^T (\beta_t^{(n)})^2\,dt\right)\right]^{-2},$$

and, by the Fatou lemma, we infer that (7.95) is also valid without an assumption on the continuity in the mean square of the process $\beta_t, 0 \le t \le T$.

From this inequality it follows that

$$P\left(\int_0^T \beta_s^2\,ds < \infty\right) = 1 \Rightarrow \int_0^T M\beta_s^2\,ds < \beta. \qquad \square$$

Note. In proving Lemma 7.2 we made use of the fact that the random variables $\alpha = \int_0^T \beta_t\varphi(t)dt$ are Gaussian.[10] The Gaussian behavior of the variable α can be proved in the following way. Denote

$$\eta(t) = \beta_t\varphi(t), \qquad \eta_\varepsilon(t) = \frac{\eta(t)}{1 + \varepsilon\sqrt{M\eta^2(t)}} \quad (\varepsilon > 0), \qquad \alpha^\varepsilon = \int_0^T \eta_\varepsilon(t)dt.$$

Then (P-a.s.)

$$|\alpha - \alpha^\varepsilon| = \left|\int_0^T [\eta(t) - \eta_\varepsilon(t)]dt\right| \le \int_0^T |\eta(t)| \frac{\varepsilon\sqrt{M\eta^2(t)}}{1 + \varepsilon\sqrt{M\eta^2(t)}}\,dt \to 0, \qquad \varepsilon \downarrow 0.$$

since, for each t,

$$1 > \frac{\varepsilon\sqrt{M\eta^2(t)}}{1 + \varepsilon\sqrt{M\eta^2(t)}} \to 0, \qquad \varepsilon \downarrow 0,$$

$$\int_0^T |\eta(t)|dt = \int_0^T |\beta_t\varphi(t)|dt \le \left(\int_0^T \beta_t^2\,dt \cdot \int_0^T \varphi^2(t)dt\right)^{1/2} < \infty \qquad (P\text{-a.s.})$$

and the theorem on bounded convergence (Theorem 1.4) can be applied.

For proving the Gaussian behavior of the variable α, if suffices to check that a distribution of the variables α^ε for $\varepsilon > 0$ is Gaussian.

It is easy to calculate that due to the Gaussian behavior of the process $\eta_\varepsilon(t), 0 \le t \le T$, for each n, $n = 1, 2, \ldots$, with $\varepsilon > 0$,

$$\int_0^T M|\eta_\varepsilon(t)|^n\,dt < \infty.$$

[10] Note that this is a nontrivial fact, since the integral $\int_0^T \beta_t\varphi(t)dt$ is a Lebesgue integral for fixed ω, and not a Riemann integral.

It is well known[11] that, in satisfying the condition

$$\int_0^T M|\eta_\varepsilon(t)|^k\, dt < \infty,$$

the kth semi-invariant $S_{\alpha^\varepsilon}^{(k)}$ of the random variable $\alpha^\varepsilon = \int_0^T \eta_\varepsilon(t)dt$ is expressed in terms $S_{\eta_\varepsilon}^{(k)}(t_1, \ldots, t_k)$ of the vector $(\eta_\varepsilon(t_1), \ldots, \eta_\varepsilon(t_k))$ by the formula

$$S_{\alpha^\varepsilon}^{(k)} = \int_0^T \cdots \int_0^T S_{\eta_\varepsilon}^{(k)}(t_1, \ldots, t_k)dt_1, \ldots, dt_k.$$

But the random vector $(\eta_\varepsilon(t_1), \ldots, \eta_\varepsilon(t_k))$ is Gaussian and therefore $S_\eta^{(k)}(t_1, \ldots, t_k) = 0$, $k \geq 3$. Therefore only the first two semi-invariants $S_{\alpha^\varepsilon}^{(1)}$, $S_{\alpha^\varepsilon}^{(2)}$ of the variable α^ε can be different from zero and henceforth the random variable α^ε has a Gaussian distribution.

PROOF OF THEOREM 7.16. From (7.90) and Lemma 7.2 it follows that

$$\int_0^T M\beta_t^2\, dt < \infty.$$

Hence (7.91) and (7.92) follow immediately from Theorem 7.14.

Let us pass to the proof of Theorem 7.16 (2). Let the functional $\alpha = (\alpha_t(x), \mathscr{B}_{t+})$, $0 \leq t \leq T$, $x \in C_T$, be such that $\alpha_t(\xi) = M(\beta_t(\omega)|\mathscr{F}_t^\xi)$ (P-a.s.). Then, because of (7.73),

$$\xi_t = \int_0^t \alpha_s(\xi)ds + \overline{W}_t. \tag{7.99}$$

It is clear that $\mathscr{F}_t^\xi \supseteq \mathscr{F}_t^{\overline{W}}$. We will now show the validity of the inverse inclusions $\mathscr{F}_t^\xi \subseteq \mathscr{F}_t^{\overline{W}}$. For this purpose it will be noted that for each t, $0 \leq t \leq T$, the random variable $\eta = \alpha_t(\xi)$ is $\mathscr{F}_t^\xi$-measurable and, by the theorem on the normal correlation (Theorem 13.1), the system (η, W, ξ) is Gaussian. Then, by Corollary 2 of Theorem 5.21,

$$\alpha_t(\xi) = M\alpha_t(\xi) + \int_0^t G(t, s)d\overline{W}_s,$$

where the deterministic function $G(t, s)$ is such that $\int_0^t G^2(t, s)ds < \infty$. Consequently[12] $\alpha_t(\xi)$ is $\mathscr{F}_t^{\overline{W}}$-measurable. From this it follows that the integral $\int_0^t \alpha_s(\xi)ds$ is also $\mathscr{F}_t^{\overline{W}}$-measurable, and, because of (7.99), $\mathscr{F}_t^\xi \subseteq \mathscr{F}_t^{\overline{W}}$. Thus for all t, $0 \leq t \leq T$, the σ-algebras $\mathscr{F}_t^\xi$ and $\mathscr{F}_t^{\overline{W}}$ coincide. The feasibility of the representation given by (7.93) follows from Theorem 5.21. □

[11] See [103], [164].

[12] All the σ-algebras under investigation are considered to be augmented by sets from $\mathscr{F}$ of zero probability.

Corollary. *If $\eta = \eta(\omega)$ is an $\mathscr{F}_T^\xi$-measurable Gaussian random variable and the system (η, β, W) is Gaussian, then*

$$\eta = M\eta + \int_0^T f_T(s)d\overline{W}_s,$$

where the function $f_T(s)$, $0 \le s \le T$, is such that $\int_0^T f_T^2(s)ds < \infty$.

7.5.3

Note. If the mutual distribution of the processes β and W is Gaussian, then from (7.90) it follows that the measures μ_ξ and μ_W are equivalent ($\mu_\xi \sim \mu_W$). Indeed, in this case the process ξ is Gaussian. And for Gaussian processes their measures are either equivalent or singular (see [57]). But $\mu_\xi \ll \mu_W$, hence $\mu_\xi \sim \mu_W$. This result could be also obtained in a direct way, since in the case considered it is not difficult to check that not only $\int_0^T M\alpha_t^2(\xi)dt < \infty$, but also $\int_0^T M\alpha_t^2(W)dt < \infty$. Hence the equivalence of the measures μ_ξ and μ_W follows from Theorem 7.7 and the density of the measures

$$\frac{d\mu_\xi}{d\mu_W}(t, W) \quad \text{and} \quad \frac{d\mu_W}{d\mu_\xi}(t, \xi)$$

are given by (7.87) and (7.88).

7.6 The absolute continuity of measures of the Ito processes with respect to measures corresponding to processes of the diffusion type

7.6.1

The results of the previous sections permit extension to wider classes of Ito processes and processes of the diffusion type.

In accordance with Definition 6, given in Section 4.2, the process $\xi = (\xi_t, \mathscr{F}_t)$, $0 \le t \le T$, is an Ito process if for any t, $0 \le t \le T$, (P-a.s.)

$$\xi_t = \xi_0 + \int_0^t A_s(\omega)ds + \int_0^t B_s(\omega)dW_s, \tag{7.100}$$

where the processes $A = (A_s(\omega), \mathscr{F}_s)$ and $B = (B_s(\omega), \mathscr{F}_s)$ are such that (P-a.s.)

$$\int_0^T |A_s(\omega)|ds < \infty, \tag{7.101}$$

$$\int_0^T B_s^2(\omega)ds < \infty. \tag{7.102}$$

In the case where for almost all $s \le T$ the values $A_s(\omega)$ and $B_s(\omega)$ are $\mathscr{F}_s^\xi$-measurable, the Ito process is called a process of the diffusion type (Definition 7, Section 4.2). For the case $B_s(\omega) \equiv 1$ in Theorem 7.12 the conditions were given under which the Ito process was at the same time a process of the diffusion type (with respect to a Wiener process $\overline{W} = (\overline{W}_t, \mathscr{F}_t^\xi)$). For the processes given by (7.100) this result can be generalized as follows.

Theorem 7.17. *Let $\xi = (\xi_t, \mathscr{F}_t)$ be an Ito process given by* (7.100) *and let $v = (v_t, \mathscr{F}_t)$, $0 \le t \le T$, be some Wiener process independent of a Wiener process W and processes A and B. Let the following conditions be satisfied*:

$$\int_0^T M|A_t(\omega)|dt < \infty, \tag{7.103}$$

Then there will be:

(1) *measurable functionals $\bar{A} = (\bar{A}_t(x), \mathscr{B}_{t+})$ and $\bar{B} = (\bar{B}_t(x), \mathscr{B}_{t+})$, $0 \le t \le T$, satisfying* (P-a.s.) *for almost all t, $0 \le t \le T$, the equalities*

$$\bar{A}_t(\xi) = M(A_t(\omega)|\mathscr{F}_t^\xi), \tag{7.104}$$

$$\bar{B}_t(\xi) = \sqrt{B_t^2(\omega)}; \tag{7.105}$$

(2) *a Wiener process $\overline{W} = (\overline{W}_t, \mathscr{F}_t^{\xi, v})$, $0 \le t \le T$, such that the process ξ permits the representation*

$$\xi_t = \xi_0 + \int_0^t \bar{A}_s(\xi)ds + \int_0^t \bar{B}_s(\xi)d\overline{W}_s. \tag{7.106}$$

If, in addition, $B_t^2(\omega) > 0$ (P-a.s.) for almost all t, $0 \le t \le T$, then the Wiener process $\overline{W}$ is adapted to the family $\mathscr{F}^\xi = (\mathscr{F}_t^\xi)$, $0 \le t \le T$.

Proof. Because of (7.103), $M|A_t(\omega)| < \infty$ for almost all t. (Without loss of generality it can be assumed that $M|A_t(\omega)| < \infty$ for all t, substituting, if necessary, $A_t(\omega)$ for a corresponding modification). Then the existence of the required functional $\bar{A}$ follows from Lemma 4.9.

For proving the validity of Equation (7.105) it suffices to make sure that the values $B_t^2(\omega)$ for almost all t, $0 \le t \le T$, are $\mathscr{F}_t^\xi$-measurable. For this purpose we shall decompose the length $[0, t]$ into n segments, $0 \equiv t_0^{(n)} < t_1^{(n)} < t_n^{(n)} \equiv t$, so that $\max_j [t_{j+1}^{(n)} - t_j^{(n)}] \to 0, n \to \infty$.

Consider the sum

$$\sum_{j=0}^{n-1} [\xi_{t_{j+1}^{(n)}} - \xi_{t_j^{(n)}}]^2 = \sum_{j=0}^{n-1} \left(\int_{t_j^{(n)}}^{t_{j+1}^{(n)}} A_s(\omega)ds\right)^2 + 2\sum_{j=0}^{n-1} \left(\int_{t_j^{(n)}}^{t_{j+1}^{(n)}} B_s(\omega)dW_s\right)\left(\int_{t_j^{(n)}}^{t_{j+1}^{(n)}} A_s(\omega)ds\right) + \sum_{j=0}^{n-1} \left(\int_{t_j^{(n)}}^{t_{j+1}^{(n)}} B_s(\omega)dW_s\right)^2. \tag{7.107}$$

The first two items in the right-hand side of (7.107) tend to zero with $n \to \infty$ with probability one since

$$\sum_{j=0}^{n-1}\left(\int_{t_j^{(n)}}^{t_{j+1}^{(n)}} A_s(\omega)ds\right)^2 \le \max_j \int_{t_j^{(n)}}^{t_{j+1}^{(n)}} |A_s(\omega)|ds \cdot \int_0^T |A_s(\omega)|ds \to 0, \qquad n \to \infty,$$

and, similarly,

$$\left|\sum_{j=0}^{n-1}\left(\int_{t_j^{(n)}}^{t_{j+1}^{(n)}} B_s(\omega)dW_s\right)\left(\int_{t_j^{(n)}}^{t_{j+1}^{(n)}} A_s(\omega)ds\right)\right| \le \max_j \left|\int_{t_j^{(n)}}^{t_{j+1}^{(n)}} B_s(\omega)dW_s\right|$$

$$\cdot \int_0^T |A_s(\omega)|ds \to 0, \qquad n \to 0.$$

The last item in the right-hand side of (7.107) can be rewritten with the help of the Ito formula in the following form:

$$\sum_{j=0}^{n-1}\left(\int_{t_j^{(n)}}^{t_{j+1}^{(n)}} B_s(\omega)dW_s\right)^2 = \sum_{j=0}^{n-1}\int_{t_j^{(n)}}^{t_{j+1}^{(n)}} B_s^2(\omega)ds$$

$$+ 2\sum_{j=0}^{n-1}\int_{t_j^{(n)}}^{t_{j+1}^{(n)}}\left(\int_{t_j^{(n)}}^{s} B_u(\omega)dW_u\right)B_s(\omega)dW_s$$

$$= \int_0^t B_s^2(\omega)ds + 2\int_0^t f_n(s)B_s(\omega)dW_s,$$

where

$$f_n(s) = \int_{t_j^{(n)}}^{s} B_u(\omega)dW_u, \qquad t_j^{(n)} \le s < t_{j+1}^{(n)}.$$

Because

$$\int_0^T f_n^2(s)B_s^2(\omega)ds \le \left(\max_j \sup_{t_j^{(n)} \le s < t_{j+1}^{(n)}} f_n^2(s)\right) \cdot \int_0^T B_s^2(\omega)ds \to 0,$$

the $P\text{-}\lim_n \int_0^t f_n(s)B_s(\omega)dW_s = 0$ and the last item in the right-hand side of (7.107) tends in probability to $\int_0^t B_s^2(\omega)ds$ with $n \to \infty$. The left-hand side of Equation (7.107) for each n, $n = 1, 2, \ldots,$ is $\mathcal{F}_t^\xi$-measurable. Therefore, the $\int_0^t B_s^2(\omega)ds$ are $\mathcal{F}_t^\xi$-measurable for each t, $0 \le t \le T$. From this (see proof of Lemma 5.2) there follows the existence of a process $\tilde{B}^2 = (\tilde{B}_s^2(\omega), \mathcal{F}_s^\xi)$, $0 \le s \le t$, such that for almost all s, $0 \le s \le t$, $\tilde{B}_s^2(\omega) = B_s^2(\omega)$ (P-a.s.). Then the existence of the sought functional $\bar{B}$ follows from Lemma 4.9.

Let us consider now the random process $\eta = (\eta_t, \mathcal{F}_t^{\xi\,\nu})$ defined by the equation

$$\eta_t = \xi_t - \xi_0 - \int_0^t \bar{A}_s(\xi)ds \tag{7.108}$$

and show that the process $\eta^{(n)} = (\eta_t^{(n)}, \mathscr{F}_t^{\xi,\nu}), t \le T$ with $\eta_t^{(n)} = \eta_{t \wedge \tau_n}$, where $\tau_n = \inf(t: \int_0^t \bar{B}_s^2(\xi)ds + \sup_{s \le t}|\eta_s| \ge \eta)$ and $\tau_n = \infty$ if

$$\int_0^T \bar{B}_s^2(\xi)ds + \sup_{s \le t}|\eta_s| < n,$$

is a square integrable martingale.

Indeed, from (7.100), (7.108) and by independence of ν and (W, A, B) we obtain

$$\begin{aligned} M(\eta_t^{(n)} - \eta_s^{(n)} | \mathscr{F}_s^{\xi,\nu}) &= M(\eta_t^{(n)} - \eta_s^{(n)} | \mathscr{F}_s^{\xi}) \\ &= M\left(\int_s^t \chi(\tau_n \ge u)[A_u(\omega) - \bar{A}_u(\xi)]du \,|\, \mathscr{F}_s^{\xi}\right) \\ &\quad + M\left(\int_s^t \chi(\tau_n \ge u)B_u(\omega)dW_u | \mathscr{F}_s^{\xi}\right) \\ &= M\left(\int_s^t \chi(\tau_n \ge u)M[A_u(\omega) - \bar{A}_u(\xi) | \mathscr{F}_u^{\xi}]du \,|\, \mathscr{F}_s^{\xi}\right) \\ &\quad + M\left[M\left(\int_s^t \chi(\tau_n \ge uB_u(\omega)dW_u | \mathscr{F}_s\right) | \mathscr{F}_s^{\xi}\right] = 0. \end{aligned}$$

Further, by the Ito formula

$$(\eta_t^{(n)})^2 = 2\int_0^{t \wedge \tau_n} \eta_s d\eta_s + \int_0^{t \wedge \tau_n} B_s^2(\omega)ds.$$

From this and (7.105) it follows that

$$\langle \eta^{(n)} \rangle_t = \int_0^{t \wedge \tau_n} \bar{B}_s^2(\xi)\, ds.$$

Let us consider the process $\bar{W} = (\bar{W}_t, \mathscr{F}_t^{\xi,\nu}), t \le T$, with

$$\bar{W}_t = \int_0^t \bar{W}_s^+(\xi)d\eta_s + \int_0^t [1 - \bar{B}_s^+(\xi)\bar{B}_s(\xi)]d\nu_s \tag{7.109}$$

and show that $\bar{W}$ is a Wiener process. By the Ito formula

$$\begin{aligned} e^{iz\bar{W}_{t \wedge \tau_n}} &= e^{iz\bar{W}_{s \wedge \tau_n}} + iz\int_s^t \chi(\tau_n \ge u)e^{iz\bar{W}_u}\bar{B}_u^+(\xi)d\eta_u \\ &\quad + iz\int_s^t \chi(\tau_n \ge u)e^{iz\bar{W}_u}[1 - \bar{B}_u^+(\xi)\bar{B}_u(\xi)]d\nu_u - \frac{z^2}{2}\int_s^t \chi(\tau_n \ge u)e^{iz\bar{W}_u}du. \end{aligned}$$

As in the proof of Theorem 7.12, computing the conditional expectation $M(\cdot | \mathscr{F}_s^{\xi,\nu})$ from the left-hand and right-hand sides of this equation and passing to the limit as $n \to \infty$ we obtain

$$M(e^{iz\bar{W}_t} | \mathscr{F}_s^{\xi,\nu}) = e^{iz\bar{W}_s} - \frac{z^2}{2}\int_s^t M(e^{iz\bar{W}_u} | \mathscr{F}_s^{\xi,\nu})du,$$

$$M(e^{iz(\bar{W}_t - \bar{W}_s)} | \mathscr{F}_s^{\xi,\nu}) = e^{-(z^2/2)(t-s)},$$

which proves the Wiener behavior of the process $\bar{W}$.

For proving (7.106) it will be noted that, because of (7.108).

$$\int_0^t \bar{B}_s(\xi)d\bar{W}_s = \int_0^t \bar{B}_s(\xi)\bar{B}_s^+(\xi)d\xi_s - \int_0^t \bar{B}_s(\xi)\bar{B}_s^+(\xi)\bar{A}_s(\xi)ds$$

$$= \xi_t - \xi_0 - \int_0^t \bar{A}_s(\xi)ds + \zeta_t, \tag{7.110}$$

where

$$\zeta_t = \int_0^t [1 - \bar{B}_s(\xi)\bar{B}_s^+(\xi)]\, d\eta_t.$$

The process $(\zeta_{t \wedge \tau_n}, \mathscr{F}^{\xi,\nu})$, $t \le T, n = 1, 2, \ldots$ is a square-integrable martingale and by (3.8)

$$M \sup_{\tau \le T} \zeta_{t\wedge\tau_n}^2 \le 4M \int_0^{T\wedge\tau_n} (1 - \bar{B}_s^+(\xi))\bar{B}_s^2(\xi)ds = 0. \tag{7.111}$$

Therefore,

$$P\left\{\sup_{t \le T} |\zeta_t| = 0\right\} \le P\{\tau_n < T\} \to 0, \qquad n \to \infty. \tag{7.112}$$

From (7.112) and (7.110) follows (7.106) for the process ξ. But if $B_t^2(\omega) > 0$ (P-a.s.) for almost all t, $0 \le t \le T$, then, from the definition of the process $\bar{W}$, it follows that the $\bar{W}_t$ are $\mathscr{F}_t^\xi$-measurable for each $0 \le t \le T$. □

7.6.2

An immediate generalization of Theorem 7.1 is the following:

Theorem 7.18. *Let $\xi = (\xi_t, \mathscr{F}_t)$ be an Ito process with the differential*

$$d\xi_t = A_t(\omega)dt + b_t(\xi)dW_t, \tag{7.113}$$

where $\eta = (\eta_t, \mathscr{F}_t)$ is a process of the diffusion type with

$$d\eta_t = a_t(\eta)dt + b_t(\eta)dW_t, \qquad \eta_0 = \xi_0, \tag{7.114}$$

and ξ_0 is an $\mathscr{F}_0$-measurable random variable with $P(|\xi_0| < \infty) = 1$. Let the following assumptions be fulfilled:

(I) *the nonanticipative functionals $a_t(x)$ and $b_t(x)$ satisfy (4.110) and (4.111), providing the existence and uniqueness of a strong solution of Equation (7.114);*

(II) *for any t, $0 \le t \le T$, the equation*

$$b_t(\xi)\alpha_t(\omega) = A_t(\omega) - a_t(\xi) \tag{7.115}$$

has (with respect to $\alpha_t(\omega)$) (P-a.s.) a bounded solution;

(III)

$$P\left(\int_0^T \alpha_t^2(\omega)dt < \infty\right) = 1; \tag{7.116}$$

(IV)

$$M \exp\left(-\int_0^T \alpha_t(\omega)dW_t - \frac{1}{2}\int_0^T \alpha_t^2(\omega)dt\right) = 1. \tag{7.117}$$

Then $\mu_\xi \sim \mu_\eta$ *and* (P-a.s.)

$$\frac{d\mu_\eta}{d\mu_\xi}(\xi) = M\left\{\exp\left(-\int_0^T \alpha_t(\omega)dW_t - \frac{1}{2}\int_0^T \alpha_t^2(\omega)dt\right)\Big| \mathscr{F}_T^\xi\right\}. \tag{7.118}$$

PROOF. Note first of all that the solution of Equation (7.115) can be represented as

$$\alpha_t(\omega) = b_t^+(\xi)[A_t(\omega) - a_t(\xi)], \tag{7.119}$$

where

$$b_t^+(\xi) = \begin{cases} b_t^{-1}(\xi), & b_t(\xi) \neq 0, \\ 0, & b_t(\xi) = 0. \end{cases} \tag{7.120}$$

Denote

$$\varkappa_t = \exp\left(-\int_0^t \alpha_s(\omega)dW_s - \frac{1}{2}\int_0^t \alpha_s^2(\omega)ds\right),$$

$$d\tilde{P}(\omega) = \varkappa_T(\omega)dP(\omega).$$

By Theorem 6.3 the process

$$\tilde{W}_t = W_t + \int_0^t \alpha_s(\omega)ds$$

is a Wiener process (with respect to the system $(\mathscr{F}_t)$, $0 \le t \le T$, and measure $\tilde{P}$). We have ($\tilde{P}$-a.s.)

$$\begin{aligned}\eta_0 + \int_0^t a_s(\xi)ds + \int_0^t b_s(\xi)d\tilde{W}_s &= \eta_0 + \int_0^t a_s(\xi)ds + \int_0^t b_s(\xi)\alpha_s(\omega)ds + \int_0^t b_s(\xi)dW_s \\ &= \eta_0 + \int_0^t a_s(\xi)ds + \int_0^t b_s(\xi)b_s^+(\xi) \\ &\quad \times [A_s(\omega) - a_s(\xi)]ds + \int_0^t b_s(\xi)dW_s \\ &= \eta_0 + \int_0^t A_s(\omega)ds + \int_0^t b_s(\xi)dW_s = \xi_t.\end{aligned}$$

In other words, the process $\xi = (\xi_t, \mathscr{F}_t)$, considered on a probability space $(\Omega, \mathscr{F}, \tilde{P})$, satisfies the same equation that the process $\eta = (\eta_t, \mathscr{F}_t)$ on $(\Omega, \mathscr{F}, P)$ does. Hence, because of assumption (I), $\tilde{P}(\xi \in A) = P\{\eta \in A\}$, and therefore,

$$\mu_\eta(A) = P\{\eta \in A\} = \tilde{P}\{\xi \in A\} = \int_{\{\omega : \xi \in A\}} \varkappa_T(\omega) dP(\omega)$$

$$= \int_A M(\varkappa_T | \mathscr{F}_T^\xi)_{\xi = x} \, d\mu_\xi(x). \qquad (7.121)$$

From this it follows that $\mu_\eta \ll \mu_\xi$, and also (7.118) holds. The absolute continuity of the measure μ_ξ w.r.t. μ_η can be proved in the same way as in Theorem 7.1. □

Corollary. *Let* $\xi = (\xi_t, \mathscr{F}_t)$, $0 \le t \le T$, *be a process of the diffusion type with*

$$d\xi_t = A_t(\xi)dt + b_t(\xi)dW_t, \qquad \xi_0 = \eta_0 \qquad (7.122)$$

(i.e., in (7.113) *let* $A_t(\omega) = A_t(\xi(\omega))$*). If the assumptions* (I), (II), *and* (IV) *of Theorem* 7.18 *are fulfilled and*

$$P\left\{\int_0^T [b_s^+(\xi)A_s(\xi)]^2 \, ds < \infty\right\} = P\left\{\int_0^T [b_s^+(\xi)a_s(\xi)]^2 \, ds < \infty\right\} = 1, \qquad (7.123)$$

then (compare with the corollary of Theorem (7.1)) (P-a.s.)

$$\frac{d\mu_\eta}{d\mu_\xi}(\xi) = \exp\left[-\int_0^T (b_s^+(\xi))^2 [A_s(\xi) - a_s(\xi)] d\xi_s + \frac{1}{2}\int_0^T (b_s^+(\xi))^2 [A_s^2(\xi) - a_s^2(\xi)] ds\right], \qquad (7.124)$$

$$\frac{d\mu_\xi}{d\mu_\eta}(\eta) = \exp\left[\int_0^T (b_s^+(\eta))^2 [A_s(\eta) - a_s(\eta)] d\eta_s - \frac{1}{2}\int_0^T (b_s^+(\eta))^2 [A_s^2(\eta) - a_s^2(\eta)] ds\right]. \qquad (7.125)$$

It will be noted that the stochastic integrals in (7.124) and (7.125) are defined due to the equivalence of the measures μ_ξ and μ_η, (7.123), and the fact that (P-a.s.)

$$\int_0^T (b_s^+(\xi))^4 (A_s(\xi) - a_s(\xi))^2 b_s^2(\xi) ds \le \int_0^T \alpha_s^2(\xi) ds < \infty.$$

EXAMPLE. Let $\xi = (\xi_t)$ and $\eta = (\eta_t)$ be two processes of the diffusion type with the differentials

$$d\xi_t = \xi_t dt + \xi_t dW_t, \qquad \xi_0 = \eta_0, \qquad d\eta_t = \eta_t dW_t,$$

where $P(\eta_0 = 0) > 0$.

With the help of the Ito formula we convince ourselves that solutions of these equations are given by the formulas

$$\xi_t = \eta_0 \exp\left(W_t + \frac{t}{2}\right), \qquad \eta_t = \eta_0 \exp\left(W_t - \frac{t}{2}\right).$$

The conditions of Theorem 7.18, as it is easy to check, are fulfilled. Hence, $\mu_\xi \sim \mu_\eta$ and (P-a.s.)

$$\begin{aligned}\frac{d\mu_\xi}{d\mu_\eta}(\eta) &= \exp\left[\int_0^T \eta_s(\eta_s^+)^2\, d\eta_s - \frac{1}{2}\int_0^T (\eta_s\eta_s^+)^2\, ds\right] \\ &= \exp\left[\int_0^T \eta_s^+\, d\eta_s - \frac{1}{2}\int_0^T \eta_s\eta_s^+\, ds\right] \\ &= \exp\left[\int_0^T (\eta_s\eta_s^+) dW_s - \frac{1}{2}\int_0^T (\eta_s\eta_s^+) ds\right] = \exp\left[\eta_0\eta_0^+\left(W_T - \frac{T}{2}\right)\right].\end{aligned} \tag{7.126}$$

But (P-a.s.)

$$\begin{aligned}\exp\left[\eta_0\eta_0^+\left(W_T - \frac{T}{2}\right)\right] &= (1 - \eta_0\eta_0^+) + \eta_0\eta_0^+ \exp\left[\eta_0\eta_0^+\left(W_T - \frac{T}{2}\right)\right] \\ &= (1 - \eta_0\eta_0^+) + \eta_0\eta_0^+ \exp\left(W_T - \frac{T}{2}\right) \\ &= (1 - \eta_0\eta_0^+) + \eta_0^+\eta_T.\end{aligned}$$

Thus (P-a.s.)

$$\frac{d\mu_\xi}{d\mu_\eta}(\eta) = (1 - \eta_0\eta_0^+) + \eta_0^+\eta_T \tag{7.127}$$

and, similarly,

$$\frac{d\mu_\eta}{d\mu_\xi}(\xi) = \frac{1}{(1 - \xi_0\xi_0^+) + \xi_0^+\xi_T}. \tag{7.128}$$

From (7.127) it is seen that on the set $\{\omega: \xi_0 = \eta_0 = 0\}$

$$\frac{d\mu_\xi}{d\mu_\eta}(\eta) = 1,$$

and on $\{\omega: \xi_0 = \eta_0 \neq 0\}$

$$\frac{d\mu_\xi}{d\mu_\eta}(\eta) = \frac{\eta_T}{\eta_0}.$$

7.6.3

For the processes of the diffusion type under consideration we shall give analogs of some statements of Theorems 7.5–7.7.

Theorem 7.19. *Let $\xi = (\xi_t)$ and $\eta = (\eta_t)$, $0 \le t \le T$, be two processes of the diffusion type with*

$$d\xi_t = A_t(\xi)dt + b_t(\xi)dW_t, \qquad \xi_0 = \eta_0, \tag{7.129}$$

$$d\eta_t = a_t(\eta)dt + b_t(\eta)dW_t. \tag{7.130}$$

Let the assumptions (I) *and* (II) *of Theorem* 7.18 *be fulfilled* (*with* $A_t(\omega) = A_t(\xi(\omega))$). *If*

$$P\left\{\int_0^T (b_s^+(\xi))^2[A_s^2(\xi) + a_s^2(\xi)]ds < \infty\right\}$$
$$= P\left\{\int_0^T (b_s^+(\eta))^2[A_s^2(\eta) + a_s^2(\eta)]ds < \infty\right\} = 1, \tag{7.131}$$

then $\mu_\xi \sim \mu_\eta$ and the densities $d\mu_\eta/d\mu_\xi$ and $d\mu_\xi/d\mu_\eta$ are given by (7.124) *and* (7.125).

PROOF. Set

$$\tau_n(x) = \begin{cases} \inf\left[t \le T: \int_0^t (b_s^+(x)[A_s(x) - a_s(x)])^2\, ds \ge n\right], \\ T, \qquad \text{if } \int_0^T (b_s^+(x)[A_s(x) - a_s(x)])^2\, ds < n, \end{cases}$$

$$\chi_t^{(n)}(x) = \chi_{\{\tau_n(x) \ge t\}}, \qquad A_t^{(n)}(x) = a_t(x) + \chi_t^{(n)}(x)[A_t(x) - a_t(x)].$$

Let us consider the process $\xi^{(n)} = (\xi_t^{(n)}, \mathscr{F}_t)$, $0 \le t \le T$, being defined by the equalities

$$\xi_t^{(n)} = \xi_{t \wedge \tau_n(\xi)} + \int_0^t [1 - \chi_s^{(n)}(\xi)]a_s(\xi^{(n)})ds + \int_0^t [1 - \chi_s^{(n)}(\xi)]b_s(\xi^{(n)})dW_s. \tag{7.132}$$

By Theorem 4.8, Equation (7.132) has a unique strong solution, with $\xi_t^{(n)} = \xi_t$ for $t \le \tau_0(\xi)$. Taking this fact into account, with the help of the Ito formula we find that

$$d\xi_t^{(n)} = A_t^{(n)}(\xi^{(n)})dt + b_t(\xi^{(n)})dW_t, \qquad \xi_0^{(n)} = \xi_0. \tag{7.133}$$

Since

$$A_t^{(n)}(x) - a_t(x) = \chi_t^{(n)}(x)[A_t(x) - a_t(x)],$$

then (P-a.s.)

$$\int_0^T (b_t^+(\xi^{(n)})[A_t^{(n)}(\xi^{(n)}) - a_t(\xi^{(n)})])^2\, dt \le n,$$

and, according to Theorem 6.1,

$$M\exp\left\{-\int_0^T b_t^+(\xi^{(n)})[A_t^{(n)}(\xi^{(n)}) - a_t(\xi^{(n)})]dW_t - \frac{1}{2}\int_0^T (b_t^+(\xi^{(n)})[A_t^{(n)}(\xi^{(n)}) - a_t(\xi^{(n)})])^2\,dt\right\} = 1.$$

Taking into account Theorem 7.18, we conclude that $\mu_{\xi(n)} \sim \mu_\eta$ and

$$\begin{aligned}\frac{d\mu_{\xi(n)}}{d\mu_\eta}(\eta) &= \exp\left\{\int_0^T (b_t^+(\eta))^2[A_t^{(n)}(\eta) - a_t(\eta)]d\eta_t - \frac{1}{2}\int_0^T (b_t^+(\eta))^2[(A_t^{(n)}(\eta))^2 - (a_t(\eta))^2]dt\right\}\\ &= \exp\left\{\int_0^{T\wedge\tau_n(\eta)} (b_t^+(\eta))^2[A_t(\eta) - a_t(\eta)]d\eta_t - \frac{1}{2}\int_0^{T\wedge\tau_n(\eta)} (b_t^+(\eta))^2[A_t^2(\eta) - a_t^2(\eta)]dt\right\} = \varkappa_{T\wedge\tau_n(\eta)}(\eta).\end{aligned}$$

Let now $\Gamma \in \mathscr{B}_T$. Then, because of (7.131),

$$\begin{aligned}\mu_\xi(\Gamma) &= \lim_n \mu_{\xi(n)}(\Gamma \cap (\tau_n(x) = T)) = \lim_n \int_{\Gamma\cap(\tau_n(x)=T)} \varkappa_{T\wedge\tau_n(x)}(x)d\mu_\eta(x)\\ &= \lim_n \int_{\Gamma\cap(\tau_n(x)=T)} \varkappa_T(x)d\mu_\eta(x) = \int_\Gamma \varkappa_T(x)d\mu_\eta(x).\end{aligned}$$

Therefore

$$\mu_\xi \ll \mu_\eta \quad\text{and}\quad \frac{d\mu_\xi}{d\mu_\eta}(x) = \varkappa_T(x).$$

But since $\mu_\eta(x:\varkappa_T(x) = 0) = 0$, by Lemma 6.8 $\mu_\eta \ll \mu_\xi$ and

$$\frac{d\mu_\eta}{d\mu_\xi}(x) = \varkappa_T^{-1}(x).$$ □

Theorem 7.20. *Let the assumptions of Theorem* 7.19 *be fulfilled with the exception of* (7.131), *which is replaced with the condition*

$$P\left\{\int_0^T (b_s^+(\xi))^2[A_s^2(\xi) + a_s^2(\xi)]ds < \infty\right\} = 1. \tag{7.134}$$

Then $\mu_\xi \ll \mu_\eta$, *the density*

$$\varkappa_t(\eta) = \frac{d\mu_\xi}{d\mu_\eta}(t, \eta)$$

is a unique (continuous) solution of the equation

$$\varkappa_t(\eta) = 1 + \int_0^t \varkappa_s(\eta)(b_s^+(\eta))^2[A_s(\eta) - a_s(\eta)]d\eta_s, \tag{7.135}$$

and $\varkappa_t(\xi)$ *can be defined by the formula*

$$\varkappa_t(\xi) = \exp\left\{\int_0^t (b_s^+(\xi))^2[A_s(\xi) - a_s(\xi)]d\xi_s - \frac{1}{2}\int_0^t (b_s^+(\xi))^2[A_s^2(\xi) - a_s^2(\xi)]ds\right\}. \tag{7.136}$$

Proof of this theorem is analogous to those of Theorems 7.19, 7.2, and 7.6.

7.6.4

We shall consider, finally, multi-dimensional analogs of Theorems 7.19 and 7.20, restricting ourselves to their formulations.

Let $\xi = (\xi_t)$ and $\eta = (\eta_t)$, $0 \le t \le T$, be vector processes, $\xi_t = (\xi_1(t), \dots, \xi_n(t))$, $\eta_t = (\eta_1(t), \dots, \eta_n(t))$, having the differentials

$$d\xi_t = A_t(\xi)dt + b_t(\xi)dW_t, \qquad \xi_0 = \eta_0,$$
$$d\eta_t = a_t(\eta)dt + b_t(\eta)dW_t,$$

where: $W_t = (W_1(t), \dots, W_k(t))$ is a k-dimensional Wiener process with respect to the system $(\mathscr{F}_t)$, $0 \le t \le T$; $A_t(x) = (A_1(t, x), \dots, A_n(t, x))$; $a_t(x) = (a_1(t, x), \dots, a_n(t, x))$; $b_t(x) = \|b_{ij}(t, x)\|$ is a matrix of the order $n \times k$; and $\eta_0 = (\eta_1(0), \dots, \eta_n(0))$ is a vector of initial values with $P(\sum_{i=1}^n |\eta_i(0)| < \infty) = 1$.

It will be assumed that a system of algebraic equations

$$b_t(x)\alpha_t(x) = [A_t(x) - a_t(x)]$$

has (with respect to $\alpha_t(x)$) a bounded solution for each t, $0 \le t \le T$, $x \in C$.

The functionals $a_t(x)$ and $b_t(x)$ satisfy (by components) (4.100) and (4.111).

If μ_ξ (P a.s.)

$$\int_0^T [A_t^*(x)(b_t(x)b_t^*(x))^+ A_t(x) + a_t^*(x)(b_t(x)b(x))^+ a_t(x)]dt < \infty, \tag{7.137}$$

then[13] $\mu_\xi \ll \mu_\eta$. If, in addition, (7.137) is satisfied and μ_η (P-a.s.) then $\mu_\xi \sim \mu_\eta$ and

$$\frac{d\mu_\xi}{d\mu_\eta}(t, \eta) = \exp\left\{\int_0^t (A_s(\eta) - a_s(\eta))^*(b_s(\eta)b_s^*(\eta))^+ \, d\eta_s - \frac{1}{2}\int_0^t (A_s(\eta) - a_s(\eta))^*(b_s(\eta)b_s^*(\eta))^+(A_s(\eta) + a_s(\eta))ds\right\}, \tag{7.138}$$

$$\frac{d\mu_\eta}{d\mu_\xi}(t, \xi) = \exp\left\{-\int_0^t (A_s(\xi) - a_s(\xi))^*(b_s(\xi)b_s^*(\xi))^+ \, d\xi_s + \frac{1}{2}\int_0^t (A_s(\xi) - a_s(\xi))^*(b_s(\xi)b_s^*(\xi))^+(A_s(\xi) + a_s(\xi))ds\right\}. \tag{7.139}$$

[13] The matrix R^+ is pseudo-inverse with respect to the matrix R (see Section 13.1).

7.7 The Cameron–Martin formula

7.7.1

Let $W = (W_t, \mathscr{F}_t)$ be an n-dimensional Wiener process, $W_t = (W_1(t), \ldots, W_n(t))$, and let $Q(t)$ be a symmetric nonnegative definite matrix whose elements $q_{ij}(t), i, j = 1, \ldots, n$, satisfy the condition

$$\int_0^T \sum_{i,j=1}^n |q_{ij}(t)| dt < \infty. \tag{7.140}$$

Making use of the results of Subsection 7.2.7, let us establish the following result, known as the *Cameron–Martin formula.*

Theorem 7.21. *Let* (7.140) *be fulfilled. Then*

$$M \exp\left[-\int_0^T (W_t, Q(t)W_t)dt\right] = \exp\left[\frac{1}{2}\int_0^T \mathrm{Sp}\, \Gamma(t)dt\right], \tag{7.141}$$

where $(W_t, Q(t)W_t)$ is a scalar product equal to $W_t^* Q(t)W_t$, and $\Gamma(t)$ is a symmetric nonpositive definite matrix, being a unique solution of the Ricatti matrix equation

$$\frac{d\Gamma(t)}{dt} = 2Q(t) - \Gamma^2(t); \tag{7.142}$$

$\Gamma(T) = 0$ is a zero matrix.

PROOF. Consider the Ricatti equation

$$\frac{d\tilde{\Gamma}(s)}{ds} = 2Q(T - s) - \tilde{\Gamma}^2(s) \tag{7.143}$$

with a zero matrix $\tilde{\Gamma}(0)$. The uniqueness of a solution of this equation in a class of nonnegative definite matrices is proved in Theorem 10.2. The existence of a continuous solution $\tilde{\Gamma}(t) = \|\tilde{\gamma}_{ij}(t)\|$ can be deduced, for example, from the solution of some auxiliary filtering problem (see Section 10.3).

Assume $\Gamma(t) = -\tilde{\Gamma}(T - t)$. It can be immediately confirmed that $\Gamma(t)$ satisfies Equation (7.142), the solution of which is unique due to the uniqueness of a solution of Equation (7.143).

Let now $\xi_t = (\xi_1(t), \ldots, \xi_n(t))$ be a random process with the differential

$$d\xi_t = \Gamma(t)\xi_t\, dt + dW_t, \qquad \xi_0 = 0. \tag{7.144}$$

According to Theorem 4.10 a strong solution of Equation (7.144) exists, is unique and defined by (4.158), and

$$P\left\{\int_0^T \xi_t^* \Gamma^2(t)\xi_t\, dt < \infty\right\} = P\left\{\int_0^T W_t^* \Gamma^2(t) W_t\, dt < \infty\right\} = 1.$$

Making use of a multi-dimensional analog of Theorem 7.7 (see also Subsection 7.2.7), we find that $\mu_\xi \sim \mu_W$ and

$$\frac{d\mu_\xi}{d\mu_W}(t, W) = \exp\left\{\int_0^t W_s^*\Gamma(s)dW_s - \frac{1}{2}\int_0^t W_s^*\Gamma^2(s)W_s\, ds\right\}.$$

Hence,

$$M \exp\left\{\int_0^t W_s^*\Gamma(s)dW_s - \frac{1}{2}\int_0^t W_s^*\Gamma^2(s)W_s\, ds\right\} = 1. \tag{7.145}$$

By the Ito formula (see Chapter 4, Example 2, (4.102))

$$\begin{aligned} 0 &= \frac{1}{2}[W_T^*\Gamma(T)W_T - W_0^*\Gamma(0)W_0] \\ &= \frac{1}{2}\int_0^T\left(W_t^*\frac{d\Gamma(t)}{dt}W_t\right)dt + \int_0^T W_t^*\Gamma(t)dW_t + \frac{1}{2}\int_0^T \mathrm{Sp}\,\Gamma(t)dt. \end{aligned}$$

From this we find

$$\int_0^T W_t^*\Gamma(t)dW_t = \frac{1}{2}\int_0^T W_t^*\frac{d\Gamma(t)}{dt}W_t\, dt - \frac{1}{2}\int_0^T \mathrm{Sp}\,\Gamma(t)dt.$$

Substituting this expression into (7.145) and taking into account that, because of (7.142),

$$\frac{1}{2}\left[\frac{d\Gamma(t)}{dt} + \Gamma^2(t)\right] = Q(t),$$

we obtain

$$\begin{aligned} 1 &= \exp\left\{-\frac{1}{2}\int_0^T \mathrm{Sp}\,\Gamma(t)dt\right\}M \exp\left\{\frac{1}{2}\int_0^T W_t^*\left[\frac{d\Gamma(t)}{dt} + \Gamma^2(t)\right]W_t\, dt\right\} \\ &= \exp\left\{-\frac{1}{2}\int_0^T \mathrm{Sp}\,\Gamma(t)dt\right\}M \exp\left\{\frac{1}{2}\int_0^T W_t^*Q(t)W_t\, dt\right\}, \end{aligned} \tag{7.146}$$

which proves (7.141). □

Example 1. Let $n = 1$, $Q(t) = \frac{1}{2}$. In this case the equation

$$\frac{d\Gamma(t)}{dt} = 1 - \Gamma^2(t), \qquad \Gamma(T) = 0,$$

has the solution

$$\Gamma(t) = \frac{e^{2(t-T)} - 1}{e^{2(t-T)} + 1}.$$

From this we obtain

$$\frac{1}{2}\int_0^T \Gamma(t)dt = \ln(\text{ch } T)^{-1/2},$$

and, therefore,

$$M \exp\left\{-\frac{1}{2}\int_0^T W_t^2\, dt\right\} = \frac{1}{\sqrt{\text{ch } T}}. \tag{7.147}$$

7.8 The Rao–Cramer–Wolfovitz inequality

7.8.1

In parameter estimation problems an essential role is played by the Rao–Cramer inequality and the generalization given by Wolfovitz for a case where the observation time is random also.

In this section it will be shown in what way the formulas for densities of the processes of a diffusion type obtained above can be applied to search for the lower bounds of mean square errors in some problems of parameter estimation.

7.8.2

It will be assumed that θ is an unknown parameter, $-\infty < \theta < \infty$, and $f = f(\theta)$ is a function estimated on the basis of results of observation of the random process $\xi = (\xi_t)$, $t \geq 0$, having the differential

$$d\xi_t = a_t(\theta, \xi)dt + dW_t, \qquad \xi_0 = 0. \tag{7.148}$$

The measurable functional $\{a_t(\theta, x), t \geq 0, -\infty < \theta < \infty, x \in C\}$ is assumed (for each fixed θ) to be nonanticipative, i.e., $\mathscr{B}_t$-measurable for each $t \geq 0$, where the $\mathscr{B}_t = \sigma\{x: x_s, s \leq t\}$ are sub-σ-algebras in a measurable space $(C, \mathscr{B})$ of continuous functions $x = (x_t)$, $t \geq 0$, with $x_0 = 0$.

Let $\tau = \tau(x)$ be a Markov time with respect to the system $(\mathscr{B}_t)$, $t \geq 0$, and let $\delta = (\delta(t, x))$ be a progressively measurable (and therefore non-anticipative) real process defined on $(C, \mathscr{B})$.

Later on the value $\delta(t, x)$ will be regarded as an *estimate* of the function $f(\theta)$ on the basis of observations of the trajectory x over a time interval $[0, t]$. If $\tau = \tau(x)$ is Markov time, then the value $\delta(\tau(x), x)$ will specify the estimate of the function $f(\theta)$ on the basis of the results of observations of the trajectory x over the time interval $[0, \tau(x)]$.

The pair of functions $\Delta = (\tau, \delta)$ prescribes a *sequential estimation plan.* With some assumptions on regularity formulated further on (Theorem 7.22), for the sequential plans $\Delta = (\tau, \sigma)$ there will be obtained (with each θ, $-\infty < \theta < \infty$) an inequality analogous to the Rao–Cramer–Wolfovitz inequality which provides a lower bound for the value $M[f(\theta) - \delta(\tau, \xi)]^2$.

7.8.3

Consider first the notation and assumptions used from now on.

Let μ_W and μ_ξ^θ denote measures in the space $(C, \mathscr{B})$, corresponding respectively to a Wiener process W and a process ξ with the differential given by (7.148) for a given θ, $-\infty < \theta < \infty$.

Let

(a) $\int_0^\infty |a_t(\theta, x)| dt < \infty$ (μ_W- and μ_ξ^θ-a.s.). $\quad -\infty < \theta < \infty$;
(b) $\int_0^{\tau(x)} a_t^2(\theta, x) dt < \infty$ (μ_W- and μ_ξ^θ-a.s.), $\quad -\infty < \theta < \infty$.

From (a), (b) and Theorem 7.10 it follows that for each θ the measures μ_ξ^θ and μ_W are equivalent and the density

$$\varphi(\theta, W) = \frac{d\mu_\xi^\theta}{d\mu_W}(\tau(W), W)$$

is given by the formula

$$\varphi(\theta, W) = \exp\left\{\int_0^{\tau(W)} a_t(\theta, W) dW_t - \frac{1}{2}\int_0^{\tau(W)} a_t^2(\theta, W) dt\right\}. \tag{7.149}$$

This representation will play a central role for obtaining a lower bound for $M[f(\theta) - \delta(\tau, \xi)]^2$.

Assume also that

(c) for each $t \geq 0$ and $x \in C$ the function $a_t(\theta, x)$ is differentiable over θ and

$$\int_0^{\tau(W)} \left[\frac{\partial}{\partial\theta} a_t(\theta, W)\right]^2 dt < \infty \qquad (P\text{-a.s.}), \qquad -\infty < \theta < \infty,$$

$$0 < M\int_0^{\tau(\xi)} \left[\frac{\partial}{\partial\theta} a_t(\theta, \xi)\right]^2 dt < \infty, \qquad -\infty < \theta < \infty;$$

(d)

$$\frac{\partial}{\partial\theta}\int_0^{\tau(W)} a_t(\theta, W) dW_t = \int_0^{\tau(W)} \frac{\partial}{\partial\theta} a_t(\theta, W) dW_t \qquad (P\text{-a.s.}),$$

$$-\infty < \theta < \infty,$$

$$\frac{\partial}{\partial\theta}\int_0^{\tau(W)} a_t^2(\theta, W) dt = 2\int_0^{\tau(W)} a_t(\theta, W)\frac{\partial}{\partial\theta} a_t(\theta, W) dt \qquad (P\text{-a.s.}),$$

$$-\infty < \theta < \infty;$$

(e) the functions $f(\theta)$ and $b(\theta) = M\delta(\tau(W), W)\varphi(\theta, W) - f(\theta)$ are differentiable over θ and

$$\frac{d}{d\theta}[b(\theta) + f(\theta)] = M\delta(\tau(W), W)\frac{\partial\varphi(\theta, W)}{\partial\theta}.$$

7.8.4

Theorem 7.22. *Let* $\Delta = (\tau, \delta)$ *be a sequential scheme of estimation with* $M\delta^2(\tau, \xi) < \infty$ *for each* θ, $-\infty < \theta < \infty$. *If the regularity conditions* (a)–(e) *are fulfilled, then for each* θ, $-\infty < \theta < \infty$

$$M[f(\theta) - \delta(\tau, \xi)]^2 \geq \frac{(d/d\theta)[f(\theta) + b(\theta)])^2}{M \int_0^{\tau(\xi)} [(\partial/\partial\theta)a_t(\theta, \xi)]^2 \, dt} + b^2(\theta). \quad (7.150)$$

PROOF. According to assumptions (d) and (e) and (7.149), for each θ, $-\infty < \theta < \infty$,

$$\begin{aligned}
\frac{d}{d\theta}[b(\theta) + f(\theta)] &= M\delta(\tau(W), W)\frac{\partial \varphi(\theta, W)}{\partial \theta} \\
&= M\delta(\tau(W), W)\left\{\int_0^{\tau(W)} \frac{\partial}{\partial\theta}[a_t(\theta, W)]dW_t \right. \\
&\qquad \left. - \int_0^{\tau(W)} a_t(\theta, W)\frac{\partial}{\partial\theta}[a_t(\theta, W)]dt\right\}\varphi(\theta, W) \\
&= M\delta(\tau(\xi), \xi)\left\{\int_0^{\tau(\xi)} \frac{\partial}{\partial\theta}[a_t(\theta, \xi)]d\xi_t \right. \\
&\qquad \left. - \int_0^{\tau(\xi)} a_t(\theta, \xi)\frac{\partial}{\partial\theta}[a_t(\theta, \xi)]dt\right\} \\
&= M\delta(\tau(\xi), \xi)\int_0^{\tau(\xi)} \frac{\partial}{\partial\theta}[a_t(\theta, \xi)]dW_t. \qquad (7.151)
\end{aligned}$$

Further, because of (c),

$$M\delta(\tau(\xi), \xi) \cdot M\int_0^{\tau(\xi)} \frac{\partial}{\partial\theta}[a_t(\theta, \xi)]dW_t = 0, \quad (7.152)$$

which together with (7.151) leads to the relation

$$\frac{d}{d\theta}[b(\theta) + f(\theta)] = M[\delta(\tau(\xi), \xi) - M\delta(\tau(\xi), \xi)]\int_0^{\tau(\xi)} \frac{\partial}{\partial\theta}[a_t(\theta, \xi)]dW_t.$$

From this, according to the Cauchy–Buniakowski inequality, assumption (c), and the property of stochastic integrals, we obtain

$$\begin{aligned}
\left(\frac{d}{d\theta}[b(\theta) + f(\theta)]\right)^2 &\leq M[\delta(\tau(\xi), \xi) - M\delta(\tau(\xi), \xi)]^2 M\int_0^{\tau(\xi)}\left[\frac{\partial}{\partial\theta}a_t(\theta, \xi)\right]^2 dt \\
&= M\int_0^{\tau(\xi)}\left[\frac{\partial}{\partial\theta}a_t(\theta, \xi)\right]^2 dt \\
&\qquad \cdot \{M[\delta(\tau(\xi), \xi) - f(\theta)]^2 - b^2(\theta)\}
\end{aligned}$$

which, because of the assumption

$$M \int_0^{\tau(\xi)} \left[\frac{\partial}{\partial \theta} a_t(\theta, \xi) \right]^2 dt > 0,$$

leads to the required inequality, (7.150). □

Corollary. *If the plan* $\Delta = (\tau, \delta)$ *is unbiased, i.e.,* $b(\theta) = M\delta(\tau, \xi) - f(\theta) \equiv 0$ *for all* θ, $-\infty < \theta < \infty$, *then*

$$M[\delta(\tau, \xi) - f(\theta)]^2 \geq \frac{[(d/d\theta)f(\theta)]^2}{M \int_0^{\tau(\xi)} [(\partial/\partial\theta)a_t(\theta, \xi)]^2 \, dt}. \tag{7.153}$$

In particular, if $f(\theta) \equiv \theta$, *then*

$$M[\delta(\tau, \xi) - \theta]^2 \geq \frac{1}{M \int_0^{\tau(\xi)} [(\partial/\partial\theta)a_t(\theta, \xi)]^2 \, dt}. \tag{7.154}$$

EXAMPLE. Let there be observed a random process

$$\xi_t = \theta t + W_t, \qquad t \geq 0, \qquad -\infty < \theta < \infty.$$

Then for unbiased sequential estimation schemes:

$$M[\delta(\tau, \xi) - \theta]^2 \geq \frac{1}{M\tau(\xi)}. \tag{7.155}$$

In particular, the scheme $\Delta^0 = (\tau^0, \delta^0)$ with $\tau^0(x) \equiv T$ and $\delta^0(T, x) = x_T/T$ is unbiased. For this scheme all the conditions of Theorem 7.22 are fulfilled, and, hence,

$$M[\delta^0(T, \xi) - \theta]^2 \geq \frac{1}{T}.$$

Now note that the left-hand side is equal to $1/T$, since $M[(\xi_T/T) - \theta]^2 = M(W_T/T)^2 = 1/T$. This implies that among all the unbiased sequential plans $\Delta = (\tau, \delta)$ with $M\tau(\xi) \leq T$ (for all $-\infty < \theta < \infty$) and satisfying the condition of Theorem 7.22, the scheme Δ^0 is optimal: for all θ, $-\infty < \theta < \infty$,

$$M[\delta(\tau, \xi) - \theta]^2 \geq M[\delta^0(T, \xi) - \theta]^2.$$

Other examples of the application of (7.150) to the problems of sequential estimation will be discussed in Chapter 17.

7.9 An abstract version of the Bayes formula

7.9.1

Let $(\Omega, \mathscr{F}, P)$ be some probability space, and let $\theta = \theta(\omega)$ and $\xi = \xi(\omega)$ be random elements with values in the measurable spaces $(\Theta, \mathscr{B}_\Theta)$, $(X, \mathscr{B}_X)$. Further, let $\mathscr{F}_\theta = \sigma\{\omega: \theta(\omega)\}$, $\mathscr{F}_\xi = \sigma\{\omega: \xi(\omega)\}$, and let Q be a restriction of the measure P on $(\Omega, \mathscr{F}_\xi)$. Denote by $Q(A; \omega) = M[\chi_A(\omega) | \mathscr{F}_\theta](\omega)$ a conditional probability of the event $A \in \mathscr{F}_\xi$. It is clear that for a given $A \in \mathscr{F}_\xi$,

$$Q(A) = \int_\Omega Q(A; \omega)P(d\omega). \tag{7.156}$$

If θ and ξ are random variables taking on only discrete values and $M|g(\theta)| < \infty$, then a conditional expectation $M[g(\theta)|\xi]$ is given by the *Bayes formula*

$$M[g(\theta)|\xi] = \frac{\sum_i g(a_i)p(\xi|a_i)P(a_i)}{\sum_i p(\xi|a_i)P(a_i)}, \tag{7.157}$$

where

$$p(b|a) = P\{\xi = b|\theta = a\}, P(a) = P(\theta = a).$$

For the case where θ and ξ are random variables whose distribution functions have densities, the Bayes formula becomes

$$M[g(\theta)|\xi] = \frac{\int_{-\infty}^{\infty} g(a)p(\xi|a)p(a)da}{\int_{-\infty}^{\infty} p(\xi|a)p(a)da}, \tag{7.158}$$

where

$$p(b|a) = \frac{dP(\xi \le b|\theta = a)}{db}, \qquad p(a) = \frac{dP(\theta \le a)}{da}.$$

Later on, we shall often deal with the abstract version of the Bayes formula generalizing (7.157) and (7.158).

Let $\theta = \theta(\omega)$, $\xi = \xi(\omega)$ be random elements with values in $(\Theta, \mathscr{B}_\Theta)$, $(X, \mathscr{B}_X)$ where $M|g(\theta)| < \infty$. For $A \in \mathscr{F}_\xi$, let

$$G(A) = \int_\Omega g(\theta(\tilde{\omega}))Q(A; \tilde{\omega})P(d\tilde{\omega}). \tag{7.159}$$

Lemma 7.3

(1) *The function $G = G(A)$, $A \in \mathscr{F}_\xi$, defined in (7.159), is a generalized measure (countable-additive function of the sets $A \in \mathscr{F}_\xi$ taking on perhaps values of different signs).*
(2) *The generalized measure G is absolutely continuous over the measure $Q: G \ll Q$.*
(3) *There is the Bayes formula*

$$M[g(\theta)|\mathscr{F}_\xi](\omega) = \frac{dG}{dP}(\omega). \tag{7.160}$$

PROOF. The first two properties follow immediately from (7.159). We shall now prove (7.160). Since $M[g(\theta)|\mathscr{F}_\xi]$ is an $\mathscr{F}_\xi$-measurable function, then one needs only to check the equality

$$M\{\chi_A(\omega)M[g(\theta)|\mathscr{F}_\xi]\} = G(A), \qquad A \in \mathscr{F}_\xi. \tag{7.161}$$

We have

$$\begin{aligned} M\{\chi_A(\omega)M[g(\theta)\mid\mathscr{F}_\xi]\} &= M\{M[\chi_A(\omega)g(\theta)\mid\mathscr{F}_\xi]\} \\ &= M\chi_A(\omega)g(\theta) = M\{g(\theta)M[\chi_A(\omega)\mid\mathscr{F}_\theta]\} \\ &= M\{g(\theta)Q(A,\omega)\} = G(A). \end{aligned}$$ □

Lemma 7.4. *Assume that the conditional probability $Q(A;\tilde{\omega})$ is regular,*[14] *the σ-algebra $\mathscr{F}_\xi$ is separable,*[15] *and that there exists a measure $\lambda = \lambda(A)$ on $(\Omega, \mathscr{F}_\xi)$, such that for almost all $\tilde{\omega} \in \Omega$*

$$Q(\cdot, \tilde{\omega}) \ll \lambda(\cdot). \tag{7.162}$$

Then, $Q \ll \lambda$, $G \ll \lambda$, and on the space $(\Omega \times \Omega, \mathscr{F}_\xi \times \mathscr{F}_\theta)$ there exists a nonnegative measurable function $q(\omega, \tilde{\omega})$ such that (P-a.s.)

$$Q(A;\tilde{\omega}) = \int_A q(\omega, \tilde{\omega})d\lambda(\omega), \tag{7.163}$$

$$\frac{dG}{\partial\lambda}(\omega) = \int_\Omega g(\theta(\tilde{\omega}))q(\omega,\tilde{\omega})P(d\tilde{\omega}), \tag{7.164}$$

$$\frac{dQ}{d\lambda}(\omega) = \int_\Omega q(\omega,\tilde{\omega})P(d\tilde{\omega}), \tag{7.165}$$

$$0 < \frac{dQ}{d\lambda}(\omega) < \infty, \tag{7.166}$$

$$M[g(\theta)\mid\mathscr{F}_\xi](\omega) = \frac{\int_\Omega g(\theta(\tilde{\omega}))q(\omega,\tilde{\omega})P(d\tilde{\omega})}{\int_\Omega q(\omega,\tilde{\omega})P(d\tilde{\omega})}. \tag{7.167}$$

PROOF. The existence of the measurable function $q(\omega, \tilde{\omega})$ satisfying (7.163) follows from regularity of the conditional probability $Q(A;\tilde{\omega})$ and separability of the σ-algebra $\mathscr{F}_\xi$.[16] For proving (7.164) and (7.165) it suffices to apply the Fubini theorem.

Next, let

$$A_0 = \left\{\omega: \frac{dQ}{d\lambda}(\omega) = 0\right\}.$$

Since $A_0 \in \mathscr{F}_\xi$,

$$P(A_0) = Q(A_0) = \int_{A_0} \frac{dQ}{d\lambda}(\omega)d\lambda(\omega) = 0.$$

Consequently,

$$\frac{dQ}{d\lambda}(\omega) > 0 \qquad (P\text{-a.s.}).$$

[14] See Subsection 1.1.4.

[15] See [46], p. 555.

[16] The proof of this fact is given in [46], Ex. 2.7 in Supplement.

For proving (7.167) it will be noted that, since $G \ll Q$, $G \ll \lambda$, and $Q \ll \lambda$, therefore

$$\frac{dG}{d\lambda} = \frac{dG}{dQ} \cdot \frac{dQ}{d\lambda}.$$

But $dQ/d\lambda > 0$ (P-a.s.); hence

$$\frac{dG}{dQ}(\omega) = \frac{dG}{d\lambda}(\omega) \Big/ \frac{dQ}{d\lambda}(\omega),$$

which together with (7.160), (7.164), and (7.165) proves (7.167). □

Note 1. If the function $g = g(\xi, \theta)$ is such that $M|g(\xi, \theta)| < \infty$, then

$$M[g(\xi, \theta)|\mathscr{F}_\xi](\omega) = \frac{\int_\Omega g(\xi(\omega), \theta(\tilde{\omega}))q(\omega, \tilde{\omega})P(d\tilde{\omega})}{\int_\Omega q(\omega, \tilde{\omega})P(d\tilde{\omega})}. \tag{7.168}$$

Indeed, if the function $g(\xi, \theta)$ can be represented in the form

$$g(\xi, \theta) = \sum_{k=1}^{n} \varphi_k(\xi) g_k(\theta),$$

then (7.168) follows immediately from (7.167). As the obvious passage to the limit (7.168) extends also to the arbitrary (measurable) functions $g(\theta, \xi)$ with $M|g(\xi, \theta)| < \infty$.

Note 2. Having denoted

$$\rho(\omega, \tilde{\omega}) = \frac{q(\omega, \tilde{\omega})}{\int_\Omega q(\omega, \tilde{\omega})P(d\tilde{\omega})},$$

we obtain for the Bayes formula given by (7.168) the following convenient representation:

$$M[g(\xi, \theta)|\mathscr{F}_\xi](\omega) = \int_\Omega g(\xi(\omega), \theta(\tilde{\omega}))\rho(\omega, \tilde{\omega})P(d\tilde{\omega}). \tag{7.169}$$

7.9.2

Let us consider in more detail the structure of the Bayes formula given by (7.169) for the case where ξ is an Ito process.

We shall assume as given a probability space $(\Omega, \mathscr{F}, P)$ with the distinguished family of sub-σ-algebras $(\mathscr{F}_t)$, $t \leq T$. Let $W = (W_t(\omega), \mathscr{F}_t)$ be a Wiener process and let $\alpha = (\alpha_t(\omega), \mathscr{F}_t)$ be some process independent on it whose trajectories $a = (a_t), 0 \leq t \leq T$, belong to the measure space $(A_T, \mathscr{B}_{A_T})$.

Consider a continuous random process $\xi = (\xi_t, \mathscr{F}_t)$, $0 \leq t \leq T$, having the differential

$$d\xi_t = A(t, \alpha, \xi)dt + B(t, \xi)dW_t, \qquad \xi_0 = 0. \tag{7.170}$$

We shall assume the following conditions as satisfied:

(A) the random process $\xi = (\xi_t, (\omega))$, $0 \le t \le T$, is a strong ($\mathcal{F}_t^{\alpha, W}$-measurable) solution of Equation (7.170).

(B) the functionals $A(t, a, x)$, $B(t, x)$ are nonanticipative and, for each $a \in A_T$ and $x \in C_T$ (note that $(C_T, \mathcal{B}_T)$ is a measure space of functions $x = (x_t)$, $0 \le t \le T$, continuous on $[0, T]$),

$$\int_0^T |A(t, a, x)| dt < \infty, \qquad \int_0^T B^2(t, x) dt < \infty. \tag{7.171}$$

(C) for any x and $\tilde{x}$ from C_T,

$$|B(t, x) - B(t, \tilde{x})|^2 \le L_1 \int_0^t |x_s - \tilde{x}_s|^2 \, dK(s) + L_2 |x_t - \tilde{x}_t|^2, \tag{7.172}$$

$$B^2(t, x) \le L_1 \int_0^t (1 + x_s^2) dK(s) + L_2(1 + x_t^2), \tag{7.173}$$

$$B^2(t, x) \ge C > 0, \tag{7.174}$$

where $K(t)$ is a nondecreasing right continuous function, $0 \le K(t) \le 1$, and C, L_1, L_2 are constants.

(D)

$$P\left(\int_0^T A^2(t, \alpha, \xi) dt < \infty\right) = P\left(\int_0^T A^2(t, \alpha, \eta) dt < \infty\right) = 1, \tag{7.175}$$

where $\eta = (\eta_t, \mathcal{F}_t^W)$ is a strong solution of the equation

$$d\eta_t = B(t, \eta) dW_t, \qquad \eta_0 = 0, \tag{7.176}$$

existing because of Theorem 4.6 and assumption (C).

(E)

$$\int_0^T M|A(t, \alpha, \xi)| dt < \infty, \qquad P\left(\int_0^T \bar{A}^2(t, \xi) dt < \infty\right) = 1, \tag{7.177}$$

where $\bar{A}(t, \xi) = M[A(t, \alpha, \xi) | \mathcal{F}_t^\xi]$.

Denote by μ_ξ, μ_η and μ_α measures corresponding to the processes ξ, η and α. Also, let $\mu_{\alpha, \xi}$ be a distribution of the probabilities in the space $(A_T \times C_T, \mathcal{B}_{A_T} \times \mathcal{B}_T)$, induced by the pair of processes (α, ξ), and let $\mu_\alpha \times \mu_\xi$ be a Cartesian product of the measures μ_α and μ_ξ.

Theorem 7.23. *Let $g_T(\alpha, \xi)$ be an $\mathcal{F}_T^{\alpha, \xi}$-measurable functional with $M|g_T(\alpha, \xi)| < \infty$. If the processes α and W are independent and the conditions* (A)–(E) *are fulfilled, then* (P-a.s.)

$$M[g_T(\alpha, \xi) | \mathcal{F}_T^\xi] = \int_{A_T} g_T(a, \xi) \rho_T(a, \xi) d\mu_\alpha(a), \tag{7.178}$$

where

$$\rho_T(a, x) = \frac{d\mu_{\alpha, \xi}}{d[\mu_\alpha \times \mu_\eta]}(a, x) \bigg/ \frac{d\mu_\xi}{d\mu_\eta}(x); \tag{7.179}$$

here (P-a.s.)

$$\rho_T(\alpha, \xi) = \exp\left\{\int_0^T \frac{A(t, \alpha, \xi) - \bar{A}(t, \xi)}{B(t, \xi)} d\overline{W}_t - \frac{1}{2}\int_0^T \frac{[A(t, \alpha, \xi) - \bar{A}(t, \xi)]^2}{B^2(t, \xi)} dt\right\}, \tag{7.180}$$

where the functional $\bar{A} = (\bar{A}(t, x), \mathscr{B}_{t+})$, $0 \le t \le T$, *is such that, for almost all* t, $0 \le t \le T$, (P-a.s.)

$$\bar{A}(t, \xi) = M[A(t, \alpha, \xi) \mid \mathscr{F}_t^\xi], \tag{7.181}$$

and $\overline{W} = (\overline{W}_t, \mathscr{F}_t^\xi)$ *is a Wiener process with*

$$\overline{W}_t = \int_0^t \frac{d\xi_s - \bar{A}(s, \xi)ds}{B(s, \xi)}. \tag{7.182}$$

For proving (7.178), which is actually no more than another way of writing the Bayes formula given by (7.169) (with substitution of the integration over the space of elementary events for integration in a function space), a number of auxiliary statements will be needed.

7.9.3

According to assumption (A) the continuous random process $\xi = (\xi_t)$, $0 \le t \le T$, is a strong solution of Equation (7.170). Let t be fixed. Since for fixed t the value ξ_t is $\mathscr{F}_t^{\alpha, W}$-measurable, then there will be (for given t) a measurable functional $Q_t(a, x)$, such that (P-a.s.)

$$\xi_t(\omega) = Q_t(\alpha(\omega), W(\omega)). \tag{7.183}$$

Consider for $a \in A_T$ the processes $\xi^a = (\xi_t^a(\omega))$, $0 \le t \le T$, defined by the equations

$$d\xi_t^a = A(t, a, \xi^a)dt + B(t, \xi^a)dW_t, \qquad \xi_0^a = 0. \tag{7.184}$$

We shall now show that at fixed t, (($\mu_a \times P$)-a.s.)

$$\xi_t^a(\omega) = Q_t(a, W(\omega)). \tag{7.185}$$

Let us consider an initial probability space $(\Omega, \mathscr{F}, P)$ such that $\Omega = A_T \times C_T$, $\mathscr{F} = \mathscr{B}_{A_T} \times \mathscr{B}_T$, $P = \mu_a \times \mu_W$. (This assumption does not restrict the generality but simplifies its consideration.) Then assuming $\omega = (\alpha, W)$, $\alpha(\omega) = \alpha$, and $W(\omega) = W$, we can see that Equation (7.185) is valid $\mu_\alpha \times \mu_W$-a.s. because of (7.170).

Introduce along with the initial space an identical space $(\tilde{\Omega}, \tilde{\mathscr{F}}, \tilde{P})$. Let $\xi(\tilde{\omega})$, $W(\tilde{\omega})$, $\alpha(\tilde{\omega})$ be processes considered on $(\tilde{\Omega}, \tilde{\mathscr{F}}, \tilde{P})$ and having the same (mutual) distributions as the processes $\xi(\omega)$, $W(\omega)$, $\alpha(\omega)$.

Consider on $(\Omega \times \tilde{\Omega}, \mathscr{F} \times \tilde{\mathscr{F}}, P \times \tilde{P})$ the process $\xi^{\alpha(\omega)}(\tilde{\omega}) = (\xi_t^{\alpha(\omega)}(\tilde{\omega}))$, $0 \le t \le T$, with

$$d\xi_t^{\alpha(\omega)}(\tilde{\omega}) = A(t, \alpha(\omega), \xi^{\alpha(\omega)}(\tilde{\omega}))dt + B(t, \xi^{\alpha(\omega)}(\tilde{\omega}))dW_t(\tilde{\omega}), \qquad \xi_0^{\alpha(\omega)}(\tilde{\omega}) = 0.$$

Then, because of (7.185), $P \times$ ($\tilde{P}$-a.s.)

$$\xi_t^{\alpha(\omega)}(\tilde{\omega}) = Q_t(\alpha(\omega), W(\tilde{\omega})). \tag{7.186}$$

Lemma 7.5. *Let the processes $\alpha(\omega)$ and $W(\omega)$ be independent. Then for any $A \in \mathscr{B}_T$*

$$P\{\xi(\omega) \in A \mid \mathscr{F}_T^\alpha\} = \tilde{P}\{\xi^{\alpha(\omega)}(\tilde{\omega}) \in A\} \qquad (P\text{-a.s.}). \tag{7.187}$$

PROOF. From the Fubini theorem it follows that the probability

$$\tilde{P}\{\xi_t^a(\tilde{\omega}) \le b\} = \tilde{P}\{Q_t(a, W(\tilde{\omega})) \le b\} = \mu_W\{x: Q_t(a, x) \le b\}, \tag{7.188}$$

regarded as a function $a \in A_T$ is $\mathscr{B}_{A_T}$-measurable. Consequently, $\tilde{P}\{\xi_t^a(\tilde{\omega}) \le b\}$ is an $\mathscr{F}_T^\alpha$-measurable function from (A).

It will be shown that (P-a.s.)

$$P\{\xi_t(\omega) \le b \mid \mathscr{F}_T^\alpha\} = \tilde{P}\{\xi^{\alpha(\omega)}(\tilde{\omega}) \le b\}. \tag{7.189}$$

Let $\lambda(\alpha(\omega))$ be an $\mathscr{F}_T^\alpha$-measurable bounded random variable. Then, by the Fubini theorem,

$$\begin{aligned} M\lambda(\alpha(\omega))\chi_{\{\xi_t(\omega) \le b\}}(\omega) &= \int_{A_T}\int_{C_T} \lambda(a)\chi_{\{Q_t(a, x) \le b\}}(x)d\mu_\alpha(a)d\mu_W(x) \\ &= \int_{A_T} \lambda(a)\left[\int_{C_T} \chi_{\{Q_t(a, x) \le b\}}(x)d\mu_W(x)\right]d\mu_\alpha(a) \\ &= \int_{A_T} \lambda(a)\tilde{P}\{\xi_t^a(\tilde{\omega}) \le b\}d\mu_\alpha(a) \\ &= M[\lambda(\alpha(\omega))\tilde{P}\{\xi_t^{\alpha(\omega)}(\tilde{\omega}) \le b\}], \end{aligned}$$

where we make use of Equation (7.183).

Consequently,

$$M[\lambda(\alpha(\omega))\tilde{P}\{\xi_t^{\alpha(\omega)}(\tilde{\omega}) \le b\}] = M\lambda(\alpha(\omega))\chi_{\{\xi_t(\omega) \le b\}}(\omega),$$

which proves (7.189).

Analogously, it is proved that for any b_i, $-\infty < b_i < \infty$, $i = 1, \ldots, n$, $0 \le t_1 < t_2 \cdots < t_n \le T$, ($P$-a.s.)

$$P\{\xi_{t_1}(\omega) \le b_1, \ldots, \xi_{t_n}(\omega) \le b_n \mid \mathscr{F}_T^\alpha\} = \tilde{P}\{\xi_{t_1}^{\alpha(\omega)}(\tilde{\omega}) \le b_1, \ldots, \xi_{t_n}^{\alpha(\omega)}(\tilde{\omega}) \le b_n\}, \tag{7.190}$$

from which follows (7.187). □

In the next two lemmas it will be shown that $\mu_{\alpha,\xi} \sim \mu_\alpha \times \mu_\eta$, $\mu_\xi \sim \mu_\eta$ and densities of these measures will be found.

Lemma 7.6. *Let the processes α and W be independent. Then in assumptions* (A)–(D):

$$\mu_{\alpha,\xi} \sim \mu_\alpha \times \mu_\eta, \tag{7.191}$$

and (P-a.s.)

$$\frac{d\mu_{\alpha,\xi}}{d[\mu_\alpha \times \mu_\eta]}(\alpha, \eta) = \exp\left[\int_0^T \frac{A(t, \alpha, \eta)}{B^2(t, \eta)} d\eta_t - \frac{1}{2}\int_0^T \frac{A^2(t, \alpha, \eta)}{B^2(t, \eta)} dt\right]. \tag{7.192}$$

PROOF. Let us consider the processes introduced above, $\xi^a(\omega) = \{\xi_t^a(\omega), 0 \le t \le T\}$, and show that $\mu_\alpha \times \mu_\eta$-a.s.

$$\frac{d\mu_{\alpha,\xi}}{d[\mu_\alpha \times \mu_\eta]}(a, x) = \frac{d\mu_{\xi^a}}{d\mu_\eta}(x). \tag{7.193}$$

Let the set $\Gamma = \Gamma_1 \times \Gamma_2$, $\Gamma_1 \in \mathscr{B}_{A_T}$, $\Gamma_2 \in \mathscr{B}_T$. Then, because of the preceding lemma,

$$\begin{aligned}\mu_{\alpha,\xi}(\Gamma) &= P\{\omega:\alpha(\omega) \in \Gamma_1, \xi(\omega) \in \Gamma_2\}\\ &= \int_{\{\omega:\alpha(\omega)\in\Gamma_1\}} P\{\xi(\omega) \in \Gamma_2 | \mathscr{F}_T^\alpha\} dP(\omega) = \int_{\Gamma_1} \tilde{P}\{\xi^a(\tilde{\omega}) \in \Gamma_2\} d\mu_\alpha(a)\\ &= \int_{\Gamma_1} \mu_{\xi^a}(\Gamma_2) d\mu_\alpha(a).\end{aligned} \tag{7.194}$$

According to (7.185), assumptions (A)–(E) and Theorem 7.19, $\mu_{\xi^a} \sim \mu_\eta$ for μ_α-almost all a, where

$$\frac{d\mu_{\xi^a}}{d\mu_\eta}(\eta) = \exp\left(\int_0^T \frac{A(t, a, \eta)}{B^2(t, \eta)} d\eta_t - \frac{1}{2}\int_0^T \frac{A^2(t, a, \eta)}{B^2(t, \eta)} dt\right) \quad (\mu_\eta\text{-a.s.}). \tag{7.195}$$

Hence, from (7.194) it follows that

$$\begin{aligned}\mu_{\alpha,\xi}(\Gamma) = \int_{\Gamma_1} \mu_{\xi^a}(\Gamma_2) d\mu_\alpha(a) &= \int_{\Gamma_1}\left[\int_{\Gamma_2} \frac{d\mu_{\xi^a}}{d\mu_\eta}(x) d\mu_\eta(x)\right] d\mu_\alpha(a)\\ &= \int_{\Gamma_1\times\Gamma_2} \frac{d\mu_{\xi^a}}{d\mu_\eta} d[\mu_\alpha \times \mu_\eta](a, x).\end{aligned}$$

Consequently, $\mu_{\alpha,\xi} \ll \mu_\alpha \times \mu_\eta$ and $\mu_\alpha \times \mu_\eta$-a.s. Equation (7.193) holds. Finally, according to (7.195),

$$\mu_\alpha \times \mu_\eta\left\{a, x: \frac{d\mu_{\xi^a}}{d\mu_\eta}(x) = 0\right\} = 0;$$

hence, by Lemma 6.8, $\mu_\alpha \times \mu_\eta \ll \mu_{\alpha,\xi}$. □

Lemma 7.7. *Let the processes* α *and* W *be independent. Then in assumptions* (A)–(E), $\mu_\xi \sim \mu_\eta$ *and*

$$\frac{d\mu_\xi}{d\mu_\eta}(x) = \int_{A_T} \frac{d\mu_{\alpha,\xi}}{d[\mu_\alpha \times \mu_\eta]}(a, x) d\mu_\alpha(a) \qquad (\mu_\eta\text{-a.s.}), \tag{7.196}$$

$$\frac{d\mu_\xi}{d\mu_\eta}(\eta) = \exp\left\{\int_0^T \frac{\bar{A}(t, \eta)}{B^2(t, \eta)} d\eta_t - \frac{1}{2}\int_0^T \frac{\bar{A}^2(t, \eta)}{B^2(t, \eta)} dt\right\}, \tag{7.197}$$

where $\bar{A}(t, x) = M[A(t, \alpha, \xi) | \mathscr{F}_t^\xi]_{\xi = x}$.

PROOF. Denoting

$$\varphi(a, x) = \frac{d\mu_{\alpha,\xi}}{d[\mu_\alpha \times \mu_\xi]}(a, x),$$

we find that for $\Gamma \in \mathscr{B}_T$

$$\mu_\xi(\Gamma) = \int_{A_T}\int_\Gamma d\mu_{\alpha,\xi}(a, x) = \int_{A_T \times \Gamma} \varphi(a, x) d[\mu_\alpha \times \mu_\eta](a, x)$$

$$= \int_\Gamma\left[\int_{A_T} \varphi(a, x) d\mu_\alpha(a)\right] d\mu_\eta(x).$$

Hence, $\mu_\xi \ll \mu_\eta$. Similarly one can show that $\mu_\eta \ll \mu_\xi$ where

$$\frac{d\mu_\eta}{d\mu_\xi}(x) = M\left\{\frac{d[\mu_\alpha \times \mu_\eta]}{d\mu_{\alpha\xi}}(\alpha, \xi) | \mathscr{F}_T^\xi\right\}_{\xi = x}.$$

From the equivalence of the measures μ_ξ and μ_η and the assumption

$$P\left(\int_0^T \bar{A}^2(t, \xi) dt < \infty\right) = 1$$

it follows also that

$$P\left(\int_0^T \bar{A}^2(t, \eta) dt < \infty\right) = 1.$$

Applying Theorem 7.19, we obtain (7.197) as well as the representation

$$\frac{d\mu_\eta}{d\mu_\xi}(\xi) = \exp\left\{-\int_0^T \frac{\bar{A}(t, \xi)}{B^2(t, \xi)} d\xi_t + \frac{1}{2}\int_0^T \frac{\bar{A}^2(t, \xi)}{B^2(t, \xi)} dt\right\}. \quad \square \tag{7.198}$$

7.9.4

PROOF OF THEOREM 7.23

From the continuity of the process ξ it follows that the σ-algebra $\mathscr{F}_T^\xi$ is separable. Next, since the process ξ is continuous, the conditional probability $M[\chi_A(\omega) | \mathscr{F}_T^\alpha](\omega)$ has[17] a regular version (which we denote by $Q(A, \omega)$). Let

[17] See, for example, [13] and [37].

the sets $A \in \mathscr{F}_T^\xi$ and $B \in \mathscr{B}_T$ be related as $A = \{\tilde{\omega}:\xi(\tilde{\omega}) \in B\}$. Then ($P$-a.s.)

$$Q(A, \tilde{\omega}) = \tilde{P}\{A|\mathscr{F}_T^\alpha\}(\tilde{\omega}) = \tilde{P}\{\tilde{\omega}:\xi(\tilde{\omega}) \in B|\mathscr{F}_T^\alpha\}(\tilde{\omega})$$

$$= P\{\omega: \xi^{\alpha(\tilde{\omega})}(\omega) \in B\} = \int_B \frac{d\mu_\xi\alpha(\tilde{\omega})}{d\mu_\xi}(x)d\mu_\xi(x) = \int_A \frac{d\mu_\xi\alpha(\tilde{\omega})}{d\mu_\xi}(\xi(\omega))dP(\omega).$$

Denote

$$q(\omega, \tilde{\omega}) = \frac{d\mu_\xi\alpha(\tilde{\omega})}{d\mu_\xi}(\xi(\omega)).$$

Then according to (7.168),

$$M[g_T(\alpha, \xi)|\mathscr{F}_T^\xi] = \frac{\int_\Omega g_T(\alpha(\tilde{\omega}), \xi(\omega))d\mu_\xi\alpha(\tilde{\omega})/d\mu_\xi(\xi(\omega))dP(\tilde{\omega})}{\int_\Omega d\mu_\xi\alpha(\tilde{\omega})/d\mu_\xi(\xi(\omega))dP(\tilde{\omega})}. \tag{7.199}$$

But $\mu_\xi \sim \mu_\eta$ and (P-a.s.) $\mu_\xi\alpha(\tilde{\omega}) \sim \mu_\xi$. Hence

$$\frac{d\mu_\xi\alpha(\tilde{\omega})}{d\mu_\xi}(\xi(\omega)) = \frac{d\mu_\xi\alpha(\tilde{\omega})}{d\mu_\eta}(\xi(\omega)) \cdot \frac{d\mu_\eta}{d\mu_\xi}(\xi(\omega)),$$

which after substitution in (7.199) yields

$$M[g_T(\alpha, \xi)|\mathscr{F}_T^\xi = \frac{\int_\Omega g_T(\alpha(\tilde{\omega}), \xi(\omega))d\mu_\xi\alpha(\tilde{\omega})/d\mu_\eta(\xi(\omega))dP(\tilde{\omega})}{\int_\Omega d\mu_\xi\alpha(\tilde{\omega})/d\mu_\eta(\xi(\omega))dP(\tilde{\omega})}.$$

Taking into account Equation (7.193) and (7.179), we find that

$$M[g_T(\alpha, \xi)|\mathscr{F}_T^\xi] = \frac{\int_\Omega g_T(\alpha(\tilde{\omega}), \xi(\omega))d\mu_{\alpha,\xi}/d[\mu_\alpha \times \mu_\eta](\alpha(\tilde{\omega}), \xi(\omega))dP(\tilde{\omega})}{\int_\Omega d\mu_{\alpha,\xi}/d[\mu_\alpha \times \mu_\eta](\alpha(\tilde{\omega}), \xi(\omega))dP(\tilde{\omega})}$$

$$= \int_\Omega g_T(\alpha(\tilde{\omega}), \xi(\omega))\rho_T(\alpha(\tilde{\omega}), \xi(\omega))dP(\tilde{\omega})$$

$$= \int_{A_T} g_T(\alpha, \xi(\omega))\rho_T(\alpha, \xi(\omega))d\mu_\alpha(\alpha).$$

This proves the Bayes formula given by (7.178). (7.180), yielding a representation for $\rho_T(a, x)$, follows from (7.192), (7.196) and (7.197).

Let us note that the validity of the Bayes formula given by (7.178) can be established by direct calculation without reference to (7.168). Indeed, first the random variable $\int_A g_T(a, \xi(\omega))\rho_T(a, \xi(\omega))d\mu_\alpha(a)$ is $\mathscr{F}_T^\xi$-measurable. Next, let $\lambda(\xi) = \lambda(\xi(\omega))$ be an $\mathscr{F}_T^\xi$-measurable bounded variable. Then

$$M[g_T(\alpha, \xi)\lambda(\xi)] = M\left[\frac{d\mu_{\alpha,\xi}}{d[\mu_\alpha \times \mu_\eta]}(\alpha, \eta)g_T(\alpha, \eta)\lambda(\eta)\right]$$

$$= M\left[\lambda(\eta)M\left\{\frac{d\mu_{\alpha,\xi}}{d[\mu_\alpha \times \mu_\eta]}(\alpha, \eta)g_T(\alpha, \eta)|\mathscr{F}_T^\eta\right\}\right].$$

But the processes α and η are independent. Hence

$$M\left\{\frac{d\mu_{\alpha,\xi}}{d[\mu_\alpha \times \mu_\eta]}(\alpha, \eta)g_T(\alpha, \eta)|\mathscr{F}_T^\eta\right\} = \int_{A_T} \frac{d\mu_{\alpha,\xi}}{d[\mu_\alpha \times \mu_\eta]}(a, \eta)g_T(a, \eta)d\mu_\alpha(a),$$

and, therefore,

$$M[g_T(\alpha, \xi)\lambda(\xi)] = M\left[\lambda(\eta)\int_{A_T} \frac{d\mu_{\alpha,\xi}}{d[\mu_\alpha \times \mu_\eta]}(a, \eta)g_T(a, \eta)d\mu_\alpha(a)\right]$$

$$= M\left[\lambda(\xi)\int_{A_T} \frac{d\mu_{\alpha,\xi}}{d[\mu_\alpha \times \mu_\eta]}(a, \xi)g_T(a, \xi)d\mu_\alpha(a)\frac{d\mu_\eta}{d\mu_\xi}(\xi)\right]$$

$$= M\left\{\lambda(\xi)\int_{A_T}\left[\frac{d\mu_{\alpha,\xi}}{d[\mu_\alpha \times \mu_\xi]}(a, \xi)\Big/\frac{d\mu_\xi}{d\mu_\eta}(\xi)\right]g_T(a, \xi)d\mu_\alpha(a)\right\}$$

$$= M\left\{\lambda(\xi)\int_{A_T} g_T(a, \xi)\rho_T(a, \xi)d\mu_\alpha(a)\right\},$$

from which follows (7.178). □

Note 1. From proof of Theorem 7.23 it is seen that one can omit conditions (A) and (C) in the formulation of the theorem, if there is equivalence of the measures $\mu_{\alpha,\xi}$ and $\mu_\alpha \times \mu_\eta$ and (7.192) is valid for the density.

Note 2. Let there exist a regular conditional probability $\mu_{\alpha|\xi_0}$ corresponding to the process α for the given ξ_0. If, in (7.170) and (7.176), $\xi_0 = \eta_0 = \zeta$, where $P(|\zeta| < \infty) = 1$, then in similar fashion one can prove that

$$M[g_T(\alpha, \xi)|\mathscr{F}_T^\xi] = \int_{A_T} g_T(a, \xi)\rho_T(a, \xi)d\mu_{\alpha|\xi_0}(a) \tag{7.200}$$

(compare with (7.178)).

These notes will be recalled in Lemma 11.5.

Note 3. (7.178) with $\rho_T(a, \xi)$ from (7.180) remains valid if, instead of condition (D), we specify that $P\{\int_0^T A^2(t, \alpha, \xi)dt < \infty\} = 1$.

7.9.5

With (7.179) in mind, we find that

$$P(\alpha_T \le b|\mathscr{F}_T^\xi) = \int_{A_T} \chi_{\{a_T \le b\}}(a)\rho_T(a, \xi)d\mu_\alpha(a). \tag{7.201}$$

Note that[18]

$$\int_A \chi_{\{a_T \le b\}}\rho_T(a, \xi)d\mu_\alpha(a) = \tilde{M}[\chi_{\{\alpha_T(\tilde{\omega}) \le b\}}\rho_T(\alpha(\tilde{\omega})), \xi(\omega)]$$

$$= \tilde{M}[\chi_{\{\alpha_T(\tilde{\omega}) \le b\}}\tilde{M}(\rho_T(\alpha(\tilde{\omega}), \xi(\omega))|\alpha_T(\tilde{\omega}))].$$

Hence, from (7.201) we find

$$P(\alpha_T \le b|\mathscr{F}_T^\xi) = \int_{-\infty}^b \tilde{M}[\rho_T(\alpha(\tilde{\omega}), \xi(\omega))|\alpha_T(\tilde{\omega}) = a]dF_{\alpha_T}(a), \tag{7.202}$$

where $F_{\alpha_T}(a) = P(\alpha_T \le a)$, $a \in \mathbb{R}^1$.

[18] $\tilde{M}$ is an averaging over the measure $\tilde{P}$, identical to measure P but defined on $(\tilde{\Omega}, \tilde{\mathscr{F}})$.

If, in particular, $F_{\alpha_T}(a)$ has the density $p_{\alpha_T}(a)$, then

$$P(\alpha_T \le b | \mathscr{F}_T^{\xi}) = \int_{-\infty}^{b} \tilde{M}[\rho_T(\alpha(\tilde{\omega}), \xi(\omega)) | \alpha_T(\tilde{\omega}) = a] p_{\alpha_T}(a) da. \quad (7.203)$$

Corollary 1. *If the random variable α_T has the density of probability distribution $P_{\alpha_T}(a)$, then the a posteriori distribution $P(\alpha_T \le b | \mathscr{F}_T^{\xi})$ also has* (P-a.s.) *the density*

$$\frac{dP(\alpha_T \le b | \mathscr{F}_T^{\xi})}{db} = p_{\alpha_T}(b) \tilde{M}[\rho_T(\alpha(\tilde{\omega}), \xi(\omega)) | \alpha_T(\tilde{\omega}) = b]. \quad (7.204)$$

If the random variable α_T takes on a finite or countable set of values b_1, $b_2, \ldots$, then

$$P(\alpha_T = b_k | \mathscr{F}_T^{\xi}) = p_{\alpha_T}(b_k) M[\rho_T(\alpha(\tilde{\omega}), \xi(\omega)) | \alpha_T(\tilde{\omega})) = b_k], \quad (7.205)$$

where $p_{\alpha_T}(b_k) = P(\alpha_T = b_k)$.

Corollary 2. *If $\alpha = \alpha(\omega)$ is a random variable with the distribution function $F_\alpha(a) = P(\alpha(\omega) \le \alpha)$, then*

$$P(\alpha \le b | \mathscr{F}_T^{\xi}) = \int_{-\infty}^{b} \rho_T(a, \xi(\omega)) dF_\alpha(a). \quad (7.206)$$

Notes and references

7.1,7.2. Some general problems of absolute continuity of the measures in function spaces are dealt with in Gykhman and Skorokhod's paper [35]. Absolute continuity of the Wiener measure under various transformations was discussed in Cameron and Martin [80], [81], and Prokhorov [134]. The results related to these sections are due to Yershov [53], Liptser and Shiryayev [118], Kadota, Shepp [66].

7.3. The structure of the processes whose measure is absolutely continuous and equivalent to Wiener measure was dicussed in Hitsuda [158], Liptser and Shiryayev [118], Yershov [53], and Kailath [69].

7.4. Representation (7.73) for Ito processes involving the innqvation process $\overline{W}$ is due to Shiryayev [166] and Kailath [67]. See also the papers of Yershov [52], Lyptser and Shiryayev [111], Fujisaki, Kallianpur, Kunita [156].

7.5. Lemma 7.2 for the case of Gaussian processes with the zero mean was proved in Kadota [64]. The proof of the reduction of the general case ($M\beta_t \not\equiv 0$) to a case of the processes with zero mean ($M\beta_t \equiv 0$) was noted by A. S. Kholevo. Representations of the type (7.99) were examined in Hitsuda [158]. In proving Gaussianness of the (Lebesgue) integral $\int_0^T \alpha(t)dt$ from the Gaussian process $\alpha(t)$, $0 \le t \le T$, representations for semi-invariants were exploited, see Leonov and Shiryayev [103] and Shiryayev [164]. Another proof of Gaussianness can be obtained with the aid of Theorem 2.8 presented in Doob [46].

7.6. The results related to this subsection are due to the authors.

7.7. Theorem 7.21 generalizes a well-known result due to Cameron and Martin [80], [81].

7.8. Theorem 7.22 generalizes a well-known Rao–Cramer inequality [90] and Wolfowitz inequality [22].

7.9. Lemmas 7.3 and 7.4 are contained in Kallianpur and Striebel [74].

General equations of optimal nonlinear filtering, interpolation and extrapolation of partially observable random processes

8

8.1 Filtering: the main theorem

8.1.1

Let $(\Omega, \mathscr{F}, P)$ be a complete probability space, and let $(\mathscr{F}_t)$, $0 \leq t \leq T$, be a nondecreasing family of right continuous σ-algebras $\mathscr{F}$ augmented by sets from $\mathscr{F}$ of zero probability.

Let (θ, ξ) be a two-dimensional partially observable random process where $\theta = (\theta_t, \mathscr{F}_t)$, $0 \leq t \leq T$, is an *unobservable* component, and $\xi = (\xi_t, \mathscr{F}_t)$, $0 \leq t \leq T$, is an observable component. The problem of optimal filtering for a partially *observable* process (θ, ξ) consists in the construction for each instant t, $0 \leq t \leq T$, of an optimal mean square estimate of some $\mathscr{F}_t$-measurable function h_t of (θ, ξ) on the basis of observation results ξ_s, $s \leq t$.

If $Mh_t^2 < \infty$, then the optimal estimate evidently is the a posteriori mean $\pi_t(h) = M(h_t | \mathscr{F}_t^\xi)$. Without special assumptions on the structure of the processes (h, ξ), $\pi_t(h)$ is difficult to determine. However under the assumption that the components of the process (h, ξ) are processes of the type (8.1) and (8.2) we can characterize $\pi_t(h)$ by the stochastic differential equation given by (8.10) and called the *optimal nonlinear filtering equation*. Following chapters will be devoted to the application of these equations for efficient construction of optimal "filters."

8.1.2

Let us begin by formulating the basic assumptions on the structure of the process (h, ξ). It will be assumed that the process $h = (h_t, \mathscr{F}_t)$, $t \leq T$, can be represented as follows:

$$h_t = h_0 + \int_0^t H_s \, ds + x_t, \tag{8.1}$$

where $X = (x_t, \mathscr{F}_t)$, $t \leq T$, is a martingale, and $H = (H_t, \mathscr{F}_t)$, $t \leq T$, is a random process with $\int_0^T |H_s|\, ds < \infty$ (P-a.s.). Because of the right continuity of the σ-algebras $\mathscr{F}_t$ and Theorem 3.1, the martingale $X = (x_t, \mathscr{F}_t)$, $t \leq T$, has a right continuous modification which will be treated further on.

As to the observable process $\xi = (\xi_t, \mathscr{F}_t)$ it will be assumed that it is an Ito process,

$$\xi_t = \xi_0 + \int_0^t A_s(\omega)ds + \int_0^t B_s(\xi)dW_s, \tag{8.2}$$

where $W = (W_t, \mathscr{F}_t)$ is a Wiener process. The processes $A = (A_t(\omega), \mathscr{F}_t)$, and $B = (B_t(\xi)\mathscr{F}_t)$ are assumed to be such that

$$P\left(\int_0^T |A_t(\omega)|dt < \infty\right) = 1, \qquad P\left(\int_0^T B_t^2(\xi)dt < \infty\right) = 1, \tag{8.3}$$

where the measurable functional $B_t(x)$, $0 \leq t \leq T$, $x \in C_T$, is $\mathscr{B}_t$-measurable for each $t \leq T$.

Next it will be assumed that the functional $B_t(x)$, $x \in C_T$, $0 \leq t \leq T$, satisfies the following conditions:

$$|B_t(x) - B_t(y)|^2 \leq L_1 \int_0^t [x_s - y_s]^2\, dK(s) + L_2[x_t - y_t]^2, \tag{8.4}$$

$$B_t^2(x) \leq L_1 \int_0^t (1 + x_s^2)dK(s) + L_2(1 + x_t^2), \tag{8.5}$$

where L_1, L_2 are nonnegative constants and $K(t)$, $0 \leq K(t) \leq 1$, is a non-decreasing right continuous function ($x, y \in C_T$).

If $g_t = g_t(\omega)$, $0 \leq t \leq T$, is some measurable random process with $M|g_t| < \infty$, then the conditional expectation $M(g_t|\mathscr{F}_t^\xi)$ has a measurable modification (see [52], [126]) which will be denoted by $\pi_t(g)$.

8.1.3

The main result of this chapter can be formulated as follows.

Theorem 8.1. *Let the partially observable random process* (h, ξ) *permit the representation given by* (8.1)–(8.2). *Let* (8.3)–(8.5) *be satisfied and let*

$$\sup_{0 \leq t \leq T} Mh_t^2 < \infty, \tag{8.6}$$

$$\int_0^T MH_t^2\, dt < \infty, \tag{8.7}$$

$$\int_0^T MA_t^2\, dt < \infty, \tag{8.8}$$

$$B_t^2(x) \geq C > 0. \tag{8.9}$$

Then for each t, $0 \leq t \leq T$, (P-a.s.)

$$\pi_t(h) = \pi_0(h) + \int_0^t \pi_s(H)ds + \int_0^t \{\pi_s(D) + [\pi_s(hA) - \pi_s(h)\pi_s(A)]B_s^{-1}(\xi)\}d\overline{W}_s, \tag{8.10}$$

where

$$\overline{W}_t = \int_0^t \frac{d\xi_s - \pi_s(A)ds}{B_s(\xi)}$$

is a Wiener process (with respect to the system $(\mathscr{F}_t^\xi)$, $0 \leq t \leq T$), *and* $D = (D_t, \mathscr{F}_t)$ *is a process with*[1]

$$D_t = \frac{d\langle x, W\rangle_t}{dt}. \tag{8.11}$$

Equation (8.10) will be the basic equation of (optimal nonlinear) filtering.

8.2 Filtering: proof of the main theorem

8.2.1

Proof of Theorem 8.1. The proof will be based essentially on the results of Chapters 5 and 7.

From (8.8) and (8.9) it follows that

$$\int_0^T M|A_t|dt < \infty, \qquad |B_t(x)| \geq \sqrt{C} > 0, \tag{8.12}$$

Consequently, $M|A_t| < \infty$ for almost all t, $0 \leq t \leq T$. Without restricting generality it can be assumed that $M|A_t| < \infty$ for all t, $0 \leq t \leq T$. Then, by Theorem 7.17, $\overline{W} = (\overline{W}_t, \mathscr{F}_t^\xi)$ is a Wiener process and the process $\xi = (\xi_t, \mathscr{F}_t)$, defined in (8.2) permits the differential

$$d\xi_t = \pi_t(A)dt + B_t(\xi)d\overline{W}_t, \tag{8.13}$$

where

$$\pi_t(A) = M[A_t(\omega)|\mathscr{F}_t^\xi].$$

Because of Jensen's inequality and (8.8),

$$\int_0^T M\pi_t^2(A)dt \leq \int_0^T MA_t^2\, dt < \infty, \tag{8.14}$$

[1] The definition of the process $\langle x, W\rangle_t$ is given in Subsection 5.1.2.

which, together with (8.4), (8.5) and (8.9), yields the applicability of Theorem 5.18. According to this theorem and Lemma 4.9, any martingale $Y = (y_t, \mathscr{F}_t^\xi)$, $0 \le t \le T$, has a continuous modification permitting the representation[2]

$$y_t = y_0 + \int_0^t f_s(\xi) d\overline{W}_s, \tag{8.15}$$

where $P(\int_0^T f_s^2(\xi) ds < \infty) = 1$ and, in the case of the square integrable martingale, $\int_0^T M f_s^2(\xi) ds < \infty$ (compare with (5.122)).

From (8.1), (8.6), and (8.7) it follows that the martingale $X = (x_t, \mathscr{F}_t)$ is square integrable. Taking on both sides of (8.1) the conditional expectation $M(\cdot | \mathscr{F}_t^\xi)$, we find that

$$\pi_t(h) = M(h_0 | \mathscr{F}_t^\xi) + M\left(\int_0^t H_s \, ds | \mathscr{F}_t^\xi\right) + M(x_t | \mathscr{F}_t^\xi). \tag{8.16}$$

8.2.2

We shall formulate now as lemmas a number of auxiliary statements which will enable us to transform the right side in (8.16) to the expression in the right side of (8.10).

Lemma 8.1. *The process $(M(h_0 | \mathscr{F}_t^\xi), \mathscr{F}_t^\xi)$, $0 \le t \le T$, is a square integrable martingale yielding the representation*

$$M(h_0 | \mathscr{F}_t^\xi) = \pi_0(h) + \int_0^t g_s^h(\xi) d\overline{W}_s, \tag{8.17}$$

where $M \int_0^T [g_s^h(\xi)]^2 \, ds < \infty$.

PROOF. This follows in the obvious way from Theorem 5.18 and Theorem 1.6; also according to the latter, the martingale $M(h_0 | \mathscr{F}_t^\xi)$ has (P-a.s.) the limits to the right for each t, $0 \le t \le T$. □

Lemma 8.2. *The process $(M(x_t | \mathscr{F}_t^\xi), \mathscr{F}_t^\xi)$, $0 \le t \le T$, is a square integrable martingale with the representation*

$$M(x_t | \mathscr{F}_t^\xi) = \int_0^t g_s^x(\xi) d\overline{W}_s \tag{8.18}$$

with $M \int_0^T (g_s^x(\xi))^2 \, ds < \infty$

PROOF. The fact that this process is a martingale can be verified in the same way as in Lemma 5.7. The square integrability follows from the square

[2] In (8.15) the measurable functional $f_s(x)$ is $\mathscr{F}_{s+}$-measurable for each s, $0 \le s \le T$.

integrability of the martingale $X = (x_t, \mathscr{F}_t)$. The existence of $\lim_{s \downarrow t} M(x_s | \mathscr{F}_s^\xi)$ follows from Theorem 3.1. Hence the conclusion of the lemma is a direct corollary of Theorem 5.18. □

Lemma 8.3. *Let $\alpha = (\alpha_t, \mathscr{F}_t)$, $0 \le t \le T$, be some random process with $\int_0^T M|\alpha_t|dt < \infty$, and let $\mathscr{G}$ be some sub-σ-algebra of $\mathscr{F}$. Then*

$$M\left(\int_0^t \alpha_s\, ds | \mathscr{G}\right) = \int_0^t M(\alpha_s | \mathscr{G})ds \qquad (P\text{-a.s.}), \qquad 0 \le t \le T. \quad (8.19)$$

PROOF. Let $\lambda = \lambda(\omega)$ be a bounded $\mathscr{G}$-measurable random variable. Then, using the Fubini theorem, we find that

$$M\left[\lambda \int_0^t \alpha_s\, ds\right] = \int_0^t M[\lambda\alpha_s]ds = \int_0^t M\{\lambda M(\alpha_s | \mathscr{G})\}ds = M\left[\lambda \int_0^t M(\alpha_s | \mathscr{G})ds\right].$$

On the other hand,

$$M\left[\lambda \int_0^t \alpha_s\, ds\right] = M\left[\lambda M\left(\int_0^t \alpha_s\, ds | \mathscr{G}\right)\right].$$

Hence,

$$M\left[\lambda \int_0^t M(\alpha_s | \mathscr{G})ds\right] = M\left[\lambda M\left(\int_0^t \alpha_s\, ds | \mathscr{G}\right)\right].$$

From this, because of the arbitrariness of $\lambda = \lambda(\omega)$, we obtain (8.19). □

Lemma 8.4. *The random process*

$$\left(M\left(\int_0^t H_s\, ds | \mathscr{F}_t^\xi\right) - \int_0^t \pi_s(H)ds,\ \mathscr{F}_t^\xi\right), \qquad 0 \le t \le T, \qquad (8.20)$$

is a square integrable martingale having the representation

$$M\left(\int_0^t H_s\, ds | \mathscr{F}_t^\xi\right) - \int_0^t \pi_s(H)ds = \int_0^t g_s^H(\xi)d\overline{W}_s \qquad (8.21)$$

with $\int_0^T M(g_s^H(\xi))^2\, ds < \infty$.

PROOF. The existence (P-a.s.) of

$$\lim_{s \downarrow t}\left[M\left(\int_0^s H_u\, du | \mathscr{F}_s^\xi\right) - \int_0^s \pi_u(H)du\right] \qquad (8.22)$$

follows from Theorem 1.6. Hence the statement of the lemma will follow immediately from Theorem 5.18, if it is shown that the process given by (8.20) is a martingale (the square integrability follows from the assumption (8.7)).

Let $s \le t$. Then, because of Lemma 8.3,

$$M\left\{M\left[\int_0^t H_u\,du \mid \mathscr{F}_t^\xi\right] - \int_0^t \pi_u(H)du \mid \mathscr{F}_s^\xi\right\}$$

$$= M\left[\int_0^t H_u\,du \mid \mathscr{F}_s^\xi\right] - \int_0^t M[\pi_u(H) \mid \mathscr{F}_s^\xi]du$$

$$= M\left[\int_0^s H_u\,du \mid \mathscr{F}_s^\xi\right] + M\left[\int_s^t H_u\,du \mid \mathscr{F}_s^\xi\right]$$

$$- \int_0^s M[\pi_u(H) \mid \mathscr{F}_s^\xi]du - \int_s^t M[\pi_u(H) \mid \mathscr{F}_s^\xi]du. \tag{8.23}$$

Here

$$\int_0^s M[\pi_u(H) \mid \mathscr{F}_s^\xi]du = \int_0^s \pi_u(H)du \qquad (P\text{-a.s.}) \tag{8.24}$$

and, for $u \ge s$,

$$M[\pi_u(H) \mid \mathscr{F}_s^\xi] = M\{M(H_u \mid \mathscr{F}_u^\xi) \mid \mathscr{F}_s^\xi\} = M\{H_u \mid \mathscr{F}_s^\xi\}.$$

Hence, by Lemma 8.3,

$$M\left[\int_s^t H_u\,du \mid \mathscr{F}_s^\xi\right] = \int_s^t M[\pi_u(H) \mid \mathscr{F}_s^\xi]du. \tag{8.25}$$

From (8.23)–(8.25) it follows that the process given by (8.20) is a martingale. □

8.2.3

Let us return again to proving the theorem. From (8.16), Lemmas 8.1, 8.2 and 8.4 we find that

$$\pi_t(h) = \pi_0(h) + \int_0^t \pi_s(H)ds + \int_0^t g_s(\xi)d\overline{W}_s, \tag{8.26}$$

where

$$g_s(\xi) = g_s^h(\xi) + g_s^x(\xi) + g_s^H(\xi), \tag{8.27}$$

with

$$\int_0^T Mg_s^2(\xi)ds < \infty. \tag{8.28}$$

It will be shown now that for almost all t, $0 \le t \le T$,

$$g_s(\xi) = \pi_s(D) + [\pi_s(hA) - \pi_s(h)\pi_s(A)]B_s^{-1}(\xi) \qquad (P\text{-a.s.}). \tag{8.29}$$

Let us do this as follows. Let $y_t = \int_0^t g_s(\xi)d\overline{W}_s$ and $z_t = \int_0^t \lambda_s(\xi)d\overline{W}_s$, where $\lambda = (\lambda_s(\xi), \mathscr{F}_s^\xi)$ is some bounded random process with $|\lambda_s(\xi)| \le C < \infty$. By properties of stochastic integrals,

$$My_t z_t = M\int_0^t \lambda_s(\xi)g_s(\xi)ds. \tag{8.30}$$

Compute now $My_t z_t$ in the other way, taking into account that, according to (8.26),

$$y_t = \pi_t(h) - \pi_0(h) - \int_0^t \pi_s(H)ds. \tag{8.31}$$

It will be noted that

$$Mz_t\pi_0(h) = M\{\pi_0(h)M(z_t|\mathscr{F}_0^\xi)\} = 0$$

and

$$M\left[z_t\int_0^t \pi_s(H)ds\right] = \int_0^t M[z_t\pi_s(H)]ds$$

$$= \int_0^t M[M(z_t|\mathscr{F}_s^\xi)\pi_s(H)]ds = \int_0^t M[z_s\pi_s(H)]ds.$$

Hence, taking into account that the random variables z_t are $\mathscr{F}_t^\xi$-measurable, we find

$$My_t z_t - Mz_t\pi_t(h) - \int_0^t Mz_s\pi_s(H)ds$$

$$= M[z_t M(h_t|\mathscr{F}_t^\xi)] - \int_0^t M[z_s M(H_s|\mathscr{F}_s^\xi)]ds$$

$$= M\left[z_t h_t - \int_0^t z_s H_s\,ds\right]. \tag{8.32}$$

Let us now make use of

$$\overline{W}_t = \int_0^t \frac{d\xi_s - \pi_s(A)ds}{B_s(\xi)} = W_t + \int_0^t \frac{A_s(\omega) - \pi_s(A)}{B_s(\xi)}ds. \tag{8.33}$$

We obtain

$$z_t = \tilde{z}_t + \int_0^t \lambda_s(\xi)\frac{A_s(\omega) - \pi_s(A)}{B_s(\xi)}ds, \tag{8.34}$$

where

$$\tilde{z}_t = \int_0^t \lambda_s(\xi)dW_s. \tag{8.35}$$

From (8.32) and (8.34) we find that

$$
\begin{aligned}
My_tz_t &= M\left[z_th_t - \int_0^t z_sH_s\,ds\right] \\
&= M\left[\tilde{z}_th_t - \int_0^t \tilde{z}_sH_s\,ds\right] + M\left[h_t\int_0^t \lambda_s(\xi)\frac{A_s(\omega) - \pi_s(A)}{B_s(\xi)}\,ds\right. \\
&\quad \left. - \int_0^t\left(\int_0^s \lambda_u(\xi)\frac{A_u(\omega) - \pi_u(A)}{B_u(\xi)}\right)H_s\,ds\right]. \qquad (8.36)
\end{aligned}
$$

The process $\tilde{z} = (\tilde{z}_t, \mathscr{F}_t)$ is a square integrable martingale. Hence

$$
M\tilde{z}_th_0 = M(h_0M(\tilde{z}_t|\mathscr{F}_0)) = Mh_0\tilde{z}_0 = 0
$$

and

$$
M\int_0^t \tilde{z}_sH_s\,ds = M\int_0^t [M(\tilde{z}_t|\mathscr{F}_s)H_s]ds = M\tilde{z}_t\int_0^t H_s\,ds.
$$

Therefore, because of (8.1) and Theorem 5.2,

$$
M\left[\tilde{z}_th_t - \int_0^t \tilde{z}_sH_s\,ds\right] = M\left[\tilde{z}_t\left(h_t - h_0 - \int_0^t H_s\,ds\right)\right] = M\tilde{z}_tx_t = M\langle\tilde{z}, x\rangle_t.
$$

By Lemma 5.1,

$$
\langle\tilde{z}, x_t\rangle = \int_0^t \lambda_s(\xi)D_s\,ds \qquad (P\text{-a.s.}) \qquad (8.37)
$$

and hence

$$
M\left[\tilde{z}_th_t - \int_0^t \tilde{z}_sH_s\,ds\right] = M\langle\tilde{z}, x\rangle_t = M\int_0^t \lambda_s(\xi)D_s\,ds = M\int_0^t \lambda_s(\xi)\pi_s(D)ds. \qquad (8.38)
$$

Computing now the second item in the right side of (8.36), we get

$$
\begin{aligned}
M\left[h_t\int_0^t \lambda_s(\xi)\frac{A_s(\omega) - \pi_s(A)}{B_s(\xi)}\,ds\right] &= M\int_0^t \lambda_s(\xi)\frac{h_sA_s - h_s\pi_s(A)}{B_s(\xi)}\,ds \\
&\quad + M\int_0^t \lambda_s(\xi)[h_t - h_s]\frac{A_s - \pi_s(A)}{B_s(\xi)}\,ds \\
&= M\int_0^t \lambda_s(\xi)\frac{\pi_s(hA) - \pi_s(h)\pi_s(A)}{B_s(\xi)}\,ds \\
&\quad + M\int_0^t \lambda_s(\xi)[h_t - h_s]\frac{A_s - \pi_s(A)}{B_s(\xi)}\,ds. \qquad (8.39)
\end{aligned}
$$

Note that

$$h_t - h_s = \int_s^t H_u\,du + (x_t - x_s)$$

and $M(x_t - x_s|\mathscr{F}_s) = 0$. Hence

$$M\int_0^t \lambda_s(\xi)[h_t - h_s]\frac{A_s - \pi_s(A)}{B_s(\xi)}\,ds = M\int_0^t \lambda_s(\xi)[x_t - x_s]\frac{A_s - \pi_s(A)}{B_s(\xi)}\,ds$$

$$+ M\int_0^t \lambda_s(\xi)\frac{A_s - \pi_s(A)}{B_s(\xi)}\left(\int_s^t H_u\,du\right)ds$$

$$= M\int_0^t\left[\int_0^s \lambda_u(\xi)\frac{A_u - \pi_u(A)}{B_u(\xi)}\,du\right]H_s\,ds.$$

From this and (8.39) it follows that

$$Mh_t\int_0^t \lambda_s(\xi)\frac{A_s(\omega) - \pi_s(A)}{B_s(\xi)}\,ds = M\int_0^t\left[\int_0^s \lambda_u(\xi)\frac{A_u(\omega) - \pi_u(A)}{B_u(\xi)}\,du\right]H_s\,ds$$

$$+ M\int_0^t \lambda_s(\xi)\frac{\pi_s(hA) - \pi_s(h)\pi_s(A)}{B_s(\xi)}\,ds. \tag{8.40}$$

From (8.36), (8.38) and (8.40) we find that

$$My_tz_t = M\int_0^t \lambda_s(\xi)\left[\pi_s(D) + \frac{\pi_s(hA) - \pi_s(h)\pi_s(A)}{B_s(\xi)}\right]ds.$$

By comparing this expression with (8.30) we readily convince ourselves of the validity of (8.29) (P-a.s.) for almost all t, $0 \le t \le T$. Since the value of the integral $\int_0^t g_s(\xi)d\overline{W}_s$ entering into (8.26) does not change with the change of the function $g_t(\xi)$ on the set of Lebesgue measure zero, then Equation (8.29) can be regarded as satisfied (P-a.s.) for all t, $0 \le t \le T$. Therefore, Theorem 8.1 is proved. □

8.2.4

Note. From the proof of Theorem 8.1 it follows that

$$\int_0^T M\left\{\pi_t(D) + \frac{\pi_t(hA) - \pi_t(h)\pi_t(A)}{B_t(\xi)}\right\}^2 dt < \infty. \tag{8.41}$$

8.2.5

Let us point out one particular case of Theorem 8.1, i.e., when $A_t \equiv 0$, $B_t \equiv 1$, $\xi_0 \equiv 0$.

Theorem 8.2. *Let $W = (W_t, \mathscr{F}_t)$, $0 \le t < T$, be a Wiener process and let the process $h_t = h_0 + \int_0^t H_s\,ds + x_t$, where $X = (x_t, \mathscr{F}_t)$, be a martingale. If*

(I) $\sup_{0 \le t \le T} Mh_t^2 < \infty$,

(II) $\int_0^T MH_t^2\,dt < \infty$,

then

$$\pi_t^W(h) = \pi_0^W(h) + \int_0^t \pi_s^W(H)ds + \int_0^t \pi_s^W(D)dW_s, \tag{8.42}$$

where

$$\pi_t^W(g) = M[g_t | \mathscr{F}_t^W],$$

and

$$D_t = \frac{d\langle x, W\rangle_t}{dt}.$$

PROOF. The representation given by (8.42) follows from (8.10) by noting that in the case under consideration

$$\xi_t = W_t = \overline{W}_t. \quad \square$$

Corollary. *Let $X = (x_t, \mathscr{F}_t)$ be a square integrable martingale. Then*

$$M(x_t | \mathscr{F}_t^W) = Mx_0 + \int_0^t M(a_s | \mathscr{F}_s^W)dW_s, \tag{8.43}$$

where $\langle x, W\rangle_t = \int_0^t a_s\,ds$.

(8.43) was obtained in Section 5.5 as (5.85).

8.3 Filtering of diffusion Markov processes

As an illustration of the basic theorem given and proved in Section 8.2 (i.e., Theorem 8.1) we consider the problem of estimating the unobservable component θ_t of the two-dimensional diffusion Markov process (θ_t, ξ_t), $0 \le t \le T$, on the basis of results of the observations ξ_s, $s \le t$.

We shall give the exact statement of the problem.

On the probability space $(\Omega, \mathscr{F}, P)$ let there be specified independent Wiener processes $W_i = (W_i(t))$, $i = 1, 2$, $0 \le t \le T$, and a random vector $(\tilde{\theta}_0, \tilde{\xi}_0)$ independent of W_1, W_2. Denote

$$\mathscr{F}_t = \sigma\{\omega : \tilde{\theta}_0, \tilde{\xi}_0, W_1(s), W_2(s), s \le t\}.$$

According to Theorem 4.3 the (augmented) σ-algebras $\mathcal{F}_t^W$ are continuous. Similarly it is proved that the (augmented) σ-algebras $\mathcal{F}_t$ are also continuous.

Let $(\theta, \xi) = (\theta_t, \xi_t)$, $0 \le t \le T$, be a random process with

$$\begin{aligned} d\theta_t &= a(t, \theta_t, \xi_t)dt + b_1(t, \theta_t, \xi_t)dW_1(t) + b_2(t, \theta_t, \xi_t)dW_2(t), \\ d\xi_t &= A(t, \theta_t, \xi_t)dt + B(t, \xi_t)dW_2(t), \\ \theta_0 &= \tilde{\theta}_0, \xi_0 = \tilde{\xi}_0, P(|\tilde{\theta}_0| < \infty) = P(|\tilde{\xi}_0| < \infty) = 1. \end{aligned} \tag{8.44}$$

If $g(t, \theta, x)$ denotes any of the functions $a(t, \theta, x)$, $A(t, \theta, x)$, $b_1(t, \theta, x)$, $b_2(t, \theta, x)$, $B(t, x)$, then it will be assumed that

$$\begin{gathered} |g(t, \theta', x'') - g(t, \theta'', x'')|^2 \le K(|\theta' - \theta''|^2 + |x' - x''|^2), \\ g^2(t, \theta, x) \le K(1 + \theta^2 + x^2). \end{gathered} \tag{8.45}$$

From these assumptions, Theorem 4.6 and the note to this theorem, it follows that the system of equations given by (8.44) has a unique (strong) solution which is a Markov process. Hence, if

$$M(\tilde{\theta}_0^2 + \tilde{\xi}_0^2) < \infty, \tag{8.46}$$

then

$$\sup_{t \le T} M(\theta_t^2 + \xi_t^2) < \infty, \tag{8.47}$$

and, because of (8.45),

$$\sup_{t \le T} M[A^2(t, \theta_t, \xi_t) + B^2(t, \xi_t)] < \infty. \tag{8.48}$$

Let $h - h(t, \theta_t, \xi_t)$ be a measurable function, such that $M|h(t, \theta_t, \xi_t)| < \infty$. Making use of Theorem 8.1, we shall find an equation for $\pi_t(h) = M[h(t, \theta_t, \xi_t)|\mathcal{F}_t^\xi]$.

Along with the assumptions given by (8.45) and (8.46) made above, we shall assume that

$$B^2(t, x) \ge C > 0 \tag{8.49}$$

and that the following conditions are satisfied:
the function $h = h(t, \theta, x)$ is continuous together with its partial derivatives

$$h_t', h_\theta', h_x', h_{\theta\theta}'', h_{\theta x}'', h_{xx}''; \tag{8.50}$$

$$\sup_{t \le T} Mh^2(t, \theta_t, \xi_t) < \infty; \tag{8.51}$$

$$\int_0^T M[\mathfrak{L}h(t, \theta_t, \xi_t)]^2 \, dt < \infty, \tag{8.52}$$

where

$$\begin{aligned} \mathfrak{L}h(t, \theta, x) = {} & h_t'(t, \theta, x) + h_\theta'(t, \theta, x)a(t, \theta, x) + h_x'(t, \theta, x)A(t, \theta, x) \\ & + \tfrac{1}{2}h_{\theta\theta}''(t, \theta, x)[b_1^2(t, \theta, x) + b_2^2(t, \theta, x)] + \tfrac{1}{2}h_{xx}''(t, \theta, x)B^2(t, x) \\ & + h_{\theta x}''(t, \theta, x)b_2(t, \theta, x)\, B(t, x). \end{aligned} \tag{8.53}$$

Finally, it will be assumed that

$$\int_0^T M\{[h'_\theta(t, \theta_t, \xi_t)]^2[b_1^2(t, \theta_t, \xi_t) + b_2^2(t, \theta_t, \xi_t)]\}dt < \infty, \tag{8.54}$$

$$\int_0^T M[h'_x(t, \theta_t, \xi_t)]^2 B^2(t, \xi_t)dt < \infty. \tag{8.55}$$

Theorem 8.3. *If the assumptions given by* (8.45), (8.46), (8.49)–(8.52), (8.54), *and* (8.55) *are fulfilled, then* (P-a.s.)

$$\pi_t(h) = \pi_0(h) + \int_0^t \pi_s(\mathfrak{L}h)ds + \int_0^t \left[\pi_s(\mathcal{N}h) + \frac{\pi_s(Ah) - \pi_s(A)\pi_s(h)}{B(s, \xi_s)}\right]d\overline{W}_s, \tag{8.56}$$

where

$$\overline{W}_t = \int_0^t \frac{d\xi_s - \pi_s(A)ds}{B(s, \xi_s)}$$

is a Wiener process (with respect to $\mathcal{F}_t^\xi$*), $0 \le t \le T$) and*

$$\mathcal{N}h(t, \theta, x) = h'_\theta(t, \theta, x)b_2(t, \theta, x) + h'_x(t, \theta, x)B(t, x). \tag{8.57}$$

Proof. By the Ito formula

$$h(t, \theta_t, \xi_t) = h(0, \theta_0, \xi_0) + \int_0^t \mathfrak{L}h(s, \theta_s, \xi_s)ds + x_t \qquad (P\text{-a.s.}),$$

where

$$x_t = \sum_{i=1}^2 \int_0^t h'_\theta(s, \theta_s, \xi_s)b_i(s, \theta_s, \xi_s)dW_i(s) + \int_0^t h'_x(s, \theta_s, \xi_s)B(s, \xi_s)dW_2(s).$$

According to the assumptions made, the process $X = (x_t, \mathcal{F}_t)$ is a square integrable martingale.

Next we will establish that

$$\langle x, W_2\rangle_t = \int_0^t [h'_\theta(s, \theta_s, \xi_s)\, b_2(s, \theta_s, \xi_s) + h'_x(s, \theta_s, \xi_s)B(s, \xi_s)]ds. \tag{8.58}$$

With the help of the Ito formula, it can easily be shown that

$$\begin{aligned} x_t W_2(t) = {} & \int_0^t W_2(s)h'_\theta(s, \theta_s, \xi_s)b_1(s, \theta_s, \xi_s)dW_1(s) \\ & + \int_0^t [x_s + W_2(s)h'_\theta(s, \theta_s, \xi_s)b_2(s, \theta_s, \xi_s) \\ & + W_2(s)h'_x(s, \theta_s, \xi_s)B(s, \xi_s)]dW_2(s) \\ & + \int_0^t [h'_\theta(s, \theta, \xi_s)b_2(s, \theta_s, \xi_s) + h'_x(s, \theta_s, \xi_s)B(s, \xi_s)]ds. \end{aligned} \tag{8.59}$$

Hence, it immediately follows that the process $Y = (y_t, \mathscr{F}_t)$ with

$$y_t = x_t W_2(t) - \int_0^t [h'_\theta(s, \theta_s, \xi_s)b_2(s, \theta_s, \xi_s) + h'_x(s, \theta_s, \xi_s)B(s, \xi_s)]ds$$

is a martingale.

From this, as in Example 3 in Chapter 5, follows the validity of (8.58), and to obtain (8.56) it remains only to make use of Theorem 8.1. □

8.4 Equations of optimal nonlinear interpolation

8.4.1

As in Section 8.1, it will be assumed that the two-dimensional process $(h, \xi) = (h_t, \xi_t)$, $0 \leq t \leq T$, is such that

$$h_t = h_0 + \int_0^s H_s\, ds + x_t, \tag{8.60}$$

$$\xi_t = \xi_0 + \int_0^t A_s\, ds + \int_0^t B_s(\xi) dW_s, \tag{8.61}$$

and satisfies the assumptions of Theorem 8.1.

The problem of optimal interpolation is to find the optimal (in the mean square sense) estimate h_s on the basis of results of the observation ξ_u, $u \leq t$, where $t \geq s$. If $Mh_s^2 < \infty$, then such an estimate is the a posteriori mean

$$\pi_{s,t}(h) = M[h_s | \mathscr{F}_t^\xi]. \tag{8.62}$$

For $\pi_{s,t}(h)$ one can obtain equations of two types: "forward" (over t for fixed s) and "backward" (over s for fixed t).

In the present section we shall deduce forward equations analogous to Equation (8.10) for $\pi_t(h) = \pi_{t,t}(h)$.

Theorem 8.4. *Let the assumptions of Theorem* 8.1 *be satisfied. Then for* $0 \leq s \leq t \leq T$,

$$\pi_{s,t}(h) = \pi_s(h) + \int_s^t \frac{M[h_s A_u | \mathscr{F}_u^\xi] - M[h_s | \mathscr{F}_u^\xi] M[A_u | \mathscr{F}_u^\xi]}{B_u(\xi)} d\overline{W}_u. \tag{8.63}$$

Proof. First of all note that (8.63) can be rewritten as follows:

$$\pi_{s,t} = \pi_s(h) + \int_s^t \frac{M[h_s A_u | \mathscr{F}_u^\xi] - \pi_{s,u}(h)\pi_u(A)}{B_u(\xi)} d\overline{W}_u;$$

or,

$$d_t \pi_{s,t}(h) = \frac{M[h_s A_t | \mathscr{F}_t^\xi] - \pi_{s,t}(h)\pi_t(A)}{B_t(\xi)} d\overline{W}_t,$$

where $t \geq s$ and $\pi_{s,s}(h) = \pi_s(h)$.

Consider now the square integrable martingale $Y = (y_t, \mathscr{F}_t^\xi)$ with

$$y_t = M(h_s | \mathscr{F}_t^\xi), \qquad t \geq s. \tag{8.64}$$

According to Theorem 5.18, y_t for $t \geq s$ permits the representation

$$y_t = \pi_s(h) + \int_s^t g_{s,u}(\xi) d\overline{W}_u \tag{8.65}$$

and the Wiener process $\overline{W} = (\overline{W}_u, \mathscr{F}_u^\xi)$, and the process $(g_{s,u}(\xi), \mathscr{F}_u^\xi)$, $u \geq s$, satisfying the condition

$$M \int_s^t g_{s,u}^2(\xi) du < \infty.$$

As in the proof of Theorem 8.1, let us introduce the square integrable martingale $Z = (z_t, \mathscr{F}_t^\xi)$ where

$$z_t = \int_s^t \lambda_{s,u}(\xi) d\overline{W}_u$$

and $|\lambda_{s,u}(\xi)| \leq C \leq \infty$. It is not difficult to show that

$$M y_t z_t = M \int_s^t \lambda_{s,u}(\xi) g_{s,u}(\xi) du. \tag{8.66}$$

On the other hand, taking into consideration that

$$\overline{W}_t = W_t + \int_0^t \frac{A_u - \pi_u(A)}{B_u(\xi)} du,$$

we find that

$$\begin{aligned} M y_t z_t &= M M(h_s | \mathscr{F}_t^\xi) z_t = M h_s z_t \\ &= M h_s \int_s^t \lambda_{s,u}(\xi) dW_u + M h_s \int_s^t \lambda_{s,u}(\xi) \frac{A_u - \pi_u(A)}{B_u(\xi)} du \\ &= M\left[h_s M\left(\int_s^t \lambda_{s,u}(\xi) dW_u \Big| \mathscr{F}_s \right)\right] + M \int_s^t \lambda_{s,u}(\xi) \frac{h_s A_u - h_s \pi_u(A)}{B_u(\xi)} du \\ &= M \int_s^t \lambda_{s,u}(\xi) \frac{M(h_s A_u | \mathscr{F}_u^\xi) - M(h_s | \mathscr{F}_u^\xi) \pi_u(A)}{B_u(\xi)} du. \end{aligned} \tag{8.67}$$

Comparing the right-hand sides in (8.66) and (8.67) we find that for almost all $u \geq s$

$$g_{s,u}(\xi) = \frac{M(h_s A_u | \mathscr{F}_u^\xi) - M(h_s | \mathscr{F}_u^\xi) \pi_u(A)}{B_u(\xi)} \qquad (P\text{-a.s.}). \tag{8.68}$$

As in Theorem 8.1, without restricting the generality we may take the function $g_{s,u}(\xi)$ to be defined by Equation (8.68) (P-a.s.) for all $u \geq s$. □

Note. From the proof of the theorem it follows that

$$M\int_s^T \left[\frac{M(h_s A_u|\mathscr{F}_u^\xi) - M(h_s|\mathscr{F}_u^\xi)\pi_u(A)}{B_u(\xi)}\right]^2 du < \infty.$$

8.4.2

Applying Theorem 8.4 to the process (θ, ξ) considered in Section 8.3, we find that (under the assumptions of Theorem 8.3) for $t \geq s$

$$M[h(s, \theta_s, \xi_s)|\mathscr{F}_t^\xi]$$
$$= \pi_s(h) + \int_s^t \frac{M\{h(s, \theta_s, \xi_s)[A(u, \theta_u, \xi_u) - \pi_u(A)]|\mathscr{F}_u^\xi\}}{B_u(\xi)} d\overline{W}_u. \quad (8.69)$$

8.5 Equations of optimal nonlinear extrapolation

8.5.1

We shall assume again that the process (h, ξ) can be described by (8.60) and (8.61) and that the conditions of Theorem 8.1 are satisfied.

Let $t \geq s$ and

$$\pi_{t,s}(h) = M[h_t|\mathscr{F}_s^\xi]. \quad (8.70)$$

It is obvious that, if $Mh_t^2 < \infty$, then $\pi_{t,s}(h)$ is an optimal (generally speaking, nonlinear) estimate of the "extrapolatable" value of h_t over the observations ξ_u, $u \leq s \leq t$. The ideas applied in deducing Equations (8.10) for $\pi_t(h)$ allow us to obtain equations also for $\pi_{t,s}(h)$ over $s \leq t$ for fixed t. These equations can naturally be called backward equations of the extrapolation as opposed to the forward equations (over $t \geq s$ for fixed s).

Theorem 8.5. *Let the assumptions of Theorem* 8.1 *be fulfilled. Then for fixed t and s, $s \leq t$,*

$$\pi_{t,s}(h) = \pi_{t,0}(h) + \int_0^s \left\{\pi_u(D) + \frac{M[M(h_t|\mathscr{F}_u)(A_u - \pi_u(A))|\mathscr{F}_u^\xi]}{B_u(\xi)}\right\} d\overline{W}_u, \quad (8.71)$$

where $D_s = d\langle \tilde{x}, W\rangle_s/ds$ and $\tilde{X} = (\tilde{x}_s, \mathscr{F}_s)$, $s \leq t$, is a square integrable martingale with $\tilde{x}_s = M(h_t|\mathscr{F}_s)$.

PROOF. Let t be fixed and $s \leq t$. Denote $y_t = M(h_t|\mathscr{F}_s^\xi)$. The process $Y = (y_s, \mathscr{F}_s^\xi)$, $s \leq t$, is a square integrable martingale and, by Theorem 5.18,

$$y_s = M(h_t|\mathscr{F}_0^\xi) + \int_0^s g_{u,t}(\xi) d\overline{W}_u \quad (8.72)$$

with

$$M\int_0^t g_{u,t}^2(\xi) du < \infty.$$

As in the proof of the preceding theorem, it will be assumed that

$$z_s = \int_0^s \lambda_u(\xi) d\overline{W}_u,$$

where $|\lambda_u(\xi)| \leq C \leq \infty$. Then

$$My_s z_s = M \int_0^s \lambda_u(\xi) g_{u,t}(\xi) du. \tag{8.73}$$

Let us now compute $My_s z_s$ in another way. It is clear that

$$My_s z_s = MM(h_t | \mathscr{F}_s^\xi) z_s = Mh_t z_s = MM(h_t | \mathscr{F}_s) z_s = M\tilde{x}_s z_s.$$

The process $\tilde{X} = (\tilde{x}_s, \mathscr{F}_s), 0 \leq s \leq t$, is a square integrable martingale and, by Theorem 5.3,

$$\langle \tilde{x}, W \rangle_s = \int_0^s D_u \, du,$$

where $M \int_0^t D_u^2 \, du < \infty$.

Let $\tilde{Z} = (\tilde{z}_s, \mathscr{F}_s)$, $s \leq t$, be a square integrable martingale with $\tilde{z}_s = \int_0^s \lambda_u(\xi) dW_u$. Since

$$\overline{W}_s = W_s + \int_0^s \frac{A_u - \pi_u(A)}{B_u(\xi)} du,$$

it follows that

$$z_s = \tilde{z}_s + \int_0^s \lambda_u(\xi) \frac{A_u - \pi_u(A)}{B_u(\xi)} du. \tag{8.74}$$

Consequently,

$$\begin{aligned} My_s z_s = M\tilde{x}_s z_s &= M\tilde{x}_s \tilde{z}_s + M\tilde{x}_s \int_0^s \lambda_u(\xi) \frac{A_u - \pi_u(A)}{B_u(\xi)} du \\ &= M \int_0^s \lambda_u(\xi) \pi_u(D) du + M\tilde{x}_s \int_0^s \lambda_u(\xi) \frac{A_u - \pi_u(A)}{B_u(\xi)} du, \end{aligned} \tag{8.75}$$

where

$$M\tilde{x}_s \tilde{z}_s = M \int_0^s \lambda_u(\xi) D_u \, du = M \int_0^s \lambda_u(\xi) \pi_u(D) du \tag{8.76}$$

by Lemma 5.1

Analogously we find that

$$\begin{aligned} M\tilde{x}_s \int_0^s \lambda_u(\xi) \frac{A_u - \pi_u(A)}{B_u(\xi)} du &= M \int_0^s \lambda_u(\xi) \tilde{x}_s \frac{A_u - \pi_u(A)}{B_u(\xi)} du \\ &= M \int_0^s \lambda_u(\xi) M(\tilde{x}_s | \mathscr{F}_u) \frac{A_u - \pi_u(A)}{B_u(\xi)} du. \end{aligned} \tag{8.77}$$

But $\tilde{x}_s = M(h_t|\mathscr{F}_s)$ and therefore, with $u \le s \le t$,

$$M(\tilde{x}_s|\mathscr{F}_u) = M(h_t|\mathscr{F}_u),$$

which together with (8.77) yields the relation

$$M\tilde{x}_s \int_0^s \lambda_u(\xi) \frac{A_u - \pi_u(A)}{B_u(\xi)}\, du = M \int_0^s \lambda_u(\xi) \frac{M\{M(h_t|\mathscr{F}_u)(A_u - \pi_u(A))|\mathscr{F}_u^\xi\}}{B_u(\xi)}\, du. \tag{8.78}$$

From (8.76)–(8.78) we obtain

$$My_s z_s = M \int_0^s \lambda_u(\xi)\left\{\pi_u(D) + \frac{M[M(h_t|\mathscr{F}_u)(A_u - \pi_u(A))|\mathscr{F}_u^\xi]}{B_u(\xi)}\right\} du. \tag{8.79}$$

Comparing (8.79) with (8.73) we find that for almost all $u \le s$,

$$g_u(\xi) = \pi_u(D) + \frac{M[M(h_t|\mathscr{F}_u)(A_u - \pi_u(A))|\mathscr{F}_u^\xi]}{B_u(\xi)} \quad (P\text{-a.s.}). \tag{8.80}$$

Without restricting the generality we may take the function $g_u(\xi)$ to be defined by Equation (8.80) for all $u \le s$. Together with (8.72) this proves (8.71). □

Note. From the proof of the theorem it follows that

$$M \int_0^t \left\{\pi_u(D) + \frac{M[M(h_t|\mathscr{F}_u)(A_u - \pi_u(A))|\mathscr{F}_u^\xi]}{B_u(\xi)}\right\}^2 du < \infty.$$

8.5.2

Consider the representation given by (8.71) for the diffusion process (θ, ξ) studied in Section 8.3. Let t be fixed and for $s < t$,

$$g(s, \theta, x) = M\{h(t, \theta_t, \xi_t)|\theta_s = \theta, \xi_s = x\}.$$

Assume that this function satisfies (8.50) and

$$\mathfrak{L}g(s, \theta, x) = 0, \tag{8.81}$$

where the operator $\mathfrak{L}$ is defined in (8.53).

It will also be assumed that

$$M \int_0^T \left\{(g'_\theta(s, \theta_s, \xi_s))^2 \sum_{i=1}^2 b_i^2(s, \theta_s, \xi_s) + (g'_x(s, \theta_s, \xi_s))^2 B^2(s, \xi_s)\right\} ds < \infty. \tag{8.82}$$

By the Ito formula for $s \le t$,

$$g(s, \theta_s, \xi_s) = g(0, \theta_0, \xi_0) + \int_0^s \mathfrak{L} g(u, \theta_u, \xi_u)du$$
$$+ \sum_{i=1}^{2} \int_0^s g'_\theta(u, \theta_u, \xi_u) b_i(u, \theta_u, \xi_u) dW_i(u)$$
$$+ \int_0^s g'_x(u, \theta_u, \xi_u) B(u, \xi_u) dW_2(u).$$

From this one can see that by (8.81) the process $Y = (y_s, \mathscr{F}_s)$, $s \le t$, with $y_s = g(s, \theta_s, \xi_s)$, is a square integrable martingale and that

$$\langle y, W_2 \rangle_s = \int_0^s [g'_\theta(u, \theta_u, \xi_u) b_2(u, \theta_u, \xi_u) + g'_x(u, \xi_u, \xi_u)\, B(u, \xi_u)] du.$$

Hence, by (8.71) and the fact that

$$M(h(t, \theta_t, \xi_t) | \mathscr{F}_s) = M(h(t, \theta_t, \xi_t) | \theta_s, \xi_s) = g(s, \theta_s, \xi_s),$$

we obtain

$$M\{h(t, \theta_t, \xi_t) | \mathscr{F}_s^\xi\} = M\{h(t, \theta_t, \xi_t) | \mathscr{F}_0^\xi\}$$
$$+ \int_0^s \left\{ \pi_u(\mathscr{N} g) + \frac{\pi_u(gA) - \pi_u(g)\pi_u(A)}{B(u, \xi_u)} \right\} dW_u, \quad (8.83)$$

where

$$\mathscr{N} g(u, \theta_u, \xi_u) = g'_\theta(u, \theta_u, \xi_u) b_2(u, \theta_u, \xi_u) + g'_x(u, \theta_u, \xi_u) B(u, \xi_u).$$

8.6 Stochastic differential equations with partial derivatives for the conditional density (the case of diffusion Markov processes)

8.6.1

Let us consider the two-dimensional diffusion Markov process $(\theta, \xi) = (\theta_t, \xi_t)$, $0 \le t \le T$, determined by Equations (8.44) with $B(t, \xi_t) \equiv 1$. If the function $h = h(x)$, $x \in \mathbb{R}^1$, is finite together with its derivatives $h'(x)$, $h''(x)$ and (8.45) and (8.46) are fulfilled, then, according to (8.56), the process $\pi_t(h) = M[h(\theta_t) | \mathscr{F}_t^\xi]$ permits the representation

$$\pi_t(h) = \pi_0(h) + \int_0^t \pi_s(\mathfrak{L} h) ds + \int_0^t [\pi_s(\mathscr{N} h) + \pi_s(Ah) - \pi_s(A)\pi_s(h)] d\overline{W}_s \quad (8.84)$$

where $\overline{W} = (\overline{W}_t, \mathscr{F}_t^\xi)$ is a Wiener process with

$$d\overline{W}_t = d\xi_t - \pi_t(A)dt,$$

and

$$\mathfrak{L}h(\theta_t) = h'(\theta_t)a(t, \theta_t, \xi_t) + \tfrac{1}{2}h''(\theta_t)\sum_{i=1}^{2} b_i(t, \theta_t, \xi_t),$$

$$\mathcal{N}h(\theta_t) - h'(\theta_t)b_2(t, \theta_t, \xi_t).$$

We shall assume now that the conditional distribution $P(\theta_t \le x | \mathcal{F}_t^\xi)$, $0 \le t \le T$, has the density

$$\rho_x(t) = \frac{dP(\theta_t \le x | \mathcal{F}_t^\xi)}{dx},$$

which is a measurable function from (t, x, ω). Starting from (8.84), let us find the equation satisfied by this density. Introduce the following notation:

$$\mathfrak{L}^*\rho_x(t) = -\frac{\partial}{\partial x}[a(t, x, \xi_t)\rho_x(t)] + \frac{1}{2}\frac{\partial^2}{\partial x^2}\left[\sum_{i-1}^{2} b_i^2(t, x, \xi_t)\rho_x(t)\right],$$

$$\mathcal{N}^*\rho_x(t) = -\frac{\partial}{\partial x}[b_2(t, x, \xi_t)\rho_x(t)].$$

Theorem 8.6. *Let the following conditions hold:*

(I) *with probability one for each $t, 0 \le t \le T$, there exist the derivatives*

$$\frac{\partial}{\partial x}[a[t, x, \xi_t)\rho_x(t)], \qquad \frac{\partial}{\partial x}[b_2(t, x, \xi_t)\rho_x(t)],$$

$$\frac{\partial^2}{\partial x^2}\left[\sum_{i=1}^{2} b_i^2(t, x, \xi_t)\rho_x(t)\right];$$

(II) *for any continuous and finite function $h = h(x)$,*

$$\int_0^T \int_{-\infty}^{\infty} |h(x)\mathfrak{L}^*\rho_x(t)|dx\,dt < \infty \tag{8.85}$$

and

$$M\int_0^T \int_{-\infty}^{\infty} h^2(x)[\mathcal{N}^*\rho_x(t) + \rho_x(t)(A(t, x, \xi_t) - \pi_t(A))]^2\,dx\,dt < \infty. \tag{8.86}$$

Then the conditional density $\rho_x(t)$, $x \in \mathbb{R}'$, $0 \le t \le T$, satisfies the stochastic differential equation (with partial derivatives)

$$\begin{aligned} d_t\rho_x(t) &= \mathfrak{L}^*\rho_x(t)dt \\ &\quad + \left\{\mathcal{N}^*\rho_x(t) + \rho_x(t)\left[A(t, x, \xi_t) - \int_{-\infty}^{\infty} A(t, y, \xi_t)\rho_y(t)dy\right]\right\} \\ &\quad \times \left[d\xi_t - \left(\int_{-\infty}^{\infty} A(t, y, \xi_t)\rho_y(t)dy\right)dt\right]. \end{aligned} \tag{8.87}$$

PROOF. It will be shown first that in (8.86)

$$\int_0^t \int_{-\infty}^{\infty} h(x)\{\mathcal{N}^*\rho_x(s) + \rho_x(s)[A(s, x, \xi_s) - \pi_s(A)]\}dx\, d\overline{W}_s$$

$$= \int_{-\infty}^{\infty} h(x)\left(\int_0^t \{\mathcal{N}^*\rho_x(s) + \rho_x(s)[A(s, x, \xi_s) - \pi_s(A)]\}d\overline{W}_s\right)dx \qquad (P\text{-a.s.}). \tag{8.88}$$

For brevity, set

$$\alpha_s(x, \xi) = \mathcal{N}^*\rho_x(s) + \rho_x(s)[A(s, x, \xi_s) - \pi_s(A)].$$

Then for proving (8.86) one needs to show that (P-a.s.)

$$\chi_t(\xi) = \int_0^t\left[\int_{-\infty}^{\infty} h(x)\alpha_s(x, \xi)dx\right]d\overline{W}_s - \int_{-\infty}^{\infty} h(x)\left[\int_0^t \alpha_s(x, \xi)d\overline{W}_s\right]dx = 0. \tag{8.89}$$

The variable $\chi_t(\xi)$ is $\mathcal{F}_t^\xi$-measurable and, according (8.86), $M\chi_t(\xi) = 0$ and $M\chi_t^2(\xi) < \infty$. Hence for proving (8.89) it suffices to establish that $M[\chi_t(\xi)\lambda_t(\xi)] = 0$ for any $\mathcal{F}_t^\xi$-measurable values $\lambda_t(\xi)$ with $|\lambda_t(\xi)| \le 1$.

By Theorem 5.18,

$$\lambda_t(\xi) = M\lambda_t(\xi) + \int_0^t g_s(\xi)d\overline{W}_s,$$

where the process $g = (g_s(\xi), \mathcal{F}_s^\xi)$, $s \le t$, is such that $\int_0^t Mg_s^2(\xi)ds < \infty$. Hence, by the Fubini theorem,

$$M\chi_t(\xi)\lambda_t(\xi) = M\chi_t(\xi)\int_0^t g_s(\xi)d\overline{W}_s = \int_0^t M\left\{g_s(\xi)\int_{-\infty}^{\infty} h(x)\alpha_s(x, \xi)dx\right\}ds$$

$$- \int_{-\infty}^{\infty} h(x)\left\{\int_0^t Mg_s(\xi)\alpha_s(x, \xi)ds\right\}dx = 0.$$

Thus $\chi_t(\xi) = 0$ (P-a.s.) for any t ($0 \le t \le T$), which proves (8.88).

Let us proceed now directly to the deduction of Equation (8.87). For this purpose let us note that

$$\pi_s(Ah) - \pi_s(A)\pi_s(h) = M\{h(\theta_s)[A(s, \theta_s, \xi_s) - \pi_s(A)] \mid \mathcal{F}_s^\xi\}.$$

Hence, according to (8.84),

$$\pi_t(h) = \pi_0(h) + \int_0^t \pi_s(\mathfrak{L}h)ds$$

$$+ \int_0^t M\{\mathcal{N}h(\theta_s) + h(\theta_s)[A(s, \theta_s, \xi_s) - \pi_s(A)] \mid \mathcal{F}_s^\xi\}d\overline{W}_s,$$

and, if there exists the density $\rho_x(t)$, then

$$\int_{-\infty}^{\infty} h(x)\rho_x(t)dx = \int_{-\infty}^{\infty} h(x)\rho_x(0)dx$$
$$+ \int_0^t \int_{-\infty}^{\infty} \left[h'(x)a(s, x, \xi_s) + \tfrac{1}{2}h''(x) \sum_{i=1}^{2} b_i^2(s, x, \xi_s) \right] \rho_x(s)dx\, ds$$
$$+ \int_0^t \int_{-\infty}^{\infty} [h'(x)b_2(s, x, \xi_s) + h(x)(A(s, x, \xi_s)$$
$$- \pi_s(A))]\rho_x(s)dx\, d\overline{W}_s. \tag{8.90}$$

By integrating in (8.90) by parts and changing the order of integration (which is permitted by (8.88), (8.85) and the Fubini theorem) we obtain

$$\int_{-\infty}^{\infty} h(x)\{\rho_x(t) - \rho_x(0) - \int_0^t \mathfrak{L}^* \rho_x(s)ds$$
$$- \int_0^t [\mathcal{N}^* \rho_x(s) + \rho_x(s)(A(s, x, \xi_s) - \pi_s(A))]d\overline{W}_s\}dx = 0.$$

From this, because of the arbitrariness of the finite function $h(x)$, we arrive at Equation (8.87). □

8.6.2

The assumptions of Theorem 8.6 are usually difficult to check. The case of the conditional Gaussian processes (θ, ξ) examined further in Chapters 10 and 11 is an exception. Hence, next we shall discuss in detail a fairly simple (but nevertheless nontrivial!) case of the processes (θ, ξ) for which the conditional density $\rho_x(t)$ exists and is a unique solution of Equation (8.87).

It will be assumed that the random process $(\theta, \xi) = [(\theta_t, \xi_t), \mathcal{F}_t]$, $0 \leq t \leq T$, satisfies the stochastic differential equations

$$d\theta_t = a(\theta_t)dt + dW_1(t), \tag{8.91}$$

$$d\xi_t = A(\theta_t)dt + dW_2(t), \tag{8.92}$$

where the random variable θ_0 and the Wiener processes $W_i = (W_i(t), \mathcal{F}_t)$, $i = 1, 2$, are independent, $P(\xi_0 = 0) = 1$, $M\theta_0^2 < \infty$.

Theorem 8.7. *Let:*

(I) *the functions $a(x)$, $A(x)$ be uniformly bounded together with their derivatives $a'(x)$, $a''(x)$, $a'''(x)$, $A'(x)$ and $A''(x)$ (by a constant K);*
(II) $|A''(x) - A''(y)| \leq K|x - y|$, $|a'''(x) - a'''(y)| \leq K|x - y|$;
(III) *the distribution function $F(x) = P(\theta_0 \leq x)$ has twice continuously differentiable density $f(x) = dF(x)/dx$.*

Then there exists (P-a.s.) *for each* t, $0 \le t \le T$,

$$\rho_x(t) = \frac{dP(\theta_t \le x | \mathscr{F}_t^\xi)}{dx},$$

which is an $\mathscr{F}_t^\xi$*-measurable* (*for each* t, $0 \le t \le T$) *solution of the equation*

$$\begin{aligned} d_t \rho_x(t) &= \mathfrak{L}^* \rho_x(t) dt \\ &+ \rho_x(t) \left[A(x) - \int_{-\infty}^{\infty} A(y) \rho_y(t) dy \right] \left[d\xi_t - \left(\int_{-\infty}^{\infty} A(y) \rho_y(t) dy \right) dt \right] \end{aligned} \tag{8.93}$$

with $\rho_x(0) = f(x)$ *and*

$$\mathfrak{L}^* \rho_x(t) = -\frac{\partial}{\partial x}[a(x)\rho_x(t)] + \frac{1}{2}\frac{\partial^2}{\partial x}[\rho_x(t)].$$

In the class of measurable (t, x, ω) *twice continuously differentiable functions* $U_x(t)$ *over* x, $\mathscr{F}_t^\xi$*-measurable for each* t, $0 \le t \le T$, *and satisfying the condition*

$$P\left\{ \int_0^T \left(\int_{-\infty}^{\infty} A(x) U_x(t) dx \right)^2 dt < \infty \right\} = 1, \tag{8.94}$$

the solution to Equation (8.93) *is unique in the following sense: if* $U_x^{(1)}(t)$ *and* $U_x^{(2)}(t)$ *are two such solutions, then*

$$P\left\{ \sup_{0 \le t \le T} |U_x^{(1)}(t) - U_x^{(2)}(t)| > 0 \right\} = 0, \qquad -\infty < x < \infty. \tag{8.95}$$

8.6.3

Before proving Theorem 8.7 let us make a number of auxiliary propositions.

Let $(\tilde{\Omega}, \tilde{\mathscr{F}}, \tilde{P})$ be a probability space identical to $(\Omega, \mathscr{F}, P)$ on which there is specified the random variable $\tilde{\theta}_0$ with $\tilde{P}(\tilde{\theta}_0 \le x) = P(\theta_0 \le x)$ and the independent Wiener process $\tilde{W} = (\tilde{W}_t)$, $0 \le t \le T$.

Introduce also the following variables:

$$\tilde{W}_t^y = y + \tilde{W}_t \quad (-\infty < y < \infty), \qquad \bar{A}_s(\xi) = M[A(\theta_s) | \mathscr{F}_s^\xi],$$

$$\bar{W}_t = \xi_t - \int_0^t \bar{A}_s(\xi) ds, \qquad D(x) = \int_0^x a(y) dy$$

and

$$\psi_t(\xi) = \exp\left\{ \int_0^t \bar{A}_s(\xi) d\bar{W}_s - \frac{1}{2} \int_0^t \bar{A}_s^2(\xi) ds \right\}, \tag{8.96}$$

$$\begin{aligned} \rho_t(y, \tilde{W}, \xi) = \exp\Bigg\{ &\int_0^t A(y + \tilde{W}_s) d\bar{W}_s - \frac{1}{2} \int_0^t [a^2(y + \tilde{W}_s) \\ &+ a^2(y + \tilde{W}_s) - a'(y + \tilde{W}_s) - 2A(y + \tilde{W}_s)\bar{A}_s(\xi)] ds \Bigg\}, \end{aligned} \tag{8.97}$$

where $\int_0^t A(y + \tilde{W}_s)d\overline{W}_s$ is defined for each $\tilde{\omega} \in \Omega$ as a stochastic integral from the determined function $A(y + \tilde{W}_s(\tilde{\omega}))$.

Lemma 8.5. *Under the conditions of Theorem* 8.7 *there exists* (P-a.s.) *the density*

$$\rho_x(t) = \frac{dP(\theta_t \le x | \mathscr{F}_t^\xi)}{dx}, \qquad 0 \le t \le T,$$

defined by the formulas

$$\rho_x(0) = f(x),$$

and for t, $0 < t \le T$,

$$\rho_x(t) = \frac{1}{\sqrt{2\pi t}\,\psi_t(\xi)} \int_{-\infty}^{\infty} \exp\left\{-\frac{(x-y)^2}{2t} + D(x) - D(y)\right\}$$

$$\times \tilde{M}(\rho_t(y, \tilde{W}, \xi) | \tilde{W}_t = x - y) f(y) dy, \tag{8.98}$$

where $\tilde{M}$ *is an averaging* w.r.t. *measure* $\tilde{P}$.

PROOF. Consider on $(\tilde{\Omega}, \tilde{\mathscr{F}}, \tilde{P})$ a process $\tilde{\theta} = (\tilde{\theta}_t)$, $0 \le t \le T$, with the differential

$$d\tilde{\theta}_t = a(\tilde{\theta}_t)dt + d\tilde{W}_t. \tag{8.99}$$

The conditions of Theorem 8.7 guarantee the existence and uniqueness of a strong solution to Equation (8.99) with the initial value $\tilde{\theta}_0$. Hence the measures μ_θ and $\mu_{\tilde{\theta}}$ corresponding to the processes θ and $\tilde{\theta}$ coincide.

Consider now the equation

$$d\tilde{\theta}_t^y = a(\tilde{\theta}_t^y)dt + d\tilde{W}_t, \qquad \tilde{\theta}_0^y = y. \tag{8.100}$$

This equation also has a unique strong solution and

$$\tilde{P}(\tilde{\theta} \in \Gamma | \tilde{\theta}_0 = y) = \tilde{P}(\tilde{\theta}^y \in \Gamma), \qquad \Gamma \in \mathscr{B}.$$

Therefore,

$$\tilde{P}(\tilde{\theta} \in \Gamma) = \int_{-\infty}^{\infty} \tilde{P}(\tilde{\theta}^y \in \Gamma) f(y) dy,$$

which will be symbolically denoted by

$$d\mu_{\tilde{\theta}} = d\mu_{\tilde{\theta}^y} f(y) dy, \tag{8.101}$$

where $\mu_{\tilde{\theta}^y}$ is a measure corresponding to the process $\tilde{\theta}^y$. Denote by $\mu_{\tilde{W}^y}$ a measure of the process $\tilde{W}^y$; according to Theorem 7.7, $\mu_{\tilde{\theta}^y} \sim \mu_{\tilde{W}^y}$ and

$$\frac{d\mu_{\tilde{\theta}^y}}{d\mu_{\tilde{W}^y}}(t, \tilde{W}^y) = \exp\left\{\int_0^t a(y + \tilde{W}_s)d\tilde{W}_s - \frac{1}{2}\int_0^t a^2(y + \tilde{W}_s)ds\right\}. \tag{8.102}$$

Employing the Ito formula we find that

$$D(y+\tilde{W}_t)=D(y)+\int_0^t a(y+\tilde{W}_s)d\tilde{W}_s+\frac{1}{2}\int_0^t a'(y+\tilde{W}_s)ds.$$

Hence (8.102) can be rewritten as follows:

$$\frac{d\mu_{\tilde{\theta}^y}}{d\mu_{\tilde{W}^y}}(t,\tilde{W}^y)=\exp\left\{D(y+\tilde{W}_t)-D(y)-\frac{1}{2}\int_0^t[a^2(y+\tilde{W}_s)+a'(y+\tilde{W}_s)]ds\right\}. \tag{8.103}$$

From (8.101) and (8.103) it is not difficult to deduce that

$$\begin{aligned}\frac{d\mu_{\tilde{\theta}}}{d\mu_{\tilde{W}^y}\times dy}&(t,\tilde{W},y)\\&=f(y)\exp\left\{D(y+\tilde{W}_t)-D(y)-\frac{1}{2}\int_0^t[a^2(y+\tilde{W}_s)+a'(y+\tilde{W}_s)]ds\right\}.\end{aligned} \tag{8.104}$$

Employing this representation and the Bayes' formula (Theorem 7.23) we obtain

$$\begin{aligned}&P(\theta_t\le x\mid\mathscr{F}_t^\xi)\\&=M[\chi_{(\theta_t\le x)}\mid\mathscr{F}_t^\xi]\\&=\tilde{M}\chi_{(\theta_t\le x)}\exp\left\{\int_0^t[A(\tilde{\theta}_s)-\bar{A}_s(\xi)]d\overline{W}_s-\frac{1}{2}\int_0^t[A(\tilde{\theta}_s)-\bar{A}_s(\xi)]^2\,ds\right\}\\&=\int_{C_T}\chi_{(c_t\le x)}\exp\left\{\int_0^t[A(c_s)-\bar{A}_s(\xi)]d\overline{W}_s-\frac{1}{2}\int_0^t[A(c_s)-\bar{A}_s(\xi)]^2\,ds\right\}d\mu_\theta(c)\\&=\int_{C_T}\chi_{(c_t\le x)}\exp\left\{\int_0^t[A(c_s)-\bar{A}_s(\xi)]d\overline{W}_s\right.\\&\qquad\left.-\frac{1}{2}\int_0^t[A(c_s)-\bar{A}_s(\xi)]^2\,ds\right\}\frac{d\mu_{\tilde{\theta}}}{d\mu_{\tilde{W}^y}\times dy}(t,c,y)d\mu_{\tilde{W}^y}\times dy\\&=\frac{1}{\psi_t(\xi)}\int_{-\infty}^{\infty}\tilde{M}\chi_{(y+\tilde{W}_t\le x)}\exp\{D(y+\tilde{W}_t)-D(y)\}\rho_t(y,\tilde{W},\xi)f(y)dy.\end{aligned} \tag{8.105}$$

But

$$\begin{aligned}&\tilde{M}\chi_{(y+\tilde{W}_t\le x)}\exp\{D(y+\tilde{W}_t)-D(y)\}\rho_t(y,\tilde{W},\xi)\\&=\tilde{M}\{\chi_{(y+\tilde{W}_t\le x)}\exp[D(y+\tilde{W}_t)-D(y)]\tilde{M}[\rho_t(y,\tilde{W},\xi)\mid\tilde{W}_t]\}\\&=\frac{1}{\sqrt{2\pi t}}\int_{-\infty}^{x-y}\exp[D(y+z)-D(y)]\tilde{M}[\rho_t(y,\tilde{W},\xi)\mid\tilde{W}_t=z]\exp\left\{-\frac{z^2}{2t}\right\}dz\\&=\frac{1}{\sqrt{2\pi t}}\int_{-\infty}^{x}\exp\left[-\frac{(z-y)^2}{2t}+D(z)-D(y)\right]\tilde{M}[\rho_t(y,\tilde{W},\xi)\mid\tilde{W}_t=z-y]dz.\end{aligned} \tag{8.106}$$

From the Fubini theorem, (8.105) and (8.106), for $t > 0$ we obtain

$$P(\theta_t \le x \mid \mathscr{F}_t^{\xi}) = \frac{1}{\sqrt{2\pi t}\,\psi_t(\xi)} \int_{-\infty}^{x} \int_{-\infty}^{\infty} \exp\left[-\frac{(z-y)^2}{2t} + D(z) - D(y)\right] \times \tilde{M}[\rho_t(y, \tilde{W}, \xi) \mid \tilde{W}_t = z - y] f(y)\,dy\,dz, \tag{8.107}$$

proving (8.98). The formula $\rho_x(0) = f(x)$ is clear. □

To formulate the next statement, we shall denote

$$B_{t,\xi}(x) = a^2(x) + A^2(x) - a'(x) - 2A(x)\bar{A}_t(\xi), \tag{8.108}$$

$$\tilde{\eta}_s = \tilde{W}_s - \frac{s}{t}\tilde{W}_t, \qquad s \le t, \tag{8.109}$$

and

$$\bar{\rho}_t(y, x - y, \tilde{\eta}, \xi) = \exp\left\{\int_0^t A\left(y\frac{t-s}{t} + \tilde{\eta}_s + \frac{s}{t}x\right)d\overline{W}_s - \frac{1}{2}\int_0^t B_{s,\xi}\left(y\frac{t-s}{t} + \tilde{\eta}_s + \frac{s}{t}x\right)ds\right\}. \tag{8.110}$$

Lemma 8.6. *From the assumptions of Theorem* 8.7 *it follows that for any* x, y $(-\infty < x < \infty, -\infty < y < \infty)$

$$\tilde{M}[\rho_t(y, \tilde{W}, \xi) \mid \tilde{W}_t = x - y] = \tilde{M}\bar{\rho}(y, x - y, \tilde{\eta}, \xi) \qquad (\tilde{P}\text{-a.s.}). \tag{8.111}$$

PROOF. Employing (8.108), the function $\rho_t(y, \tilde{W}, \xi)$ defined in (8.97) can be represented as follows:

$$\rho_t(y, \tilde{W}, \xi) = \exp\left\{\int_0^t A(y + \tilde{W}_s)d\overline{W}_s - \frac{1}{2}\int_0^t B_{s,\xi}(y + \tilde{W}_s)ds\right\}. \tag{8.112}$$

Proceeding from the theorem of normal correlation (Theorem 13.1) it will not be difficult to show that the conditional (under the condition $\tilde{W}_t$) distribution of the process $\tilde{\eta} = (\tilde{\eta}_s)$, $s \le t$, with $\tilde{\eta}_s = \tilde{W}_s - (s/t)\tilde{W}_t$ does not depend on $\tilde{W}_t$ ($\tilde{P}$-a.s.). Hence, if $\Phi_s(\eta, \tilde{W}_t)$ is a $\mathscr{G}_s^{\tilde{\eta}, \tilde{W}_t}$-measurable functional ($\mathscr{G}_s^{\tilde{\eta}, \tilde{W}_t} = \sigma\{\omega : \tilde{\eta}_u, u \le s; \tilde{W}_t\}$, $s \le t$) with $\tilde{M}|\Phi_s(\tilde{\eta}, \tilde{W}_t)| < \infty$, then

$$\tilde{M}(\Phi_s(\tilde{\eta}, \tilde{W}_t) \mid \tilde{W}_t = x) = \tilde{M}\Phi_s(\tilde{\eta}, x). \tag{8.113}$$

Substituting $\tilde{W}_s = \tilde{\eta}_s + (s/t)\tilde{W}_t$ into $\rho_t(y, \tilde{W}, \xi)$ and applying (8.113), from (8.109) and (8.110) we obtain (8.111). □

Corollary. *From* (8.98) *and* (8.111) *it follows that*

$$\rho_x(t) = \frac{1}{\sqrt{2\pi t}\,\psi_t(\xi)} \int_{-\infty}^{\infty} \exp\left\{-\frac{(x-y)^2}{2t} + D(x) - D(y)\right\} \times \tilde{M}\bar{\rho}_t(y, x - y, \tilde{\eta}, \xi) f(y)\,dy. \tag{8.114}$$

Lemma 8.7. *From the assumptions of Theorem* 8.7,

$$\sup_{0 \le t \le T} M\rho_x^2(t) < \infty, \qquad -\infty < x < \infty. \tag{8.115}$$

PROOF. Set $z = (x - y)/\sqrt{t}$. Then by (8.114),

$$\rho_x(t) = -\frac{\psi_t^{-1}(\xi)}{\sqrt{2\pi}} \int_{-\infty}^{\infty} g(x, z, t)\tilde{M}\bar{\rho}_t[x - z\sqrt{t}, z\sqrt{t}, \tilde{\eta}, \xi]dz, \tag{8.116}$$

where

$$g(x, z, t) = \exp\left\{-\frac{z^2}{2} + D(x) - D(x - z\sqrt{t})\right\} f(x - z\sqrt{t}).$$

But $|D(x)| \le \int_0^{|x|} |\alpha(y)|dy \le K|x|$. Hence, for each x, $-\infty < x < \infty$,

$$|g(x, z, t)| \le \exp\left\{-\frac{z^2}{2} + D(x) + K|x| + K|z|\sqrt{T}\right\} \sup_{-\infty \le y \le \infty} f(y)$$

$$= d(x)\exp\left\{-\frac{z^2}{2} + K|z|\sqrt{T}\right\},$$

where

$$d(x) = \exp\{D(x) + K|x|\} \sup_{-\infty \le y \le \infty} f(y).$$

Next, from (8.108) and (8.110) we find

$$0 \le \bar{\rho}_t[x - z\sqrt{t}, z\sqrt{t}, \tilde{\eta}, \xi]$$

$$\le K_1 \exp\left\{\int_0^t A\left(x - z\sqrt{t} + \frac{sz}{\sqrt{t}} + \tilde{\eta}_s\right)d\overline{W}_s\right\},$$

where K_1 is a certain constant. From this, by the Jensen inequality, Lemma 6.1, and the Fubini theorem we obtain

$$M(\tilde{M}\bar{\rho}_t[x - z\sqrt{t}, z\sqrt{t}, \tilde{\eta}, \xi])^{2n} \le M\tilde{M}\bar{\rho}_t^{2n}[x - z\sqrt{t}, z\sqrt{t}, \tilde{\eta}, \xi]$$

$$\le C_1(n)M\tilde{M} \exp\left\{2n \int_0^t A\left(x - z\sqrt{t} - \frac{sz}{\sqrt{t}} + \tilde{\eta}_s\right)d\overline{W}_s\right.$$

$$\left. - \frac{(2n)^2}{2} \int_0^t A^2\left(x - z\sqrt{t} + \frac{sz}{\sqrt{t}} + \tilde{\eta}_s\right)ds\right\} \le C_1(n),$$

where $C_1(n)$ is a certain constant. In a similar way it is also shown that

$$M\psi_t^{-2n}(\xi) \le C_2(n), \qquad n = 1, 2, \ldots.$$

Making use of these estimates, the integrability of the function $\exp\{-z^2/2 + K\sqrt{T}|z|\}$, and the Cauchy–Buniakowski inequalities, from (8.116) we obtain (8.115). □

Lemma 8.8. *If the assumptions of Theorem* 8.7 *are fulfilled, then the conditional density* $\rho_x(t)$, $0 \le t \le T$, *is twice differentiable over* x *and*

$$\sup_{t \le T} M\left[\frac{\partial \rho_x(t)}{\partial x}\right]^2 < \infty, \qquad \sup_{t \le T} M\left[\frac{\partial^2 \rho_x(t)}{\partial x^2}\right]^2 < \infty. \tag{8.117}$$

PROOF. For $t > 0$, denote

$$\Phi_{t, y, \tilde{\eta}, \xi}(x) = \exp\left\{-\frac{(x-y)^2}{2t} + D(x) - D(y)\right\}\tilde{\rho}_t(y, x - y, \tilde{\eta}, \xi).$$

Then, by (8.114),

$$\rho_x(t) = \frac{1}{\sqrt{2\pi t \psi_t(\xi)}} \int_{\tilde{\Omega} \times \mathbb{R}^1} \Phi_{t, y, \tilde{\eta}, \xi}(x)\tilde{P}(d\tilde{\omega})f(y)dy, \tag{8.118}$$

and for the existence of the derivatives $\partial^i \rho_x(t)/\partial x^i$ it suffices to establish that

$$V(x) = \int_{\tilde{\Omega} \times \mathbb{R}^1} \Phi_{t, y, \tilde{\eta}, \xi}(x)\tilde{P}(d\tilde{\omega})f(y)dy$$

is twice differentiable with respect to x.

Assume that with fixed $t, y, \tilde{\eta}, \xi$ the function $\Phi_{t, y, \tilde{\eta}, \xi}(x)$ is twice differentiable over x. Then, for any x', x'' $(-\infty < x' < x'' < \infty)$,

$$V(x'') - V(x') = \int_{\tilde{\Omega} \times \mathbb{R}^1} \left[\int_{x'}^{x''} \frac{\partial}{\partial z} \Phi_{t, y, \tilde{\eta}, \xi}(z)dz\right]\tilde{P}(d\tilde{\omega})f(y)dy, \tag{8.119}$$

and if (P-a.s.)

$$\int_{\tilde{\Omega} \times \mathbb{R}^1} \int_{x'}^{x''} \left|\frac{\partial}{\partial z} \Phi_{t, y, \tilde{\eta}, \xi}(z)\right| \tilde{P}(d\tilde{\omega})f(y)dy < \infty, \tag{8.120}$$

then, by the Fubini theorem, in (8.119) the change of the orders of integration is permitted and

$$V(x) = V(0) + \int_0^x \left[\int_{\tilde{\Omega} \times \mathbb{R}^1} \frac{\partial}{\partial z} \Phi_{t, y, \tilde{\eta}, \xi}(z)\tilde{P}(d\tilde{\omega})f(y)dy\right]dz.$$

Hence, if, in addition, the function

$$R(x) = \int_{\tilde{\Omega} \times \mathbb{R}^1} \frac{\partial}{\partial x} \Phi_{t, y, \tilde{\eta}, \xi}(x)\tilde{P}(d\tilde{\omega})f(y)dy$$

is continuous in x ((P-a.s.) for t, $0 \le t \le T$), then the function $V(x)$ will be differentiable in x and $dV(x)/dx = R(x)$.

Let us establish first that the function $\partial/\partial x\, \Phi_{t, y, \tilde{\eta}, \xi}(x)$ is continuous in x. Since the function $D(x)$ is continuously differentiable, it suffices to show that the functions

$$\int_0^t A\left(y\frac{t-s}{t} + \tilde{\eta}_s + \frac{s}{t}x\right)d\overline{W}_s, \qquad \int_0^t B_{s, \xi}\left(y\frac{t-s}{t} + \tilde{\eta}_s + \frac{s}{t}x\right)ds$$

are continuously differentiable in x.

The derivatives

$$\frac{\partial}{\partial x} A\left(y\frac{t-s}{t} + \tilde{\eta}_s + \frac{s}{t}x\right), \qquad \frac{\partial}{\partial x} B_{s,\xi}\left(y\frac{t-s}{t} + \tilde{\eta}_s + \frac{s}{t}x\right)$$

exist and are uniformly bounded under the assumption of Theorem 8.7. Repeating the considerations above, we can see that

$$\frac{\partial}{\partial x}\int_0^t B_{s,\xi}\left(y\frac{t-s}{t} + \tilde{\eta}_s + \frac{s}{t}x\right)ds = \int_0^t \frac{\partial}{\partial x} B_{s,\xi}\left(y\frac{t-s}{t} + \tilde{\eta}_s + \frac{s}{t}x\right)ds, \tag{8.121}$$

if the function

$$\int_0^t \frac{\partial}{\partial x} B_{s,\xi}\left(y\frac{t-s}{t} + \tilde{\eta}_s + \frac{s}{t}x\right)ds$$

(in terms of a function of x for fixed $t, \xi, \tilde{\eta}$) is continuous. But the function

$$\frac{\partial}{\partial x} B_{s,\xi}\left(y\frac{t-s}{t} + \tilde{\eta}_s + \frac{s}{t}x\right)$$

is uniformly bounded and continuous, which implies (8.121) and the continuity in x of the function

$$\frac{\partial}{\partial x}\int_0^t B_{s,\xi}\left(y\frac{t-s}{t} + \tilde{\eta}_s + \frac{s}{t}x\right)ds.$$

Next let us establish the differentiability of the function

$$\int_0^t A\left(y\frac{t-s}{t} + \tilde{\eta}_s + \frac{s}{t}x\right)d\overline{W}_s$$

and the equality

$$\frac{\partial}{\partial x}\int_0^t A\left(y\frac{t-s}{t} + \tilde{\eta}_s + \frac{s}{t}x\right)d\overline{W}_s = \int_0^t \frac{\partial}{\partial x} A\left(y\frac{t-s}{t} + \tilde{\eta}_s + \frac{s}{t}x\right)d\overline{W}_s. \tag{8.122}$$

It will be noted that the function

$$\lambda(x) = \int_0^t \frac{\partial}{\partial x} A\left(y\frac{t-s}{t} + \tilde{\eta}_s + \frac{s}{t}x\right)d\overline{W}_s$$

(for fixed $t, \tilde{\eta}, \xi$) is continuous in x. Indeed, by the assumptions of Theorem 8.7,

$$\begin{aligned} M|\lambda(x') - \lambda(x'')|^2 &= M\left\{\int_0^t \left[\frac{\partial}{\partial x'} A\left(y\frac{t-s}{t} + \tilde{\eta}_s + \frac{s}{t}x'\right)\right.\right. \\ &\qquad \left.\left. - \frac{\partial}{\partial x''} A\left(y\frac{t-s}{t} + \tilde{\eta}_s + \frac{s}{t}x''\right)\right]d\overline{W}_s\right\}^2 \\ &\le KT|x' - x''|^2. \end{aligned}$$

Hence the continuity of $\lambda(x)$ follows from Kolmogorov's continuity criterion (Theorem 1.10).

Next, by the Fubini theorem for stochastic integrals (Theorem 5.15), with $-\infty < x' < x'' < \infty$,

$$\int_{x'}^{x''}\left\{\int_0^t \frac{\partial}{\partial z} A\left(y\frac{t-s}{t} + \tilde{\eta}_s + \frac{s}{t}z\right)d\overline{W}_s\right\}dz$$

$$= \int_0^t\left\{\int_{x'}^{x''} \frac{\partial}{\partial z} A\left(y\frac{t-s}{t} + \tilde{\eta}_s + \frac{s}{t}z\right)dz\right\}d\overline{W}_s$$

$$= \int_0^t A\left(y\frac{t-s}{t} + \tilde{\eta}_s + \frac{s}{t}x''\right)d\overline{W}_s - \int_0^t A\left(y\frac{t-s}{t} + \tilde{\eta}_s + \frac{s}{t}x'\right)d\overline{W}_s.$$

From this, because of the continuity of the function $\lambda(x)$, it follows that the derivative

$$\frac{\partial}{\partial x}\int_0^t A\left(y\frac{t-s}{t} + \tilde{\eta}_s + \frac{s}{t}x\right)d\overline{W}_s$$

exists and (8.122) is satisfied.

Thus, the function

$$\frac{\partial}{\partial x}\Phi_{t,y,\tilde{\eta},\xi}(x)$$

is continuous over x. Therefore the density $\rho_x(t)$ is differentiable in x (for almost all ω and t, $0 \le t \le T$), and hence

$$\frac{\partial \rho_x(t)}{\partial x} = \frac{1}{\sqrt{2\pi t}\,\psi_t(\xi)}\int_{\Omega\times\mathbb{R}^1}\frac{\partial}{\partial x}\Phi_{t,y,\tilde{\eta},\xi}(x)\tilde{P}(d\tilde{\omega})f(y)dy. \tag{8.123}$$

In a similar way one can establish the existence for $t > 0$ of the derivative $\partial^2\rho_x(t)/\partial x^2$ and the formula

$$\frac{\partial^2 \rho_x(t)}{\partial x^2} = \frac{1}{\sqrt{2\pi t}\,\psi_t(\xi)}\int_{\Omega\times\mathbb{R}^1}\frac{\partial^2}{\partial x^2}\Phi_{t,y,\tilde{\eta},\xi}(x)\tilde{P}(d\tilde{\omega})f(y)dy. \tag{8.124}$$

The inequalities given by (8.117) are shown in the same way as in Lemma 8.7. □

8.6.4

Proof of Theorem 8.7. The validity of Equation (8.93) for $\rho_x(t)$ follows from Theorem 8.6 and Lemmas 8.5–8.8 (which guarantee that the conditions of Theorem 8.6 are satisfied).

Let us now prove the uniqueness of a solution of this equation for the class of functions defined under the conditions of the theorem.

Let $U_x(t)$, $x \in \mathbb{R}^1$, $0 \le t \le T$, be some solution of Equation (8.93) from the given class, with $U_x(0) = f(x)$ (P-a.s.). Set

$$\varkappa_t = \exp\left\{\int_0^t \left(\int_{-\infty}^{\infty} A(y)U_y(s)dy\right)d\xi_s - \frac{1}{2}\int_0^t \left(\int_{-\infty}^{\infty} A(y)U_y(s)dy\right)^2 ds\right\} \quad (8.125)$$

and

$$Q_x(t) = U_x(t)\varkappa_t. \quad (8.126)$$

By the Ito formula,

$$d_t Q_x(t) = \left\{-\frac{\partial}{\partial x}[a(x)U_x(t)] + \frac{1}{2}\frac{\partial^2}{\partial x^2}[U_x(t)]\right\}\varkappa_t\, dt + U_x(t)\varkappa_t A(x)d\xi_t \quad (8.127)$$

or, equivalently,

$$d_t Q_x(t) = \left\{-\frac{\partial}{\partial x}[a(x)Q_x(t)] + \frac{1}{2}\frac{\partial^2}{\partial x^2}[Q_x(t)]\right\}dt + Q_x(t)A(x)d\xi_t, \quad (8.128)$$

where

$$Q_x(0) = U_x(0) = f(x). \quad (8.129)$$

Therefore Equation (8.128) with the initial condition given by (8.129) has the strong (i.e., $\mathscr{F}_t^\xi$-measurable for each t, $0 \le t \le T$) solution $Q_x(t) = U_x(t)\varkappa_t$.

By the Ito formula,

$$\varkappa_t = 1 + \int_0^t \varkappa_s\left(\int_{-\infty}^{\infty} A(y)U_y(s)dy\right)d\xi_s = 1 + \int_0^t \left(\int_{-\infty}^{\infty} A(y)Q_y(s)dy\right)d\xi_s, \quad (8.130)$$

and obviously $P\{0 < \varkappa_t < \infty, 0 \le t \le T\} = 1$. Hence, from (8.126) and (8.130) it follows that

$$U_x(t) = \frac{Q_x(t)}{1 + \int_0^t \left(\int_{-\infty}^{\infty} A(y)Q_y(s)dy\right)d\xi_s}, \quad (8.131)$$

where $Q_x(t)$ satisfies Equation (8.128).

(8.126) and (8.131) determine a one-to-one correspondence between the solutions of Equation (8.93) and those of (8.128). Hence, for proving uniqueness of a solution of Equation (8.93) it suffices to establish uniqueness of a solution of Equation (8.128) in a class of the functions $Q_x(t)$ satisfying the condition

$$P\left\{\int_0^T \left(\int_{-\infty}^{\infty} A(x)Q_x(t)dx\right)^2 dt < \infty\right\} = 1$$

(see (8.131)).

Assume

$$\psi_x(t) = \exp\{A(x)\xi_t - \tfrac{1}{2}A^2(x)t\} \tag{8.132}$$

and

$$R_x(t) = \frac{Q_x(t)}{\psi_x(t)}. \tag{8.133}$$

By the Ito formula, from (8.128), (8.132) and (8.133) we find that

$$d_t R_x(t) = \left\{ -\frac{\partial}{\partial x}[a(x)Q_x(t)] + \frac{1}{2}\frac{\partial^2}{\partial x^2}[Q_x(t)] \right\} \psi_x^{-1}(t)dt. \tag{8.134}$$

The multiplier for dt in (8.134) is a continuous function over t and hence

$$\begin{aligned}
\frac{\partial R_x(t)}{dt} &= \left\{ -\frac{\partial}{\partial x}[a(x)Q_x(t)] + \frac{1}{2}\frac{\partial^2}{\partial x^2}[Q_x(t)] \right\} \psi_x^{-1}(t) \\
&= \left\{ -\frac{\partial}{\partial x}[a(x)R_x(t)\psi_x(t)] + \frac{1}{2}\frac{\partial^2}{\partial x^2}[R_x(t)\psi_x(t)] \right\} \psi_x^{-1}(t) \\
&= -a'(x)R_x(t) - a(x)\frac{\partial R_x(t)}{\partial x} + \frac{1}{2}\frac{\partial^2}{\partial x^2}R_x(t) \\
&\quad - a(x)R_x(t)\frac{\partial \psi_x(t)}{\partial x}\psi_x^{-1}(t) + \frac{\partial R_x(t)}{\partial x}\frac{\partial \psi_x(t)}{\partial x}\psi_x^{-1}(t) \\
&\quad + \frac{1}{2}R_x(t)\frac{\partial^2 \psi_x(t)}{\partial x^2}\psi_x^{-1}(t),
\end{aligned} \tag{8.135}$$

where

$$\frac{\partial \psi_x(t)}{\partial x}\psi_x^{-1}(t) = A'(x)[A(x)t - \xi_t],$$

$$\frac{\partial^2 \psi_x(t)}{\partial x^2}\psi_x^{-1}(t) = (A'(x))^2[A'(x)t - \xi_t]^2 + A''(x)[A(x)t - \xi_t] + (A'(x))^2. \tag{8.136}$$

Denoting

$$\bar{a}(t, x) = -a(x) + A'(x)[A(x)t - \xi_t], \tag{8.137}$$

$$\begin{aligned}
\bar{c}(t, x) = {} & -a'(x) - a(x)A'(x)[A(x)t - \xi_t] \\
& + \tfrac{1}{2}(A'(x))^2(1 + [A(x)t - \xi_t]^2) + A''(x)[A(x)t - \xi_t],
\end{aligned} \tag{8.138}$$

from (8.134)–(8.138) we obtain for $R_x(t)$ the equation

$$\frac{\partial R_x(t)}{\partial t} = \frac{1}{2}\frac{\partial^2 R_x(t)}{\partial x^2} + \bar{a}(t, x)\frac{\partial R_x(t)}{\partial x} + \bar{c}(t, x)R_x(t) \tag{8.139}$$

with $R_x(0) = f(x)$.

The coefficients $\bar{a}(t, x)$, $\bar{c}(t, x)$ are continuous (P-a.s.) over all the variables and uniformly bounded. Hence, from the known results of the theory of differential equations with partial derivatives[3] it follows that Equation (8.139) has (P-a.s.) a unique solution with $R_x(0) = f(x)$ in the class of the functions $R_x(t)$ satisfying the condition (for each ω)

$$R_x(t) \leq c_1(\omega)\exp(c_2(\omega)x^2),$$

where $c_i(\omega)$, $i = 1, 2$, are such that

$$R_x(t) \leq c_1(\omega)\exp(c_2(\omega)x^2).$$

But $P(\inf_{t \leq T} \psi_x(t) > 0) = 1$, $-\infty < x < \infty$. Hence the solution of Equation (8.128) is also unique in the given class.

From this follows the uniqueness of a solution of Equation (8.93) in the class of the random functions $\{U_x(t), -\infty < x < \infty, 0 \leq t \leq T\}$ satisfying the condition

$$\int_0^T \left(\int_{-\infty}^{\infty} A(x)U_x(t)dx\right)^2 dt < \infty \qquad (P\text{-a.s.}). \tag{8.140}$$

To complete the proof it remains only to note that the function $\rho_x(t)$ satisfies (8.140) since

$$\int_0^T M\left(\int_{-\infty}^{\infty} A(x)\rho_x(t)dx\right)^2 dt = \int_0^T M[M(A(\theta_t)|\mathscr{F}_t^{\xi})]^2\, dt \leq KT. \qquad \square$$

Notes and references

8.1,8.2. Many works deal with the deduction of representations for conditional mathematical expectations $\pi_t(h)$ under various assumptions on (θ, ξ, h). First of all the classical works of Kolmogorov [87], and Wiener [21] should be noted where the problems of constructing optimal estimates for a case of stationarily associated processes were examined within a linear theory.

More extended discussion of the results obtained by them as well as the latest advances in this field in recent years can be found in Yaglom [172], Rozanov [139], Prokhorov and Rozanov [135]. For the results concerning nonlinear filtering see, for example, Stratanovich [146], [147], Wentzel [19], Wonham [25], Kushner [98], [99], Shiryayev [165], [166], [170], Liptser and Shiryayev[111], [114]–[116], Liptser [108]–[110], Kailath [67], [70], Frost, Kailath [155], Striebel [148], Kallianpur and Striebel [74], [75], Yershov [50], [51], and Grigelionis [41]. The deduction presented here principally follows Fujisaki, Kallianpur, and Kunita [156]. The first general results on the construction of optimal nonlinear estimates for Markov processes were obtained by Stratonovich [146], [147], within the theory of conditional Markov processes.

8.3. Representation (8.56) for $\pi_t(h)$ in a case of diffusion type processes is due to Shiryayev [165] and Liptser and Shiryayev [111].

8.4,8.5. Theorems 8.4 and 8.5 have been first presented here. Particular cases are due to Stratonovich [147], Liptser and Shiryayev [112]–[116], and Liptser [108]–[110].

8.6. The stochastic differential equations with partial derivatives for conditional density considered here are due to Liptser and Shiryayev [111]. The results on uniqueness of the solution are due to Rozovsky [140].

[3] See, for example, [154], Theorem 10, Chapter II, Section 4.

9 Optimal filtering, interpolation and extrapolation of Markov processes with a countable number of states

9.1 Equations of optimal nonlinear filtering

9.1.1

The present chapter will be concerned with a pair of random processes $(\theta, \xi) = (\theta_t, \xi_t)$, $0 \leq t \leq T$, where the unobservable component θ is a Markov process with a finite or countable number of states, and the observable process ξ permits the stochastic differential

$$d\xi_t = A_t(\theta_t, \xi)dt + B_t(\xi)dW_t, \tag{9.1}$$

where W_t is a Wiener process.

Many problems of random processes statistics lead to such a scheme where an unobservable process takes discrete values, and the noise is of the nature of "white" Gaussian noise.

In this section, which draws on the results of the previous chapter, equations of optimal nonlinear filtering will be deduced and studied. Interpolation and extrapolation (phenomena) will be treated in Sections 9.2 and 9.3.

9.1.2

Let $(\Omega, \mathscr{F}, P)$ be a complete probability space with a nondecreasing family of right continuous sub-σ-algebras $\mathscr{F}_t$, $0 \leq t \leq T$. Let $\theta = (\theta_t, \mathscr{F}_t)$, $0 \leq t \leq T$, be a real right continuous Markov process with values in the countable set $E = \{\alpha, \beta, \gamma, \dots\}$; let $W = (W_t, \mathscr{F}_t)$, $0 \leq t \leq T$, be a standard Wiener process independent of θ, and let ξ_0 be an $\mathscr{F}_0$-measurable random variable independent of θ. It will be assumed that the nonanticipative functionals $A_t(\varepsilon, x)$ and

$B_t(x)$ entering in (9.1) satisfy the following conditions.

$$A_t^2(\varepsilon_t, x) \le L_1 \int_0^t (1 + x_s^2) dK(s) + L_2(1 + \varepsilon_t^2 + x_t^2), \tag{9.2}$$

$$0 < C \le B_t^2(x) \le L_1 \int_0^t (1 + x_s^2) dK(s) + L_2(1 + x_t^2), \tag{9.3}$$

$$|A_t(\varepsilon_t, x) - A_t(\varepsilon_t, y)|^2 + |B_t(x) - B_t(y)|^2$$
$$\le L_1 \int_0^t (x_s - y_s)^2 \, dK(s) + L_2(x_t - y_t)^2, \tag{9.4}$$

where C, L_1, L_2 are certain constants, $K(s)$ is a nondecreasing right continuous function, $0 \le K(s) \le 1$, $x \in C_T$, $y \in C_T$, $\varepsilon_t \in E$, $0 \le t \le T$.

Along with (9.2)–(9.4) it will be also assumed that

$$M\xi_0^2 < \infty \tag{9.5}$$

and

$$M \int_0^t \theta_t^2 \, dt < \infty. \tag{9.6}$$

By Theorem 4.6[1] (9.2)–(9.6) provide Equation (9.1) with existence and uniqueness of the (strong) solution

$$\xi = (\xi_t, \mathscr{F}_t^{\xi_0, \theta, W}), \; 0 \le t \le T,$$

with $\sup_{0 \le t \le T} M\xi_t^2 < \infty$.

Let the realization $\xi_0^t = \{\xi_s, s \le t\}$ of the observable process ξ be known for $0 \le t \le T$. The filtering problem for an unobservable process θ is the construction of estimates of the value θ_t on the basis of ξ_0^t. The most convenient criterion for optimality for estimating θ_t is the a posteriori probability

$$\pi_\beta(t) = P(\theta_t = \beta \,|\, \mathscr{F}_t^\xi), \qquad \beta \in E.$$

Indeed, with the help of $\pi_\beta(t)$, $\beta \in E$, the most various estimates of the value θ_t can be obtained. In particular, the conditional expectation

$$M(\theta_t \,|\, \mathscr{F}_t^\xi) = \sum_{\beta \in E} \beta \pi_\beta(t) \tag{9.7}$$

is the optimal mean square estimate. The estimate $\beta_t(\xi)$, obtained from the condition

$$\max_\beta P(\theta_t = \beta \,|\, \mathscr{F}_t^\xi) = \pi_{\beta_t(\xi)}(t), \tag{9.8}$$

is an estimate maximizing the a posteriori probability.

[1] More precisely, because of an obvious extension of this theorem to the case where the functionals $a(t, x)$ in (4.112) are replaced by the functionals $A_t(\varepsilon_t, x)$.

9.1.3

We shall formulate a number of auxiliary statements with respect to the processes θ and ξ which will be employed in proving the main result (Theorem 9.1).

Denote

$$p_\beta(t) = P(\theta_t = \beta),$$

$$p_{\beta\alpha}(t, s) = P(\theta_t = \beta | \theta_s = \alpha), \qquad 0 \le s < t \le T, \quad \beta, \alpha \in E.$$

Lemma 9.1. *Let there exist a function $\lambda_{\alpha\beta}(t)$, $0 \le t \le T$, α, $\beta \in E$, such that (uniformly over α, β) it is continuous over t, $|\lambda_{\alpha\beta}(t)| \le K$, and*

$$|p_{\beta\alpha}(t + \Delta, t) - \delta(\beta, \alpha) - \lambda_{\alpha\beta}(t) \cdot \Delta| \le o(\Delta), \tag{9.9}$$

where $\delta(\beta, \alpha)$ is a Kronecker's symbol, and the value $0(\Delta)/\Delta \to 0(\Delta \to 0)$ uniformly over α, β, t. Then $p_{\beta\alpha}(t, s)$ satisfies the forward Kolmogorov equation

$$p_{\beta\alpha}(t, s) = \delta(\beta, \alpha) + \int_s^t \mathfrak{L}^* p_{\beta\alpha}(u, s) du, \tag{9.10}$$

where

$$\mathfrak{L}^* p_{\beta\alpha}(u, s) = \sum_{\gamma \in E} \lambda_{\gamma\beta}(u) p_{\gamma\alpha}(u, s). \tag{9.11}$$

The probabilities $p_\beta(t)$ satisfy the equation

$$p_\beta(t) = p_\beta(0) + \int_0^t \mathfrak{L}^* p_\beta(u) du, \tag{9.12}$$

where

$$\mathfrak{L}^* p_\beta(u) = \sum_{\gamma \in E} \lambda_{\gamma\beta}(u) p_\gamma(u).$$

PROOF. Let $s = t_0^{(n)} < t_1^{(n)} < \cdots < t_n^{(n)} = t$ and let $\max_j |t_{j+1}^{(n)} - t_j^{(n)}| \to 0$, $n \to \infty$. Because of the Markov behavior of the process θ,

$$\begin{aligned} p_{\beta\alpha}(t_{j+1}^{(n)}, s) &= P(\theta_{t_{j+1}^{(n)}} = \beta | \theta_s = \alpha) \\ &= M\{P(\theta_{t_{j+1}^{(n)}} = \beta | \theta_{t_j^{(n)}}, \theta_s = \alpha) | \theta_s = \alpha\} \\ &= M\{P(\theta_{t_{j+1}^{(n)}} = \beta | \theta_{t_j^{(n)}}) | \theta_s = \alpha\}, \end{aligned}$$

or

$$p_{\beta\alpha}(t_{j+1}^{(n)}, s) = \sum_{\gamma \in E} p_{\beta\gamma}(t_{j+1}^{(n)}, t_j^{(n)}) p_{\gamma\alpha}(t_j^{(n)}, s). \tag{9.13}$$

Denote

$$r_{\beta\gamma}(t_{j+1}^{(n)}, t_j^{(n)}) = p_{\beta\gamma}(t_{j+1}^{(n)}, t_j^{(n)}) - \delta(\beta, \gamma) - \lambda_{\gamma\beta}(t_j^{(n)})(t_{j+1}^{(n)} - t_j^{(n)}).$$

Then from (9.13) we find that

$$p_{\beta\alpha}(t_{j+1}^{(n)}, s) = \sum_{\gamma\in E}[\delta(\beta,\gamma) + \lambda_{\gamma\beta}(t_j^{(n)})(t_{j+1}^{(n)} - t_j^{(n)}) + r_{\beta\gamma}(t_{j+1}^{(n)}, t_j^{(n)})]p_{\gamma\alpha}(t_j^{(n)}, s)$$

$$= p_{\beta\alpha}(t_j^{(n)}, s) + \left(\sum_{\gamma\in E}\lambda_{\gamma\beta}(t_j^{(n)})p_{\gamma\alpha}(t_j^{(n)}, s)\right)[t_{j+1}^{(n)} - t_j^{(n)}]$$

$$+ \sum_{\gamma\in E} r_{\beta\gamma}(t_{j+1}^{(n)}, t_j^{(n)})p_{\gamma\alpha}(t_j^{(n)}, s). \tag{9.14}$$

From the conditions of the lemma and this equality it follows that the function $p_{\beta\alpha}(t, s)$ is continuous over t (uniformly over α, β, s). Next, again by (9.14),

$$p_{\beta\alpha}(t, s) - \delta(\beta, \alpha) = \sum_{j=0}^{n-1}[p_{\beta\alpha}(t_{j+1}^{(n)}, s) - p_{\beta\alpha}(t_j^{(n)}, s)]$$

$$= \int_s^t \sum_{\gamma\in E}\lambda_{\gamma\beta}(\varphi_n(u))p_{\gamma\alpha}(\varphi_n(u), s)du$$

$$+ \sum_{j=0}^{n-1}\sum_{\gamma\in E} r_{\beta\gamma}(t_{j+1}^{(n)}, t_j^{(n)})p_{\gamma\alpha}(t_j^{(n)}, s), \tag{9.15}$$

where $\varphi_n(u) = t_j^{(n)}$ when $t_j^{(n)} \le u < t_{j+1}^{(n)}$.

According to the assumptions of the lemma,

$$\overline{\lim_{n\to\infty}} \sum_{j=0}^{n-1}\sum_{\gamma\in E}|r_{\beta\gamma}(t_{j+1}^{(n)}, t_j^{(n)})|p_{\gamma\alpha}(t_j^{(n)}, s) = 0,$$

and

$$\sum_{\gamma\in E}|\lambda_{\gamma\beta}(\varphi_n(u))|p_{\gamma\alpha}(\varphi_n(u), s) \le K < \infty.$$

Taking this fact into account as well as the continuity of $\lambda_{\alpha\beta}(t)$ and $p_{\beta\alpha}(t, s)$ over t (uniformly over α, β, s), from (9.15) (after the passage to the limit with $n \to \infty$) we obtain the Equation (9.10); Equation (9.12) is easily deduced from (9.10). □

Note. The function $\lambda_{\alpha\beta}(t)$ is the density of transition probabilities from α into β at time t.

Lemma 9.2. *Let the conditions of Lemma* 9.1 *be satisfied. For each* $\beta \in E$ *we set*

$$x_t^\beta = \delta(\beta, \theta_t) - \delta(\beta, \theta_0) - \int_0^t \lambda_{\theta_s\beta}(s)ds. \tag{9.16}$$

The random process $X^\beta = (x_t^\beta, \mathscr{F}_t)$, $0 \le t \le T$, *is a square integrable martingale with right continuous trajectories.*

PROOF. The process x_t^β, $0 \le t \le T$, is bounded ($|x_t^\beta| \le 2 + KT$) and right continuous because of right continuity of the trajectories of the process θ_t, $0 \le t \le T$.

Let us show that $X^\beta = (x_t^\beta, \mathscr{F}_t)$, $0 \le t \le T$, is a martingale. Let $t > s$. Then

$$x_t^\beta = x_s^\beta + \left[\delta(\beta, \theta_t) - \delta(\beta, \theta_s) - \int_s^t \lambda_{\theta_u \beta}(u)du\right]$$

and therefore,

$$M(x_t^\beta | \mathscr{F}_s) = x_s^\beta + M\left[\delta(\beta, \theta_t) - \delta(\beta, \theta_s) - \int_s^t \lambda_{\theta_u \beta}(u)du \,|\, \mathscr{F}_s\right].$$

Because of the Markov behavior of the process $\theta = (\theta_t)$, $0 \le t \le T$, and Equation (9.10),

$$M\left[\delta(\beta, \theta_t) - \delta(\beta, \theta_s) - \int_s^t \lambda_{\theta_u \beta}(u)du \,|\, \mathscr{F}_s\right]$$

$$= M\left[\delta(\beta, \theta_t) - \delta(\beta, \theta_s) - \int_s^t \lambda_{\theta_u \beta}(u)du \,|\, \theta_s\right]$$

$$= p_{\beta\theta_s}(t, s) - \delta(\beta, \theta_s) - \int_s^t \sum_{\gamma \in E} \lambda_{\gamma\beta}(u) p_{\gamma\theta_s}(u, s) = 0. \quad \square$$

9.1.4

Theorem 9.1. *Let the conditions of Lemma 9.1 and (9.2)–(9.6) be fulfilled. Then the a posteriori probabilities $\pi_\beta(t)$, $\beta \in E$, satisfy a system of the equations*

$$\pi_\beta(t) = p_\beta(0) + \int_0^t \mathfrak{L}^* \pi_\beta(u)du + \int_0^t \pi_\beta(u) \frac{A_u(\beta, \xi) - \bar{A}_u(\xi)}{B_u(\xi)} d\overline{W}_u, \tag{9.17}$$

where

$$\mathfrak{L}^* \pi_\beta(u) = \sum_{\gamma \in E} \lambda_{\gamma\beta}(u) \pi_\beta(u), \tag{9.18}$$

$$\bar{A}_u(\xi) = \sum_{\gamma \in E} A_u(\gamma, \xi) \pi_\gamma(u), \tag{9.19}$$

and $\overline{W} = (\overline{W}_t, \mathscr{F}_t)$ is a Wiener process with

$$\overline{W}_t = \int_0^t \frac{d\xi_u - \bar{A}_u(\xi)du}{B_u(\xi)}. \tag{9.20}$$

PROOF. By Lemma 9.2,

$$\delta(\beta, \theta_t) = \delta(\beta, \theta_0) + \int_0^t \lambda_{\theta_u\beta}(u)du + x_t^\beta, \tag{9.21}$$

where $X^\beta = (x_t^\beta, \mathcal{F}_t)$ is a square integrable martingale. Since the processes X^β and W are independent, $\langle x^\beta, W\rangle_t \equiv 0$ (P-a.s.), $0 \le t \le T$.

The assumptions given by (9.2)–(9.6) make possible the application (to $h_t = \delta(\beta, \theta_t)$) of Theorem 8.1, according to which

$$\pi_t^\beta(\delta) = \pi_0^\beta(\delta) + \int_0^t \pi_s^\beta(\lambda)ds + \int_0^t \frac{\pi_s^\beta(\delta A) - \pi_s^\beta(\delta)\pi_s^\beta(A)}{B_u(\xi)} d\overline{W}_u, \tag{9.22}$$

where

$$\pi_t^\beta(\delta) = M[\delta(\beta, \theta_t)|\mathcal{F}_t^\xi] = \pi_\beta(t),$$

$$\pi_s^\beta(\lambda) = M[\lambda_{\theta_s\beta}(s)|\mathcal{F}_s^\xi] = \sum_{\gamma\in E}\lambda_{\gamma\beta}(s)\pi_\gamma(s) = \mathfrak{L}^*\pi_\beta(s),$$

$$\pi_s^\beta(\delta A) = M[\delta(\beta, \theta_s)A_s(\theta_s, \xi)|\mathcal{F}_s^\xi] = A_s(\beta, \xi)\pi_\beta(s),$$

$$\pi_s^\beta(A) = M[A_s(\theta_s, \xi)|\mathcal{F}_s^\xi] = \bar{A}_s(\xi) = \sum_{\gamma\in E} A_s(\gamma, \xi)\pi_\gamma(s).$$

Using this notation we can see that (9.22) coincides with (9.17). □

Note. If in (9.1) the coefficients $A_t(\theta_t, \xi)$ do not depend upon θ_t, then $\pi_\beta(t) = p_\beta(t)$ and the equations given by (9.17) become the (forward) Kolmogorov equations ((9.12)).

9.1.5

From (9.17) we can see that countably-valued process $\Pi = \{\pi_\beta(t), \beta \in E\}$, $0 \le t \le T$, is a solution of the following infinite system of stochastic differential equations

$$\begin{aligned} dz_\beta(t, \xi) = \Bigg[&\sum_{\gamma\in E}\lambda_{\gamma\beta}(t)z_\gamma(t, \xi) - z_\beta(t, \xi)\frac{A_t(\beta, \xi) - \sum_{\gamma\in E} A_t(\gamma, \xi)z_\gamma(t, \xi)}{B_t^2(\xi)} \\ &\times \sum_{\gamma\in E} A_t(\gamma, \xi)z_\gamma(t, \xi)\Bigg]dt \\ &+ z_\beta(t, \xi)\frac{A_t(\beta, \xi) - \sum_{\gamma\in E} A_t(\gamma, \xi)z_\gamma(t, \xi)}{B_t^2(\xi)} d\xi_t, \qquad \beta \in E, \end{aligned} \tag{9.23}$$

to be solved under the conditions $z_\beta(0, \xi) = p_\beta(0)$, $\beta \in E$.

An important question is the uniqueness of solution of this (nonlinear) system of equations.

Theorem 9.2. *Let the conditions of Lemma* 9.1 *and* (9.2)–(9.6) *be fulfilled. Then in the class of the nonnegative continuous processes* $Z = \{z_\beta(t, \xi), \beta \in E\}$, $0 \le t \le T$, $\mathcal{F}_t^\xi$*-measurable for each t and satisfying the conditions*

$$P\left\{\sup_{0 \le t \le T} \sum_{\beta \in E} z_\beta(t, \xi) \le C\right\} = 1 \qquad (C = \text{const.}), \tag{9.24}$$

$$P\left\{\int_0^T \left(\sum_{\gamma \in E} \frac{|A_t(\gamma, \xi)| z_\gamma(t, \xi)}{B_t(\xi)}\right)^2 dt < \infty\right\} = 1, \tag{9.25}$$

the system of equations given by (9.23) *has a unique solution in the following sense: if Z and Z′ are two solutions, then* $P\{\sup_{0 \le t \le T} |z_\beta(t, \xi) - z'_\beta(t, \xi)| > 0\} = 0$, $\beta \in E$.

Proof. Note first of all that the a posteriori probabilities $\Pi = \{\pi_\beta(t), \beta \in E\}$, $0 \le t \le T$, belong to a class of the processes satisfying (9.24) and (9.25). Hence, from the statement of the theorem it follows that in the class under consideration the process Π is a unique solution of (9.23). Let us note also that (9.24), (9.25), and the assumed continuity of the component trajectories of the processes Z provide the existence of the corresponding integrals (over dt and $d\xi_t$) in (9.23).

Let $Z = \{z_\beta(t, \xi), \beta \in E\}$, $0 \le t \le T$, be some solution of (9.23), with $z_\beta(0, \xi) = p_\beta(0)$, $\sum_{\beta \in E} p_\beta(0) = 1$. Denote

$$I_Z(t, \xi) = \exp\left\{\int_0^T \frac{\sum_{\gamma \in E} A_s(\gamma, \xi) z_\gamma(s, \xi)}{B_s^2(\xi)} d\xi_s - \frac{1}{2}\int_0^t \left[\frac{\sum_{\gamma \in E} A_s(\gamma, \xi) z_\gamma(s, \xi)}{B_s(\xi)}\right]^2 ds\right\} \tag{9.26}$$

and

$$\varkappa_\beta(t, \xi) = z_\beta(t, \xi) I_Z(t, \xi). \tag{9.27}$$

(By (9.25), (9.2) and (9.3) the integrals in (9.26) are defined.)

From (9.26), (9.27) and (9.23), with the help of the Ito formula we find that

$$I_Z(t, \xi) = 1 + \int_0^t I_Z(s, \xi) \frac{\sum_{\gamma \in E} A_s(\gamma, \xi) z_\gamma(s, \xi)}{B_s^2(\xi)} d\xi_s \tag{9.28}$$

and

$$d\varkappa_\beta(t, \xi) = \sum_{\gamma \in E} \lambda_{\gamma\beta}(t) \varkappa_\gamma(t, \xi) dt + \varkappa_\beta(t, \xi) \frac{A_t(\beta, \xi)}{B_t^2(\xi)} d\xi_t. \tag{9.29}$$

Comparing (9.27) with (9.28) we note that

$$I_Z(t, \xi) = 1 + \int_0^t \frac{\sum_{\gamma \in E} A_s(\gamma, \xi) \varkappa_\gamma(s, \xi)}{B_s^2(\xi)} d\xi_s. \tag{9.30}$$

Since $P\{0 < I_Z(t, \xi) < \infty, 0 \le t \le T\} = 1$, then, by (9.27) and (9.30),

$$z_\beta(t, \xi) = \frac{\varkappa_\beta(t, \xi)}{1 + \int_0^t \sum_{\gamma \in E} (A_s(\gamma, \xi) \varkappa_\gamma(s, \xi)/B_s^2(\xi)) d\xi_s}. \tag{9.31}$$

If the process $\varkappa = \{\varkappa_\beta(t, \xi), \beta \in E\}$, $0 \leq t \leq T$, is a solution of (9.29), then applying the Ito formula to the right-hand side of (9.31) it is not difficult to show that the process $Z = \{z_\beta(t, \xi), \beta \in E\}$, $0 \leq t \leq T$, satisfies the system of equations given by (9.23).

Thus (9.27) and (9.31) determine a one-to-one correspondence between the processes Z which are solutions of (9.23) and the processes $\varkappa$ which are solutions of (9.29).

Let

$$\varphi(t) = \exp\left\{-\int_0^t \frac{A_s(\theta_s, \xi)}{B_s^2(\xi)}\, d\xi_s + \frac{1}{2}\int_0^t \frac{A_s^2(\theta_s, \xi)}{B_s^2(\xi)}\, ds\right\}. \tag{9.32}$$

If the process Z satisfies (9.24), then the process $\varkappa$ corresponding to it satisfies the condition

$$\sup_{0 \leq t \leq T} M\left(\sum_{\beta \in E} \varkappa_\beta(t, \xi)\varphi(t)\right) < \infty. \tag{9.33}$$

Indeed, by (9.24),

$$M \sum_{\beta \in E} \varkappa_\beta(t, \xi)\varphi(t) = MI_Z(t, \xi)\varphi(t) \cdot \sum_{\beta} z_\beta(t, \xi)$$
$$\leq MI_Z(t, \xi)\varphi(t) \sup_{0 \leq t \leq T} \sum_{\beta \in E} z_\beta(t, \xi) \leq CMI_Z(t, \xi)\varphi(t) \leq C < \infty,$$

where we have made use of the fact that

$$MI_Z(t, \xi)\varphi(t) = M \exp\left\{\int_0^t \frac{\sum_{\gamma \in E} A_s(\gamma, \xi) z_\gamma(s, \xi) - A_s(\theta_s, \xi)}{B_s(\xi)}\, dW_s\right.$$
$$\left. - \frac{1}{2}\int_0^t \left[\frac{\sum_{\gamma \in E} A_s(\gamma, \xi) z_\gamma(s, \xi) - A_s(\theta_s, \xi)}{B_s(\xi)}\right]^2 ds\right\} \leq 1 \tag{9.34}$$

(see Lemma 6.1).

Because of the above mentioned one-to-one correspondence between the processes Z and the processes $\varkappa$, for proving uniqueness of a solution to the (nonlinear) system of equations in (9.23) in the class of processes satisfying (9.24) and (9.25), it suffices to establish the uniqueness of a solution of the (linear) system of equations in (9.29) in a class of processes satisfying (9.33).

Setting

$$\psi_s^t(\beta) = \exp\left\{\int_s^t \lambda_{\beta\beta}(u)du + \int_s^t \frac{A_u(\beta, \xi)}{B_u^2(\xi)}\, d\xi_u - \frac{1}{2}\int_s^t \frac{A_u^2(\beta, \xi)}{B_u^2(\xi)}\, du\right\},$$

we can show that the system of equations in (9.29) is equivalent to the system of equations

$$\varkappa_\beta(t, \xi) = \psi_0^t(\beta)p_\beta(0) + \int_0^t \psi_s^t(\beta) \sum_{\gamma \neq \beta} \lambda_{\gamma\beta}(s)\varkappa_\gamma(s, \xi)ds. \tag{9.35}$$

The fact that any solution of the system of equations in (9.35) is at the same time a solution of the system of equations in (9.29) can be checked by the Ito formula.

On the other hand, we shall rewrite the system of equations in (9.29) as follows:

$$d\varkappa_\beta(t, \xi) = [\lambda_{\beta\beta}(t)\varkappa_\beta(t, \xi) + \alpha_\beta(t, \xi)]dt + \varkappa_\beta(t, \xi)\frac{A_t(\beta, \xi)}{B_t^2(\xi)}d\xi_t, \tag{9.36}$$

where

$$\alpha_\beta(t, \xi) = \sum_{\gamma\neq\beta}\lambda_{\gamma\beta}(t)\varkappa_\gamma(t, \xi).$$

Equation (9.36) is (with the given process $\alpha_\beta(t, \xi)$) linear with respect to $\varkappa_\beta(t, \xi)$. According to the note to Theorem 4.10, the solution of this equation can be represented in the form

$$\varkappa_\beta(t, \xi) = \psi_0^t(\beta)p_\beta(0) + \int_0^t \psi_s^t(\beta)\alpha_\beta(s, \xi)ds. \tag{9.37}$$

Thus the problem reduces to establishing the uniqueness of a solution of the system of integral equations given by (9.35), which no longer involves stochastic integrals (over $d\xi_s$).

Let $\Delta_\beta(t, \xi) = \varkappa'_\beta(t, \xi) - \varkappa''_\beta(t, \xi)$ be a difference of two solutions of (9.35) satisfying (9.33). Then

$$\Delta_\beta(t, \xi) = \int_0^t \psi_s^t(\beta)\sum_{\gamma\neq\beta}\lambda_{\gamma\beta}(s)\Delta_\gamma(s, \xi)ds \tag{9.38}$$

and

$$\varphi(t)|\Delta_\beta(t, \xi)| \leq \int_0^t \psi_s^t(\beta)\varphi(t)\sum_{\gamma\neq\beta}\lambda_{\gamma\beta}(s)|\Delta_\gamma(s, \xi)|ds.$$

Hence,

$$M\varphi(t)|\Delta_\beta(t, \xi)| \leq \int_0^t \sum_{\gamma\neq\beta}\lambda_{\gamma\beta}(s)M(\psi_s^t(\beta)\varphi(t)|\Delta_\gamma(s, \xi)|)ds.$$

But

$$M(\psi_s^t(\beta)\varphi(t)|\Delta_\gamma(s, \xi)||\mathscr{F}_s^{\theta,\xi}) = |\Delta_\gamma(s, \xi)|\varphi(s)M\left[\psi_s^t\frac{\varphi(t)}{\varphi(s)}\bigg|\mathscr{F}_s^{\theta,\xi}\right] \leq |\Delta_\gamma(s, \xi)|\varphi(s),$$

since

$$M\left(\psi_s^t(\beta)\frac{\varphi(t)}{\varphi(s)}\bigg|\mathscr{F}_s^{\theta,\xi}\right) \leq 1,$$

which can be established in the same way as (9.34) by taking into consideration that $\lambda_{\beta\beta}(u) \leq 0$.

Consequently,

$$M(\varphi(t)|\Delta_\beta(t, \beta)|) \le \int_0^t \sum_{\gamma \neq \beta} \lambda_{\gamma\beta}(s) M(\varphi(s)|\Delta_\gamma(s, \xi)|)ds$$

and

$$\begin{aligned}\sum_{\beta \in E} M(\varphi(t)|\Delta_\beta(t, \xi)|) &\le \int_0^t \sum_{\beta \in E} \sum_{\gamma \neq \beta} \lambda_{\gamma\beta}(s) M(\varphi(s)|\Delta_\gamma(s, \xi)|)ds \\ &\le \int_0^t \sum_{\gamma \in E} M(\varphi(s)|\Delta_\gamma(s, \xi)|) \sum_{\beta \in E} |\lambda_{\gamma\beta}(s)| ds \\ &\le 2K \int_0^t \sum_{\beta \in E} M(\varphi(s)|\Delta_\beta(s, \xi)| ds), \end{aligned} \tag{9.39}$$

where we have made use of the fact that

$$\sum_{\beta \in E} |\lambda_{\gamma\beta}(s)| = \sum_{\beta \neq \gamma} \lambda_{\gamma\beta}(s) + |\lambda_{\gamma\gamma}(s)| = 2|\lambda_{\gamma\gamma}(s)| \le 2K.$$

From (9.39) it follows that

$$\sum_{\beta \in E} M\{\varphi(t)|\Delta_\beta(t, \xi)|\} \le 2K \int_0^t \sum_{\beta \in E} M\{\varphi(s)|\Delta_\beta(s, \xi)|\} ds.$$

According to Lemma 4.13, it follows from this that $\sum_{\beta \in E} M\{\varphi(t)|\Delta_\beta(t, \xi)|\} = 0$. But $P\{\varphi(t) > 0\} = 1$; therefore $P\{|\Delta_\beta(t, \xi)| > 0\} = 0$.

Hence, because of the continuity of the processes $\mathscr{z}'$ and $\mathscr{z}''$ and the countability of the set E,

$$P\{|\mathscr{z}'_\beta(t, \xi) - \mathscr{z}''_\beta(t, \xi)| = 0, \qquad 0 \le t \le T, \beta \in E\} = 1.$$

Thus the uniqueness of a solution (in the class of processes satisfying (9.33)) of the system of equations in (9.29) is established. From the uniqueness of the solution of (9.29) the uniqueness of a solution of (9.23) (in the class of processes satisfying (9.24) and (9.25)) follows as well. □

Note. If $A_t(\theta_t, \xi) \equiv A_t(\theta_t, \xi_t)$, $B_t(\xi) \equiv B_t(\xi_t)$, then the two-dimensional process (θ_t, ξ_t), $0 \le t \le T$, is a Markov process (with respect to the system $(\mathscr{F}_t)$, $0 \le t \le T$):

$$P\{\theta_t \in A, \xi_t \in B | \mathscr{F}_s\} = P\{\theta_t \in A, \xi_t \in B | \theta_s, \xi_s\}. \tag{9.40}$$

Employing Theorem 9.2, which establishes the uniqueness of a solution to the system of equations in (9.23), it can be shown that in this case the (infinite-dimensional) process $\{\xi_t, \pi_\beta(t), \beta \in E\}$, $0 \le t \le T$, is a Markov process with respect to the system $(\mathscr{F}_t^\xi)$, $0 \le t \le T$:

$$P\{\xi_t \in B, \pi_\beta(t) \in A_\beta, \beta \in E | \mathscr{F}_t^\xi\} = P\{\xi_t \in B, \pi_\beta(t) \in A_\beta, \beta \in E | \xi_s, \pi_\beta(s), \beta \in E\}. \tag{9.41}$$

9.1.6

In a number of problems in statistics (in particular, in problems of interpolation which will be discussed later on) there arises the necessity for considering equations satisfied by the conditional probabilities

$$\omega_{\beta\alpha}(t, s) = P(\theta_t = \beta | \mathscr{F}_t^\xi, \theta_s = \alpha), \tag{9.42}$$

where $0 \le s \le t \le T$. It is clear that if $p_\alpha(0) = 1$, then $\omega_{\beta\alpha}(t, 0) = \pi_\beta(t)$, with $\pi_\alpha(0) = p_\alpha(0) = 1$ and $\pi_\beta(0) = 0$ for all $\beta \neq \alpha$.

Theorem 9.3. *Let the conditions of Lemma* 9.1 *and also* (9.2)–(9.6) *be satisfied. Then the conditional probabilities* $\{\omega_{\beta\alpha}(t, s), \beta \in E\}$, $s \le t \le T$, *satisfy (with the given* $\alpha \in E$ *and* $s \ge 0$) *the system* ($\beta \in E$) *of equations*

$$\begin{aligned}\omega_{\beta\alpha}(t, s) = \delta(\beta, \alpha) &+ \int_0^t \mathfrak{L}^*\omega_{\beta\alpha}(u, s)du \\ &- \int_s^t \omega_{\beta\alpha}(u, s)\frac{A_u(\beta, \xi) - \sum_{\gamma \in E} A_u(\gamma, \xi)\omega_{\gamma\alpha}(u, s)}{B_u^2(\xi)} \\ &\times \sum_{\gamma \in E} A_u(\gamma, \xi)\omega_{\gamma\alpha}(u, s)du \\ &+ \int_s^t \omega_{\beta\alpha}(u, s)\frac{A_u(\beta, \xi) - \sum_{\gamma \in E} A_u(\gamma, \xi)\omega_{\gamma\alpha}(u, s)}{B_u^2(\xi)} d\xi_u.\end{aligned} \tag{9.43}$$

In the class of the nonnegative continuous functions $\{\omega_{\beta\alpha}(t, s), \beta \in E, s \le t \le T\}$ *satisfying the conditions*

$$P\left\{\sup_{s \le t \le T} \sum_{\beta \in E} \omega_{\beta\alpha}(t, s) \le C\right\} = 1 \qquad (C = \text{const.}), \tag{9.44}$$

$$P\left\{\int_s^T \left(\sum_{\gamma \in E} \frac{|A_u(\gamma, \xi)|\omega_{\gamma\alpha}(u, s)}{B_u(\xi)}\right)^2 du < \infty\right\} = 1, \tag{9.45}$$

the system of equations in (9.43) *has a unique solution.*

PROOF. Let (θ_u^α), $s \le u \le T$, be a Markov process which has the same transition probabilities as the initial process θ, and which satisfies the condition $\theta_s^\alpha = \alpha$. Hence, in particular,

$$P\{\theta_t = \beta | \theta_s = \alpha\} = P\{\theta_t^\alpha = \beta\}, \qquad t \ge s. \tag{9.46}$$

Let next $(\xi_u^{(\alpha, \xi_0^s)})$, $0 \le u \le T$, be a random process, such that

$$\xi_u^{(\alpha, \xi_0^s)} = \xi_u, \qquad u \le s, \tag{9.47}$$

and, with $u > s$,

$$\xi_u^{(\alpha,\,\xi_0^s)} = \xi_s + \int_s^u A_v(\theta_v^\alpha, \xi^{(\alpha,\,\xi_0^s)})dv + \int_s^u B_v(\xi^{(\alpha,\,\xi_0^s)})dW_v. \tag{9.48}$$

By (9.2)–(9.4), Equation (9.48) has a unique strong solution (see Theorem 4.6)[2] and with probability one

$$\xi_u^{(\theta_s,\,\xi_0^s)} = \xi_u, \qquad u \le s.$$

Let us show that (P-a.s.)[3]

$$P\{\xi_t \le y \mid \theta_s = \alpha, \xi_0^s\} = P\{\xi_t^{(\alpha,\,\xi_0^s)} \le y\}. \tag{9.49}$$

For this purpose it will be noted that for each $t \ge s$ there will be a (measurable) functional $Q_t(\cdot,\cdot,\cdot)$, defined on $C_{[0,s]} \times E_{[s,t]} \times C_{[s,t]}$, where $C_{[0,s]}$ and $C_{[s,t]}$ are the spaces of functions continuous on $[0, s]$ and $[s, t]$, and $E_{[s,t]}$ is the space of right continuous functions defined on $[s, t]$, such that

$$\xi_t = Q_t(\xi_0^s, \theta_s^t, W_s^t) \qquad (P\text{-a.s.}). \tag{9.50}$$

Because of the uniqueness of a strong solution of Equation (9.48) (see above), for each $t \ge s$ (P-a.s.)

$$\xi_t^{(\alpha,\,\xi_0^s)} = Q_t(\xi_0^s, (\theta^\alpha)_s^t, W_s^t). \tag{9.51}$$

From (9.49), (9.50), the independence of the processes θ and W, the Markov behavior of the process θ, and (9.46), it follows that

$$\begin{aligned}
P\{\xi_t \le x \mid \theta_s = \alpha, \xi_0^s = x_0^s\} &= P\{Q_t(\xi_0^s, \theta_s^t, W_s^t) \le x \mid \theta_s = \alpha, \xi_0^s = x_0^s\}\\
&= P\{Q_t(x_0^s, \theta_s^t, W_s^t) \le x \mid \theta_s = \alpha, \xi_0^s = x_0^s\}\\
&= P\{Q_t(x_0^s, \theta_s^t, W_s^t) \le x \mid \theta_s = \alpha\}\\
&= P\{Q_t(x_0^s, (\theta^\alpha)_s^t, W_s^t) \le x\}.
\end{aligned}$$

Together with (9.51) this proves (9.49).

Similarly it can be shown that for $s \le s_1 \le \cdots \le s_n \le t$ and $x_1, \ldots, x_n \in \mathbb{R}^1$,

$$P\{\theta_t = \beta, \xi_{s_1} \le x_1, \ldots, \xi_{s_n} \le x_n \mid \theta_s = \alpha, \xi_0^s\}$$
$$= P\{\theta_t^\alpha = \beta, \xi_{s_1}^{(\alpha,\,\xi_0^s)} \le x_1, \ldots, \xi_{s_n}^{(\alpha,\,\xi_0^s)} \le x_n\}. \tag{9.52}$$

From this it is not difficult to deduce that for $s \le t$

$$\omega_{\beta\alpha}(t, s) = P(\theta_t = \beta \mid \mathcal{F}_t^\xi, \theta_s = \alpha) = P(\theta_t^\alpha = \beta \mid \mathcal{F}_t^{\xi(\alpha,\,\xi_0^s)}).$$

Applying Theorem 9.1 to the process $(\theta_t^\alpha, \xi_t^{(\alpha,\,\xi_0^s)})$, $t \ge s$, (taking into account the obvious changes in the notation) we infer that the $\omega_{\beta\alpha}(t, s)$ satisfy (for fixed α and s) the system of equations in (9.43). The uniqueness of a continuous solution satisfying (9.44) and (9.45) follows from Theorem 9.2. □

[2] See also Footnote 1.

[3] As to the notation for conditional probabilities used here and from now on, see Section 1.2.

9.2 Forward and backward equations of optimal nonlinear interpolation

9.2.1

Let $(\theta, \xi) = (\theta_t, \xi_t)$, $0 \le t \le T$, be the random process introduced in the preceding section.

Denote

$$\pi_\beta(s, t) = P(\theta_s = \beta | \mathcal{F}_t^\xi), \qquad s \le t. \tag{9.53}$$

Knowing the a posteriori probabilities $\{\pi_\beta(s, t), \beta \in E\}$, one can solve various problems of the interpolation of an unobservable component on the basis of the observations $\xi_0^t = \{\xi_u, u \le t\}$, $s < t$. In the present section forward (over t for fixed s) and backward (over s for fixed t) equations will be deduced for $\pi_\beta(s, t)$.

Theorem 9.4. *Let the conditions of Lemma* 9.1 *and* (9.2)–(9.6) *be fulfilled. Then, for all* s, t $(0 \le s \le t \le T)$ *the conditional probabilities* $\pi_\beta(s, t)$ *satisfy the (forward) equations* $(\pi_\beta(s, s) = \pi_\beta(s))$

$$d_t \pi_\beta(s, t) = \pi_\beta(s, t) B_t^{-2}(\xi) \sum_{\gamma \in E} A_t(\gamma, \xi)[\omega_{\gamma\beta}(t, s) - \pi_\gamma(t)] \times \left[d\xi_t - \sum_{\gamma \in E} A_t(\gamma, \xi)\pi_\gamma(t) dt \right] \tag{9.54}$$

and can be represented as follows:

$$\pi_\beta(s, t) = \pi_\beta(s) \exp\left\{ \int_s^t B_s^{-2}(\xi) \sum_{\gamma \in E} A_u(\gamma, \xi)[\omega_{\gamma\beta}(u, s) - \pi_\gamma(u)] d\xi_u - \frac{1}{2} \int_s^t B_s^{-2}(\xi) \left\{ \left[\sum_{\gamma \in E} A_u(\gamma, \xi)\omega_{\gamma\beta}(u, s) \right]^2 - \left[\sum_{\gamma \in E} A_u(\gamma, \xi)\pi_\gamma(u) \right]^2 \right\} du \right\}. \tag{9.55}$$

PROOF. Since

$$\pi_\beta(s, t) = M[\delta(\theta_s, \beta) | \mathcal{F}_t^\xi],$$

then, by Theorem 8.4,

$$\pi_\beta(s, t) = M[\delta(\theta_s, \beta) | \mathcal{F}_t^\xi] = M[\delta(\theta_s, \beta) | \mathcal{F}_s^\xi] + \int_s^t [B_u(\xi)]^{-1} \{ M[\delta(\theta_s, \beta) A_u(\theta_u, \xi) | \mathcal{F}_u^\xi] - M[\delta(\theta_s, \beta) | \mathcal{F}_u^\xi] M[A_u(\theta_u, \xi) | \mathcal{F}_u^\xi] \} d\overline{W}_u, \tag{9.56}$$

where $\overline{W} = (\overline{W}_t, \mathscr{F}_t^\xi)$ is a Wiener process with

$$\overline{W}_t = \int_0^t \frac{d\xi_s - M[A_s(\theta_s, \xi) | \mathscr{F}_s^\xi] ds}{B_s(\xi)}.$$

Here

$$M[A_u(\theta_u, \xi) | \mathscr{F}_u^\xi] = \sum_\gamma A_u(\gamma, \xi)\pi_\gamma(u),$$

$$\begin{aligned} M[\delta(\theta_s, \beta)A_u(\theta_u, \xi) | \mathscr{F}_u^\xi] &= M[\delta(\theta_s, \beta)M(A_u(\theta_s, \xi) | \mathscr{F}_u^\xi, \theta_s) | \mathscr{F}_u^\xi] \\ &= \pi_\beta(s, u) \sum_{\gamma \in E} A_u(\gamma, \xi)\omega_{\gamma\beta}(u, s). \end{aligned}$$

Taking this into account, Equation (9.54) follows from (9.56); (9.55) follows from (9.54) and the Ito formula. □

Note. From (9.54) and (9.55) we can see that in the problems of interpolation involving computation of the conditional probabilities $\pi_\beta(s, t)$, $\beta \in E$, it is necessary to solve two auxiliary problems of filtering (for finding $\pi_\gamma(u)$ and $\omega_{\gamma\beta}(u, s)$, $u \geq s$).

9.2.2

For deducing the backward equations of interpolation we shall need a number of auxiliary results related to the conditional probability

$$\rho_{\alpha\beta}(s, t) = P(\theta_s = \alpha | \mathscr{F}_t^\xi, \theta_t = \beta).$$

Lemma 9.3. *For given $\beta \in E$, let any of the following two conditions be satisfied:*

(1) $p_\beta(0) > 0$,
(2) $\inf_{0 \leq t \leq T} \inf_{\gamma \neq \beta} \lambda_{\gamma\beta}(t) \geq \varepsilon_\beta > 0$.

Then for each t, $0 \leq t \leq T$,

$$p\{\pi_\beta(t) > 0\} = 1. \tag{9.57}$$

Proof. From the Bayes formula given by (7.205) it follows that $\pi_\beta(t)$ turns into zero (P-a.s.) simultaneously with $p_\beta(t)$. From (9.12), for $p_\beta(t)$, $T \geq t \geq s \geq 0$, we obtain the representation

$$p_\beta(t) = \exp\left\{\int_s^t \lambda_{\beta\beta}(u)du\right\}\left\{p_\beta(s) + \int_s^t \exp\left[-\int_s^u \lambda_{\beta\beta}(v)dv\right] \sum_{\gamma \neq \beta} \lambda_{\gamma\beta}(u)p_\gamma(u)du\right\}. \tag{9.58}$$

Since $0 \leq \lambda_{\gamma\beta}(t) \leq K$ with $\gamma \neq \beta$, then from (9.58) it follows that for all $t \geq s$

$$p_\beta(t) \geq \exp(-K(t-s))\left\{p_\beta(s) + \varepsilon_\beta \int_s^t [1 - p_\beta(u)]du\right\}. \tag{9.59}$$

From this it is seen that if $p_\beta(0) > 0$, then $\inf_{0 \leq t \leq T} p_\beta(t) > 0$. But if $p_\beta(0) = 0$ and $\varepsilon_\beta > 0$, then

$$p_\beta(t) \geq \varepsilon_\beta \int_0^t [1 - p_\beta(s)]ds. \tag{9.60}$$

Hence, because of the continuity of $p_\beta(s)$, $s \geq 0$, from (9.60) it follows that $p_\beta(t) > 0$, at least for sufficiently small positive t. This fact together with (9.59) proves that $p_\beta(t) > 0$ for each $t > 0$. □

Lemma 9.4. *If $P\{\pi_\beta(t) > 0\} = 1$, then for $t \geq s$*

$$\rho_{\alpha\beta}(s, t) = \frac{\omega_{\beta\alpha}(t, s)\pi_\alpha(s, t)}{\pi_\beta(t)}. \tag{9.61}$$

Proof. If $t \geq s$ then

$$\begin{aligned} M[\delta(\theta_s, \alpha)\delta(\theta_t, \beta) | \mathscr{F}_t^\xi] &= M[\delta(\theta_t, \beta)M(\delta(\theta_s, \alpha) | \mathscr{F}_t^\xi, \theta_t) | \mathscr{F}_t^\xi] \\ &= M[\delta(\theta_t, \beta)\rho_{\alpha\theta_t}(s, t) | \mathscr{F}_t^\xi] = \rho_{\alpha\beta}(s, t)\pi_\beta(t). \end{aligned} \tag{9.62}$$

On the other hand,

$$\begin{aligned} M[\delta(\theta_s, \alpha)\delta(\theta_t, \beta) | \mathscr{F}_t^\xi] &= M[\delta(\theta_s, \alpha)M(\delta(\theta_t, \beta) | \mathscr{F}_t^\xi, \theta_s) | \mathscr{F}_t^\xi] \\ &= M[\delta(\theta_s, \alpha)\omega_{\beta\theta_s}(t, s) | \mathscr{F}_t^\xi] = \pi_\alpha(s, t)\omega_{\beta\alpha}(t, s). \end{aligned} \tag{9.63}$$

Comparing (9.62) and (9.63) and taking into consideration that $p\{\pi_\beta(t) > 0\} = 1$, we obtain (9.61). □

Note. (9.61) holds if any of the conditions of Lemma 9.3 is fulfilled.

Lemma 9.5. *Let $p_\beta(0) > 0$. Then the process $\rho_{\alpha\beta}(s, t)$, where $\alpha \in E$, $0 \leq s \leq t \leq T$, permits the stochastic differential*

$$d_t\rho_{\alpha\beta}(s, t) = \frac{1}{\pi_\beta(t)} \sum_{\gamma \in E} \lambda_{\gamma\beta}(t)\pi_\gamma(t)[\rho_{\alpha\gamma}(s, t) - \rho_{\alpha\beta}(s, t)]dt \tag{9.64}$$

and

$$\rho_{\alpha\beta}(s, s) = \delta(\alpha, \beta).$$

Proof. By the condition $p_\beta(0) > 0$ and by Lemma 9.3, it follows that $P(\pi_\beta(t) > 0) = 1$. Hence, (9.61) is valid. Applying the Ito formula to the right-hand side of (9.61), and taking into account that $\omega_{\beta\alpha}(t, s)$, $\pi_\alpha(s, t)$ and $\pi_\beta(t)$ permit the representations given by (9.43), (9.54), and (9.17), respectively, we arrive at (9.64) after some arithmetic. □

9.2.3

Let us deduce next the backward equations of interpolation, considering here only the case where the set E is finite.

Theorem 9.5. *Let the set E be finite and let $p_\alpha(0) > 0$ for all $\alpha \in E$. Then the conditional probabilities $\pi_\alpha(s, t) = P(\theta_s = \alpha | \mathscr{F}_t^\xi)$, $s < t$, $\alpha \in E$, satisfy the system of equations*

$$-\frac{\partial \pi_\alpha(s, t)}{\partial s} = \pi_\alpha(s)\mathfrak{L}\left(\frac{\pi_\alpha(s, t)}{\pi_\alpha(s)}\right) - \frac{\pi_\alpha(s, t)}{\pi_\alpha(s)} \mathfrak{L}^*\pi_\alpha(s), \tag{9.65}$$

where

$$\mathfrak{L}\left(\frac{\pi_\alpha(s, t)}{\pi_\alpha(s)}\right) = \sum_{\gamma \in E} \lambda_{\alpha\gamma}(s) \frac{\pi_\gamma(s, t)}{\pi_\gamma(s)}, \tag{9.66}$$

$$\mathfrak{L}^*\pi_\alpha(s) = \sum_{\gamma \in E} \lambda_{\gamma\alpha}(s)\pi_\gamma(s). \tag{9.67}$$

Proof. First of all let us note that

$$\begin{aligned}\pi_\alpha(s, t) &= M[\delta(\theta_s, \alpha) | \mathscr{F}_t^\xi] = M[M(\delta(\theta_s, \alpha) | \mathscr{F}_t^\xi, \theta_t) | \mathscr{F}_t^\xi] \\ &= M[\rho_{\alpha\theta_t}(s, t) | \mathscr{F}_t^\xi] = \sum_{\gamma \in E} \rho_{\alpha\gamma}(s, t)\pi_\gamma(t).\end{aligned} \tag{9.68}$$

Hence, if we establish that

$$-\frac{\partial \rho_{\alpha\gamma}(s, t)}{\partial s} = \pi_\alpha(s)\mathfrak{L}\left(\frac{\rho_{\alpha\gamma}(s, t)}{\pi_\alpha(s)}\right) - \frac{\rho_{\alpha\gamma}(s, t)}{\pi_\alpha(s)} \mathfrak{L}^*\pi_\alpha(s), \tag{9.69}$$

then (9.65) will follow from (9.68).

By Lemma 9.5, the probabilities $\rho_{\alpha\beta}(s, t)$ have a derivative over t:

$$\frac{\partial \rho_{\alpha\beta}(s, t)}{\partial t} = \frac{1}{\pi_\beta(t)} \sum_{\gamma \in E} \lambda_{\gamma\beta}(t)\pi_\gamma(t)[\rho_{\alpha\gamma}(s, t) - \rho_{\alpha\beta}(s, t)]. \tag{9.70}$$

Let $R(s, t) = \|\rho_{\alpha\beta}(s, t)\|$, $\alpha, \beta \in E$. The matrix $R(s, t)$ is fundamental: $R(s, s)$ is a unit matrix and

$$\frac{\partial R(s, t)}{\partial t} = R(s, t)C(t, \omega), \tag{9.71}$$

where $C(t, \omega)$ is a matrix with the elements

$$c_{\alpha\alpha}(t, \omega) = \frac{\lambda_{\alpha\alpha}(t)\pi_\alpha(t) - \sum_{\gamma \in E} \lambda_{\gamma\alpha}(t)\pi_\gamma(t)}{\pi_\alpha(t)},$$

$$c_{\alpha\beta}(t, \omega) = \frac{\lambda_{\alpha\beta}(t)\pi_\alpha(t)}{\pi_\beta(t)},$$

and is (P-a.s.) a continuous function since $\pi_\gamma(t)$, $\lambda_{\gamma\alpha}(t)$ ($\gamma, \alpha \in E$) are continuous over t and the set E is finite.

If $s < u < t$ then, because of the properties of fundamental matrices,

$$R(s, t) = R(s, u)R(u, t).$$

Since the matrix $R(s, u)$ is (P-a.s.) nonsingular

$$R(u, t) = R^{-1}(s, u)R(s, t). \tag{9.72}$$

From (9.71) and the explicit identity

$$0 = \frac{\partial}{\partial u}(R(s, u)R^{-1}(s, u)),$$

it follows that

$$\frac{\partial}{\partial u} R^{-1}(s, u) = -C(u, \omega)R^{-1}(s, u).$$

Hence,

$$-\frac{\partial}{\partial u} R(u, t) = \frac{\partial}{\partial u} R^{-1}(s, u)R(s, t) = C(u, \omega)R^{-1}(s, u)R(s, t) = C(u, \omega)R(u, t)$$

and, therefore (for $s < t$),

$$-\frac{\partial}{\partial s} R(s, t) = C(s, \omega)R(s, t).$$

Writing this system by coordinates we arrive at the system of equations in (9.69), from which, as was noted above; there follow the equations in (9.65). □

Note. If in (9.1) the coefficients $A_t(\theta_t, \xi)$ do not depend on θ_t, then $\rho_{\alpha\beta}(s, t) = P(\theta_s = \alpha | \theta_t = \beta, \mathscr{F}_t^\xi) = P(\theta_s = \alpha | \theta_t = \beta) = p_{\alpha\beta}(s, t)$. Hence, if the set E is finite and $p_\beta(0) > 0$, $\beta \in E$, then

$$-\frac{\partial p_{\alpha\beta}(s, t)}{\partial s} = p_\alpha(s)\mathfrak{L}\left(\frac{p_{\alpha\beta}(s, t)}{p_\alpha(s)}\right) - \frac{p_{\alpha\beta}(s, t)}{p_\alpha(s)} \mathfrak{L}^* p_\alpha(s). \tag{9.73}$$

9.3 Equations of optimal nonlinear extrapolation

9.3.1

For $s < t < T$ let us denote

$$\pi_\beta(t, s) = P(\theta_t = \beta | \mathscr{F}_s^\xi), \qquad \beta \in E.$$

The knowledge of these probabilities enables us to solve various problems related to predicting θ_t on the basis of the observations $\xi_0^s = \{\xi_u, u \le s\}$. Thus, if $M\theta_t^2 < \infty$, then $\sum_{\beta \in E} \beta\pi_\beta(t, s)$ is an optimal (in the mean square sense) estimate θ_t over ξ_0^s.

For the probabilities $\pi_\beta(t, s)$ one can obtain equations both over t (for fixed s) and over s (for fixed t). The first of these equations (which it is natural to call forward equations) allow us to understand how the prediction of θ from ξ_0^s deteriorates when t increases. From the equations over s (t is fixed) one can judge the degree to which the prediction improves with increase in "the number of observations" (i.e., as $s \uparrow t$).

9.3.2

Theorem 9.6. *Let the conditions of Lemma* 9.1 *and* (9.2)–(9.6) *be fulfilled. Then for each fixed s the conditional probabilities* $\{\pi_\beta(t, s), t \geq s, \beta \in E\}$ *satisfy the (forward) equations*

$$\pi_\beta(t, s) = \pi_\beta(s) + \int_s^t \mathfrak{L}^*\pi_\beta(u, s)du, \tag{9.74}$$

where

$$\mathfrak{L}^*\pi_\beta(u, s) = \sum_{\gamma \in E} \lambda_{\gamma\beta}(u)\pi_\gamma(u, s).$$

The system of equations in (9.74) *has a unique solution (in the class of non-negative continuous solutions)* $x_\beta(t, s)$ *with* $\sup_{s \leq t \leq T} \sum_\beta x_\beta(t, s) < \infty$ *(P-a.s.). For fixed t the conditional probabilities* $\{\pi_\beta(t, s), s \leq t, \beta \in E\}$ *permit the representation*

$$\pi_\beta(t, s) = \pi_\beta(t, 0) + \int_0^t B_u^{-2}(\xi)\Bigg\{ \sum_{\gamma \in E} p_{\beta\gamma}(t, u)\pi_\gamma(u)\Bigg[A_u(\gamma, \xi) - \sum_{\gamma \in E} A_u(\gamma, \xi)\pi_\gamma(u)\Bigg]\Bigg\}\Bigg[d\xi_u - \sum_{\gamma \in E} A_u(\gamma, \xi)\pi_\gamma(u)du\Bigg]. \tag{9.75}$$

PROOF. For deducing (9.74) let us use the fact that for $t \geq s$

$$\pi_\beta(t, s) = P(\theta_t = \beta | \mathscr{F}_s^\xi) = M[P(\theta_t = \beta | \mathscr{F}_t^\xi) | \mathscr{F}_s^\xi] = M[\pi_\beta(t) | \mathscr{F}_s^\xi] \tag{9.76}$$

and, according to (9.17),

$$\pi_\beta(t) = \pi_\beta(s) + \int_s^t \mathfrak{L}^*\pi_\beta(u)du + \int_s^t \pi_\beta(u)\frac{A_u(\beta, \xi) - \bar{A}_u(\xi)}{B_u(\xi)}d\overline{W}_u. \tag{9.77}$$

Then, taking the conditional expectation $M[\cdot | \mathscr{F}_s^\xi]$, on both sides of (9.77), we obtain

$$\pi_\beta(t) = \pi_\beta(s) + M\left(\int_s^t \mathfrak{L}^*\pi_\beta(u)du | \mathscr{F}_s^\xi\right) + M\left(\int_s^t \pi_\beta(u)\frac{A_u(\beta, \xi) - \bar{A}_u(\xi)}{B_u(\xi)}d\overline{W}_u | \mathscr{F}_s^\xi\right). \tag{9.78}$$

But

$$M\left(\int_s^t \mathfrak{L}^*\pi_\beta(u)du \,|\, \mathcal{F}_s^\xi\right) = \int_s^t \sum_{\gamma\in E} \lambda_{\gamma\beta}(u) M[\pi_\gamma(u)\,|\,\mathcal{F}_s^\xi]du$$

$$= \int_s^t \sum_{\gamma\in E} \lambda_{\gamma\beta}(u)\pi_\gamma(u, s)du = \int_s^t \mathfrak{L}^*\pi_\beta(u, s)du. \quad (9.79)$$

Next, in deducing the basic theorem of filtering (see the note to Theorem 8.1) it was established that the random process

$$\left(\int_0^t \pi_\beta(u)\frac{A_u(\beta, \xi) - \bar{A}_u(\xi)}{B_u(\xi)} d\overline{W}_u, \mathcal{F}_t^\xi\right), \qquad 0 \le t \le T,$$

is a square integrable martingale. Therefore

$$M\left(\int_s^t \pi_\beta(u)\frac{A_u(\beta, \xi) - \bar{A}_u(\xi)}{B_u(\xi)} d\overline{W}_u \,|\, \mathcal{F}_s^\xi\right) = 0 \qquad (P\text{-a.s.})$$

which together with (9.78) and (9.79) proves the validity of (9.74).

Let $x_\beta(t, s)$ and $x'_\beta(t, s)$ be two solutions of the system of equations in (9.74). Then

$$x_\beta(t, s) - x'_\beta(t, s) = \int_s^t \sum_{\gamma\in E} \lambda_{\gamma\beta}(u)[x_\gamma(u, s) - x'_\gamma(u, s)]du$$

and, therefore,

$$\sum_{\beta\in E} |x_\beta(t, s) - x'_\beta(t, s)| \le \int_s^t \sum_{\gamma\in E}\sum_{\beta\in E} |\lambda_{\gamma\beta}(u)|\,|x_\gamma(u, s) - x'_\gamma(u, s)|du.$$

Note that

$$\sum_{\beta\in E} |\lambda_{\gamma\beta}(u)| = \sum_{\beta\ne\gamma} \lambda_{\gamma\beta}(u) - \lambda_{\gamma\gamma}(u) = -2\lambda_{\gamma\gamma}(u) \le 2K.$$

Hence,

$$\sum_{\beta\in E} |x_\beta(t, s) - x'_\beta(t, s)| \le 2K \int_s^t \sum_{\beta\in E} |x_\beta(u, s) - x'_\beta(u, s)|du,$$

and, by Lemma 4.13,

$$\sum_{\beta\in E} |x_\beta(t, s) - x'_\beta(t, s)| = 0 \qquad (P\text{-a.s.}).$$

This proves the uniqueness of solutions of the forward equations in (9.74).

Let us next establish (9.75). For this purpose we will consider the random process $Y = (y_s, \mathcal{F}_s)$, $0 \le s \le t$, with $y_s = p_{\beta\theta_s}(t, s)$. Because of the Markov behavior of the process $\theta = (\theta_s, \mathcal{F}_s)$,

$$M(y_s|\mathcal{F}_u) = M[p_{\beta\theta_s}(t, s)|\mathcal{F}_u] = M[p_{\beta\theta_s}(t, s)|\theta_u]$$

$$= \sum_{\gamma\in E} p_{\beta\gamma}(t, s)p_{\gamma\theta_u}(s, u) = p_{\beta\theta_u}(t, u) = y_u \qquad (P\text{-a.s.}), \qquad u \ge s.$$

Hence, the process $Y = (y_s, \mathscr{F}_s)$, $0 \le s \le t$, is a square integrable martingale.

Since for $t \ge s$

$$\begin{aligned}\pi_\beta(t, s) &= M[\delta(\theta_t, \beta)|\mathscr{F}_s^\xi] = M[M(\delta(\theta_t, \beta)|\mathscr{F}_s)|\mathscr{F}_s^\xi] \\ &= M[M(\delta(\theta_t, \beta)|\theta_s)|\mathscr{F}_s^\xi] = M(p_{\beta\theta_s}(t, s)|\mathscr{F}_s^\xi) = M[y_s|\mathscr{F}_s^\xi],\end{aligned}$$

then, by Theorem 8.5,

$$\pi_\beta(t, s) = \pi_\beta(t, 0) + \int_0^s \alpha_u(\xi)d\overline{W}_u,$$

where

$$\begin{aligned}\alpha_u(\xi) = B_u^{-1}(\xi)[&M(p_{\beta\theta_u}(t, u)A_u(\theta_u, \xi)|\mathscr{F}_u^\xi) \\ &- M(p_{\beta\theta_u}(t, u)|\mathscr{F}_u^\xi)M(A_u(\theta_u, \xi)|\mathscr{F}_u^\xi)]. \quad \square\end{aligned}$$

9.4 Examples

EXAMPLE 1. Let $\theta = \theta(\omega)$ be a random variable taking either of two values β and α with the probabilities p and $1-p$ respectively. The random process ξ_t, $t \ge 0$, with

$$d\xi_t = \theta\, dt + dW_t, \qquad \xi_0 = 0,$$

is observed. Then the a posteriori probability $\pi(t) = P\{\theta = \beta|\mathscr{F}_t^\xi\}$ satisfies, according to (9.17), the equation

$$d\pi(t) = (\beta - \alpha)\pi(t)(1 - \pi(t))[d\xi_t - (\alpha + \pi(t)(\beta - \alpha))dt], \qquad \pi(0) = p. \tag{9.80}$$

In particular, if $\beta = 1$, $\alpha = 0$, then

$$d\pi(t) = \pi(t)(1 - \pi(t))[d\xi_t - \pi(t)dt] \tag{9.81}$$

with $\pi(0) = p$.

If

$$\varphi(t) = \frac{d\mu_1}{d\mu_0}(t, \xi)$$

is the density of the Radon–Nikodym derivative μ_1, corresponding to the process ξ with $\theta = 1$ w.r.t. the measure μ_0, corresponding to the process ξ with $\theta = 0$, then from the Bayes formula it follows that with $p < 1$

$$\pi(t) = \frac{p}{1-p}\varphi(t)\bigg/\left(1 + \frac{p}{1-p}\varphi(t)\right). \tag{9.82}$$

In the case under consideration "the likelihood functional" (see Theorem 7.7) $\varphi(t) = \exp\{\xi_t - t/2\}$ and, therefore,

$$d\varphi(t) = \varphi(t)d\xi_t. \tag{9.83}$$

(9.81) could also be easily deduced from (9.82) and (9.83). And, vice versa, (9.83) follows easily from (9.81) and (9.82).

It will be noted that the a posteriori probability $\pi(t)$ (as well as $\varphi(t)$) is a sufficient statistic in the problem of testing two simple hypotheses[4] $H_0:\theta = 0$ and $H_1:\theta = 1$.

EXAMPLE 2. Let $\theta_t, t \geq 0$, be a Markov process with the two states 0 and 1 with $P(\theta_0 = 1) = p$, $P(\theta_0 = 0) = 1 - p$, and the single transition from 0 into 1:

$$\lambda_{00} = -\lambda, \qquad \lambda_{01} = \lambda, \qquad \lambda_{10} = 0, \qquad \lambda_{11} = 0.$$

Let there be observed the random process

$$\xi_t = \int_0^t \theta_s \, ds + W_t$$

to this scheme we reduce the so-called "disruption" problem (see [169]) of the earliest detection of the time θ after the drift coefficient of the observable process has been changed under the condition that

$$P\{\theta \geq t \,|\, \theta > 0\} = e^{-\lambda t}, \qquad P(\theta = 0) = p.$$

In the case under consideration the a posteriori probability

$$\pi(t) - P(\theta_t - 1 \,|\, \mathscr{F}_t^\xi) \qquad (-P\{\theta \leq t \,|\, \mathscr{F}_t^\xi\})$$

satisfies (according to (9.17)) the equation

$$d\pi(t) = \lambda(1 - \pi(t))dt + \pi(t)(1 - \pi(t))(d\xi_t - \pi(t)dt) \tag{9.84}$$

with $\pi(0) = p$.

Note that $\pi(t) = M(\theta_t \,|\, \mathscr{F}_t^\xi)$. Hence $\pi(t)$ is the optimal (in the mean square sense) estimate of θ_t on the basis of the observation $\xi_0^t = \{\xi_s, s \leq t\}$.

EXAMPLE 3. Let θ_t, $t \geq 0$, be a Markov process with two states 0 and 1. Assume $P(\theta_0 = 0) = P(\theta_0 = 1) = \frac{1}{2}$, the densities of the transition probabilities $\lambda_{\alpha\beta}(t)$ do not depend on t, and

$$\lambda_{00} = -\lambda, \qquad \lambda_{01} = \lambda, \qquad \lambda_{10} = \lambda, \qquad \lambda_{11} = -\lambda.$$

(The process $\theta_t, t \geq 0$, is called a "telegraph signal.")

Let the observable process ξ_t, $t \geq 0$, permit the differential

$$d\xi_t = \theta_t \, dt + dW_t, \qquad \xi_0 = 0. \tag{9.85}$$

The a posteriori probability $\pi(t) = P(\theta_t = 1 \,|\, \mathscr{F}_t^\xi)$, being in this case an optimal (in the mean square sense) estimate of the values θ_t, satisfies the stochastic equation

$$d\pi(t) = \lambda(1 - 2\pi(t))dt + \pi(t)(1 - \pi(t))(d\xi_t - \pi(t)dt) \tag{9.86}$$

with $\pi(0) = \frac{1}{2}$.

[4] For more detail see, for example, [169], Chapter 4.

Analogously, $\omega_1(t, s) = P(\theta_t = 1 | \theta_s = 1, \mathscr{F}_t^\xi)$ satisfies the equation

$$\omega_{11}(t, s) = 1 + \lambda \int_s^t [1 - 2\omega_{11}(u, s)]du$$
$$+ \int_s^t \omega_{11}(u, s)[1 - \omega_{11}(u, s)][d\xi_u - \omega_{11}(u, s)du]. \tag{9.87}$$

For $s \leq t$, denote $\pi_1(s, t) = P(\theta_s = 1 | \mathscr{F}_t^\xi)$. Then, from (9.55), it is seen that $\pi_1(s, t)$ is the optimal (in the mean square sense) estimate of θ_s on the basis of $\xi_0^t, s \leq t$:

$$\pi_1(s, t) = \pi_1(s)\exp\left\{\int_s^t [\omega_{11}(u, s) - \pi(u)]d\xi_u - \frac{1}{2}\int_s^t [\omega_{11}^2(u, s) - \pi^2(u)]du\right\}.$$

For $t \geq s$, now let $\pi_1(t, s) = P(\theta_t = 1 | \mathscr{F}_s^\xi)$. Then, according to (9.74),

$$\pi_1(t, s) = \pi(s) + \lambda \int_s^t [1 - 2\pi_1(u, s)]du.$$

From this we find

$$\pi_1(t, s) = \pi(s)e^{-2\lambda(t-s)} + \tfrac{1}{2}(1 - e^{-2\lambda(t-s)}). \tag{9.88}$$

By (9.75),

$$\pi_1(t, s) = \pi_1(t, 0) + \int_0^s [p_{11}(t, u) - p_{10}(t, u)]\pi(u)(1 - \pi(u))[d\xi_u - \pi(u)du].$$

It is not difficult to infer from (9.12) that

$$p_{11}(t, u) = \tfrac{1}{2}(1 + e^{-2\lambda(t-u)}), \qquad p_{10}(t, u) = \tfrac{1}{2}(1 - e^{-2\lambda(t-u)}).$$

Hence,

$$\pi_1(t, s) = \frac{1}{2} + \frac{1}{2}\int_0^s \pi(u)(1 - \pi(u))e^{-2\lambda(t-u)}[d\xi_u - \pi(u)du]. \tag{9.89}$$

$\pi_1(t, s)$ is an extrapolating estimate of θ_t on the basis of $\xi_0^s, s \leq t$.

Notes and references

9.1–9.3. Particular cases of Theorem 9.1 have been published by Wonham [25], Shiryayev [166], Liptser and Shiryayev [166], and Stratonovich [147]. The martingale deduction here is new. The uniqueness of the solution to a nonlinear system of equations (9.23) has been studied by Rozovsky and Shiryayev [141]. The deductions of forward and backward interpolation equations have been dealt with in Stratonovich [147] and in Liptser and Shiryayev [116].

Optimal linear nonstationary filtering

10

10.1 The Kalman–Bucy method

10.1.1

On the probability space $(\Omega, \mathscr{F}, P)$ with a distinguished family of the σ-algebras $(\mathscr{F}_t)$, $t \leq T$, we shall consider the two-dimensional Gaussian random process $(\theta_t, \mathscr{F}_t)$, $0 \leq t \leq T$, satisfying the stochastic differential equations

$$d\theta_t = a(t)\theta_t\, dt + b(t)dW_1(t), \tag{10.1}$$

$$d\xi_t = A(t)\theta_t\, dt + B(t)dW_2(t), \tag{10.2}$$

where $W_1 = (W_1(t), \mathscr{F}_t)$ and $W_2 = (W_2(t), \mathscr{F}_t)$ are two independent Wiener processes and θ_0, ξ_0 are $\mathscr{F}_0$-measurable.

It will be assumed that the measurable functions $a(t)$, $b(t)$, $A(t)$, $B(t)$ are such that

$$\int_0^T |a(t)|dt < \infty, \qquad \int_0^T b^2(t)dt < \infty, \tag{10.3}$$

$$\int_0^T |A(t)|dt < \infty, \qquad \int_0^T B^2(t)dt < \infty. \tag{10.4}$$

From Theorem 4.10 it follows that the linear equation given by (10.1) has a unique, continuous solution, given by the formula

$$\theta_t = \exp\left[\int_0^t a(u)du\right]\left[\theta_0 + \int_0^t \exp\left\{-\int_0^s a(u)du\right\}b(s)dW_1(s)\right]. \tag{10.5}$$

The problem of *optimal linear nonstationary filtering* (θ_t on ξ_0^t) examined by Kalman and Bucy consists in the following. Suppose the process θ_t, $0 \leq t \leq T$, is inaccessible for observation, and one can observe only the values ξ_t, $0 \le t \le T$, containing incomplete (due to the availability in (10.2) of the multiplier $A(t)$ and the noise $\int_0^t B(s)dW_2(s)$) information on the values θ_t. It is required at each moment t to estimate (to filter) in the "optimal" way the values θ_t on the basis of the observed process: $\xi_0^t = \{\xi_s, 0 \le s \le t\}$.

If we take the optimality of estimation in the mean square sense, then the optimal (at t) estimate for θ_t given $\xi_0^t = \{\xi_s, 0 \le s \le t\}$ coincides with the conditional expectation[1]

$$m_t = M(\theta_t | \mathscr{F}_t^\xi) \tag{10.6}$$

(in the notation of Chapter 8, $m_t = \pi_t(\theta)$). An error of estimation (of filtering) we denote by

$$\gamma_t = M(\theta_t - m_t)^2. \tag{10.7}$$

The method employed by Kalman and Bucy to find m_t and γ_t yields a *closed* system of dynamic equations (see (10.10)–(10.11)) for the estimate in a form convenient for instrumentation of optimal "filter."

The process (θ_t, ξ_t), $0 \le t \le T$, studied by Kalman and Bucy is Gaussian. As a consequence, the optimal estimate $m_t = M(\theta_t | \mathscr{F}_t^\xi)$ turns out to be *linear* (see Lemma 10.1). The next chapter contains an essential generalization of the Kalman–Bucy scheme. It will be shown there that in the so-called conditionally Gaussian case for $m_t = M(\theta_t | \mathscr{F}_t^\xi)$ and $\gamma_t = M[(\theta_t - m_t)^2 | \mathscr{F}_t^\xi]$ a closed system of equations can also be obtained (see (12.29), (12.30)), although the estimate m_t will then be, generally speaking, *nonlinear*.

In the case of (10.1) and (10.2) the equations for m_t and γ_t can be easily deduced from the general equations of filtering obtained in Chapter 8. It will be done in Sections 10.2 and 10.3.

In Subsections 10.1.2–10.1.4 the filtering equations for m_t and γ_t will be deduced (with some modifications and refinements) following the scheme originally suggested by Kalman and Bucy. As noted in the introduction, (10.24) is the basis of this deduction (in the case $m_0 = 0$). In Subsection 10.1.5 another (simpler) deduction of the same equations will be given, employing the fact that m_t can be represented in the form of (10.52), where $\overline{W}$ is an innovation process.

10.1.2

Theorem 10.1. *Let (θ_t, ξ_t), $0 \le t \le T$, be a two-dimensional Gaussian process satisfying the system of the equations in* (10.1) *and* (10.2). *Let* (10.3) *and* (10.4) *also be satisfied, and require, further, that*

$$\int_0^T A^2(t)dt < \infty, \tag{10.8}$$

$$B^2(t) \ge C > 0, \qquad 0 \le t \le T. \tag{10.9}$$

[1] Henceforth only the measurable modifications of conditional expectations will be taken.

Then the conditional expectation $m_t = M(\theta_t | \mathscr{F}_t^\xi)$ *and the mean square error of filtering* $\gamma_t = M(\theta_t - m_t)^2$ *satisfy the system of equations*

$$dm_t = a(t)m_t\,dt + \frac{\gamma_t A(t)}{B^2(t)}(d\xi_t - A(t)m_t\,dt), \tag{10.10}$$

$$\dot{\gamma}_t = 2a(t)\gamma_t - \frac{A^2(t)\gamma_t^2}{B^2(t)} + b^2(t), \tag{10.11}$$

with $m_0 = M(\theta_0 | \xi_0)$, $\gamma_0 = M(\theta_0 - m_0)^2$. *The system of equations in* (10.10) *and* (10.11) *has a unique continuous solution (for* γ_t, *in a class of nonnegative functions).*

10.1.3

As a preliminary we shall prove a number of auxiliary statements.

Lemma 10.1. *Let* $\xi = (\xi_t, \mathscr{F}_t)$, $0 \le t \le T$, *be a Gaussian random process with*

$$\xi_t = \xi_0 + \int_0^t \alpha_s\,ds + \int_0^t B(s)dW_s, \qquad B^2(s) \ge C > 0, 0 \le s \le T, \tag{10.12}$$

where the Wiener process $W = (W_t, \mathscr{F}_t)$ *does not depend on the Gaussian process* $\alpha = (\alpha_t, \mathscr{F}_t)$, $0 \le t \le T$, *with* $M(\alpha_t | \xi_0) \equiv 0$ *and*

$$P\left(\int_0^T \alpha_s^2\,ds < \infty\right) = 1. \tag{10.13}$$

If the random variable $\eta = \eta(\omega)$ *and the process* $\xi = (\xi_t)$, $0 \le t \le T$, *form a Gaussian system, then for each t,* $0 \le t \le T$, *we can find a function* $G(t, s)$, $0 \le s \le t$, *with*

$$\int_0^t G^2(t, s)ds < \infty \tag{10.14}$$

such that (P-a.s.)

$$M(\eta | \mathscr{F}_t^\xi) = M(\eta | \xi_0) + \int_0^t G(t, s)d\xi_s. \tag{10.15}$$

Proof. First of all, it will be noted that from (10.13) it follows that $\int_0^T M\alpha_t^2\,dt < \infty$ (Lemma 7.2). Let $0 = t_0^{(n)} < t_1^{(n)} < \cdots < t_{2^n}^{(n)} = t$ be a binary rational decomposition of the interval $[0, t]$, $t_k^{(n)} = (k/2^n)t$. Denote

$$\mathscr{F}_{t,n}^\xi = \sigma\{\omega : \xi_{t_0^{(n)}}, \ldots, \xi_{t_{2^n}^{(n)}}\} = \sigma\{\omega : \xi_{t_0^{(n)}}, \xi_{t_1^{(n)}} - \xi_{t_0^{(n)}}, \ldots, \xi_{t_{2^n}^{(n)}} - \xi_{t_{2^n-1}^{(n)}}\}.$$

Then, since $\mathscr{F}_{t,n}^\xi \uparrow \mathscr{F}_t^\xi$, by Theorem 1.5 with probability one

$$M(\eta | \mathscr{F}_{t,n}^\xi) \to M(\eta | \mathscr{F}_t^\xi). \tag{10.16}$$

The sequence of random variables $\{(M(\eta | \mathscr{F}_{t,n}^\xi))^2, n = 1, 2, \ldots\}$ is uniformly integrable. Hence from (10.16) it follows that

$$\underset{n\to\infty}{\text{l.i.m.}}\, M(\eta | \mathscr{F}_{t,n}^\xi) = M(\eta | \mathscr{F}_t^\xi). \tag{10.17}$$

By the theorem of normal correlation (Theorem 13.1), for each n, $n = 1, 2, \ldots$, (P-a.s.)

$$M(\eta \mid \mathscr{F}^{\xi}_{t,n}) = M(\eta \mid \xi_0) + \sum_{j=0}^{2^{n-1}} G_n(t, t_j^{(n)})[\xi_{t_{j+1}^{(n)}} - \xi_{t_j^{(n)}}] \tag{10.18}$$

for a certain (nonrandom) function $G_n(t, t_j^{(n)})$, $0 \le j \le 2^{n-1}$

Denote

$$G_n(t, s) = G_n(t, t_j^{(n)}), \qquad t_j^{(n)} \le s < t_{j+1}^{(n)}.$$

Then Equation (10.18) can be rewritten as follows:

$$M(\eta \mid \mathscr{F}^{\xi}_{t,n}) = M(\eta \mid \xi_0) + \int_0^t G_n(t, s) d\xi_s. \tag{10.19}$$

From (10.19) and the independence of the processes α and W it follows that

$$\begin{aligned}
M[M(\eta \mid \mathscr{F}^{\xi}_{t,n}) - M(\eta \mid \mathscr{F}^{\xi}_{t,m})]^2 &= M\left\{\int_0^t [G_n(t, s) - G_m(t, s)] d\xi_s\right\}^2 \\
&= M\left\{\int_0^t [G_n(t, s) - G_m(t, s)]\alpha_s \, ds\right\}^2 \\
&\quad + M\left\{\int_0^t [G_n(t, s) - G_m(t, s)]B(s) dW_s\right\}^2 \\
&= M\left\{\int_0^t [G_n(t, s) - G_m(t, s)]\alpha_s \, ds\right\}^2 \\
&\quad + \int_0^t [G_n(t, s) - G_m(t, s)]^2 B^2(s) ds. \qquad (10.20)
\end{aligned}$$

But by (10.17), $\lim_{n,m\to\infty} M[M(\eta \mid \mathscr{F}^{\xi}_{t,n}) - M(\eta \mid \mathscr{F}^{\xi}_{t,m})]^2 = 0$. Hence, according to (10.20) and the inequality $B^2(s) \ge C > 0$, $0 \le s \le T$,

$$\lim_{n,m\to\infty} \int_0^t [G_n(t, s) - G_m(t, s)]^2 \, ds = 0.$$

In other words, the sequence of functions $\{G_n(t, s), n = 1, 2, \ldots\}$ is fundamental in $L_2[0, t]$. Because of the completeness of this space there exists (at the given t) a measurable (over s, $0 \le s \le t$), function

$$G(t, s) \in L_2[0, T]$$

such that

$$\lim_{n\to\infty} \int_0^t [G(t, s) - G_n(t, s)]^2 \, ds = 0,$$

$$\lim_{n\to\infty} \int_0^t [G(t, s) - G_n(t, s)]^2 B^2(s) ds = 0. \tag{10.21}$$

Since $M \int_0^t \alpha_t^2\, ds < \infty$, from (10.21) it follows also that

$$\lim_{n \to \infty} M\left\{ \int_0^t [G_n(t, s) - G(t, s)]\alpha_s\, ds \right\}^2 = 0.$$

Consequently,

$$\underset{n}{\text{l.i.m.}} \int_0^t G_n(t, s) d\xi_s = \int_0^t G(t, s) d\xi_s,$$

which together with (10.17) and (10.19) proves (10.15). □

Corollary 1. *Let $W = (W_t, \mathscr{F}_t)$, $0 \le t \le T$, be a Wiener process and let $\eta = \eta(W)$ be a (Gaussian) random variable, such that (η, W) forms a Gaussian system. Then, (P-a.s.) for any t, $0 \le t \le T$,*

$$M(\eta | \mathscr{F}_t^W) = M\eta + \int_0^t G(t, s) dW_s, \tag{10.22}$$

where $G(t, s)$, $0 \le s \le t$, is a deterministic function with $\int_0^t G^2(t, s)ds < \infty$ (compare with (5.16)). In particular, if the random variable η is $\mathscr{F}_t^W$-measurable, then

$$\eta = M\eta + \int_0^t G(t, s) dW_s.$$

Corollary 2. *Let the conditions of Theorem 10.1 be satisfied ($m_0 = 0$). Then, for each t, $0 \le t \le T$, there exists a function $G(t, s)$, $0 \le s \le t$, such that*

$$\int_0^t G^2(t, s) ds < \infty, \qquad \int_0^t G^2(t, s) B^2(s) ds < \infty,$$

$$\int_0^t \int_0^t G(t, u) G(t, v) A(u) A(v) M(\theta_u, \theta_v) du\, dv < \infty, \tag{10.23}$$

and $m_t = M(\theta_t | \mathscr{F}_t^\xi)$ is given by:

$$m_t = \int_0^t G(t, s) d\xi_s. \tag{10.24}$$

From Lemma 10.3 it will follow that the function $G(t, s)$ entering into (10.24) has a modification which is measurable over a pair of variables.

Lemma 10.2. *Let the assumptions of Theorem 10.1 be fulfilled and $m_0 = 0$ (P-a.s.). Then for each t, $0 \le t \le T$, the function $G(t, s)$, $0 \le s \le t$, satisfies a Wiener–Hopf integral equation: for almost all u, $0 \le u \le t$,*

$$K(t, u) A(u) = \int_0^t G(t, s) A(s) K(s, u) A(u) ds + G(t, u) B^2(u), \tag{10.25}$$

where $K(t, u) = M\theta_t \theta_u$.

PROOF. First of all, note that from the assumption $m_0 = 0$ (P-a.s.) it follows that $M\theta_0 = Mm_0 = 0$, and by (10.5) $M\theta_t \equiv 0, 0 \le t \le T$.

Next, the integral $\int_0^t G(t, s)A(s)K(s, u)ds$ exists and is finite, since $\int_0^t G^2(t, s)ds < \infty$, $\int_0^T A^2(s)ds < \infty$, and $K(s, u)$ is a bounded function, continuous over a pair of variables, which, according to (10.5), may be represented as follows:

$$K(s, u) = \exp\left[\int_0^s a(z)dz + \int_0^u a(z)dz\right] \times \left[M\theta_0^2 + \int_0^{s\wedge u} \exp\left(-2\int_0^z a(y)dy\right)b^2(z)dz\right], \tag{10.26}$$

where $s \wedge u = \min(s, u)$.

Pass now to the deduction of Equation (10.25). Let $t \in [0, T]$ and let $f(t, s)$, $0 \le s \le t$, be a bounded measurable (w.r.t. s) function. Consider the integral $I(t) = \int_0^t f(t, s)d\xi_s$. This random variable is $\mathscr{F}_t^\xi$-measurable, and it is not difficult to show that

$$M\left[\int_0^t f(t, s)d\xi_s\right]^2 < \infty.$$

Hence,

$$M(\theta_t - m_t)\int_0^t f(t, s)d\xi_s = 0,$$

i.e.,

$$M\theta_t \int_0^t f(t, s)d\xi_s = Mm_t \int_0^t f(t, s)d\xi_s. \tag{10.27}$$

Since the random variables θ_t and $\int_0^t f(t, s)B(s)dW_2(s)$ are independent, then

$$\begin{aligned} M\theta_t \int_0^t f(t, s)d\xi_s &= M\theta_t \int_0^t f(t, s)A(s)\theta_s\, ds + M\theta_t \int_0^t f(t, s)B(s)dW_2(s) \\ &= M\theta_t \int_0^t f(t, s)A(s)\theta_s\, ds \\ &= \int_0^t f(t, s)A(s)M\theta_t\theta_s\, ds = \int_0^t f(t, s)A(s)K(t, s)ds. \end{aligned} \tag{10.28}$$

On the other hand, using (10.24), we find that

$$\begin{aligned} Mm_t \int_0^t f(t, s)d\xi_s &= M\int_0^t G(t, s)d\xi_s \int_0^t f(t, s)d\xi_s \\ &= M\left[\int_0^t G(t, s)A(s)\theta_s\, ds + \int_0^t G(t, s)B(s)dW_2(s)\right] \\ &\quad \times \left[\int_0^t f(t, s)A(s)\theta_s\, ds + \int_0^t f(t, s)B(s)dW_2(s)\right]. \end{aligned} \tag{10.29}$$

Let us again make use of the independence of $\int_0^t G(t, s)A(s)\theta_s\,ds$ and $\int_0^t f(t, s)B(s)dW_2(s)$, $\int_0^t f(t, s)A(s)\theta_s\,ds$ and $\int_0^t G(t, s)B(s)dW_2(s)$. Then from (10.29) we obtain

$$\begin{aligned} Mm_t \int_0^t f(t, s)d\xi_s &= M \int_0^t \int_0^t G(t, s)A(s)\theta_s\theta_u A(u)f(t, u)du \\ &\quad + M \int_0^t G(t, s)B(s)dW_2(s) \int_0^t f(t, s)B(s)dW_2(s) \\ &= \int_0^t \int_0^t G(t, s)A(s)K(s, u)A(u)f(t, u)ds\,du \\ &\quad + \int_0^t G(t, u)B^2(u)f(t, u)du. \end{aligned} \tag{10.30}$$

Comparing (10.27), (10.28) and (10.30), and also taking into account the arbitrariness of the function $f(t, u)$, we obtain (10.25). ☐

Lemma 10.3. *Let $t \in [0, T]$ be fixed. The solution $G(t, s)$, $0 \le s \le t$, of Equation* (10.25) *is unique*[2] *(in the class of the functions satisfying* (10.23)) *and is given by the formula*

$$G(t, s) = \varphi_s^t G(s, s), \tag{10.31}$$

where

$$G(s, s) = \frac{\gamma_s A(s)}{B^2(s)} \tag{10.32}$$

and φ_s^t is a solution of the differential equation

$$\frac{d\varphi_s^t}{dt} = \left[a(t) - \gamma_t \frac{A^2(t)}{B^2(t)}\right]\varphi_s^t, \quad \varphi_s^s = 1. \tag{10.33}$$

PROOF. We will establish uniqueness first. Let $G_i(t, s)$, $i = 1, 2$, be two solutions of Equation (10.25), such that

$$\int_0^t G_i^2(t, s)ds < \infty, \qquad \int_0^t G_i^2(t, s)B^2(s)ds < \infty.$$

Then $\Delta(t, s) = G_1(t, s) - G_2(t, s)$ is a solution of the equation

$$\int_0^t \Delta(t, s)A(s)K(s, u)A(u)ds + \Delta(t, u)B^2(u) = 0. \tag{10.34}$$

[2] The two solutions $G_1(t, s)$ and $G_2(t, s)$ are considered to coincide if $G_1(t, s) = G_2(t, s)$ for almost all s, $0 \le s \le t$.

Multiplying both sides of this equation by $\Delta(t, u)$ and integrating over u from 0 to t, we obtain

$$\int_0^t \int_0^t \Delta(t, s)A(s)K(s, u)A(u)\Delta(t, u)ds\, du + \int_0^t \Delta^2(t, u)B^2(u)du = 0. \quad (10.35)$$

Because of the nonnegative definiteness of the correlation function $K(s, u)$,

$$\int_0^t \int_0^t [\Delta(t, s)A(s)]K(s, u)[A(u)\Delta(t, s)] \geq 0.$$

Hence,

$$\int_0^t \Delta^2(t, u)B^2(u)du = 0,$$

and, because $\inf_{0 \leq u \leq t} B^2(u) > 0$, therefore $\Delta(t, u) = 0$ for almost all u, $0 \leq u \leq t$.

It will also be noted that Equation (10.33), which defines the function φ_s^t, has a unique continuous solution. This follows from Theorem 4.10 and the fact that

$$\int_s^T \left| a(t) - \gamma_t \frac{A^2(t)}{B^2(t)} \right| dt \leq \int_s^T |a(t)|dt + \frac{\sup_{0 \leq t \leq T} M\theta_t^2}{C} \int_0^T A^2(t)dt < \infty;$$

the constant C is defined in (10.9).

Let us establish (10.32) next. From (10.25) we find

$$\begin{aligned} G(t, t)B^2(t) &= K(t, t)A(t) - \int_0^t G(t, s)A(s)K(s, t)A(t)ds \\ &= M\theta_t^2 A(t) - \int_0^t G(t, s)A(s)M\theta_s\theta_t A(t)ds \\ &= M\left[\theta_t - \int_0^t G(t, s)A(s)\theta_s\, ds\right]\theta_t A(t). \end{aligned} \quad (10.36)$$

Since $M\theta_t \int_0^t G(t, s)B(s)dW_2(s) = 0$, the right-hand side in (10.36) is equal to

$$\begin{aligned} &M\left[\theta_t - \int_0^t G(t, s)A(s)\theta_s\, ds - \int_0^t G(t, s)B(s)dW_2(s)\right]\theta_t A(t) \\ &= M\left[\theta_t - \int_0^t G(t, s)d\xi_s\right]\theta_t A(t) = M[\theta_t - m_t]\theta_t A(t) \\ &= M(\theta_t - m_t)^2 A(t) + M(\theta_t - m_t)m_t A(t). \end{aligned} \quad (10.37)$$

But $M(\theta_t - m_t)m_t A(t) = 0$, and $M(\theta_t - m_t)^2 = \gamma_t$. Therefore, by virtue of (10.36) and (10.37), $G(t, t)B^2(t) = \gamma_t A(t)$, which proves (10.32).

We shall seek a solution of Equation (10.25) on the assumption that the function $G(t, s)$ is almost everywhere differentiable over t ($s \leq t \leq T$). This assumption does not restrict the generality, because if Equation (10.25) has a solution satisfying (10.23), then by the proven uniqueness it is the required solution.

Let us establish first of all that the function $K(t, u)$ is almost everywhere differentiable in $t(t \geq u)$ and that

$$\frac{\partial K(t, u)}{\partial t} = a(t)K(t, u). \tag{10.38}$$

Indeed, by (10.1),

$$\theta_t \theta_u = \theta_u^2 + \int_u^t a(v)\theta_u \theta_v \, dv + \theta_u \int_u^t b(v) dW_1(v).$$

Taking expectations on both sides of this equality and taking into account that $M\theta_u^2 \int_u^t b^2(v)dv < \infty$, we find

$$K(t, u) = K(u, u) + \int_u^t a(v)K(u, v)dv. \tag{10.39}$$

This proves the validity of Equation (10.38).

Assuming the differentiability of the function $G(t, u)$, let us differentiate over t the left and right sides of Equation (10.25). Taking into consideration (10.38), we obtain

$$\begin{aligned} a(t)K(t, u)A(u) = {} & G(t, t)A(t)K(t, u)A(u) \\ & + \int_0^t \frac{\partial G(t, s)}{\partial t} A(s)K(s, u)A(u)ds + \frac{\partial G(t, u)}{\partial t} B^2(u). \end{aligned} \tag{10.40}$$

But, according to (10.25),

$$K(t, u)A(u) = \int_0^t G(t, s)A(s)K(s, u)A(u)ds + G(t, u)B^2(u)$$

and

$$G(t, t) = \frac{\gamma_t A(t)}{B^2(t)}.$$

Hence (10.40) can be transformed to

$$\begin{aligned} & \int_0^t \left\{ \left[a(t) - \frac{\gamma_t A^2(t)}{B^2(t)} \right] G(t, s) - \frac{\partial G(t, s)}{\partial t} \right\} A(s)K(s, u)A(u)ds \\ & + \left\{ \left[a(t) - \frac{\gamma_t A^2(t)}{B^2(t)} \right] G(t, u) - \frac{\partial G(t, u)}{\partial t} \right\} B^2(u) = 0. \end{aligned} \tag{10.41}$$

From this it is seen that the function $G(t, s)$, being a solution of the equation

$$\frac{\partial G(t, s)}{\partial t} = \left[a(t) - \frac{\gamma_t A^2(t)}{B^2(t)} \right] G(t, s), \qquad t \geq s,$$

with $G(s, s) = \gamma_s A(s)/B^2(s)$, satisfies as well Equation (10.41). □

10.1.4

Proof of Theorem 10.1. Assume first that $m_0 = 0$ (P-a.s.). Then, by Lemmas 10.1 and 10.3,

$$m_t = \int_0^t G(t, s)d\xi_s = \int_0^t G(s, s)\varphi_s^t \, d\xi_s = \varphi_0^t \int_0^t (\varphi_0^s)^{-1} \frac{\gamma_s A(s)}{B^2(s)} \, d\xi_s, \tag{10.42}$$

since $\varphi_s^t = \varphi_0^t(\varphi_0^s)^{-1}$. Taking into account that $d\xi_t = A(t)\theta_t \, dt + B(t)dW_2(t)$, from (10.42), with the help of the Ito formula, we find that

$$dm_t = \frac{d\varphi_0^t}{dt}\left[\int_0^t (\varphi_0^s)^{-1} \frac{\gamma_s A(s)}{B^2(s)} \, d\xi_s\right]dt + \frac{\gamma_t A(t)}{B^2(t)} \, d\xi_t. \tag{10.43}$$

But

$$\frac{d\varphi_0^t}{dt} = \left[a(t) - \frac{\gamma_t A^2(t)}{B^2(t)}\right]\varphi_0^t.$$

Hence,

$$\frac{d\varphi_0^t}{dt}\left[\int_0^t (\varphi_0^s)^{-1} \frac{\gamma_s A(s)}{B^2(s)} \, d\xi_s\right] = \left[a(t) - \frac{\gamma_t A^2(t)}{B^2(t)}\right]m_t,$$

which together with (10.43) leads (in the case $m_0 = 0$) to the equation

$$dm_t = \left[a(t) - \frac{\gamma_t A^2(t)}{B^2(t)}\right]m_t \, dt + \frac{\gamma_t A(t)}{B^2(t)} \, d\xi_t,$$

corresponding to Equation (10.10).

Let $P\{m_0 \neq 0\} > 0$. Introduce the process $(\tilde{\theta}_t, \tilde{\xi}_t)$, $0 \le t \le T$, with

$$\tilde{\theta}_t = \theta_t - m_0 \exp\left(\int_0^t a(s)ds\right), \tag{10.44}$$

$$\tilde{\xi}_t = \xi_t - m_0 \int_0^t A(s)\exp\left(\int_0^s a(u)du\right)ds. \tag{10.45}$$

Then

$$\begin{aligned} d\tilde{\theta}_t &= a(t)\tilde{\theta}_t \, dt + b(t)dW_1(t), & \tilde{\theta}_0 &= \theta_0 - m_0, \\ d\tilde{\xi}_t &= A(t)\tilde{\theta}_t \, dt + B(t)dW_2(t), & \tilde{\xi}_0 &= \xi_0. \end{aligned} \tag{10.46}$$

Denote $\tilde{m}_t = M(\tilde{\theta}_t | \mathscr{F}_t^{\tilde{\xi}})$ and $\tilde{\gamma}_t = M(\tilde{\theta}_t - \tilde{m}_t)^2$. Since $\tilde{\xi}_0 = \xi_0$, then, by (10.45),

$$\mathscr{F}_t^{\xi} = \mathscr{F}_t^{\tilde{\xi}}, \qquad 0 \le t \le T;$$

therefore,

$$\tilde{m}_t = M(\tilde{\theta}_t | \mathscr{F}_t^{\xi}) = M(\theta_t | \mathscr{F}_t^{\xi}) - m_0 \exp\left(\int_0^t a(s)ds\right) = m_t - m_0 \exp\left(\int_0^t a(s)ds\right). \tag{10.47}$$

Also

$$d\tilde{m}_t = \left[a(t) - \frac{\tilde{\gamma}_t A^2(t)}{B^2(t)}\right]\tilde{m}_t\,dt + \frac{\tilde{\gamma}_t A(t)}{B^2(t)}\,d\tilde{\xi}_t. \tag{10.48}$$

It will be noted that

$$\begin{aligned}\gamma_t &= M(\theta_t - m_t)^2 \\ &= M\left[\left(\theta_t - m_0 \exp\left(\int_0^t a(s)ds\right)\right) - \left(m_t - m_0 \exp\left(\int_0^t a(s)ds\right)\right)\right]^2 \\ &= M[\tilde{\theta}_t - \tilde{m}_t]^2 = \tilde{\gamma}_t.\end{aligned}$$

Hence (10.48), taking into account (10.45) and (10.47), can be rewritten as follows:

$$\begin{aligned}\left[dm_t - m_0 a(t)\exp\left(\int_0^t a(s)ds\right)dt\right] &= \left[a(t) - \frac{A^2(t)\gamma_t}{B^2(t)}\right] \\ &\quad\times\left[m_t - m_0 \exp\left(\int_0^t a(s)ds\right)\right]dt \\ &\quad+ \frac{\gamma_t A(t)}{B^2(t)}\left[d\xi_t - m_0 A(t)\exp\left(\int_0^t a(s)ds\right)dt\right]\end{aligned}$$

After simple transformations, we obtain (10.10) for $m_t = M(\theta_t | \mathscr{F}_t^\xi)$.

We shall deduce now Equation (10.11) for $\gamma_t = M[\theta_t - m_t]^2$. Denote $\delta_t = \theta_t - m_t$. From (10.1), (10.10) and (10.2) we obtain

$$d\delta_t = a(t)\delta_t\,dt + b(t)dW_1(t) - \frac{\gamma_t A^2(t)}{B^2(t)}\delta_t\,dt - \frac{\gamma_t A(t)}{B(t)}\,dW_2(t).$$

From this, with the help of the Ito formula, we find that

$$\begin{aligned}\delta_t^2 = \delta_0^2 &+ 2\int_0^t \left[a(s) - \frac{\gamma_s A^2(s)}{B^2(s)}\right]\delta_s^2\,ds + \int_0^t \left[b^2(s) + \frac{\gamma_s^2 A^2(s)}{B^2(s)}\right]ds \\ &+ 2\int_0^t \delta_s b(s)dW_1(s) - 2\int_0^t \delta_s \frac{\gamma_s A(s)}{B(s)}\,dW_2(s).\end{aligned} \tag{10.49}$$

Noting that $M\delta_t^2 = \gamma_t$ and that

$$M\int_0^t \delta_s b(s)dW_1(s) = 0, \qquad M\int_0^t \delta_s \frac{\gamma_s A(s)}{B(s)}\,dW_2(s) = 0,$$

from (10.49) we obtain

$$\gamma_t = \gamma_0 + 2\int_0^t \left[a(s) - \frac{\gamma_s A^2(s)}{B^2(s)}\right]\gamma_s\,ds + \int_0^t \left[b^2(s) + \frac{\gamma_s^2 A^2(s)}{B^2(s)}\right]ds.$$

After obvious simplifications this equation can be transformed into Equation (10.11).

Let us proceed now to the conclusion of the theorem concerning the uniqueness of the solution of the system of equations in (10.10) and (10.11).

If the solution of the Ricatti equation, (10.11), is unique, then the uniqueness of the solution of Equation (10.10) follows from its linearity, which can be proved in the same way as in Theorem 4.10.

Next let us prove the uniqueness (in the class of nonnegative functions) of the solution of Equation (10.11).

Any nonnegative solution γ_t, $0 \le t \le T$, of this equation satisfies, as can easily be checked, the integral equation

$$\gamma_t = \exp\left\{2 \int_0^t a(s)ds\right\}\left\{\gamma_0 + \int_0^t \exp\left(-2 \int_0^s a(u)du\right)\left[b^2(s) - \frac{\gamma_s^2 A^2(s)}{B^2(s)}\right]ds\right\}.$$

From this by (10.3) and the assumption $M\theta_0^2 < \infty$, we obtain

$$0 \le \gamma_t \le \exp\left\{2 \int_0^T |a(s)|ds\right\}\left\{\gamma_0 + \exp\left(2 \int_0^T |a(u)|du\right) \int_0^T b^2(u)du\right\}$$
$$\le L < \infty, \tag{10.50}$$

where L is a certain constant.

Now let $\gamma_1(t)$ and $\gamma_2(t)$ be two solutions of Equation (10.11). Assume $\Delta(t) = |\gamma_1(t) - \gamma_2(t)|$. Then, according to (10.11), (10.50), (10.3), (10.8), and (10.9),

$$\Delta(t) \le 2 \int_0^t \left\{|a(s)| + \frac{L}{C} A^2(s)\right\}\Delta(s)ds.$$

From this, by Lemma 4.13, it follows that $\Delta(t) \equiv 0$. □

10.1.5

The Kalman–Bucy method was based essentially on the possibility of representation of conditional expectations $m_t = M(\theta_t | \mathscr{F}_t^\xi)$ in the form

$$m_t = \int_0^t G(t, s)d\xi_s \tag{10.51}$$

(we assume here, and henceforth, that $m_0 = 0$; therefore, by (10.5), $M(\theta_t | \mathscr{F}_0^\xi) = 0$). In the case under consideration, however, where the process (θ, ξ) is Gaussian, the conditional expectations m_t can be represented as well in the form

$$m_t = \int_0^t F(t, s)d\overline{W}_s, \tag{10.52}$$

where $\int_0^t F^2(t, s)ds < \infty$ and the process $\overline{W} = (\overline{W}_t, \mathscr{F}_t^\xi)$, $0 \le t \le T$, is a Wiener process and is determined by the equality

$$\overline{W}_t = \int_0^t \frac{d\xi_s}{B(s)} - \int_0^t \frac{A(s)}{B(s)} m_s\, ds$$

(see Theorems 7.12, 7.16 and 7.17).

It will be shown that the deduction of Equation (10.10) for m_t, $0 \le t \le T$, becomes considerably simpler if we start not from (10.51) but from (10.52).

We shall follow the scheme adopted in proving Theorem 10.1.

Let us fix t, $0 \le t \le T$, and let $f(t, s)$, $0 \le s \le t$, be a measurable bounded function. Then

$$M(\theta_t - m_t)\int_0^t f(t, s)d\overline{W}_s = 0,$$

i.e. (compare with (10.27)),

$$M\theta_t \int_0^t f(t, s)d\overline{W}_s = \int_0^t F(t, s)f(t, s)ds.$$

By definition of the innovation process $\overline{W} = (\overline{W}_t, \mathscr{F}_t^{\xi})$,

$$\overline{W}_t = W_2(t) + \int_0^t \frac{A(s)}{B(s)}(\theta_s - m_s)ds,$$

and, therefore,

$$\begin{aligned}\int_0^t F(t, s)f(t, s)ds &= M\left[\theta_t \int_0^t f(t, s)dW_2(s)\right]\\ &\quad + M\left[\theta_t \int_0^t f(t, s)\frac{A(s)}{B(s)}(\theta_s - m_s)ds\right]\\ &= \int_0^t f(t, s)\frac{A(s)}{B(s)}M[\theta_t(\theta_s - m_s)]ds,\end{aligned}$$

where we have made use of the fact that, because of the independence of the processes θ and W_2,

$$M\theta_t \int_0^t f(t, s)dW_2(s) = M\theta_t M\int_0^t f(t, s)dW_2(s) = 0.$$

Next, by (10.5),

$$M(\theta_t | \mathscr{F}_s) = \exp\left\{\int_s^t a(u)du\right\}\theta_s \qquad (P\text{-a.s.}).$$

Hence,

$$\begin{aligned}M\theta_t(\theta_s - m_s) &= M\{M(\theta_t | \mathscr{F}_s)(\theta_s - m_s)\}\\ &= \exp\left\{\int_s^t a(u)du\right\}M\theta_s[\theta_s - m_s]\\ &= \exp\left\{\int_s^t a(u)du\right\}M[\theta_s - m_s]^2 = \exp\left\{\int_s^t a(u)du\right\}\gamma_s,\end{aligned}$$

and, therefore,

$$\int_0^t F(t, s) f(t, s) ds = \int_0^t f(t, s) \frac{A(s)}{B(s)} \exp\left\{\int_s^t a(u) du\right\} \gamma_s \, ds.$$

From this, because of the arbitrariness of the function $f(t, s)$, we obtain

$$F(t, s) = \exp\left\{\int_s^t a(u) du\right\} \gamma_s \frac{A(s)}{B(s)}.$$

Thus

$$m_t = \int_0^t F(t, s) d\overline{W}_s = \int_0^t \exp\left\{\int_s^t a(u) du\right\} \frac{A(s)}{B(s)} \gamma_s \, d\overline{W}_s$$

$$= \exp\left\{\int_0^t a(u) du\right\} \int_0^t \exp\left\{-\int_0^s a(u) du\right\} \frac{A(s)\gamma_s}{B^2(s)} [d\xi_s - m_s \, ds].$$

From this by the Ito formula for $m_t, 0 \leq t \leq T$, we obtain Equation (10.10).

10.2 Martingale proof of the equations of linear nonstationary filtering

10.2.1

As was noted in Section 10.1, Equations (10.10) and (10.11) for m_t and γ_t can be deduced from general equations of filtering obtained in Chapter 8. We shall sketch this deduction since it will also serve as a particular example of how to employ the general equations.

We shall use the notation and concepts employed in proving Theorem 10.1. Assume also

$$G_t = \sigma\{\omega : \theta_0(\omega), \xi_0(\omega); W_1(s), W_2(s), s \leq t\}, \qquad 0 \leq t \leq T,$$

and

$$\psi_s^t = \exp\left(\int_s^t a(u) du\right).$$

Then from (10.5),

$$\theta_t = \psi_0^t \left(\theta_0 + \int_0^t (\psi_0^s)^{-1} b(s) dW_1(s)\right),$$

where the process $\bar{\theta} = (\bar{\theta}_t, G_t), 0 \leq t \leq T$, with

$$\bar{\theta}_t = \theta_0 + \int_0^t (\psi_0^s)^{-1} b(s) dW_1(s), \tag{10.53}$$

is a square integrable martingale.

Let us deduce now equations for $\overline{m}_t = M(\overline{\theta}_t | \mathscr{F}_t^\xi)$ and $\overline{\gamma}_t = M(\overline{\theta}_t - \overline{m}_t)^2$, from which there will easily be found equations for

$$m_t = \psi_0^t \overline{m}_t, \qquad \gamma_t = (\psi_0^t)^2 \overline{\gamma}_t. \tag{10.54}$$

By (10.53), $\langle \theta, W_2 \rangle_t = 0$ (P-a.s.), $0 \le t \le T$. Hence, according to the general equation of filtering ((8.10)) for $\pi_t(\overline{\theta}) = M(\overline{\theta}_t | \mathscr{F}_t^\xi)$ $(= \overline{m}_t)$, we obain

$$\pi_t(\overline{\theta}) = \pi_0(\overline{\theta}) + \int_0^t \frac{\pi_s(\overline{\theta}^2)\psi_0^s A(s) - (\pi_s(\overline{\theta}))^2 \psi_0^s A(s)}{B(s)}\, d\overline{W}_s, \tag{10.55}$$

where $\pi_s(\overline{\theta}^2) = M(\overline{\theta}_s^2 | \mathscr{F}_s^\xi)$ and

$$\overline{W}_t = \int_0^t \frac{d\xi_s - A(s)m_s\, ds}{B(s)}$$

is a Wiener process (with respect to $(\mathscr{F}_t^\xi)$, $0 \le t \le T$).

Note that

$$\begin{aligned}\pi_s(\overline{\theta}^2)\psi_0^s A(s) - (\pi_s(\overline{\theta}))^2 \psi_0^s A(s) &= \psi_0^s A(s)[\pi_s(\overline{\theta}^2) - (\pi_s(\overline{\theta}))^2] \\ &= \psi_0^s A(s) M[(\overline{\theta}_s - \overline{m}_s)^2 | \mathscr{F}_s^\xi]. \end{aligned} \tag{10.56}$$

It will be shown that

$$M[(\overline{\theta}_s - \overline{m}_s)^2 | \mathscr{F}_s^\xi] = M[\overline{\theta}_s \quad \overline{m}_s]^2 \qquad (-\overline{\gamma}_s). \tag{10.57}$$

Let $\mathscr{F}_{s,n}^\xi$ be the σ-algebras introduced in proving Lemma 10.1:

$$m_s^{(n)} = M(\theta_s | \mathscr{F}_{s,n}^\xi), \qquad \gamma_s^{(n)} = M(\theta_s - m_s^{(n)})^2.$$

From the theorem of normal correlation (Chapter 13) it follows that (P-a.s.)

$$M[(\theta_s - m_s^{(n)})^2 | \mathscr{F}_{s,n}^\xi] = M[\theta_s - m_s^{(n)}]^2. \tag{10.58}$$

We shall make use of this fact to prove the equality $M[(\theta_s - m_s)^2 | \mathscr{F}_s^\xi] = M[\theta_s - m_s]^2$ (P-a.s.), from which in an obvious manner will follow (10.57) as well.

By Theorem 1.5 and (10.58),

$$\begin{aligned} M[(\theta_s - m_s)^2 | \mathscr{F}_s^\xi] &= M(\theta_s^2 | \mathscr{F}_s^\xi) - m_s^2 \\ &= \lim_n M(\theta_s^2 | \mathscr{F}_{s,n}^\xi) - \lim_n (m_s^{(n)})^2 \\ &= \lim_n M[(\theta_s - m_s^{(n)})^2 | \mathscr{F}_{s,n}^\xi] \\ &= \lim_n M[\theta_s - m_s^{(n)}]^2 = \lim_n \gamma_s^{(n)}. \end{aligned} \tag{10.59}$$

On the other hand,

$$\begin{aligned} \gamma_s = M(\theta_s - m_s)^2 &= M[(\theta_s - m_s^{(n)}) + (m_s^{(n)} - m_s)]^2 \\ &= \gamma_s^{(n)} + M(m_s^{(n)} - m_s)^2 + 2M(\theta_s - m_s^{(n)})(m_s^{(n)} - m_s), \end{aligned}$$

and, therefore, according to the proof of Lemma 10.1,

$$|\gamma_s - \gamma_s^{(n)}| \le M(m_s^{(n)} - m_s)^2 + 2\sqrt{M(\theta_s - m_s^{(n)})^2 M(m_s^{(n)} - m_s)^2}$$

$$\le M(m_s^{(n)} - m_s)^2 + 2\sqrt{M\theta_s^2 M(m_s^{(n)} - m_s)^2} \to 0, \qquad n \to \infty.$$

Together with (10.59) this proves the equality

$$M[(\theta_s - m_s)^2 | \mathscr{F}_s^\xi] = M[\theta_s - m_s]^2 \qquad (P\text{-a.s.})$$

and, therefore, Equation (10.57).

Taking into account (10.57) and (10.54), the right-hand side in (10.56) can be rewritten as follows:

$$\psi_0^s A(s) M[(\bar{\theta}_s - \bar{m}_s)^2 | \mathscr{F}_s^\xi] = \psi_0^s A(s)\bar{\gamma}(s) = A(s)\gamma_s(\psi_0^s)^{-1}.$$

Hence, according to (10.55),

$$d\bar{m}_t = \frac{A(t)\gamma_t}{B(t)\psi_0^t} d\bar{W}_t. \tag{10.60}$$

Applying now the Ito formula to the product $m_t = \psi_0^t \bar{m}_t$, we obtain Equation (10.10):

$$dm_t = \frac{d\psi_0^t}{dt}\bar{m}_t\, dt + \frac{\gamma_t A(t)}{B(t)} d\bar{W}_t = a(t)(\psi_0^t \bar{m}_t) dt + \frac{\gamma_t A(t)}{B(t)} d\bar{W}_t$$

$$= a(t) m_t\, dt + \frac{\gamma_t A(t)}{B^2(t)}(d\xi_t - A(t) m_t\, dt).$$

10.2.2

In order to deduce Equation (10.11) from (8.10) it will be noted that, according to (10.53),

$$\bar{\theta}_t^2 = \bar{\theta}_0^2 + 2\int_0^t \bar{\theta}_s(\psi_0^s)^{-1} b(s) dW_1(s) + \int_0^t b^2(s)(\psi_0^s)^{-2}\, ds.$$

Hence, by (8.10),

$$\pi_t(\bar{\theta}^2) = \pi_0(\bar{\theta}^2) + \int_0^t b^2(s)(\psi_0^s)^{-2}\, ds + \int_0^t \frac{A(s)\psi_0^s}{B(s)} M[\bar{\theta}_s^2(\bar{\theta}_s - \bar{m}_s) | \mathscr{F}_s^\xi] d\bar{W}_s.$$

Since the process $(\bar{\theta}_s, \bar{\xi}_s)$, $0 \le s \le T$, is Gaussian,

$$M[\bar{\theta}_s^2(\bar{\theta}_s - \bar{m}_s) | \mathscr{F}_s^\xi] = 2\bar{m}_s \bar{\gamma}_s.$$

Therefore,

$$\pi_t(\bar{\theta}^2) = \pi_0(\bar{\theta}^2) + \int_0^t b^2(s)(\psi_0^s)^{-2}\, ds + 2\int_0^t \frac{A(s)\psi_0^s}{B(s)} \bar{m}_s \bar{\gamma}_s\, d\bar{W}_s. \tag{10.61}$$

From (10.60) and (10.61) we obtain

$$
\begin{aligned}
d\bar{\gamma}_t &= d[\pi_t(\bar{\theta}^2) - (\overline{m}_t)^2] \\
&= b^2(t)(\psi_0^t)^{-2}\,dt + 2\frac{A(t)\psi_0^t}{B(t)}\overline{m}_t\bar{\gamma}_t\,d\overline{W}_t - 2\overline{m}_t\frac{A(t)\gamma_t}{B(t)\psi_0^t}\,d\overline{W}_t - \left(\frac{A(t)\gamma_t}{B(t)\psi_0^t}\right)^2 dt \\
&= b^2(t)(\psi_0^t)^{-2}\,dt - \frac{A^2(t)}{B^2(t)}(\psi_0^t)^{-2}\gamma_t^2\,dt.
\end{aligned}
$$

From this we find

$$
\begin{aligned}
d\gamma_t &= (\psi_0^t)^2\,d\bar{\gamma}_t + 2(\psi_0^t)^2\bar{\gamma}_t a(t)dt \\
&= b^2(t)dt - \frac{A^2(t)}{B^2(t)}(\psi_0^t)^4\bar{\gamma}_t^2\,dt + 2a(t)\bar{\gamma}_t(\psi_0^t)^2\,dt \\
&= b^2(t)dt - \frac{A^2(t)}{B^2(t)}\gamma_t^2\,dt + 2a(t)\gamma_t\,dt,
\end{aligned}
$$

which coincides with Equation (10.11).

10.2.3

Note. Equations (10.10) and (10.11) could be deduced from the equations for (θ_t, ξ_t) without introducing the process $(\bar{\theta}_t, \xi_t)$, $0 \le t \le T$, by requiring $\int_0^T a^2(t)dt < \infty$ instead of $\int_0^T |a(t)|dt < \infty$.

10.3 Equations of linear nonstationary filtering: the multi-dimensional case

10.3.1

The present section concerns the extension of Theorem 10.1 in two directions: first, linear dependence of the observable component ξ_t will be introduced into the coefficients of transfer in (10.1) and (10.2); second, the multi-dimensional processes θ_t and ξ_t will be examined.

Thus, let us consider the $k + l$-dimensional Gaussian random process $(\theta_t, \xi_t) = [(\theta_1(t), \ldots, \theta_k(t)), (\xi_1(t), \ldots, \xi_l(t)]$, $0 \le t \le T$, with

$$
d\theta_t = [a_0(t) + a_1(t)\theta_t + a_2(t)\xi_t]dt + \sum_{i=1}^{2} b_i(t)dW_i(t), \tag{10.62}
$$

$$
d\xi_t = [A_0(t) + A_1(t)\theta_t + A_2(t)\xi_t]dt + \sum_{i=1}^{2} B_i(t)dW_i(t). \tag{10.63}
$$

In (10.62) and (10.63), $W_1 = [W_{11}(t), \ldots, W_{1k}(t)$, $W_2 = [W_{21}(t), \ldots, W_{2l}(t)]$ are two independent Wiener processes. A Gaussian vector of the initial values θ_0, ξ_0 is assumed to be independent of the processes W_1 and W_2. The measurable (deterministic) vector functions

$$
a_0(t) = [a_{01}(t), \ldots, a_{0k}(t)], \qquad A_0(t) = [A_{01}(t), \ldots, A_{0l}(t)]
$$

and the matrices[3]

$$a_1(t) = \|a_{ij}^{(1)}(t)\|_{(k \times k)}, \qquad a_2(t) = \|a_{ij}^{(2)}(t)\|_{(k \times l)},$$
$$A_1(t) = \|A_{ij}^{(1)}(t)\|_{(l \times k)}, \qquad A_2(t) = \|A_{ij}^{(2)}(t)\|_{(l \times l)},$$
$$b_1(t) = \|b_{ij}^{(1)}(t)\|_{(k \times k)}, \qquad b_2(t) = \|b_{ij}^{(2)}(t)\|_{(k \times l)},$$
$$B_1(t) = \|B_{ij}^{(1)}(t)\|_{(l \times k)}, \qquad B_2(t) = \|B_{ij}^{(2)}(t)\|_{(l \times l)},$$

are assumed to have the following properties:

$$\int_0^T \left[\sum_{i=1}^k |a_{0i}(t)| + \sum_{j=1}^l (A_{0j}(t))^2\right] dt < \infty; \tag{10.64}$$

$$\int_0^T \left[\sum_{i=1}^k \sum_{j=1}^k |a_{ij}^{(1)}(t)| + \sum_{i=1}^k \sum_{j=1}^l |a_{ij}^{(2)}(t)|\right] dt < \infty; \tag{10.65}$$

$$\int_0^T \left[\sum_{i=1}^k \sum_{j=1}^l (A_{ij}^{(1)}(t))^2 + \sum_{i=1}^l \sum_{j=1}^l (A_{ij}^{(2)}(t))^2\right] dt < \infty; \tag{10.66}$$

$$\int_0^T \left[\sum_{i=1}^k \sum_{j=1}^k (b_{ij}^{(1)}(t))^2 + \sum_{i=1}^k \sum_{j=1}^l (b_{ij}^{(2)}(t))^2 + \sum_{i=1}^k \sum_{j=1}^l (B_{ij}^{(1)}(t))^2 + \sum_{i=1}^l \sum_{j=1}^l (B_{ij}^{(2)}(t))^2\right] dt < \infty; \tag{10.67}$$

for all t, $0 \le t \le T$, the matrices $B_1(t)B_1^*(t) + B_2(t)B_2^*(t)$ are uniformly nonsingular, i.e., the smallest eigenvalues of the matrices $B_1(t)B_1^*(t) + B_2(t)B_2^*(t)$, $0 \le t \le T$, are uniformly (in t) bounded away from zero.[4]

According to Theorem 4.10 the system of equations in (10.62) and (10.63) has a unique continuous solution.

Let $m_t = M(\theta_t | \mathscr{F}_t^\xi)$ be a vector of conditional expectations,

$$[m_1(t), \ldots, m_k(t)] = [M(\theta_1(t) | \mathscr{F}_t^\xi), \ldots, M(\theta_k(t) | \mathscr{F}_t^\xi)],$$

where $\|\gamma_{ij}(t)\|_{(k \times k)} = \gamma_t$ is a matrix of covariances with

$$\gamma_{ij}(t) = M[(\theta_i(t) - m_i(t))(\theta_j(t) - m_j(t))].$$

The vector $m_t = [m_1(t), \ldots, m_k(t)]$ is, evidently, an $\mathscr{F}_t^\xi$-measurable estimate of the vector $\theta_t = (\theta_1(t), \ldots, \theta_k(t))$, optimal in the sense that

$$\operatorname{Sp} \gamma_t \equiv \sum_{i=1}^k \gamma_{ii}(t) \le \operatorname{Sp} M[(\theta_t - v_t)(\theta_t - v_t)^*] \tag{10.68}$$

for any $\mathscr{F}_t^\xi$-measurable vector $v_t = [v_1(t), \ldots, v_k(t)]$ with $\sum_{i=1}^k Mv_i^2(t) < \infty$.

[3] The indices $(p \times q)$ indicate the order of the matrix, the first index (p) giving the number of rows, and the second index (q) the number of columns.

[4] It can be shown that in this case the elements $(B_1(t)B_1^*(t) + B_2(t)B_2^*(t))^{-1}$, $0 \le t \le T$, are uniformly bounded.

Because of the Gaussian behavior of the process (θ_t, ξ_t), $0 \le t \le T$, the components of the vector m_t depend linearly on the observable values $\xi_0^t = \{\xi_s, s \le t\}$ (see below, (10.73)). Hence the optimal (in terms of (10.68)) filtering (of the values θ_t on the basis of ξ_0^t) is linear, but, generally speaking, nonstationary. As for (10.1) and (10.2), in the case under consideration one can also obtain a closed system of equations for m_t and γ_t defining the optimal filter.

10.3.2

Let us begin with a particular case of the system of equations given by (10.62) and (10.63): namely, a multi-dimensional analog of the system of equations given by (10.1) and (10.2).

Theorem 10.2. *Let the $k + l$-dimensional Gaussian process (θ_t, ξ_t), $0 \le t \le T$, permit the differentials*

$$d\theta_t = a(t)\theta_t\, dt + b(t)dW_1(t), \tag{10.69}$$

$$d\xi_t = A(t)\theta_t\, dt + B(t)dW_2(t) \tag{10.70}$$

(i.e., in (10.62) and (10.63) let $a_0(t) \equiv 0$, $A_0(t) \equiv 0$, $a_1(t) = a(t)$, $A_1(t) = A(t)$, $a_2(t) \equiv 0$, $A_2(t) \equiv 0$, $b_2(t) \equiv 0$, $B_1(t) \equiv 0$, $b_1(t) = b(t)$, $B_2(t) = B(t)$). Then m_t and γ_t are the solutions of the system of equations

$$dm_t = a(t)m_t\, dt + \gamma_t A^*(t)(B(t)B^*(t))^{-1}(d\xi_t - A(t)m_t\, dt), \tag{10.71}$$

$$\dot{\gamma}_t = a(t)\gamma_t + \gamma_t a^*(t) - \gamma_t A^*(t)(B(t)B^*(t))^{-1}A(t)\gamma_t + b(t)b^*(t), \tag{10.72}$$

with the initial conditions

$$m_0 = M(\theta_0 | \xi_0), \qquad \gamma_0 = M[(\theta_0 - m_0)(\theta_0 - m_0)^*]$$

The system of equations in (10.71) and (10.72) has a unique solution (for γ_t in the class of symmetric nonnegative definite matrices).

PROOF. With $k = l = 1$, (10.71) and (10.72) coincide with Equations (10.10) and (10.11), whose validity was established in Theorem 10.1.

The Kalman–Bucy method is applicable in deducing these equations in the general case $k \ge 1$, $l \ge 1$.

As in proving Theorem 10.1, first it is shown that (in the case $m_0 = 0$) for each t, $0 \le t \le T$,

$$m_t = \int_0^t G(t, s)d\xi_s \tag{10.73}$$

with the deterministic matrix $G(t, s)$ (of the order $(k \times l)$), measurable in s and such that

$$\mathrm{Sp} \int_0^t G(t, s)G^*(t, s)ds < \infty, \tag{10.74}$$

$$\mathrm{Sp} \int_0^t G(t, s)B(s)B^*(s)G^*(t, s)ds < \infty. \tag{10.75}$$

Further, it is established that

$$G(t, s) = \varphi_s^t G(s, s), \tag{10.76}$$

where

$$G(s, s) = \gamma_s A^*(s)(B(s)B^*(s))^{-1}, \tag{10.77}$$

and the matrix φ_s^t is a solution of the differential equation

$$\frac{d\varphi_s^t}{dt} = [a(t) - \gamma_t A^*(t)(B(t)B^*(t))^{-1}A(t)]\varphi_s^t, \qquad \varphi_s^s = E_{(k \times k)}. \tag{10.78}$$

Hence, m_t permits the representation

$$m_t = \varphi_0^t \int_0^t (\varphi_0^s)^{-1}\gamma_s A^*(s)(B(s)B^*(s))^{-1}\, d\xi_s, \tag{10.79}$$

from which (in the case $m_0 = 0$) Equation (10.71) is deduced.

The case $m_0 \neq 0$ (P-a.s.) is investigated in the same way as the case $k = l = 1$.

To obtain Equation (10.72), a vector $\delta_t = \theta_t - m_t$ is introduced, and then, for $\delta_t \delta_t^*$, with the help of the Ito formula an integral representation analogous to (10.49) is found. Taking, then, the expectation we obtain an (integral) equation equivalent to Equation (10.72).

The uniqueness of a solution of the system of equations in (10.71) and (10.72) can be proved in the same way as in the scalar case. It will merely be noted that instead of the estimate given by (10.50) one should employ the estimate

$$0 \leq \mathrm{Sp}\,\gamma_t \leq \mathrm{Sp}\,\Phi_0^t\left\{\gamma_0 + \int_0^t (\Phi_0^s)^{-1}b(s)b^*(s)[(\Phi_0^s)^{-1}]^*\,ds\right\}(\Phi_0^t)^* \leq L < \infty,$$

where L is a certain constant, and Φ_0^t is a fundamental matrix solution of the matrix equation

$$\frac{d\Phi_0^t}{dt} = a(t)\Phi_0^t, \qquad \Phi_0^0 = E_{(k \times k)}. \quad \square$$

10.3.3

Let us proceed to considering the general case. We shall use the following notation:

$$\begin{aligned}(b \circ b)(t) &= b_1(t)b_1^*(t) + b_2(t)b_2^*(t);\\ (b \circ B)(t) &= b_1(t)B_1^*(t) + b_2(t)B_2^*(t);\\ (B \circ B)(t) &= B_1(t)B_1^*(t) + B_2(t)B_2^*(t).\end{aligned} \tag{10.80}$$

Theorem 10.3. *Let the coefficients of the system of equations in* (10.62) *and* (10.63) *satisfy the conditions of Subsection* 10.3.1. *Then the vector m_t and the matrix γ_t are solutions of the system of the equations*

$$\begin{aligned}dm_t = {}& [a_0(t) + a_1(t)m_t + a_2(t)\xi_t]dt + [(b \circ B)(t) + \gamma_t A_1^*(t)]((B \circ B)(t))^{-1}\\ &\times [d\xi_t - (A_0(t) + A_1(t)m_t + A_2(t)\xi_t)dt],\end{aligned} \tag{10.81}$$

$$\begin{aligned}\dot{\gamma}_t = {}& a_1(t)\gamma_t + \gamma_t a_1^*(t) + b \circ b(t)\\ &- [(b \circ B)(t) + \gamma_t A_1^*(t)]((B \circ B)(t))^{-1}[(b \circ B)(t) + \gamma_t A_1^*(t)]^*\end{aligned}$$

with the initial conditions $m_0 = M(\theta_0 | \xi_0)$ *and*

$$\gamma_0 = \|\gamma_{ij}(0)\|, \qquad \gamma_{ij}(0) = M[(\theta_i(0) - m_i(0))(\theta_j(0) - m_j(0))^*]. \quad (10.82)$$

The system of equations in (10.81) *and* (10.82) *has a unique solution (for* γ_t *in the class of symmetric nonnegative definite matrices).*

For proving this we shall need the following lemma.

Lemma 10.4. *Let* $W = ([W_1(t), \ldots, W_N(t)], \mathscr{F}_t)$, $0 \le t \le T$, *be an* N*-dimensional Wiener process, and let* $B = (B_t, \mathscr{F}_t)$ *be a matrix random process where* $B_t = \|B_{ij}(t)\|_{(n \times N)}$ *and* (P-a.s.)

$$\mathrm{Sp} \int_0^T B_t B_t^* \, dt < \infty. \quad (10.83)$$

Let the matrix process $D = (D_t, \mathscr{F}_t)$, $D_t = \|D_{ij}(t)\|_{(n \times k)}$, *be such that for almost all* t, $0 \le t \le T$, (P-a.s.)

$$D_t D_t^* = B_t B_t^*. \quad (10.84)$$

Then[5] *there is a* k*-dimensional Wiener process* $\tilde{W} = ([\tilde{W}_1(t), \ldots, \tilde{W}_k(t)], \mathscr{F}_t)$, *such that for each* t, $0 \le t \le T$, (P-a.s.)

$$\int_0^t B_s \, dW_s = \int_0^t D_s \, d\tilde{W}_s. \quad (10.85)$$

PROOF. Let the initial probability space be "rich" enough to accommodate the k-dimensional Wiener process $Z = (z_t, \mathscr{F}_t)$, $0 \le t \le T$, independent of the Wiener process W. Assume

$$\tilde{W}_t = \int_0^t D_s^+ B_s \, dW_s + \int_0^t (E - D_s^+ D_s) dz_s, \quad (10.86)$$

where E is a unit matrix of the order $(k \times k)$, and D_s^+ is a pseudoinverse matrix to D_s (see Section 13.1).

The process $\tilde{W} = (\tilde{W}_s, \mathscr{F}_t)$, $0 \le t \le T$, is a Wiener process since by Theorem 4.2 it is a (vectorial) square integrable martingale with continuous trajectories and

$$\begin{aligned} M[(\tilde{W}_t - \tilde{W}_s)(\tilde{W}_t - \tilde{W}_s)^* | \mathscr{F}_s] &= M\left[\int_s^t D_u^+ B_u B_u^* (D_u^+)^* \, du | \mathscr{F}_s\right] \\ &\quad + M\left[\int_s^t (E - D_u^+ D_u)(E - D_u^+ D_u)^* \, du | \mathscr{F}_s\right] \\ &= E(t - s) \qquad (P\text{-a.s.}), \end{aligned}$$

[5] It is also assumed that the initial probability space is sufficiently "rich."

where the last equality holds because

$$D_u^+ B_u B_u^+ (D_u^+)^* = D_u^+ D_u D_u^* (D_u^+)^* = D_u^+ D_u (D_u^+ D_u)^* = (D_u^+ D_u)^2 = D_u^+ D_u,$$

and

$$(E - D_u^+ D_u)(E - D_u^+ D_u)^* = (E - D_u^+ D_u)^2 = E - D_u^+ D_u$$

(see Subsection 13.1.3).

Let us now establish the validity of Equation (10.85). Since $D_s(E - D_s^+ D_s) = D_s - D_s D_s^+ D_s = 0$ (P-a.s.), by (10.86)

$$\int_0^t D_s \, d\tilde{W}_s = \int_0^t D_s D_s^+ B_s \, dW_s \qquad (P\text{-a.s.}).$$

Next,

$$\int_0^t D_s D_s^+ B_s \, dW_s = \int_0^t B_s \, dW_s - \int_0^t (E - D_s D_s^+) B_s \, dW_s$$

Set $x_t = \int_0^t [E - D_s^+ D_s] B_s \, dW_s$. Then

$$Mx_t x_t^* = M \int_0^t (E - D_s D_s^+) B_s B_s^* (E - D_s D_s^+)^* \, ds$$

$$= M \int_0^t (E - D_s D_s^+) D_s D_s^* (E - D_s D_s^+)^* \, ds = 0,$$

since $(E - D_s D_s^+) D_s = 0$. Consequently, $x_t = 0$ (P-a.s.) and, therefore,

$$\int_0^t D_s \, d\tilde{W}_s = \int_0^t D_s D_s^+ B_s \, dW_s = \int_0^t B_s \, dW_s. \quad \square$$

PROOF OF THEOREM 10.3. It will be shown that there is a block matrix

$$D_t = \begin{pmatrix} d_1(t) & d_2(t) \\ 0 & D_2(t) \end{pmatrix},$$

the dimensions of which coincide with those of the corresponding blocks of the matrix

$$B_t = \begin{pmatrix} b_1(t) & b_2(t) \\ B_1(t) & B_2(t) \end{pmatrix},$$

and

$$D_t D_t^* = B_t B_t^*. \tag{10.87}$$

It is clear that (10.87) is equivalent to the system of the matrix equations

$$\begin{aligned} d_1(t) d_1^*(t) + d_2(t) d_2^*(t) &= (b \circ b)(t), \\ d_2(t) D_2^*(t) &= (b \circ B)(t), \\ D_2(t) D_2^*(t) &= (B \circ B)(t). \end{aligned} \tag{10.88}$$

The matrices $D_t, 0 \le t \le T$, with the desired properties can be constructed in the following way. Assume (omitting for simplicity the index t)

$$D_2 = D_2^* = (B \circ B)^{1/2}. \tag{10.89}$$

Then, since the matrix $B \circ B$ is nonsingular, from the second equality in (10.88) we obtain

$$d_2 = (b \circ B)(B \circ B)^{-1/2}. \tag{10.90}$$

Next,

$$d_1 d_1^* = (b \circ b) - (b \circ B)(B \circ B)^{-1}(b \circ B)^*. \tag{10.91}$$

By Lemma 13.2 the matrix $(b \circ b) - (b \circ B)(B \circ B)^{-1}(b \circ B)^*$ is nonnegative definite; as d_1 one can take the matrix

$$d_1 = d_1^* = [(b \circ b) - (b \circ B)(B \circ B)^{-1}(b \circ B)^*]^{1/2}. \tag{10.92}$$

Thus, the block matrix

$$D_t = \begin{pmatrix} d_1(t) & d_2(t) \\ 0 & D_2(t) \end{pmatrix}$$

with the property given by (10.87) has been constructed.

By Lemma 10.4, for the system of equations in (10.62) and (10.63) there is also the representation

$$d\theta_t = [a_0(t) + a_1(t)\theta_t + a_2(t)\xi_t]dt + d_1(t)d\tilde{W}_1(t) + d_2(t)d\tilde{W}_2(t), \tag{10.93}$$

$$d\xi_t = [A_0(t) + A_1(t)\theta_t + A_2(t)\xi_t]dt + D_2(t)d\tilde{W}_2(t), \tag{10.94}$$

where $\tilde{W}_1$ and $\tilde{W}_2$ are new Wiener processes which are mutually independent.

Let us define now a random process $v = (v_t, \mathscr{F}_t^\xi)$, $0 \le t \le T$, which is a solution of the linear stochastic differential equation

$$\begin{aligned} dv_t = \{&[a_0(t) - d_2(t)D_2^{-1}(t)A_0(t)] + [a_1(t) - d_2(t)D_2^{-1}(t)A_1(t)]v_t \\ &+ [a_2(t) - d_2(t)D_2^{-1}(t)A_2(t)]\xi_t\}dt + d_2(t)D_2^{-1}(t)d\xi_t, \qquad v_0 = 0. \end{aligned} \tag{10.95}$$

By the assumptions, (10.89), (10.90), (10.92), and the note to Theorem 4.10 (see Subsection 4.4.6), Equation (10.95) has a unique continuous solution $v = (v_t, \mathscr{F}_t^\xi)$.

Set

$$\tilde{\theta}_t = \theta_t - v_t, \qquad \tilde{\xi}_t = \xi_t - \int_0^t [A_0(s) + A_1(s)v_s + A_2(s)\xi_s]ds. \tag{10.96}$$

By (10.94) and the nonsingularity of the matrices $D_2(t)$,

$$\tilde{W}_2(t) = \int_0^t D_2^{-1}(s)[d\xi_s - (A_0(s) + A_1(s)\theta_s + A_2(s)\xi_s)ds] \tag{10.97}$$

(compare with the proof of Theorem 5.12).

From (10.95)–(10.97) we find

$$d\tilde{\theta}_t = [a_1(t) - d_2(t)D_2^{-1}(t)A_1(t)]\tilde{\theta}_t\,dt + d_1(t)d\tilde{W}_1(t), \tag{10.98}$$

$$d\tilde{\xi}_t = A_t(t)\tilde{\theta}_t\,dt + D_2(t)d\tilde{W}_2(t). \tag{10.99}$$

From the construction of the process $\tilde{\xi} = (\tilde{\xi}_t)$, $0 \leq t \leq T$ (see (10.96)), it follows that $\mathscr{F}_t^{\xi} \supseteq \mathscr{F}_t^{\tilde{\xi}}$. It will be shown that actually the σ-algebras $\mathscr{F}_t^{\xi}$ and $\mathscr{F}_t^{\tilde{\xi}}$ coincide for all t, $0 \leq t \leq T$.

For the proof we shall consider the linear system of equations

$$d\xi_t = [A_0(t) + A_1(t)v_t + A_2(t)\xi_t]dt + d\tilde{\xi}_t, \qquad \xi_0 = \tilde{\xi}_0, \tag{10.100}$$

$$dv_t = [a_0(t) + a_1(t)v_t + a_2(t)\xi_t]dt + d_2(t)D_2^{-1}(t)d\tilde{\xi}_t, \qquad v_0 = 0, \tag{10.101}$$

obtained from (10.95) and (10.96).

This linear system of equations has a unique, strong solution (see Theorem 4.10 and the note to it), which implies $\mathscr{F}_t^{\tilde{\xi}} \supseteq \mathscr{F}_t^{\xi}, 0 \leq t \leq T$, i.e., $\mathscr{F}_t^{\xi} = \mathscr{F}_t^{\tilde{\xi}}$ and

$$\tilde{m}_t = M(\tilde{\theta}_t | \mathscr{F}_t^{\tilde{\xi}}) = M(\tilde{\theta}_t | \mathscr{F}_t^{\xi}).$$

Hence,

$$m_t = M(\theta_t | \mathscr{F}_t^{\xi}) = M[\tilde{\theta}_t + v_t | \mathscr{F}_t^{\xi}] = \tilde{m}_t + v_t \tag{10.102}$$

and

$$\tilde{\theta}_t - \tilde{m}_t = (\theta_t - v_t) - (m_t - v_t) = \theta_t - m_t.$$

From this,

$$\gamma_t = \tilde{\gamma}_t. \tag{10.103}$$

According to Theorem 10.3,

$$\begin{aligned} d\tilde{m}_t = {} & [a_1(t) - d_2(t)D_2^{-1}(t)A_1(t)]\tilde{m}_t\,dt \\ & + \tilde{\gamma}_t A_1^*(t)(D_2(t)D_2^*(t))^{-1}(d\tilde{\xi}_t - A_1(t)\tilde{m}_t\,dt), \end{aligned} \tag{10.104}$$

$$\begin{aligned} \dot{\tilde{\gamma}} = {} & [a_1(t) - d_2(t)D_2^{-1}(t)A_1(t)]\tilde{\gamma}_t + \tilde{\gamma}_t[a_1(t) - d_2(t)D_2^{-1}(t)A_1(t)]^* \\ & - \tilde{\gamma}_t A_1^*(t)(D_2(t)D_2^*(t))^{-1}A_1(t)\tilde{\gamma}_t + d_1(t)d_1^*(t). \end{aligned} \tag{10.105}$$

From this, taking into account that $m_t = \tilde{m}_t + v_t$ and $\gamma_t = \tilde{\gamma}_t$, after some simple transformations we arrive at Equations (10.81) and (10.82) for m_t and γ_t.

The uniqueness of the solution of Equation (10.82) follows from the validity of the analogous equation ((10.105)) and Theorem 10.2. The uniqueness of the solution of Equation (10.81) follows from its linearity, Theorem 4.10 and the note to this theorem. □

10.4 Equations for an almost optimal linear filter for singular $B \circ B$

10.4.1

Consider again the $k + l$-dimensional Gaussian process

$$(\theta_t, \xi_t) = [(\theta_1(t), \dots, \theta_k(t)), (\xi_1(t), \dots, \xi_l(t))], 0 \le t \le T,$$

described by Equations (10.62) and (10.63).

Assume now that the matrix $(B \circ B)(t) = B_1(t)B_1^*(t) + B_2(t)B_2^*(t)$ is singular.[6] In this case Equations (10.81) and (10.82), with the help of which the conditional expectation $m_t = M(\theta_t | \mathscr{F}_t^\xi)$ and the matrix

$$\gamma_t = M[(\theta_t - m_t)(\theta_t - m_t)^*]$$

were defined in the case where the matrix $(B \circ B)(t)$ is positive definite, lose their meaning since the matrix $[(B \circ B)(t)]^{-1}$ entering the right-hand side of these equations does not exist.

If the coefficients of Equations (10.62) and (10.63) are discontinuous, then m_t and γ_t with the singular matrice $(B \circ B)(t)$ are not necessarily continuous time functions and, therefore, are not defined by equations of the type given by (10.81) and (10.82).

In a number of cases equations for m_t and γ_t can be deduced even when $(B \circ B)(t)$ is singular (for example, when the coefficients of Equations (10.62) and (10.63) are constant or sufficiently smooth time functions).

From the point of view of the application, these equations for m_t and γ_t with singular $(B \circ B)(t)$ are not significant because, as a rule, they contain derivatives in the components of the observable process ξ_t or in their linear combination[7] which cannot be computed without great errors in a real situation.

Later, for each $\varepsilon \neq 0$ the processes m_t^ε and γ_t^ε, $0 \le t \le T$, will be constructed, which in a certain sense are close to m_t and γ_t. These processes are defined from equations of the type given by (10.81) and (10.82) for the nonnegative definite matrix $(B \circ B)(t)$, $0 \le t \le T$, and determine the filter which, following the terminology used by experts on "ill-posed" problems, could be called a "regularized" filter.

10.4.2

Let $a_2(t) \equiv 0$, $A_2(t) \equiv 0$. Along with the processes θ_t and ξ_t we shall introduce the process $\xi^\varepsilon = (\xi_t^\varepsilon)$, $0 \le t \le T$, with

$$\xi_t^\varepsilon = \xi_t + \varepsilon \tilde{W}_t, \qquad \varepsilon \neq 0, \tag{10.106}$$

[6] This case arises, for example in the problems of linear filtering of stationary processes by rational spectra (Section 15.3).

[7] See [43], [110], [172].

where $\tilde{W}_t = [\tilde{W}_1(t), \dots, \tilde{W}_l(t)]$, $t \le T$, is a Wiener process independent of (θ_0, ξ_0) and the processes W_1, W_2.

Since $a_2(t) \equiv 0$, $A_2(t) \equiv 0$, then from (10.61), (10.62) and (10.106) it follows that the process $(\theta_t, \xi_t^\varepsilon)$, $0 \le t \le T$, satisfies the system of equations

$$d\theta_t = [a_0(t) + a_1(t)\theta_t]dt + b_1(t)dW_1(t) + b_2(t)dW_2(t), \qquad (10.107)$$

$$d\xi_t^\varepsilon = [A_0(t) + A_1(t)\theta_t]dt + B_1(t)dW_1(t) + B_2(t)dW_2(t) + \varepsilon\, d\tilde{W}_t, \qquad (10.108)$$

solved under the initial conditions θ_0 and $\xi_0^\varepsilon = \xi_0$.

Denote $n_t^\varepsilon = M(\theta_t | \mathscr{F}_t^{\xi^\varepsilon})$, $\gamma_t^\varepsilon = M[(\theta_t - n_t^\varepsilon)(\theta_t - n_t^\varepsilon)^*]$. By Lemma 10.4 and Theorem 10.3, n_t^ε and γ_t^ε are defined by the equations

$$dn_t^\varepsilon = [a_0(t) + a_1(t)n_t^\varepsilon]dt + [(b \circ B)(t) + \gamma_t^\varepsilon A_1^*(t)][(B \circ B)(t) + \varepsilon^2 E]^{-1}[d\xi_t^\varepsilon - (A_0(t) + A_1(t)n_t^\varepsilon)dt], \qquad (10.109)$$

$$\dot{\gamma}_t^\varepsilon = a_1(t)\gamma_t^\varepsilon + \gamma_t^\varepsilon a_1^*(t) + (b \circ b)(t) - [(b \circ B)(t) + \gamma_t^\varepsilon A_1^*(t)][(B \circ B)(t) + \varepsilon^2 E]^{-1}[(b \circ B)(t) + \gamma_t^\varepsilon A_1^*(t)]^* \qquad (10.110)$$

with $n_0^\varepsilon = m_0 = M(\theta_0 | \xi_0)$, and $\gamma_0^\varepsilon = \gamma_0 = M[(\theta_0 - m_0)(\theta_0 - m_0)^*]$ where $E = E_{(l \times l)}$ is a unit matrix.

Let us specify the processes $\lambda_t^\varepsilon = \lambda_t^\varepsilon(\xi)$, $\Delta_t^\varepsilon = \Delta_t^\varepsilon(\tilde{W})$, $0 \le t \le T$, where $\lambda_t^\varepsilon = [\lambda_1^\varepsilon(t), \dots, \lambda_k^\varepsilon(t)]$, $\Delta_t^\varepsilon = [\Delta_1^\varepsilon(t), \dots, \Delta_k^\varepsilon(t)]$, with the help of the following differential equations:

$$d\lambda_t^\varepsilon = [a_0(t) + a_1(t)\lambda_t^\varepsilon]dt + [(b \circ B)(t) + \gamma_t^\varepsilon A_1^*(t)][(B \circ B)(t) + \varepsilon^2 E]^{-1} \times [d\xi_t - (A_0(t) + A_1(t)\lambda_t^\varepsilon)dt], \qquad \lambda_0^\varepsilon = m_0, \qquad (10.111)$$

$$d\Delta_t^\varepsilon = a_1(t)\Delta_t^\varepsilon\, dt + [(b \circ B)(t) + y_t^\varepsilon A_1^*(t)][(B \circ B)(t) + \varepsilon^2 E]^{-1} \times [\varepsilon\, d\tilde{W}_t - A_1(t)\Delta_t^\varepsilon\, dt], \qquad \Delta_0^\varepsilon = 0. \qquad (10.112)$$

It is not difficult to check, by (10.108), (10.111) and (10.112), that for each t, $0 \le t \le T$,

$$n_t^\varepsilon = n_t^\varepsilon(\xi^\varepsilon) = \lambda_t^\varepsilon(\xi) + \Delta_t^\varepsilon(\tilde{W}). \qquad (10.113)$$

Define the matrix $\delta_t^\varepsilon = \|\delta_{ij}^\varepsilon(t)\|_{(k \times k)} = M[(\theta_t - \lambda_t^\varepsilon)(\theta_t - \lambda_t^\varepsilon)^*]$.

Lemma 10.5. *Let* (10.64)–(10.67) *be satisfied. Then for any* t, $0 \le t \le T$:

(1) $M\lambda_t^\varepsilon = M\theta_t$;
(2) $\gamma_{ii}(t) \le \delta_{ii}^\varepsilon(t) \le \gamma_{ii}^\varepsilon(t)$, $\quad i = 1, \dots, k$;
(3) $\gamma_t = \lim_{\varepsilon \to 0} \delta_t^\varepsilon = \lim_{\varepsilon \to 0} \gamma_t^\varepsilon$,
(4) $\lim_{\varepsilon \to 0} M[m_i(t) - \lambda_i^\varepsilon(t)]^2 = 0$, $\quad i = 1, \dots, k$.

PROOF. We have (see (10.106)): $\xi_i^\varepsilon(s) = \xi_i(s) + \varepsilon \tilde{W}_i(s)$. For each $s \le t$, $M[\xi_i(s) | \mathscr{F}_t^{\xi^\varepsilon}]$ is an optimal, in the mean square terms, estimate for $\xi_i(s)$ on the basis of $\{\xi_u^\varepsilon, 0 \le u \le t\}$. Hence,

$$M[\xi_i(s) - M(\xi_i(s) | \mathscr{F}_t^{\xi^\varepsilon})]^2 \le M[\xi_i(s) - \xi_i^\varepsilon(s)]^2 = \varepsilon^2 M(\tilde{W}_i(s))^2 = \varepsilon^2 s \to 0, \qquad \varepsilon \to 0.$$

From this it is not difficult to deduce that for any random variable e_n that is a linear function from $\xi_{t_0}, \xi_{t_1}, \ldots, \xi_{t_n}$,

$$\lim_{\varepsilon \to 0} M[e_n - M(e_n | \mathscr{F}_t^{\xi^\varepsilon})]^2 = 0.$$

Next, if the sequence $(e_n, n = 1, 2, \ldots)$ of the random variables e_n, defined above, has the limit e in the mean square ($e = \text{l.i.m.}_n\, e_n$), then

$$\lim_{\varepsilon \to 0} M[e - M(e | \mathscr{F}_t^{\xi^\varepsilon})]^2 = 0, \tag{10.114}$$

since

$$\begin{aligned} M[e - M(e|\mathscr{F}_t^{\xi^\varepsilon})]^2 &\le 3(M[e - e_n]^2 + M[e_n - M(e_n|\mathscr{F}_t^{\xi^\varepsilon})]^2 \\ &\quad + M[M(e - e_n|\mathscr{F}_t^{\xi^\varepsilon})]^2) \\ &\le 6M[e - e_n]^2 + 3M[e_n - M(e_n|\mathscr{F}_t^{\xi^\varepsilon})]^2, \end{aligned}$$

and, consequently,

$$\overline{\lim_{\varepsilon \to 0}}\, M[e - M(e|\mathscr{F}_t^{\xi^\varepsilon})]^2 \le 6M[e - e_n]^2 \to 0, \qquad n \to \infty.$$

Note now that the components $m_i(t)$, $i = 1, \ldots, k$, of the random vector $m_t = M(\theta_t | \mathscr{F}_t^{\xi})$ are the mean square limits of sequences of random variables of the type $e_n, n = 1, 2, \ldots$. Indeed, if $\mathscr{F}_{t,n}^{\xi} = \sigma\{\omega : \xi_0, \xi_{2^{-n}}, \ldots, \xi_{k \cdot 2^{-n}}, \ldots, \xi_t\}$, then by the theorem of normal correlation (see Chapter 13) the components $m_i^{(n)}(t)$, $i = 1, \ldots, k$, of the vector $m_t^{(n)} = M(\theta_t | \mathscr{F}_{t,n}^{\xi})$ are linearly expressed via $\xi_0, \xi_{2^{-n}}, \ldots, \xi_{k \cdot 2^{-n}}, \ldots, \xi_t$. In this case, according to Theorem 1.5, $m_i^{(n)}(t) \to m_i^{(t)}$ with probability one. But $M[m_i^{(n)}(t)]^4 \le M\theta_i^4(t)$ and is uniform over all n. Hence, by Theorem 1.8, $\lim_{n \to \infty} M(m_i(t) - m_i^{(n)}(t))^2 = 0$. Thus, by (10.114), we have

$$\lim_{\varepsilon \to 0} M[m_i(t) - M(m_i(t) | \mathscr{F}_t^{\xi^\varepsilon})]^2 = 0. \tag{10.115}$$

From the definition of the process ξ^ε (see (10.106)), it follows that for any t, $0 \le t \le T$, there coincide the σ-algebras $\mathscr{F}_t^{\xi, \xi^\varepsilon}$ and $\mathscr{F}_t^{\xi, \tilde{W}}$, from which, employing the independence of the processes (θ_t, ξ_t) and $(\tilde{W}_t)$, $0 \le t \le T$, we find that (P a.s.)

$$M(\theta_t | \mathscr{F}_t^{\xi, \xi^\varepsilon}) = M(\theta_t | \mathscr{F}_t^{\xi, \tilde{W}}) = M(\theta_t | \mathscr{F}_t^{\xi}) = m_t,$$

from which, because of the property of the conditional expectation, we obtain

$$M(m_t | \mathscr{F}_t^{\xi^\varepsilon}) = M[M(\theta_t | \mathscr{F}_t^{\xi, \xi^\varepsilon}) | \mathscr{F}_t^{\xi^\varepsilon}] = n_t^\varepsilon. \tag{10.116}$$

But then, by (10.115) and (10.116),

$$\lim_{\varepsilon \to 0} M[n_i^\varepsilon(t) - m_i(t)]^2 = 0. \tag{10.117}$$

From (10.117) it can easily be deduced that

$$\lim_{\varepsilon \to 0} \gamma_t^\varepsilon = \gamma_t. \tag{10.118}$$

Let us prove now Lemma 10.5(1). For this purpose we shall consider the process $[\theta_t - \lambda_t^\varepsilon]$, defined in accord with (10.107) and (10.111) by the equation

$$[\theta_t - \lambda_t^\varepsilon] = [\theta_0 - m_0] + \int_0^t (a_1(s) - D(s)A_1(s))[\theta_s - \lambda_s^\varepsilon]ds$$

$$+ \sum_{i=1}^2 \int_0^t [b_i(s) + D(s)B_i(s)]dW_i(s),$$

where $D(s) = [b \circ b(s) + \gamma_s A_1^*(s)][B \circ B(s) + \varepsilon^2 E]^{-1}$, from which, obviously,

$$M[\theta_t - \lambda_t^\varepsilon] = \int_0^t (a_1(s) - D(s)A_1(s))M[\theta_s - \lambda_s^\varepsilon]ds, \tag{10.119}$$

and, therefore, $M[\theta_t - \lambda_t^\varepsilon] \equiv 0$.

From the unbiasedness of λ_t ($M\lambda_t^\varepsilon = M\theta_t$) and (10.113) it follows that

$$\begin{aligned}\gamma_{ii}^\varepsilon(t) &= M[\theta_i(t) - n_i^\varepsilon(t)]^2 \\ &= M[\theta_i(t) - \lambda_i^\varepsilon(t)]^2 + M[\Delta_i^\varepsilon(t)]^2 \geq M[\theta_i(t) - \lambda_i^\varepsilon(t)]^2 = \delta_{ii}^\varepsilon(t),\end{aligned}$$

which together with (10.18) proves Lemma 10.5(2)–(3).

Finally,

$$\begin{aligned}\gamma_{ii}(t) &= M[\theta_i(t) - m_i(t)]^2 = M[(\theta_i(t) - \lambda_i^\varepsilon(t)) + (\lambda_i^\varepsilon(t) - m_i(t))]^2 \\ &= \delta_{ii}^\varepsilon(t) - M[\lambda_i^\varepsilon(t) - m_i(t)]^2,\end{aligned}$$

since

$$\begin{aligned}M[(\theta_i(t) - \lambda_i^\varepsilon(t))(\lambda_i^\varepsilon(t) - m_i(t))] &= M[M(\theta_i(t) - \lambda_i^\varepsilon(t) | \mathscr{F}_t^\xi)(\lambda_i^\varepsilon(t) - m_i(t))] \\ &= -M[\lambda_i^\varepsilon(t) - m_i(t)]^2,\end{aligned}$$

which proves Lemma 10.5(4) because $\delta_{ii}^\varepsilon(t) \to \gamma_{ii}(t)$, $\varepsilon \to 0$.

10.4.3

From Lemma 10.5 it follows that, with $a_2(t) \equiv 0$, $A_2(t) \equiv 0$, λ_t^ε can be chosen as the almost optimal estimate θ_t on the basis of ξ_0^t, where the process λ_t^ε, together with γ_t^ε, can be defined from Equations (10.111) and (10.110).

But if $a_2(t) \not\equiv 0$, $A_2(t) \not\equiv 0$, then, by analogy with Equation (10.111), we define m_t^ε as a solution of the equation

$$\begin{aligned}dm_t^\varepsilon &= [a_0(t) + a_1(t)m_t^\varepsilon + a_2(t)\xi_t]dt \\ &\quad + [(b \circ B)(t) + \gamma_t^\varepsilon A_1^*(t)][(B \circ B)(t) + \varepsilon^2 E]^{-1} \\ &\quad \times [d\xi_t - (A_0(t) + A_1(t)m_t^\varepsilon + A_2(t)\xi_t)dt],\end{aligned} \tag{10.120}$$

where $m_0^\varepsilon = m_0$, and γ_t^ε is still to be found from Equation (10.110).

Theorem 10.4. *Let* (10.64)–(10.67) *be satisfied. Then the process* $(m_t^\varepsilon), 0 \le t \le T$, *defined by Equations* (10.120) *and* (10.110), *gives the estimate of the vector* θ_t *on the basis of* ξ_0^t, *and having the following properties*:

$$Mm_t^\varepsilon = M\theta_t,$$

$$\lim_{\varepsilon \to 0} M[m_i^\varepsilon(t) - m_i(t)]^2 = 0, \qquad i = 1, \dots, k,$$

$$\gamma_{ii}(t) \le M[\theta_i(t) - m_i^\varepsilon(t)]^2 \le \gamma_{ii}^\varepsilon(t), \qquad i = 1, \dots, k. \tag{10.121}$$

The matrix of the mean square errors $\Gamma_t^\varepsilon = M[(\theta_t - m_t^\varepsilon)(\theta_t - m_t^\varepsilon)^*]$ *is defined from the equation*

$$\dot{\Gamma}_t^\varepsilon = a^\varepsilon(t)\Gamma_t^\varepsilon + \Gamma_t^\varepsilon(a^\varepsilon(t))^* + \sum_{i=1}^{2} (b_i^\varepsilon(t))(b_i^\varepsilon(t))^* \tag{10.122}$$

with $\Gamma_0^\varepsilon = \gamma_0$ *and*

$$a^\varepsilon(t) = a_1(t) - [(b \circ B)(t) + \gamma_t^\varepsilon A_1^*(t)][(B \circ B)(t) + \varepsilon^2 E]^{-1} A_1(t),$$

$$b_i^\varepsilon(t) = b_i(t) - [(b \circ B)(t) + \gamma_t^\varepsilon A_1^*(t)][(B \circ B)(t) + \varepsilon^2 E]^{-1} B_i(t), \qquad i = 1, 2.$$

Proof. If $a_2(t) \equiv 0$, $A_2(t) \equiv 0$ then, obviously, $m_t^\varepsilon = \lambda_t^\varepsilon$, $0 \le t \le T$, and by Lemma 10.5 the properties given by (10.121) of the estimate m_t are satisfied. To deduce Equation (10.122) in the case under consideration, let us assume $V_t^\varepsilon = \theta_t - \lambda_t^\varepsilon$. Then

$$dV_t^\varepsilon = a^\varepsilon(t)V_t^\varepsilon\, dt + \sum_{i=1}^{2} b_i^\varepsilon(t) dW_i(t)$$

and, by the Ito formula,

$$V_t^\varepsilon(V_t^\varepsilon)^* = V_0^\varepsilon(V_0^\varepsilon)^* + \int_0^t \left[a^\varepsilon(s)V_s^\varepsilon(V_s^\varepsilon)^* + V_s^\varepsilon(V_s^\varepsilon)^*(a^\varepsilon(s))^* + \sum_{i=1}^{2} b_i^\varepsilon(s)(b_i^\varepsilon(s))^* \right] ds$$

$$+ \int_0^t V_s^\varepsilon \left(\sum_{i=1}^{2} b_i^\varepsilon(s) dW_i(s) \right)^* + \int_0^t \sum_{i=1}^{2} (b_i^\varepsilon(s) dW_i(s))(V_s^\varepsilon)^*.$$

From this, after averaging, we obtain for $\Gamma_t = MV_t^\varepsilon(V_t^\varepsilon)^*$ Equation (10.122).

Let now $a_2(t) \not\equiv 0$, $A_2(t) \not\equiv 0$. Introduce the processes $v = (v_t)$, $\tilde{\xi} = (\tilde{\xi}_t)$, $0 \le t \le T$, where

$$v_t = \int_0^t [a_1(s)v_s + a_2(s)\xi_s] ds, \tag{10.123}$$

$$\tilde{\xi}_t = \xi_t - \int_0^t [A_1(s)v_s + A_2(s)\xi_s] ds, \tag{10.124}$$

and set $\tilde{\theta}_t = \theta_t - v_t$. Then, from (10.62), (10.63), (10.123), and (10.124) we find

$$d\tilde{\xi}_t = [A_0(t) + A_1(t)\tilde{\theta}_t]dt + B_1(t)dW_1(t) + B_2(t)dW_2(t) \tag{10.125}$$

$$d\tilde{\theta}_t = [a_0(t) + a_1(t)\tilde{\theta}_t]dt + b_1(t)dW_1(t) + b_2(t)dW_2(t), \tag{10.126}$$

with $\tilde{\theta}_0 = \theta_0$, $\tilde{\xi}_0 = \xi_0$.

If one estimates $\tilde{\theta}_t$ on the basis of ξ_0^t, then, according to (10.120) (with $a_2(t) \equiv 0$, $A_2(t) = 0$), the corresponding estimate $\tilde{m}_t$ is given by the equation

$$d\tilde{m}_t^\varepsilon = [a_0(t) + a_1(t)\tilde{m}_t^\varepsilon]dt + [b \circ B(t) + \gamma_t^\varepsilon A_1^*(t)](B \circ B(t) + \varepsilon^2 E)^{-1}[d\tilde{\xi}_t - (A_0(t) + A_1(t)m_t^\varepsilon)dt],$$
$$\tilde{m}_0^\varepsilon = m_0. \tag{10.127}$$

From (10.123) and (10.124) it is not difficult to deduce (compare with the proof of Theorem 10.3), that the σ-algebras $\mathscr{F}_t^\xi$ and $\mathscr{F}_t^{\tilde{\xi}}$, $0 \le t \le T$, coincide. Hence, denoting $\tilde{m}_t = M(\tilde{\theta}_t|\mathscr{F}_t^{\tilde{\xi}})$, we find that

$$m_t = M(\theta_t|\mathscr{F}_t^\xi) = M(\tilde{\theta}_t + v_t|\mathscr{F}_t^\xi) = M(\tilde{\theta}_t|\mathscr{F}_t^\xi) + v_t = \tilde{m}_t + v_t.$$

Set $m_t^\varepsilon = \tilde{m}_t^\varepsilon + v_t$. Then $m_t - m_t^\varepsilon = \tilde{m}_t - \tilde{m}_t^\varepsilon$ and, therefore, the estimate m_t has the properties given by (10.121).

Equation (10.120) follows from the equality $m_t^\varepsilon = \tilde{m}_t^\varepsilon + v_t$ and from (10.127), (10.123), and (10.124). Equation (10.122) holds as well for the case $a_2(t) \not\equiv 0$, $A_2(t) \not\equiv 0$, since $\theta_t - m_t^\varepsilon = \tilde{\theta}_t - \tilde{m}_t^\varepsilon = \tilde{V}_t^\varepsilon$ and it is not difficult to check that $M\tilde{V}_t^\varepsilon(\tilde{V}_t^\varepsilon)^* = MV_t^\varepsilon(V_t)^*$, $0 \le t \le T$. □

Notes and references

10.1–10.3. Equations (10.10) and (10.11) determining the evolution of optimal linear filter have been obtained by Kalman and Bucy [78]. See also Chapter 9 in Stratonovich [147].

The martingale deduction of Equations (10.10) and (10.11) presented here is probably new. The proof of Lemma 10.1 is due to the authors. Another proof of Lemma 10.1 has been given in Ruymgaart [142].

10.4. The equations for an almost optimal linear filter in the case of singularity of the matrices $B \circ B$ have been first given here.

Bibliography

[1] Albert A., Sittler R. W., A method for computing least squares estimators that keep up with the data. *SIAM J. Control* **3** (1965), 384–417.

[2] Anderson T., *Introduction to the Multivariate Statistical Analysis*. Russian transl., Fizmatgiz, Moscow, 1963.

[3] Aoki M., *Optimization of Stochastic Systems*. Russian transl., "Nauka," Moscow, 1971.

[4] Arato M., Calculation of confidence bounds for the damping parameter of the complex stationary Gaussian Markov process. *Teoria Verojatn. i Primenen.* XIII, **1**, (1968), 326–333.

[5] Arato M., Kolmogorov A. N., Sinai Ja. G., On parameter estimation of a complex stationary Gaussian Markov process. *DAN SSSR* 146, **4** (1962), 747–750.

[6] Astrom K. I., Optimal control of Markov processes with incomplete state information. *J. Math. Anal. Appl.* **10** (1965), 174–205.

[7] Balakrishnan A. V., *Stochastic Differential Systems* I, Lecture notes. Dept. of Systems Science, UCLA, 1971.

[8] Balakrishnan, A. V., A martingale approach to linear recursive state estimation *SIAM J. Control* **10** (1972), 754–766.

[9] Bellman, R., Cook K. L., *Difference–Differential Equations*. Russian transl., "MIR," Moscow, 1967.

[10] Bensoussan A., *Filtrage Optimal des Systemes Lineaires*. Dunod, Paris, 1971.

[11] Blackwell D., Dubins L., Merging of opinions with increasing information. *AMS* **33** (1962), 882–886.

[12] Blumental R. M., Getoor R. K., *Markov Processes and Potential Theory*. Academic Press, N.Y., 1968.

[13] Bolshakov I. A., Repin V. G., Problems of nonlinear filtering. *Avtomatika i telemekhanika* XXII, **4** (1961), 466–478.

[14] Breiman L., *Probability*. Addison–Wesley Publ. Company, Reading 1968.

[15] Bucy R. S., Nonlinear filtering theory. *IEEE Trans. Automatic Control* AC-10 (1965), 198.

[16] Bucy R. S., Joseph P. D., *Filtering for Stochastic Processes with Application to Guidance*. Interscience, N.Y., 1968.

[17] Beckenbach A., Bellman R., *Inequalities*. Russian transl. "MIR," Moscow, 1965.

[18] Ventsel A. D., Additive functionals of a multivariate Wiener process. *DAN SSSR* 130, **1** (1961), 13–16.

[19] Ventsel A. D., On equations of the conditional Markov processes theory. *Teoria Verojatn. i Primenen.* X, **2** (1965), 390–393.

[20] Wiener N., Differential space. *J. Math. and Phys.* 58 (1923), 131–174.

[21] Wiener N., *Extrapolation, Interpolation and Smoothing of Stationary Time Series*, J. Wiley & Sons, N.Y., 1949.

[22] Wolfowitz J., On sequential binomial estimation. *AMS* **17** (1946), 489–493.

[23] Wolfowitz J., The efficiency of sequential games estimates and Wald's equation for sequential processes. *AMS* **18** (1947), 215–230.

[24] Wonham W. M., *Stochastic Problems in Optimal Control*. Tech. report 63-14, Research Institute for Advanced Study, Baltimore, 1963.

[25] Wonham W. M., Some applications of stochastic differential equations to optimal nonlinear filtering. *SIAM J. Control*, **2** (1965), 347–369.

[26] Wonham W. M., On the separation theorem of stochastic control. *SIAM J. Control* **6** (1968), 312–326.

[27] Wonham W. M., On a matrix Riccati equation of stochastic control. *SIAM J. Control* **6** (1968), 681–697.

[28] Galchuk L. I., *On One Representation of Jump Processes*. Sovetsko–Japonsky Simpozium po Teorii Verojatnostei, Khabarovsk, 1969.

[29] Galchuk L. I., Filtering of jump Markov processes. *UMN* XXV, **5** (1970), 237–238.

[30] Gantmacher F. R., *The Theory of Matrices*. "Nauka," Moscow, 1967.

[31] Girsanov I. V., On transformation of one class of random processes with the help of absolutely continuous substitution of the measure. *Teoria Verojan. i Primenen.* V, **3** (1960), 314–330.

[32] Gikhman I. I., On the theory of stochastic differential equations of random processes. *Urk. math. journal* 2, **3** (1950), 45–69.

[33] Gikhman I. I., Dorogovtsev A. Ja., On stability of solutions of stochastic differential equations. *Ukr. math. journal* 17, **6** (1965), 3–21.

[34] Gikhman I. I., Skorokhod A. V., *Introduction to Random Processes Theory*. "Nauka," Moscow, 1965.

[35] Gikhman I. I., Skorokhod A. V., On densities of probability measures on function spaces. *UMN* **21** (1966), 83–152.

[36] Gikhman I. I., Skorokhod A. V., *Stochastic Differential Equations*. "Naukova dumka," Kiev, 1968 (Ukranian).

[37] Gikhman I. I., Skorokhod A. V., *The Theory of Random Processes*. Tom I. "Nauka," Moscow, 1971.

[38] Glonti O. A., Sequential filtering and interpolation of Markov chain components. *Litovsky matem. sbornik* IX, **2** (1969), 263–279 (Russian).

[39] Glonti O. A., Extrapolation of Markov chain components. *Litovsky matem. sbornik* IX, **4** (1969), 741–754 (Russian).

[40] Glonti O. A., Sequential filtering of Markov chain components with singular diffusion matrices. *Teoria Verojatn. i Primenen.* XV, **4** (1970), 736–740.

[41] Grigelionis B., On stochastic equations of nonlinear filtering of random processes. *Litovsky matem. sbornik* XII, **4** (1972) (Russian).

[42] Grigelionis B., On the structure of densities of measures corresponding to random processes. *Litovsky matem. sbornik* XIII, **1** (1973) (Russian).

[43] Gulko F. B., Novoseltseva Zh. A., Solution of nonstationary problems of filtering and prediction by simulation. *Avtomatika i telemekhanika* **4** (1966), 122–141.

[44] Dashevsky M. L., Liptser R. S., Application of conditional semi-invariants in the problems of nonlinear filtering of Markov processes. *Avtomatika i telemekhanika* **6** (1967), 63–74.

[45] Dashevsky M. L., The method of semi-invariants in the problems of nonlinear filtering of Markov processes. *Avtomatika i telemekhanika* **7** (1968), 24–32.

[46] Doob J. L., *Probability Processes*. Russian transl., IL. Moscow, 1956.

[47] Dynkin Ye. B., *Markov Processes*. Fizmatgiz, Moscow, 1963.

[48] Djachkov A. G., Pinsker M. S., On optimal linear transmission through a memoryless Gaussian channel with complete feedback. *Problemy peredachi informatsii* 7, **2** (1971), 38–46.

[49] Djuge D., *Theoretical and Applied Statistics*. "Nauka," Moscow, 1972.

[50] Yershov M. P., Nonlinear filtering of Markov processes. *Teoria Verojatn. i Primenen.* XIV, **4** (1969), 757–758.

[51] Yershov M. P., Sequential estimation of diffusion processes. *Teoria Verojatn. i Primenen.* XV, **4** (1970), 705–717.

[52] Yershov M. P., On representations of Ito processes. *Teoria Verojatn. i Primenen.* XVII, **1** (1972), 167–172.

[53] Yershov M. P., On absolute continuity of measures corresponding to diffusion type processes. *Teoria Verojatn. i Primenen.* XVII, **1** (1972), 173–178.

[54] Yershov M. P., *Stochastic Equations*. Proc. Second Japan-USSR Sympos. Probab. Theory, Kyoto, I (1972), 101–106.

[55] Zhykovsky Ye. L., Liptser R. S., On recursive computing schemes of normal solutions of linear algebraic equations. *JVM i MF* 12, **4** (1972), 843–857.

[56] Zigangirov K. Sh., Transmission of messages through a binary Gaussian channel with feedback. *Problemy peredachi informatsii* 3, **2** (1967), 98–101.

[57] Ibragimov I. A., Rozanov Yu. A., *Gaussian Random Processes*. "Nauka," Moscow, 1970.

[58] Ibragimov I. A., Khasminsky R. Z., Information inequalities and supereffective estimates. *DAN SSSR* 204, **6** (1972).

[59] Ito K., Stochastic integrals. *Proc. Imp. Acad. Tokyo* **20** (1944), 519–524.

[60] Ito K., On one formula on stochastic differentials. *Matematika. Sbornik perevodov inostr. statei.* 3: **5** (1959), 131–141.

[61] Ito K., McKean G., Diffusion processes and their trajectories. (Russian transl.), "MIR," Moscow, 1968.

[62] Ito K., Nisio M., On stationary solutions of stochastic differential equations. *J. Math. Kyoto Univ.* 4, **1** (1964), 1–79.

[63] Kagan A. M., Linnik Yu. V., Rao S. R., Characterization problems of mathematical statistics. "Nauka," Moscow, 1972.

[64] Kadota T. T., Nonsingular Detection and Likelihood Ratio for Random Signals in White Gaussian Noise. *IEEE Trans. Inform. Theory* IT-16 (1970), 291–298.

[65] Kadota T. T., Zakai M., Ziv I., Mutual information of the white Gaussian channel with and without feedback. *IEEE Trans. Inform. Theory* IT-17, **4** (1971), 368–371.

[66] Kadota T. T., Shepp L. A., Conditions for the absolute continuity between a certain pair of probability measures. *Z. Wahrscheinlickkeitstheorie verw. Gebiete* 16, **3** (1970), 250–260.

[67] Kailath T., An innovations approach to least-squares estimation, Parts I, II. *IEEE Trans. Automatic Control* AC-13 (1968), 646–660.

[68] Kailath T., The innovations approach to detection and estimation theory. *Proc. IEEE* **58** (1970), 680–695.

[69] Kailath T., The structure of Radon–Nykodym derivatives with respect to Wiener and related measures. *AMS* **42** (1971), 1054–1067.

[70] Kailath T., Geesey R., An innovations approach to least-squares estimation, Part IV. *IEEE Trans. Automatic Control* AC-16 (1971), 720–727.

[71] Kailath T., Zakai M., Absolute continuity and Radon–Nykodym derivatives for certain measures relative to Wiener measure. *AMS* 42, **1** (1971), 130–140.

[72] Kalachev M. G., Analytical calculation of the Kalman–Bucy filter in one multivariate problem of filtering. *Avtomatika i telemekhanika* **1** (1972), 46–50.

[73] Kalachev M. G., Petrovsky A. M., Repeated differentiation of the signal with bounded spectrum. *Avtomatika i telemekhanika* **3** (1972), 28–34.

[74] Kallianpur G., Striebel C., Estimation of stochastic systems: Arbitrary system process with additive white noise observation errors. *AMS* **39** (1968), 785–801.

[75] Kallianpur G., Striebel C., Stochastic differential equations occurring in the estimation of continuous parameter stochastic processes. *Teoria Verojatn. i Primenen.* XIV, **4** (1969), 597–622.

[76] Kalman R. E., A new approach to linear filtering and prediction problems. *J. Basic. Eng.* **1** (1960), 35–45.

[77] Kalman R. E., Contributions to the theory of optimal control. *Bol. Soc. Mat. Mexicana* **5** (1960), 102–119.

[78] Kalman R. E., Bucy R. S., New results in linear filtering and the prediction theory. Russian transl. *Tekhnicheskaja mekhanika* 83, ser. D, **1** (1961), 123.

[79] Kalman R., Falb P., Arbib M., *Mathematical System Theory*. Russian translation, "MIR," Moscow, 1971.

[80] Cameron R. H., Martin W. T., Transformation of Wiener integrals under a general class of linear transformations. *Trans. Amer. Math. Soc.* **58** (1945), 184–219.

[81] Cameron R. H., Martin W. T., Transformation of Wiener integrals by nonlinear transformation. *Trans. Amer. Math. Soc.* **66** (1949), 253–283.

[82] Kitsul P. I., Nonlinear filtering of continuous and discrete observations. Papers presented at the 2nd All-Union conference on statistical methods of the control theory. Sbornik "Adaptatsia, samoorganizatsia," Tashkent. 1970, 52–57.

[83] Kitzul P. I., On continuously-discrete filtering of Markov diffusion type processes. *Avtomatika i telemekhanika* **11** (1970), 29–37.

[84] Kitsul P. I., A problem of optimal linear filtering. *Avtomatika i telemekhanika* **11** (1971), 46–52.

[85] Clark I. M. C., The representation of functionals of Brownian motion by stochastic integrals. *AMS* 41, **4** (1970), 1282–1295.

[86] Kolmogorov A. N., *The Foundation of Probability Theory*. ONTI, Moscow-Leningrad, 1936.

[87] Kolmogorov A. N., Interpolation and extrapolation of stationary random sequences. *Izv. AN SSSR*, ser. matem. 5, **5** (1941).

[88] Kolmogorov A. N., *Theory of Information Transmission*. Izd-vo AN SSSR, 1956.

[89] Kolmogorov A. N., Fomin S. V., *Elements of the Theory of Functions and Functional Analysis*. "Nauka," Moscow, 1968.

[90] Cramer G., *Mathematical Statistics Methods*. Russian transl., IL, Moscow, 1948.

[91] Cramer G., Lidbetter M., *Stationary Random Processes*. Russian transl., "MIR," Moscow, 1969.

[92] Krassovsky N. N., Lidsky A. A., Analytical design of regulators in systems with random properties. III. Optimal regulation in linear systems—The minimal meansquare error. *Avtomatika i telemekhanika* **11** (1961), 1425–1431.

[93] Krylov N. V., On Ito stochastic integral equations. *Teoria Verojatn. i Primenen.* 14, **2** (1969), 340–348.

[94] Kunita H., Asymptotic behavior of the nonlinear filtering errors of Markov processes. *J. of Multivariate Analysis* 1, **4** (1971), 365–393.

[95] Kunita H., Watanabe Sh., On square-integrable martingales. *Matematika, Sbornik perevodov inostr. statei*, 15: **1** (1971), 66–102.

[96] Courrège Ph., *Intégrales Stochastiques et Martingales de Carré Intégrable*. Seminaire Brelot–Choquet–Deny. 7-e année (1962/63).

[97] Courrège Ph., Intégrales stochastiques associées à une martingale de carré intégrable. *C. R. Acad. Sci.* **256** (1963), 867–870.

[98] Kushner H. J., On the dynamical equations of conditional probability density functions, with applications to optimal stochastic control theory. *J. Math. Anal. Appl.* **8** (1964), 332–344.

[99] Kushner H. J., Dynamical equations for optimal nonlinear filtering. *J. Differential Equations* **3** (1967), 179–190.

[100] Levy P., *Stochastic Processes and Brownian Motion*. "Nauka," Moscow, 1972.

[101] Levin B. R. *Theoretical Foundations of Statistical Radio Engineering*. "Sov. radio.", Moscow, kn. 1, 1966, kn. 2, 1968.

[102] Leman E., *Testing of Statistical Hypotheses*. "Nauka," Moscow, 1964.

[103] Leonov V. P., Shiryayev A. N., Methods for calculating semi-invariants. *Teoria Verojatn. i Primenen.* IV, **2** (1959), 342–355.

[104] Letov A. M., Analytical design of regulators. I–IV. *Avtomatika i telemekhanika* **4** (1960), 436–441; **5** (1960), 561–568; **6** (1960), 661–665; **4** (1961), 425–435.

[105] Lie R., Optimal estimates. Definition of Characteristics. Control. Russian trans., "Nauka," Moscow, 1966.

[106] Linnik Yu. V., Statistical problems with nuisance parameters. "Nauka," Moscow, 1966.

[107] Liptser R. S., The comparison of nonlinear and linear filtering of Markov processes. *Teoria Verojatn. i Primenen.* XI, **3** (1966), 528–533.

[108] Liptser R. S., On filtering and extrapolation of the components of diffusion type Markov processes. *Teoria Verojatn. i Primenen.* XII, 4 (1967), 764–765.

[109] Liptser R. S., On extrapolation and filtering of some Markov processes I. *Kibernetika* **3** (1968), 63–70.

[110] Liptser R. S., On extrapolation and filtering of some Markov processes II. *Kibernetika* **6** (1968), 70–76.

[111] Liptser R. S., Shiryayev A. N., Nonlinear filtering of diffusion type Markov processes. *Trudy matem. in-ta im. V. A. Steklova AN SSSR* **104** (1968), 135–180.

[112] Liptser R. S., Shiryayev A. N., On filtering, interpolation and extrapolation of diffusion type Markov processes with incomplete data. *Teoria Verojatn. i Primenen.* XIII, **3** (1968), 569–570.

[113] Liptser R. S., Shiryayev A. N., Extrapolation of multivariate Markov processes with incomplete data. *Teoria Verojatn. i Primenen.* XIII, **1** (1968), 17–38.

[114] Liptser R. S., Shiryayev A. N., On the cases of effective solving the problems of optimal nonlinear filtering, interpolation, and extrapolation. *Teoria Verojatn. i Primenen.* XIII, **3** (1968), 570–571.

[115] Liptser R. S., Shiryayev A. N., Nonlinear interpolation of the components of diffusion type Markov processes (forward equations, effective formulas). *Teoria Verojatn. i Primenen.* XIII, **4** (1968), 602–620.

[116] Liptser R. S., Shiryayev A. N., Interpolation and filtering of the jump component of a Markov process. *Izv. AN SSSR*, ser. matem. 33, **4** (1969), 901–914.

[117] Liptser R. S., Shiryayev A. N., On the densities of probability measures of diffusion type processes. *Izv. AN SSSR*, ser. matem. 33, **5** (1969), 1120–1131.

[118] Liptser R. S., Shiryayev A. N., On absolute continuity of measures corresponding to diffusion type processes with respect to a Wiener measure. *Izv. AN SSSR*, ser. matem. 36, **4** (1972), 874–889.

[119] Liptser R. S., Shiryayev A. N., *Statitistics of Conditionally Gaussian Random Sequences.* Proc. Sixth Berkeley Sympos. Math. Statistics and Probability (1970), Vol. II, Univ. of Calif. Press, 1972, 389–422.

[120] Loève M., *Probability Theory.* Russian transl., IL, Moscow, 1962.

[121] Laidain M., Test entre deux hypotheses pour un processus defini par une equation differentielle stochastique. *Rev. Cethedec* 8, **26** (1971), 111–121.

[122] Laning G. H., Battin R. G., *Random Processes in Automatic Control Problems.* Russian transl., IL., Moscow, 1958.

[123] McKean G., *Stochastic Integrals.* Russian transl., "MIR," Moscow, 1972.

[124] Marsaglia G., Conditional means and covariance of normal variables with singular covariance matrix. *J. Amer. Statist. Assoc.* 59, **308** (1964), 1203–1204.

[125] Meditch J. S., *Stochastic Optimal Linear Estimation and Control*, McGraw-Hill, N.Y., 1969.

[126] Meyer P. A., *Probabilitiés et Potentiel.* Herman, Paris, 1966.

[127] Nahi N. E., *Estimation Theory and Applications*, J. Wiley & Sons, N.Y., 1969.

[128] Nevelson M. B., Khasminsky R. Z., *Stochastic Approximation and Recursive Estimation.* "Nauka," Moscow, 1972.

[129] Neveu J., *Mathematical Foundations of Probability Theory.* Russian transl., "MIR," Moscow, 1969.

[130] Neveu J., *Martingales, Notes Partielles d'un Course de 3-ème Cycle.* Paris, 1970–1971.

[131] Novikov A. A., Sequential estimation of the parameters of diffusion-type processes. *Teoria Verojatn. i Primenen.* XVI, **2** (1971), 394–396.

[132] Novikov A. A., On stopping times of a Wiener process. *Teoria Verojatn. i Primenen.* XVI, **3** (1971), 548–550.

[133] Novikov A. A., On an identity for stochastic integrals. *Teoria Verojatn. i Primenen.* XVII, **4** (1972), 761–765.

[134] Prokhorov Yu. V., Convergence of random processes and limit theorems of probability theory. *Teoria Verojatn. i Primenen.* I, **2** (1956), 177–238.

[135] Prokhorov Yu. V., Rozanov Yu. A., *Probability Theory. Foundations. Limit Theorems. Random Processes.* "Nauka," Moscow, 1967.

[136] Pugachev V. S., *The Theory of Random Functions and its Application to Automatic Control Problems.* Fizmatgiz, Moscow, 1962.

[137] Rao, C. M., On decomposition theorems of Meyer. *Math. Scandinavica* 24, **1** (1969), 66–78.

[138] Rao C. R., *Linear Statistical Methods and Their Applications.* Russian transl., "Nauka," Moscow, 1968.

[139] Rozanov Yu. A., *Stationary Random Processes.* Fizmatgiz, Moscow, 1963.

[140] Rozovsky B. L., Stochastic partial differential equations arising in nonlinear filtering problems. *UMN* XXVII, **3** (1972), 213–214.

[141] Rozovsky B. L., Shiryayev A. N., On infinite systems of stochastic differential equations arising in the theory of optimal nonlinear filtering. *Teoria Verojatn. i Primenen.* XVII, **2** (1972), 228–237.

[142] Ruymgaart P. A., A note of the integral representation of the Kalman–Bucy estimate. *Indag. Math.* 33, **4** (1971), 346–360.

[143] Siegmund D., Robbins H., Wendel J., The limiting distribution of the last time $S_n \geq n\varepsilon$. *Proc. Nat. Acad. Sci.* 62, **1** (1968).

[144] Skorokhod A. V., The investigation of a random processes theory. *Izd-vo Kievsk. univ-ta*, 1961.

[145] Skorokhod A. V., *Random Processes with Independent Increments.* "Nauka," Moscow, 1964.

[146] Stratonovich R. L., Conditional Markov processes. *Teoria Verojatn. i Primenen.* V, **2** (1960), 172–195.

[147] Stratonovich R. L., Conditional Markov processes and their applications to the optimal control theory. *Izd-vo MGU*, 1966.

[148] Striebel C. T., Partial differential equations for the conditional distribution of a Markov process given noisy observations. *J. Math. Anal. Appl.* **11** (1965), 151–159.

[149] Tikhonov A. N., On stability of the algorithms for the solution of singular systems of linear algebraic equations. *JVM i MF* 5, **4** (1965), 718–722.

[150] Turin G., *Lectures on Digital Communication.* Russian transl., "MIR," Moscow, 1972.

[151] Whittle P., *Prediction and Regulation*, Van Nostrand, London, 1963.

[152] Feldbaum A. A., *The Foundations of the Theory of Optimal Automatic Systems.* "Nauka," Moscow, 1966.

[153] Ferguson T. S., *Mathematical Statistics.* Academic Press, N.Y., 1967.

[154] Friedman A., *Partial Differential Equations of the Parabolic Type.* Russian transl., "MIR," Moscow, 1968.

[155] Frost P., Kailath T., An innovations approach to least-squares estimation, Part III. *IEEE Trans. Automatic Control* AC-16 (1971), 217–226.

[156] Fujisaki M., Kallianpur G., Kunita H., Stochastic differential equations for the nonlinear filtering problem. *Osaka J. Math.* 9, **1** (1972), 19–40. (Russian transl.: *Matematika. Sbornik perevodov inostr. statei*, 17: **2** (1973), 108–128).

[157] Khasminsky R. Z., *Stability of Differential Equations Systems Given Random Disturbances of Their Parameters*. "Nauka," Moscow, 1969.

[158] Hitsuda M., Representation of Gaussian processes equivalent to Wiener processes. *Osaka J. Math.* **5** (1968), 299–312.

[159] Tsypkin Ya. Z., *Self-adaptive and Learning Systems*. "Nauka," Moscow, 1968.

[160] Chow Y. S., Robbins H., Siegmund D., *Great Expectations: The Theory of Optimal Stopping*. Houghton Mifflin Company, Boston, 1971.

[161] Shalkwijk J. P. M., Kailath T., A coding scheme for additive noise channels with feedback, Part I. *IEEE Trans. Inform. Theory* IT-12 (1966), 172–182.

[162] Shatashvili A. D., Nonlinear filtering for the solution of some stochastic differential equations. *Kibernetica* **3** (1970), 97–102.

[163] Shepp L. A., Radon–Nykodym derivatives of Gaussian measures. *AMS* **37** (1966), 321–354.

[164] Shiryayev A. N., Problems of spectral theory of higher moments I. *Teoria Verojatn. i Primenen.* V, **3** (1960), 293–313.

[165] Shiryayev A. N., On stochastic equations in the theory of conditional Markov processes. *Teoria Verojatn. i Primenen.* XI, **1** (1966), 200–206.

[166] Shiryayev A. N., Stochastic equations of nonlinear filtering of jump Markov processes. *Problemy peredachi informatsii.* II, **3** (1966), 3–22.

[167] Shiryayev A. N., New results in the theory of controlled random processes. *Trans. 4th Prague Confer. Inform. Theory* (1965), Prague, 1967, 131–203.

[168] Shiryayev A. N., Studies in the statistical sequential analysis. *Matem. zametki* 3, **6** (1968), 739–754.

[169] Shiryayev A. N., *Statistical Sequential Analysis*. "Nauka," Moscow, 1969.

[170] Shiryayev A. N., *Sur les Équations Stochastiques aux Dérivées Partielles*. Actes Congrès Intern. Math., 1970.

[171] Shiryayev A. N., Statistics of diffusion type processes. *Proc. Second Japan-USSR Sympos. Probab. Theory*, I (1971), 69–87.

[172] Yaglom A. M., Introduction to the theory of stationary random functions. *UMN* 7, 5 (1952), **3**–168.

[173] Jazwinski A. H., *Stochastic Processes and Filtering Theory*. Academic Press, N.Y., 1970.

[174] Yamada T., Watanabe Sh., On the uniqueness of solution of stochastic differential equations. *J. Math. Kyoto Univ.* 11, **1** (1971), 155–167.

[175] Yashin A. I., Filtering of a jump Markov process with unknown probability characteristics given additive noise. *Avtomatika i telemkhanika* **12** (1968), 25–30.

[176] Yashin A. I., On distinguishing jump parameters of multivariate processes. *Avtomatika i telemekhanika* **10** (1969), 60–67.

[177] Boel R., Varaiya P., Wong E., Martingles on jump processes, III. Representation results. *SIAM J. Control* vol. 13, **5** (1975), 999–1060.

[178] Bremaud P., *A Martingale Approach to Point Processes*, Univ. Calif. Berkeley, ERL Memo M 345, Aug. 1972.

[179] Bremaud P., An extension of Watanabe's theorem of characterization of a Poisson process over the positive real half line. *J. Appl. Probab.* 12, **2** (1975), 369–399.

[180] Bremaud P., Estimation de l'état d'une file d'attente et du temps de panne d'une

machine par la methode des semi-martingales. *Adv. Appl. Probab.* **7** (1975), 845–863.

[181] Bremaud P., La methode des semi-martingales en filtrage quand l'observation est un processus ponctuel marque (Preprint).

[182] Van Schuppen I. L., *Filtering for Counting Process.* a *Martingale Approach.* Proc. 4th Symp. Nonlinear Estimation and Appl., San Diego, Calif., 1973.

[183] Grigelionis B., On nonlinear filtering theory and absolute continuity of measures, corresponding to stochastic processes. In Lecture Notes in Mathematics 330, Springer-Verlag, Berlin, Heidelberg, New York, 1973.

[184] Grigelionis B., On mutual information for locally unbounded random processes. *Litovsky matem. sbornik* XIV (1974) (Russian).

[185] Grigelionis B., Random point processes and martingales. *Litovsky matem. sbornik.* XV, **4** (1975) (Russian).

[186] Dellasheric C., Integrales stochastiques par rapport aux processes de Wiener et de Poisson, Seminare de Probabilitie: VIII. In Lecture Notes in Mathematics **381**, Springer-Verlag, Berlin, Heidelberg, New York, 1974.

[187] Dellasherie C., *Capacities and Random Processes.* Russian transl., "MIR," Moscow, 1975.

[188] Doleans-Dade C., Quelques applications de la formula de changement de variables pour les semimartingales locales. *Z. Wahrscheilichkeitstheorie und verw. Gebiete.* **16** (1970), 181–194.

[189] Davis M. H. A., *The Representation of Martingales of Jump Processes.* Res. Rep. 74/78, Imperial Coll., London, 1974.

[190] Jacod J., Multivariate Point Processes: Predictable Projection, Radon–Nikodym Derivatives, Representation of Martingales. *Z. Wahrscheinlichkeitstheorie und verw. Gebiete* **31**, (1975), 235–253.

[191] Jacod J., Un theoreme de representatio pour les martingales discontinues. *Z. Wahrscheinlichkeitstheorie und verw. Gebiete*, **34** (1976), 225–244.

[192] Jacod J., *A General Theorem of Representation for Martingales.* Dep. Math. Inform. Univ. de Rennes, 1976.

[193] Jacod J., Memin J., *Caracteristiques Locales et Conditions de Continuity Absolute Pour les Semi-Martingales.* Dep. Math. Inform. Univ. de Rennes, 1975.

[194] Yor M., Representation integrale des martingales. Etude de destibution estremales, *Z. Wahrscheilichkeitstheorie und verw. Gebiete.*

[195] Kabanov Yu. M., Representation of the functionals of Wiener processes and Poisson processes as stochastic integrals. *Teoria Verojatn. i Primenen.* XVIII, **2** (1973), 376–380.

[196] Kabanov Yu. M., Integral representations of the functional of processes with independent increments. *Teoria Verojatn. i Primenen.* XIX, **4** (1974), 889–893.

[197] Kabanov Yu. M., Channel capacity of a Poisson type channel. *Teoria Verojatn. i Primenen.*

[198] Kabanov Yu. M., Liptser R. S., Shiryayev A. N., Martingale methods in the point processes theory. Trudy shcoly-seminara po teorii sluchainykh protsessov. II, *Druskeninkai*, 25–30 November, 1974, *Vilnius* 1975, 296–353.

[199] Kabanov Yu. M., Liptser R. S., Shiryayev A. N., Criteria for absolute continuity of measures of multivariate point processes. III—Sovetsko–Yaponski symposium, Tashkent, 1975, Lecture Notes, Springer-Verlag, 1976.

[200] Kerstan J., Matthes K., Mecke J., *Unbegrenzt Teilbare Punktprocesse*. Academic-Verlag, Berlin, 1974.

[201] Kutoyants Yu. A., Estimation of the intensity parameter of a nonhomogeneous Poisson process. *Problemy upravlenija i peredachi informatsii.*

[202] Liptser R. S., Optimal coding and decoding in the transmission of a Gaussian Markov signal with noiseless feedback. *Problemy peredachi informatsii* X, **4** (1974), 3–15.

[203] Liptser R. S., Gaussian martingales and generalization of the Kalman–Bucy filter. *Teoria Verojatn. i Primenen.* XX, **2** (1975), 292–308.

[204] Liptser R. S., On the representation of local martingales. *Teoria Verojatn. i Primenen.* XXI, **4** (1976).

[205] Orey S., *Radon–Nykodym Derivative of Probability Measures: Martingale Methods*. Dpt. Found. Sc., Tokyo Univ. of Eduation, 1974.

[206] Segall A., *A Martingale Approach to Modeling, Estimation and Detection of Jump Processes*. Tech. Rapport No. 7050-21, 1973.

[207] Segall A., Davis M. H. A., Kailath T., Nonlinear filtering with counting observation. *IEEE Trans. Infom. Theory* v. IT-21 (1975), 143–149.

[208] Segall A., Kailath T., The modeling of random modulated jump processes. *IEEE Trans. on Inform Theory*, v. IT-21, **2** (1975), 135–142.

[209] Snyder D. L., Filtering and detection for double stochastic Poisson processes. *IEEE Trans. Inform. Theory* v. IT-18 (1972), 91–102.

[210] Snyder D. L., Fishman P. M., How to track a swarm of fire flies by observing their flashes. *IEEE Trans. Inform. Theory* v. 21, **6** (1975), 692–695.

[211] Khadgijev D. I., On estimation of random processes from the observations of point processes. *UMN* XXXI, **2** (1976), 235–236.

[212] Khadgijev D. I., Filtering of semi-martingales with observations of point processes. *Teoria Verojatn. i Primenen.*

[213] Tsyrelson B. S., An example of the stochastic equation having no strong solution. *Teoria Verojatn. i Primenen.* XX, **2** (1975), 427–430.

[214] Chou C. S., Meyer P. A., La representation des martingales relatives a un processes punctueal discret. *C. R. Acad. Sci. Paris*, A, **278** (1974), 1561–1563.

[215] Shiryayev A. N., *Statistical Sequential Analysis*. 2nd ed., "Nauka," Moscow, 1976.

[216] Yashin A. I., On constructive algorithms of optimal nonlinear filtering, I, II. *Avtomatika i telemekhanika* 11, **12** (1975), 33–39, 108–113.

[217] Yashin A. I., Estimation of characteristics of jump random processes. *Avtomatika i telemekhanika* **4** (1976), 55–60.

Index